Research in Mathematics Education

More information about this series at http://www.springer.com/series/13030

Ji-Won Son • Tad Watanabe • Jane-Jane Lo
Editors

What Matters? Research Trends in International Comparative Studies in Mathematics Education

Springer

Editors
Ji-Won Son
Department of Learning and Instruction
University at Buffalo–The State
University of New York
Buffalo, NY, USA

Tad Watanabe
Department of Mathematics
Kennesaw State University
Kennesaw, GA, USA

Jane-Jane Lo
Department of Mathematics
Western Michigan University
Kalamazoo, MI, USA

Research in Mathematics Education
ISBN 978-3-319-51185-6 ISBN 978-3-319-51187-0 (eBook)
DOI 10.1007/978-3-319-51187-0

Library of Congress Control Number: 2017932283

Printed on acid-free paper

This Springer imprint is published by Springer Nature
The registered company is Springer International Publishing AG
The registered company address is: Gewerbestrasse 11, 6330 Cham, Switzerland

Foreword

As series editors, we hold the topic of international comparative studies near and dear to our hearts. We have both worked in and with education systems in different countries. Jinfa Cai, in particular, has studied mathematics education in China and the United States extensively (Cai, Ding, & Wang, 2014; Cai, Mok, Reedy, & Stacey, 2016). These experiences have taught us that the often subtle differences in curricular structure, relationship between teacher and student, classroom instruction, and cultural attitudes towards mathematics can together make a profound difference in patterns of learning behavior and, of course, in the outcomes of students' mathematical learning experiences. The dynamics of this interaction, how influential the system is at different levels, and what lessons we can learn from international comparative studies to improve students' learning continue to be dimensions of comparative study that merit ongoing effort. This volume provides some critical insight into these questions by examining mathematics education in the United States and several East Asian nations.

As the nations of the world grow closer in our academic and scientific and technological *capabilities*, the *expression* of those capabilities through culture, politics, and local adaptation becomes an ever more critical object of study for mathematics education. This volume examines the diversity of ways in which school mathematics is manifested in our world. Nations have evolved and developed different approaches to curriculum and learning opportunities (Part I), the preparation and continuing education of teachers (Part II), the ways in which teaching is envisioned and practiced (Part III), and the interaction with culture, politics, and economic systems to influence outcomes such as student performance and mathematical disposition (Part IV). The authors of the chapters in this volume use various research methods ranging from case studies, to larger-scale quantitative methods, to secondary data analysis to position both similarities and differences across education systems in their proper cultural context.

The result is an excellent read: Researchers from different countries, particularly those in the United States and in East Asian nations, have the opportunity to reexamine deeply held assumptions about the effects of teacher and school, community and national context, and individual student variables. But scholars from other world regions will also benefit, we think, from reading these comparisons. In particular, the studies reported in this volume will help researchers to frame new hypotheses about what factors might be innovative in improving mathematics teaching and learning in their *own* local and national contexts. This latter benefit is of particular importance: Comparative studies have been criticized in the past for providing only broad, high-level views of education systems, leaving little detail for the innovator to implement at her or his local level. This book is quite different. By examining practices at multiple levels, the authors explore nascent hypotheses about the adaptability of curriculum, teacher preparation, and pedagogical methods with some solid empirical backing. The authors acknowledge the methodological limitations of many of the studies reviewed and reported, but overall, the picture painted here is of a growing and vibrant field, making ever more important contributions to our knowledge and ability to act. It is also interesting to see a number of young researchers engaging in international comparative studies in mathematics.

Thus, in all aspects, this volume meets the overall vision of the monograph series. As we have indicated in our forewords in previous volumes, the audience for this monograph series consists of those in the intersection between researchers and mathematics education leaders—people who need the highest quality research, methodological rigor, and potentially transformative implications ready at hand to help them make decisions regarding the improvement of teaching, learning, policy, and practice. With this vision, our mission for this book series is:

1. To support the sharing of critical research findings among members of the mathematics education community.
2. To support graduate students and junior faculty and induct them into the research community by pairing them with senior faculty in the production of the highest quality, peer-reviewed, research papers.
3. To support the usefulness and widespread adoption of research-based innovation.

Finally, as series editors, we wish to thank the volume editors and authors for the quality of the research chapters and section and summative commentaries they have provided. This book will be especially useful in graduate courses on mathematics teaching and teacher education and mathematics curriculum, illustrating a broader base of what is possible than what might be assumed by looking only within one's own traditions.

Cai, J., Ding, M., & Wang, T. (2014). How do exemplary Chinese and U.S. mathematics teachers view instructional coherence? *Educational Studies in Mathematics, 85*(2), 265–280.

Cai, J., Mok, I., Reedy, V., & Stacey, K. (2016). *International comparative studies in mathematics: Lessons for improving students' learning*. New York, NY: Springer.

Jinfa Cai
James Middleton

Preface

Over the past several decades, increasing students' mathematical understanding and proficiency has been a national issue both in the United States and in many other countries, and a growing body of international comparative studies have been conducted to find ways to improve students' mathematics achievement. However, despite growing attention to international comparative studies and continued work in various aspects of education, it is not widely known what research has been done and how it was carried out.

In May 2014, Ji-Won Son approached Tad Watanabe and Jane-Jane Lo about an idea of a book on international comparative studies that would provide multiple perspectives on diverse issues and practices in mathematics education. Drawing from our own cultural backgrounds and expertise, we decided to focus this book on studies that compare data between and among the United States and five high-performing TIMSS education systems: Japan, China, Singapore, South Korea, and Taiwan. After numerous Skype meetings, we identified four main themes: (a) research on curriculum's influence on student learning, (b) research on institutional systems of mathematics teacher education, (c) research on improving teacher knowledge and pedagogical approaches, and (d) research using large-scale data. We then sent invitations to leading researchers in these areas to submit chapter abstracts to be considered for this book. The authors of those abstracts that fit one of the four themes of this book were invited to submit chapter proposals of approximately 3000 words. Detailed feedback on each proposal was provided to help the authors expand their proposals into full-length chapters. Each chapter manuscript was then reviewed by a panel of three reviewers, consisting of two editors and author(s) of another chapters and/or invited external reviewers. One to three rounds of revisions were completed before each manuscript was accepted.

This book includes 16 chapters, contributed by 15 US mathematics education researchers and 13 of their international counterparts from Australia, China, Hong Kong SAR, Japan, South Korea, and Taiwan. They are divided into four main groups. The authors of the chapters in Part I focus on curriculum-level influences on student learning by examining cross-national similarities and differences between intended and potentially implemented curricula (e.g., what is to be taught and how) and enacted curricula (e.g., student and teacher interactions and teaching approaches) that may contribute to *student achievement gaps* between and/or among educational systems. The authors of the chapters in Part II examine institutional-level influences on student learning by investigating cross-national similarities and differences in teacher education programs and in-service teacher education programs that may contribute to *teaching gaps* between and/or among countries. The authors of the chapters in Part III examine pedagogical approaches supporting preservice teachers' awareness and knowledge development between and/or across countries. While the authors in previous chapters focused, for the most part, on small samples and case studies, the authors in Part IV used large-scale data to examine the factors that explain differences in student mathematics achievement. Various factors are discussed in this section, including student-, teacher-, and school-level factors affecting mathematics achievement. The authors in Part IV also discuss inequality issues and parental influence affecting mathematics achievement, as well as professional development opportunities among different education systems.

While the authors were expanding their proposals to full manuscripts, we began to identify colleagues with research expertise in each of the four main themes to write commentary chapters. Edward Silver, Jeremy Kilpatrick, Sandra Crespo, and Sarah Lubienski accepted our invitations, and each wrote a commentary chapter on one of the four parts. Each commentary contains a brief review of the studies in that particular part, identifies important issues from each paper and across the papers, and provides thoughts on where the field should be going in that particular area of research. Furthermore, two commentaries for the entire book, one by Gabriele Kaiser and Xinrong Yang, and the other one by William Schmidt, establish the context of research in international comparative studies in mathematics education, identify important issues from each paper and across the papers, and provide thoughts on where the field should be going in research on international comparative studies.

Ji-Won Son was the leader of the editorial team. She set up the agenda for each editorial meeting and made sure that all decisions were followed through afterwards. She was also the point person for all author correspondence and promptly reminded authors when they missed a deadline. In addition, she was in charge of the editing for Parts II and III, while Tad Watanabe assumed the lead role for Part I and Jane-Jane Lo for Part IV.

We thank all the authors of this volume for their dedication in meeting the extremely tight deadlines involved in bringing this book together. Thanks also go to a group of external reviewers who took the time to help review many chapters of the book. We also thank John Acker for his attention to technical and stylistic details during the final preparation of the manuscripts, and series editors Jinfa Cai and James Middleton for their support and encouragement. We are pleased to present this volume as a timely and important resource for the mathematics education research community, to explore critical issues in the area of international comparative studies.

Buffalo, NY — Ji-Won Son
Kennesaw, GA — Tad Watanabe
Kalamazoo, MI — Jane-Jane Lo

Contents

Part I Research on Curriculum Influence on Student Learning

1 **What Can We Learn from Textbook Analysis?** 3
Ji-Won Son and Jeri Diletti

2 **Intended Treatment of Fractions and Fraction Operations in Mathematics Curricula from Japan, Korea, and Taiwan** 33
Tad Watanabe, Jane-Jane Lo, and Ji-Won Son

3 **Comparing the Difficulty Level of Junior Secondary School Mathematics Textbooks in Five Nations** 63
Yiming Cao, Libao Wu, and Lianchun Dong

4 **Uncovering the Label "Asian" in International Comparative Studies of Mathematics Education** 83
Yoshinori Shimizu

5 **Achievement Gaps in Mathematics and Opportunities to Learn: Insights from PISA 2012** 95
Yan Zhu

6 **Toward Understanding the Influence of Curriculum Resources on Students' Mathematics Learning: Cross-National Perspectives on What Matters Where, When, and for Whom** 115
Edward A. Silver

Part II Research on Institutional System of Mathematics Teacher Education

7 Knowledge Expectations Matter: Mathematics Teacher Preparation Programs in South Korea and the United States . . . 123
Rae Young Kim and Seung Hwan Ham

8 Pre-service Teacher Training for Secondary School Mathematics in Japan and Korea . . . 145
Masataka Koyama and Hee-chan Lew

9 Predictors of the Teaching Readiness of Future Secondary Mathematics Teachers: A Comparison of Singapore, Taiwan, and the United States . . . 165
Ting-Ying Wang and Feng-Jui Hsieh

10 Similarities and Differences in Programs for Prospective Secondary Mathematics Teachers . . . 187
Jeremy Kilpatrick

Part III Research on Improving Teacher Knowledge and Pedagogical Approaches

11 Cross-Cultural Lesson Planning Between the United States and South Korea . . . 199
Woong Lim and Ji-Won Son

12 The Instructional Quality of Mathematics Student Teachers in the United States and Japan: The Possible Impact of the Structure of Student Teaching . . . 215
Douglas Lyman Corey, Keith R. Leatham, and Blake E. Peterson

13 Reflective Capabilities of Mathematics Education Systems in China, Japan, and the United States . . . 237
Thomas E. Ricks

14 Research on Improving Teacher Knowledge and Pedagogical Approaches: From a Comparative to a Collaborative Perspective . . . 259
Sandra Crespo

Part IV Cross-national Comparative Studies with Large-Scale Data

15 Self-Concept, Self-Efficacy, and Mathematics Achievement: Students in 65 Regions Including the US and Asia . . . 267
Ming Ming Chiu

16 What Do TIMSS Studies Show About Math Achievement Inequality? A Sociological Perspective . . . 289
Seong Won Han, Ji-Won Son, and Chungseo Kang

17 When Knowing Basic Skills and Procedures Is Not Enough 315
Kyong Mi Choi and Dae S. Hong

18 The WIFI Study: Students' Valuing of Mathematics Learning in Hong Kong and Japan 333
Wee Tiong Seah, Takuya Baba, and Qiaoping Zhang

19 Examining the Association Between Teacher Feedback and Mathematics Instruction in Japan, Korea, Singapore, and the United States 355
Seong Won Han, Ji-Won Son, and Chungseo Kang

20 Large-Scale International Datasets—What We Can and Cannot Learn from Them, and How We Could Learn More 385
Sarah Theule Lubienski

Part V Final Commentary

21 Reflections on Research Trends in International Comparative Studies in Mathematics Education 395
Gabriele Kaiser and Xinrong Yang

22 The Missing Link—Incorporating Opportunity to Learn in Educational Research Analyses 411
William H. Schmidt, Leland S. Cogan, and Michelle L. Solorio

Index 419

Contributors

Takuya Baba is a Professor and the Dean of the Graduate School of International Development and Cooperation, Hiroshima University, Japan. He has completed both master and doctor courses in the aforementioned graduate school. He has got his Ph.D. in 2003. His interest expands from ethnomathematics, critical mathematics education and problem solving, socio-cultural aspect and values to lesson study, and endogenous development in mathematics education. He is a member of Curriculum Board of Japan Society of Mathematics Education (JSME) and the chief editor of the *Journal for Japan Academic Society of Mathematics Education* (JASME). He has extensive working experience as a mathematics teacher, developmental practitioner, and an academic researcher across countries in Asia and Africa. He has served as a project director of JICA developmental project to strengthen the quality of primary education in Bangladesh since 2004 and as a coordinator of Zambian Special Education Program since 2002. He also serves as an international expert of PISA for Development.

Graduate School for International Development and Cooperation, Hiroshima University, Hiroshima, Japan

Yiming Cao is a Professor in Mathematics Education at Beijing Normal University, China, and the president of Chinese Association of Mathematics Education. He is the member/expert of various committees in the Ministry of Education, including Primary and Secondary School Textbook Review Committee, Teacher Education and Curriculum Resources Construction Committee, and Basic Education Curriculum Textbook Commission. He has been serving on the editorial boards for several journals, such as the *Journal of Mathematics Education* (Chinese), *Research Journal of Mathematics and Technology*, and *Journal of Mathematics Education*. He is interested in international comparative study of mathematics curriculum, measurement and evaluation, and information technology integration, and hosts a number of national projects in these fields. He has published more than 150 papers in recent years.

School of Mathematical Sciences, Beijing Normal University, Beijing, China

Ming Ming Chiu is Charles Hicks Professor of Educational Psychology and Research Methodology at Purdue University. He serves on the advisory board of mainland China's Ministry of Education's National Evaluation of Primary and Secondary Schools. He invented two statistics methods: statistical discourse analysis (SDA) and multilevel diffusion analysis (MDA). SDA models online and face-to-face conversations, showing how students' social metacognitive skills enhance their micro-creativity. MDA shows how ideas spread through populations and detect corruption in the music industry. He also showed how economic growth, inequalities, and cultural values affect nearly 500,000 students learning in 65 countries. Supported by 26 grants totaling over $4.5 million, he disseminated his research through 164 publications (including 89 journal articles), 3 television broadcasts, 17 radio broadcasts, and 148 news articles in 21 countries and regions.

Department of Educational Studies, Purdue University, West Lafayette, IN, USA

Kyong Mi Choi is associate professor of mathematics education at the University of Iowa. Her research investigates how students learn mathematics better and develop higher level cognitive abilities in STEM. Choi led a Mathematics and Science Partnership 3-year grant titled I-MaP^2 (Iowa Mathematics Proficiency Project) to investigate if students learn better when the reasoning-modeling (R-M) approach is implemented to elementary and middle school mathematics classrooms in high-need school districts in rural Iowa. She has published widely and presented at national and international mathematics education conferences.

Department of Teaching and Learning, College of Education, The University of Iowa, Iowa City, IA, USA

Leland S. Cogan is a Senior Researcher in the Center for the Study of Curriculum Policy at Michigan State University. He has taught courses in educational psychology and educational research methods. He has conducted analyses of TIMSS and PISA data and coauthored technical reports, articles, and books including *Characterizing Pedagogical Flow*, *Facing the Consequences*, *Why Schools Matter*, and *The Preparation Gap: Teacher Education for Middle School Mathematics in Six Countries*. His research interests include evaluation of mathematics and science curricula, mathematics and science classroom instruction, and the preparation of mathematics and science teachers.

Center for the Study of Curriculum Policy, Michigan State University, East Lansing, MI, USA

Douglas Lyman Corey is currently an Associate Professor of mathematics education at Brigham Young University in Provo, UT. Doug earned his Ph.D. in mathematics education at the University of Michigan and has master's degrees in statistics and mathematics. Doug is interested in conceptualizations of high-quality instruction and the kind of knowledge, skills, and orientations it takes to enact high-quality lessons. To better understand these issues he is currently studying the teaching and professional learning of Japanese mathematics teachers. He is particularly interested in shared cultural factors (a teaching language, common planning practices, conceptions of high-quality teaching, etc) among Japanese mathematics teachers that seem to be a powerful resource for enabling the enactment of high-quality instruction. His other work includes an NSF-funded project (with colleagues from Harvard and the University of Michigan) to evaluate a scalable professional development model for elementary school teachers as well as to better understand the causal effect a teacher's MKT has on their instruction and student learning.

Department of Mathematics Education, Brigham Young University, Provo, UT, USA

Sandra Crespo is a professor of mathematics and teacher education and the director of the *Curriculum, Instruction, and Teacher Education* Ph.D. program at Michigan State University. She researches learning and teaching practices that support meaningful and empowering mathematical activity by disrupting and redistributing power dynamics in the mathematics classroom. Her research has been supported by the National Science Foundation grants. Her scholarship has been published in peer reviewed journals such as *Educational Studies in Mathematics* and *Journal of Mathematics Teacher Education* and in edited books such as *Mathematics in the Public Interest* and *Mathematical Problem Posing: From Research to Effective Practice*. She has participated in international research and development projects (e.g., TIMSS, TEDS-M, FirstMath) as a research assistant/consultant. She has taken leadership roles in several professional organizations that include AERA's SIG-RME, the NCTM, and TODOS-*Mathematics for All*. She is currently serving as the editor of the *Mathematics Teacher Educator* journal.

Department of Teacher Education, Michigan State University, East Lansing, MI, USA

Jeri Diletti is a doctoral student at the State University of New York at Buffalo, majoring in Curriculum, Instruction and the Science of Learning with an emphasis in Math Education. Her research interests include textbook analysis, curriculum alignment to standards, and teachers' enactment of curriculum resources. Her most recent publication includes "A Study of How Angry Birds Has Been Used in Mathematics Education," *Digital Experiences in Mathematics Education*. She is currently the president of the Math Education Student Alumni group at the University at Buffalo. Jeri has also been teaching high school mathematics at Akron High School in Akron, NY, for the last 10 years. She has taught courses in Remedial Algebra, Algebra, Algebra 2, Trigonometry, and Intermediate Algebra. She is the Math department chair as well as the chair of the Educational Study Council for Akron High School.

Department of Learning and Instruction, University at Buffalo—The State University of New York, Buffalo, NY, USA

Lianchun Dong is a Ph.D. candidate in Melbourne Graduate School of Education at the University of Melbourne, Australia. Prior to the Ph.D. research, he taught primary and secondary mathematics as a part-time teacher in various cities in China including Beijing, Shijiazhuang (Hebei Province), Baoding (Hebei Province), Cangzhou (Hebei Province), and Mianyang (Sichuan Province). He started his

Ph.D. research in the Faculty of Education at Monash University before transferring his Ph.D. candidature to the University of Melbourne. His Ph.D. research project concerns Australian and Mainland Chinese mathematics teachers' employment of questioning strategies across a unit of consecutive lessons. He has presented his Ph. D. research findings at several academic conferences including *PME*, *Mathematics Education Research Group of Australasia Annual Conference*, *The Mathematical Association of Victoria Annual Conference*, *The Australian Association for Research in Education Annual International Conference*, and *European Conference on Educational Research*. His research interest includes preservice and in-service mathematics teacher training, comparative research in mathematics teaching and learning, video analysis of mathematics classrooms, and mathematics teaching and learning in indigenous regions in Mainland China.

Melbourne Graduate School of Education, University of Melbourne, Melbourne, VIC, Australia

Seung Hwan Ham is an assistant professor of education at Hanyang University, Seoul, South Korea. His scholarly interests focus on school organization and multilayered institutional arrangements, especially in relation to their effects on educational equity and well-being for diverse students. Before joining Hanyang University, he taught teacher education courses at Michigan State University and worked as a research professor of education at Pohang University of Science and Technology.

Department of Education, Hanyang University, Seoul, South Korea

Seong Won Han is an assistant professor of educational leadership and policy at the University at Buffalo. Her research focuses on cross-national comparisons of education, gender inequality in STEM, educational policy, and teacher quality. Using large-scale international surveys and student achievement data, her cross-national comparative research investigates factors that are associated with students' expectations for STEM occupations and the teaching profession in a wide range of nations. She also investigates the factors that support improvement of instruction and student outcomes in a wide range of nations, including the United States. Her research has been published in the *International Journal of Educational Development*, *International Journal of Educational Research*, *Educational Policy*, *Journal of Educational Change*, *Teachers College Record*, among others, and as part of an edited collection on leadership and instructional change. Seong Won Han received the Thomas J. Alexander (TJA) Fellowship from the Organisation for Economic Co-operation and Development (OECD).

Department of Educational Policy and Leadership, University at Buffalo–The State University of New York, Buffalo, NY, USA

Dae S. Hong is an assistant professor of mathematics education at the University of Iowa. His interests are mathematics curriculum and textbook comparison and investigating instructional practices in mathematics classes. His previous work that compared quadratic equations lessons of different textbooks was published in *Educational Studies in Mathematics*. He has presented his curriculum-related studies at American Educational Research Association's Meetings, Psychology of Mathematics Education and Psychology of Mathematics Education—North American Chapter. His latest work on comparing function lessons of different textbooks was presented at PME 36 meeting in Vancouver. He currently serves as a Co-PI on the Mathematics and Science Partnership 3-year grant titled "I-MaP^2" Iowa Mathematics Proficiency Project.

Department of Teaching and Learning, College of Education, The University of Iowa, Iowa City, IA, USA

Feng-Jui Hsieh is a professor at National Taiwan Normal University. She is also incumbent Chair of Taiwan Association for Mathematics Education. Her research has focused on international studies of mathematics teacher education and the assessment of mathematics teaching competence, awareness, and reflection of secondary school mathematics teachers. She is currently cooperating with researchers from Mainland China to explore ideal mathematics teaching behaviors from students' perspectives and professional mathematics teaching indicators for secondary school teachers in Taiwan and China and to make comparisons of related issues between the two countries.

Department of Mathematics, National Taiwan Normal University, Taipei City, Taiwan

Gabriele Kaiser holds a master's degree as a teacher for mathematics and humanities for lower and upper secondary level. She completed her doctorate in mathematics education in 1986 with a study on applications and modeling supervised by Werner Blum and Arnold Kirsch. Based on a grant for Postdoctoral Research by the German Research Society (DFG), she undertook her postdoctoral study in pedagogy on international comparative studies at the University of Kassel, which she completed in 1997. Since 1998, she is a full professor for mathematics education at the Faculty of Education of the University of Hamburg. Since 2005 she serves as editor-in-chief of *ZDM Mathematics Education* (formerly Zentralblatt fuer Didaktik der Mathematik), published by Springer. She is Convenor of the 13th International Congress on Mathematics Education (ICME-13), which took place in 2016 at the University of Hamburg. Her areas of research include modeling and applications in school, international comparative studies, gender and cultural aspects in mathematics education, and empirical research on teacher education.

Faculty of Education, University of Hamburg, Hamburg, Germany

Chungseo Kang is a Ph.D. candidate in the Education, Culture, Policy, and Society Program at the State University of New York at Buffalo. He is interested in interdisciplinary research on the influence of societal features, policies, educational institutions, and families on learning growth gaps and outcomes in higher education, as well as quasi-experimental designs and survey methods.

Department of Educational Policy and Leadership, University at Buffalo—The State University of New York, Buffalo, NY, USA

Jeremy Kilpatrick is Regents Professor of Mathematics Education at the University of Georgia. He holds A.B. and M.A. degrees from the University of California,

Berkeley, and M.S. and Ph.D. degrees from Stanford University. Before joining the faculty at Georgia in 1975, he taught at Teachers College, Columbia University. He has taught courses in mathematics education at several European and Latin American universities and has received Fulbright awards for work in New Zealand, Spain, Colombia, and Sweden. He holds an honorary doctorate from the University of Gothenburg, is a Member of the National Academy of Education and a National Associate of the National Academy of Sciences, and received a 2003 Lifetime Achievement Award from the National Council of Teachers of Mathematics and the 2007 Felix Klein Medal honoring lifetime achievement in mathematics education from the International Commission on Mathematical Instruction. His research interests include teachers' proficiency in teaching mathematics, mathematics curriculum change and its history, assessment, and the history of research in mathematics education.

Department of Mathematics and Science Education, University of Georgia, Athens, GA, USA

Rae Young Kim is an Associate Professor in the Department of Mathematics Education at Ewha Womans University, Seoul, Korea, where she also serves as chair of the department. Her academic interests include curriculum design and development, mathematics teacher education, research methodology, international comparative study, and assessment for K-12 students in school settings. She has been engaged in several research projects related to preservice and in-service teacher knowledge and learning as well as curriculum design and assessment for K-12 school mathematics. She received a number of awards including Teacher Education Endowed Fellowship and Teaching Excellence awards. Her most recent research focuses on the development of interdisciplinary curriculum and pedagogy for teacher preparation programs and K-12 schools as well as personalized learning in technological environments for improving equity and excellence.

Department of Mathematics Education, Ewha Womans University, Seoul, South Korea

Masataka Koyama is a Professor of mathematics education and the Dean of Graduate School of Education and School of Education at Hiroshima University, Japan. He received his B.Ed., M.Ed., and Ph.D. in Education from Hiroshima University. His major scholarly interests are students' mathematical understanding, international comparative study on students' mathematical attainments, mathematics teachers' professional development, and school mathematics curricula and textbooks. He has authored or coauthored 4 books, over 40 book chapters, and over 150 journal articles, and presented at over 60 conferences including ICME, PME, and EARCME. He is also actively involved in international and national activities for improving mathematics education as editor-in-chief of the *Hiroshima Journal of Mathematics Education*, chief editor of Japanese Primary and Lower Secondary School Mathematics Textbooks, and external expert for Lesson Study in Mathematics. He will be an invited lecturer at the ICME-13 in Germany and a plenary speaker at the PME-40 in Hungary in 2016.

Department of Mathematics Education, Graduate School of Education, Hiroshima University, Higashi-Hiroshima, Japan

Keith R. Leatham is a professor of mathematics education at Brigham Young University in Provo, Utah. His research focuses on understanding how preservice secondary mathematics teachers learn to facilitate student mathematics learning. In particular he studies how preservice teachers learn to use technology in teaching and learning mathematics, how they learn to recognize and use students' mathematical thinking, and how their beliefs about mathematics, its teaching and learning are related to the learning-to-teach process. He is co-PI on the NSF-funded grant *Leveraging MOSTs: Developing a Theory of Productive Use of Student Mathematical Thinking*. He recently finished a term serving on the editorial panel for the *Journal for Research in Mathematics Education* and currently serves on the staff of the AMTE-sponsored early career development program *Service, Teaching and Research (STaR) in Mathematics Education*.

Department of Mathematics Education, Brigham Young University, Provo, UT, USA

Hee-chan Lew is a professor in the Department of Mathematics Education at Korea National University of Education since 1991. Now he serves as the president of the university since 2016. He was a researcher and a research fellow of the Korea Educational Development Institute, the President for the Korea Society of Educational Studies in Mathematics, a member of International Committee of the IGPME, and a member of the International Program Committee for ICME-12 and ICME-13. His areas of research interests in mathematics education are use of computer technology in classrooms and mathematics teacher education. He, as a chief editor, has directed various projects to develop elementary and high school textbooks and teachers' guides funded by the Ministry of Education and commercial publishing companies. He has also authored 10 books and over 100 journal articles on mathematics education.

Department of Mathematics Education, Korea National University of Education, Cheongju, South Korea

Woong Lim is an assistant professor of mathematics education at the University of New Mexico. He has a B.A. from Northwestern University and an M.A. and Ed.D from the University of Houston. His research interests include interrelations between language and mathematics, teacher preparation, and equity issues in mathematics education.

Department of Teacher Education, Educational Leadership and Policy, University of New Mexico, Albuquerque, NM, USA

Jane-Jane Lo is a professor of mathematics education at Western Michigan University, Kalamazoo, Michigan. She has a long-term research interest in studying the process of mathematical learning and concept development. This focus has been pursued in three complementary areas: rational number concepts, curriculum analysis, and international comparative studies both in the contexts of K-8 and teacher education. She has published in both research and practitioner journals, including the *Journal for Research in Mathematics Education*, *Educational Studies in Mathematics*, *Journal of Mathematics Teacher Education*, *Teaching Children*

Mathematics, *Mathematics Teaching in the Middle School*, *and Mathematics Teacher*. She and Laura Van Zoest co-chaired the 2012 PME-NA conference. Selected papers from this conference were later included in a book titled *Research Trends in Mathematics Teacher Education* co-edited by her, Keith Leatham, and Laura Zoest and published by Springer in 2014.

Department of Mathematics, Western Michigan University, Kalamazoo, MI, USA

Sarah Theule Lubienski is a professor at the University of Illinois at Urbana-Champaign. Her research focuses on inequities in students' math outcomes and the policies and practices that shape those outcomes. She conducts large-scale studies using national datasets as well as smaller, school-based studies. Dr. Lubienski is a member of the AERA Grants Governing Board, and she has chaired the National Assessment of Educational Progress (NAEP) Special Interest Group of AERA. She is Principal Investigator and Director of the University of Illinois Postdoctoral Research Training Program in Mathematics Education, funded by the U.S. Institute of Education Sciences. Dr. Lubienski's work has also been funded by the National Center of Education Statistics, the National Science Foundation, and the Fulbright Program, which supported her study of mathematics education reform in Ireland.

Department of Curriculum and Instruction, University of Illinois at Urbana-Champaign, Champaign, IL, USA

Blake E. Peterson is a professor in and chair of the Department of Mathematics Education at Brigham Young University in Provo, Utah. He is the recipient of the 2015 Excellence in Teaching in Mathematics Teacher Education award from the Association of Mathematics Teacher Educators. He has been a coauthor on the last 6 editions of a mathematics textbook titled *Mathematics for Elementary Teachers: A Contemporary Approach*. Earlier in his career, his research focused on student teaching in the United States and in Japan and how the structure of that experience influences the opportunity for preservice teachers to learn. More recently, his research has sought to understand how to help mathematics teachers recognize and effectively use student mathematical thinking as a standard method for teaching mathematics. He is a co-principal investigator on the NSF-funded MOST project, which focuses on identifying and building on high-leverage instances of student mathematical thinking.

Department of Mathematics Education, Brigham Young University, Provo, UT, USA

Qiaoping Zhang is a lecturer in the Department of Curriculum and Instruction, the Chinese University of Hong Kong. After receiving a B.Sc. degree in mathematics and a Master of Philosophy in education, he taught courses in mathematics and mathematics education in Hubei University, China. After receiving his Ph.D. degree at the Chinese University of Hong Kong, he worked in the Department of Mathematics at East China Normal University in Shanghai, China. Currently he is teaching courses in mathematics education at both the graduate and undergraduate levels at the Chinese University of Hong Kong. He also served as the program coordinator of the M.Sc. program in mathematics education. His research interests include affects in mathematics education, beliefs about mathematics, mathematics curriculum reform, and teacher education.

Department of Curriculum and Instruction, Faculty of Education, The Chinese University of Hong Kong, Shatin, NT, Hong Kong SAR

Thomas E. Ricks is an associate professor in the School of Education at Louisiana State University. His research interests include mathematics and science education teaching and learning, teacher preparation, complexity, and international cross-cultural comparisons, particularly between the United States and East Asian nations.

School of Education, Louisiana State University, Baton Rouge, LA, USA

William H. Schmidt is a University Distinguished Professor of statistics and education at Michigan State University. He serves as director of the Education Policy Center and holds faculty appointments in Statistics and Education. Previously he served as National Research Coordinator and Executive Director of the US National Center which oversaw participation of the United States in the IEA sponsored Third International Mathematics and Science Study (TIMSS). He has published in numerous journals including the *Journal of the American Statistical Association*, *Journal of Educational Statistics*, *EEPA*, *Science*, *Educational Researcher* and the *Journal of Educational Measurement*. He has coauthored ten books including *Why Schools Matter*, *Teacher Education Matters*, and *Inequality for All*. His current writing and research concerns issues of academic content in K-12 schooling including the Common Core State Standards for Mathematics, assessment theory, and the effects of curriculum on academic achievement. He is also concerned with educational policy related to mathematics, science, and testing in general. Dr. Schmidt received the 1998 Willard Jacobson Lectureship from the New York Academy of Sciences and is a member of the National Academy of Education. In 2009 he was elected in the first group of Fellows in the American Educational Research Association. Dr. Schmidt served on the Steering Committee for Review of the Evaluation Data on the Effectiveness of NSF-Supported Mathematics Curriculum Materials. He received his A.B. in mathematics from Concordia College in River Forrest, IL, and his Ph.D. from the University of Chicago in psychometrics and applied statistics. He was also awarded an honorary doctorate degree from Concordia University in 1997.

Education Policy Center, Michigan State University, East Lansing, MI, USA

Wee Tiong Seah is associate professor in mathematics education at the Melbourne Graduate School of Education, the University of Melbourne, Australia. He also heads its Mathematics Education Group. He has been teaching preservice and in-service courses at Bachelor, Masters, and Ph.D. levels since 2002, both locally and overseas. Wee Tiong was a member of the Australian federal government's Expert Advisory and Research Group in the late 2000s. Wee Tiong's research expertise is in comparative research (e.g., he currently leads a 22-nation research consortium called the "Third Wave Project"), the role of values in facilitating student learning, and the experiences of immigrant and refugee students and teachers in mathematics pedagogy. He has delivered numerous keynote or invitational speeches at different events, a recent one being at the 2016 International Congress on Mathematical Education. Wee Tiong sits on the editorial boards of several journals, including the *International Journal of Science and Mathematics Education*.

Melbourne Graduate School of Education, The University of Melbourne, Melbourne, VIC, Australia

Yoshinori Shimizu is a Professor of mathematics education and the Chair of Master's Program in Education at the University of Tsukuba in Japan. His primary research interests include international comparison of mathematics classrooms and student assessment. He was a member of Mathematics Expert Group for OECD/PISA 2003, 2006, and 2009 and a consultant for TIMSS Videotape Classroom Study. He is one of the founders of Learner's Perspective Study (LPS), an international comparative study on mathematics classrooms, and has been the Japanese team leader of the project. He serves as a member of editorial boards for international research journals, such as *ZDM Mathematics Education* and *Mathematics Education Research Journal*. He is a vice president of Japan Society of Mathematical Education and has been working closely with mathematics teachers at elementary and secondary schools as an external expert in Lesson Study.

College of Education, University of Tsukuba, Tsukuba, Japan

Edward A. Silver is currently the Senior Associate Dean for Research and Graduate Studies, the William A. Brownell Collegiate Professor of Education, and a Professor of Mathematics at the University of Michigan in Ann Arbor. He has published numerous articles, chapters, and books on these and related topics. He currently directs two projects supported by funding from the National Science Foundation—one aims to explore ways to transform PISA assessment items and results into resources for teacher professional learning, and the other involves an investigation of the beliefs and practices of mathematics teacher educators and teacher professional development specialists regarding formative assessment. He also directs the Usable Scholarship in Education (USE) Initiative at the University of Michigan, in which he and others are probing the nature of the connections and disconnections between educational scholarship and educational policy/practice. He recently completed his 3-year term as an elected member of the Board of Directors of the Association of Mathematics Teacher Educators. He has previously served on several National Research Council Committees, numerous advisory boards for projects and organizations, as one of the principal authors for *Principles and Standards for School Mathematics*, as editor of the *Journal for Research in Mathematics Education* from 2000 to 2004, and as co-editor of *The Elementary School Journal* from 2008 to 2010. He has received a number of recognitions for his work, including the 2004 Award for Outstanding Contributions of Educational Research to Practice from the American Educational Research Association, the 2007 Iris Carl Memorial Leadership and Equity Award from TODOS, the 2008

Judith Jacobs Lectureship from the Association of Mathematics Teacher Educators, the 2009 Lifetime Achievement Award from the National Council of Teachers of Mathematics, and the 2011 Senior Scholar Award from the American Educational Research Association Special Interest Group for Research in Mathematics Education. In 2016 he was selected to be a Fellow of the American Educational Research Association.

School of Education, University of Michigan, Ann Arbor, MI, USA

Michelle L. Solorio is a Ph.D. student of educational policy at Michigan State University. She holds a research assistantship under Dr. Schmidt in the Center for the Study of Curriculum at the MSU Education Policy Center and is working with Dr. Schmidt on comparative education projects related to TIMSS and PISA, the evolution of opportunity-to-learn, and measures of socioeconomic status for educational research use. As a student, she focuses on international comparative education policy as it relates to language of instruction policies and conflict education in francophone West Africa and refugee populations around the world. She is currently a Dean's Scholar at the MSU College of Education and a US Department of Education FLAS Fellow in the MSU African Studies Center, and she received a Summer Research Development Fellowship from the MSU College of Education in summer 2016. Starting in fall 2015, she actively serves as a proofreader for the *Michigan Reading Journal*. Previously, she worked in international student mobility as a study abroad advisor and incoming exchange student coordinator at the Global Education Office at Vanderbilt University.

Center for the Study of Curriculum Policy, Michigan State University, East Lansing, MI, USA

Ji-Won Son is an assistant professor of mathematics education at University at Buffalo–The State University of New York. She taught elementary and secondary school students in South Korea and was an active member of lesson study in South Korea. Her research interest includes (1) mathematics textbook analysis, (2) teachers' curriculum use, (3) teacher learning, and (4) international comparative studies. Son has published numerous research articles in nationally and internationally referred journals and in conference proceedings. She serves on the editorial boards of *Journal of Educational Studies for Educational Research Institute*, *International Journal of Education*, and *Journal of Research in STEM Education*. Son currently serves as Principal Investigator on the Spencer Foundation-funded project entitled "Core Mathematical Knowledge and Cross-Cultural Practices for Teaching Algebraic and Functional Relations" to identify, from a cross-cultural perspective, the core mathematical knowledge and practice for teaching algebraic and functional relations, to enable middle school mathematics teachers to better develop students' algebraic thinking.

Department of Learning and Instruction, University at Buffalo—The State University of New York, Buffalo, NY, USA

Ting-Ying Wang is a project assistant professor at National Taiwan Normal University. Her research has focused on international studies of mathematics teacher education and secondary school mathematics teachers' teaching competences. She is currently cooperating with researchers from Mainland China to investigate and compare preservice and in-service teachers' perspectives of ideal mathematics teaching behaviors in Taiwan and China.

Department of Mathematics, National Taiwan Normal University, Taipei City, Taiwan

Tad Watanabe is a professor of mathematics education and an assistant chair of the Department of Mathematics at Kennesaw State (GA) University. He also owns and operates an educational consultant company, Math Horizon, LLC. His recent research has examined various aspects of Japanese elementary curriculum materials, including the content of teacher's manuals, the treatment of arithmetic mean, the use of pictorial representations, and the initial treatments of multiplication and fractions. He was involved in the translation of the 2008 Japanese Course of Study for Elementary and Lower Secondary mathematics and the accompanying Teaching Guides. Watanabe is fluent in English and Japanese (both written and spoken) and knowledgeable about the United States and Japanese education. Since 2000, he has visited more than 50 Japanese schools and observed more than 100 mathematics lessons. He has taught mathematics content and mathematics methods courses for prospective elementary school teachers and worked with a number of lesson study groups within the United States. Watanabe is an expert in international level comparative studies.

Department of Mathematics, Kennesaw State University, Kennesaw, GA, USA

Libao Wu is an associate professor in mathematics education at Tianjin Normal University, China. He obtained his Ph.D. degree from Beijing Normal University, China. In the past years, he has published over 100 academic papers in a variety of education journals in China including *Curriculum, Textbooks and Pedagogy*, *China Education Journal*, *Education Journal*, *Comparative Education Research*, *China ICT education*, *Foreign School Education*, *Education Theory and Practice*, *Mathematics Education Journal*, *Mathematics Bulletin*, *Instruction and Administration* and so on. Out of his publications, 15 articles were published in the journals that are indexed by China Social Sciences Citation Index, 42 articles were published in journals that are indexed by China Core journals, and four articles were full-text cited by Information Centre for Social Sciences at the Renmin University of China. His research projects include "Investigation of school teacher's professional development in rural regions of Sichuan Province," "Improving instructional competencies of mathematics teachers in rural regions," "An international comparative study of mathematics textbooks' difficulty level (Ministry of Education, AHA120008)," "Alignment of mathematics curriculum standards and teaching practices (Ministry of Education, GOA10715)," and "Theory and practices about quality life and education (funded by Education Department in Tianjin City)." He is also the lecturer of national-level online course "Research in Middle School Mathematics Curriculum and Textbook." He participated in the preparation for provincial-level online course "Analytical Geometry." He edited two books: *Research and Cases in*

Primary Mathematics Teaching and *Mathematics Content Knowledge and Instructional Competencies in Senior High School.*

Teacher Education College, Tianjin Normal University, Tianjin, China

Xinrong Yang completed his doctoral degree at the University of Hong Kong in mathematics education in 2010 with a study on the topic of mathematics teachers' teaching expertise. After that, he started his work at Southwest University in China as an associate professor. From September 2012 to October 2014, he carried out a 1-year postdoctoral research at Umea University, Sweden, and the Institute of Education, London, UK, respectively. Now, he is undertaking his postdoctoral research at the Hamburg University with Professor Gabriele Kaiser under the support of Marie Curie Individual Fellowship on the relationship between teachers' professional competence and the progress of students' mathematics achievement in China and Germany. His research interests include mathematics teacher education, classroom environment research, and cross-cultural comparative research. He has published around 30 peer-reviewed journal articles on these themes in Chinese and international journals.

School of Mathematics and Statistics, Southwest University, Chongqing, China
Faculty of Education, University of Hamburg, Hamburg, Germany

Yan Zhu is Associate Professor in the Department of Curriculum and Instruction, East China Normal University. She received her B.Sc. and M.Ed. from East China Normal University and her Ph.D. from Nanyang Technological University, Singapore. Her research interest includes education equity, comparative studies, mathematics problem solving, mathematics assessment, and secondary data analysis.

Faculty of Education, East China Normal University, Shanghai, China

Part I
Research on Curriculum Influence on Student Learning

Chapter 1
What Can We Learn from Textbook Analysis?

Ji-Won Son and Jeri Diletti

Abstract As a fundamental resource, textbooks have the potential to shape the way we teach and learn mathematics. While a growing body of textbook analysis studies has sought better ways to improve students' mathematics achievement, no meta-analysis has yet summarized those studies and their methods. This chapter reviews international comparative studies that analyzed learning opportunities presented in mathematics textbooks in the USA and five high-achieving Asian education systems. We summarize what research studies say about learning opportunities presented in textbooks in connection to the theoretical frameworks used, and their plausible relationship with students' mathematics achievement. Following this description and analysis, we raise several questions and issues for mathematics education researchers to discuss, to promote a critical examination of what can be learned from the content of textbooks in other countries.

Keywords Textbook analysis • Literature review • Content analysis • Problem analysis

Introduction

Over the past 30 years, changes in mathematics classroom practices and teaching methodologies have led to concerns regarding the quality of mathematics textbooks. Because textbooks are often the curricular materials that are the most influential on what happens in classrooms (Kilpatrick, Swafford, & Findell, 2001), they have attracted more and more research attention from the international mathematics education community in the past three decades, particularly in connection with international assessment studies (Cai, 2010; Cai, Mok, Reedy, & Stacey, 2016). For instance, the results from international assessment studies of mathematics, including the *Trends in International Mathematics and Science Study*

J.-W. Son (✉) • J. Diletti
Department of Learning and Instruction, University at Buffalo—The State University of New York, Buffalo, NY, USA
e-mail: jiwonson@buffalo.edu; jeridile@buffalo.edu

J.-W. Son et al. (eds.), *What Matters? Research Trends in International Comparative Studies in Mathematics Education*, Research in Mathematics Education, DOI 10.1007/978-3-319-51187-0_1

(TIMSS: 1995, 1999, 2003, 2007, and 2011) and the *Program for International Student Assessment* (PISA: 2003 to 2012), offer opportunities to compare the mathematics performance of US students with that of their peers in other countries. The recent 2011 TIMSS revealed that while US fourth and eighth graders scored above the international average of the 63 TIMSS countries, they fell significantly behind the students in five Asian education systems at both grade levels: Hong Kong, Singapore, Korea, China, and Japan (Mullis, Martin, Foy, & Arora, 2012). Accordingly, researchers have been looking into the varying reasons why students in Asian countries tend to outperform their US counterparts and have identified several important factors that are linked to student achievement (Cai & Howson, 2013; Wagemaker, 2003).

Factors that potentially impact student learning include the curriculum as a whole and the curricular materials available, including textbooks. Researchers have identified multiple factors that have an impact on student learning, including student-level factors (e.g., students' home background, their socioeconomic status, and gender differences) (Bos & Kuiper, 1999), teacher- or classroom-level factors (e.g., peer influence, teacher quality, and teachers' instructional approaches) (Cai, Ding, & Wang, 2014; Kupari, 2006; Papanastasiou, 2008), and contextual or school-level factors (e.g., the location of the school, the number of desks) (Creemers, 1994). In particular, with a focus on identifying curricular influence on students' academic achievement, researchers have come to the understanding that cross-system similarities and differences in curriculum can provide *partial* explanations for cross-national discrepancies in students' mathematics performance (Cai, Ni, & Lester, 2011; Cai, Wang, Moyer, Wang, & Nie, 2011; Schmidt et al., 2001; Silver, 2009). For example, when comparing the curriculum materials from several high-achieving countries, such as Japan, South Korea, and China, researchers noted that the US curriculum materials devoted more page space to student practice than to content instruction (e.g., Carter, Li, & Ferrucci, 1997; Kim, 2012) and failed to provide challenging mathematics content and problems (e.g., Cai, Ni, & Lester, 2011; Li, 2007; Schmidt, McKnight, & Raizen, 1997; Son & Hu, 2016). Thus, by changing content presentation and organization, researchers have attempted to improve students' mathematics achievement.

However, researchers use several different methods to study the important question of how mathematics textbooks in different countries structure learning opportunities for their students. A complete framework for textbook analysis remains unavailable (Charalambous, Delaney, Hsu, & Mesa, 2010; O'Keeffe & O'Donoghue, 2015). In addition, despite the important position of curriculum materials in mathematics classrooms worldwide, researchers have expressed contrasting views about what can be learned from analyzing mathematics textbooks. Some researchers claim that textbook analysis can explain differences in students' performance in international comparative studies (Cai, Ni, & Lester, 2011; Cai, Wang, Moyer, Wang, & Nie, 2011; Fuson, Stigler, & Bartsch, 1988; Li, 2002; Son & Senk, 2010). Other researchers, however, have argued that textbooks bear little influence on instruction and on what students learn (Freeman & Porter, 1989). This line of research has viewed textbooks as a potential source for teacher learning, a goal that is frequently unfulfilled (Newton & Newton, 2007; Remillard, 2005; Son & Kim, 2015).

We generally agree that the analyses of curriculum materials reveal *nuanced insights* into variations in what is made available to students and teachers in textbooks, and also how that content is made available (Silver, 2009). In this chapter, we take a more moderate viewpoint by suggesting that textbooks afford probabilistic rather than deterministic opportunities to learn mathematics (Mesa, 2004; Valverde, Bianchi, Wolfe, Schmidt, & Houang, 2002). While the deterministic perspective toward textbook analysis links students' mathematics achievement *directly* to the content of their textbooks, the probabilistic perspective acknowledges the possible mediated effect of teachers and students on the learning opportunities presented in textbooks, because the role of textbooks in instruction depends on how students and teachers interact with them (Remillard, 2005; Son & Kim, 2015). From the probabilistic perspective, textbook analysis can only reveal different *performance expectations* made of students in different countries, the extent to which a country's textbook series prioritizes conceptual understanding or procedural fluency, and how the treatment of mathematical content and problems differs among countries. With this view, we intend to review international comparative studies that analyzed learning opportunities presented in mathematics textbooks.

This chapter presents a survey study that aims to examine, analyze, and review relevant textbook research systematically. We focus on textbook analysis research studies that perform international mathematics assessments between and among the USA and five high-performing Asian education systems: Japan, China, Singapore, South Korea, and Taiwan. We first summarize what research studies say about variations or commonalities in terms of the learning opportunities presented in mathematics textbooks across different education systems, as this might account in part for disparities in students' mathematics achievement. In doing so, we specifically look at the theoretical frameworks used in textbook analysis studies and the findings drawn from each framework. Charalambous et al. (2010) defined a "textbook signature" as "the uniform distinctive features within a particular country." By summarizing the findings from prior research, this study reports whether any unique signature represents each education system's textbooks. We then raise questions and issues for mathematics education researchers in terms of conceptualization and methodological matters, leading to a critical examination of what can be learned from textbooks from other countries. In the next section, we define curriculum and textbooks, and discuss the data and coding framework used in this study.

Methods

Assumptions and Definition of Terms

In this study, by *textbook series* we mean a set of curricular resources that teachers use for day-to-day teaching, which includes student texts, workbooks, and the teacher's guide. Drawn from TIMSS' definition (Schmidt, McKnight, Cogan, Jakwerth, & Houang, 1999), we define the *intended curriculum* as the set of standards students are required to achieve. The *potentially implemented curriculum* may include teacher manuals, students' main textbooks, and supplemental materials such as student workbooks, review materials, and assessments. An examination of textbooks informs policymakers of how societal visions and educational objectives, seen in national policies and official documents as *the intended curriculum*, are *potentially* embodied in classrooms (Schmidt et al., 1999; Valverde et al., 2002). This study focuses on research studies that analyzed the content (e.g., content coverage) and/or problems of mathematics textbooks in international comparisons, while comparing the similarities and differences of two or more series of mathematics textbooks.

Selection Search Procedures

The research review presented here is based on an analysis of peer-reviewed research articles that focused on learning opportunities presented in mathematics textbooks between and among the USA, Japan, China, Singapore, South Korea, and Taiwan. We conducted our literature review via the Education Resource and Information Center (ERIC), allegedly the world's largest digital library for education literature, as well as via Google Scholar. We obtained peer-reviewed research articles primarily using the search terms of "textbook" and "mathematics," and refined our searches further by adding several groups of terms, including "textbook research," "textbook content," and "textbook analysis." Next, we systematically examined past issues of peer-reviewed research journals in mathematics education to identify the relevant literature, including the following:

ZDM—The International Journal of Math Education
ESM—Educational Studies in Math Education
CI—Cognition and Instruction
SSM—School Science and Mathematics
JCS—Journal of Curriculum Studies
IJME—International Journal of Science and Mathematics Education
JRME—Journal for Research in Mathematics Education

These journals were selected based on two criteria. First, their scope of publication covers a great range of mathematics education research, and secondly, they are all highly ranked. Nevertheless, research articles that were published in other journals also received attention, though we mainly identified these articles through ERIC searches, not directly from the journals. Our search was also limited to peer-reviewed research articles published between 1988 and the first half of 2015, because the reform movement began with the formation of the National Council of Teachers of Mathematics (NCTM) in 1988.

It should be mentioned that our collection of the relevant literature is by no means complete, which is a limitation of our study. While the ERIC database includes a variety of sources, the main body of the literature we have identified consists of peer-reviewed journal articles based on original empirical studies. Thus, we did not pay attention to sources such as books, doctoral dissertations, and papers presented at conferences. We must point out that although we tried to make the survey as comprehensive as possible within our criteria, it is possible that some important peer-reviewed research work in this area was missed in the selection process. This is due to a variety of reasons, including the scope and focus of the study and the fact that not all research is accessible via ERIC or published in journals. These challenges are common among survey studies like this one.

Coding Framework

After we selected the literature, we first constructed a database by classifying all the articles using the six criteria established for the study, including topics, research questions, grade level, education systems analyzed, analytical framework used, and major findings reported. This analysis helped us provide an overview of general tendencies in textbook analysis studies. Next, we further analyzed each article based on the following framework for content analysis and problem analysis (see Table 1.1).

Content analysis refers to comparing learning goals, lists of topics (content coverage), topic placement, textbook size, allocation of content, allocation of time, repetition of content, development of concepts and procedures, the use of technology, and the use of worked examples. Problem analysis means classifying the textbook exercises and problems/tasks by various kinds of schemes, such as the characteristics of mathematical features, contextual features, cognitive demand, cognitive expectations, depth of knowledge required for solving problems, and the relevance of non-textual elements. For the analytical foci in problem analysis, including cognitive demand, cognitive expectations, and cognitive depth of knowledge, we referred to Stein, Grover, and Henningsen (1996), Son and Senk (2010), and Webb (1999), respectively.

Charalambous et al. (2010) called the former type of textbook analysis horizontal analysis, especially focusing on the overall structures of textbooks (i.e., what mathematics is taught at what grade level), and the latter type as vertical analysis, focusing on the treatment of a particular mathematical topic. Li, Chen, and An (2009) called the former type macroanalysis and the latter type microanalysis. In our study, content analysis exceeds the horizontal analysis or macroanalysis by including textbook size, allocation of content, allocation of time, repetition of content, development of concepts and procedures, the use of technology, and the use of worked examples. The findings reported in each article were categorized based on the sub-dimensions shown in Table 1.1 to describe how mathematics textbooks in different education systems structure learning opportunities for their

Table 1.1 A framework for classifying the literature on textbook analysis research

Analytical foci	Subcomponent
Content analysis (macro, horizontal)	• Content coverage (topic placement)
	• Size/length of book
	• Introduction and development of concepts and procedures
	• Repetition of content
	• Others (e.g., the use of technology and worked examples)
Problem analysis (micro, vertical)	• Mathematical features (number of steps required: single vs. multiple)
	• Contextual features (purely mathematical or illustrative)
	• Response type (numerical answer only or explanation required)
	• Cognitive demand (degree to which students are required to engage cognitively: high or low)
	• Cognitive expectations (kind of knowledge/process required in solving problems: conceptual knowledge, procedural knowledge, representations, mathematical reasoning, and problem-solving)
	• Depth of knowledge (the complexity of mental processing that occurs to answer a question or perform a task: level 1, 2, 3, or 4)
	• Relevance of non-textual elements (e.g., photos, pictorial illustrations, mathematical representations, and pictures)

students, and whether there is any unique signature that represents the textbooks in each education system. Each sub-dimension in both content and problem analysis will be discussed in detail in the findings section.

Results

Overall Tendencies in Textbook Analysis

Appendix gives a list of the peer-reviewed research articles analyzed in our survey. In total, we identified 31 articles that addressed international textbook comparisons between the USA and the five high-achieving Asian education systems. Figure 1.1 illustrates the frequency of the education systems surveyed in comparisons of mathematics textbooks. All but one article included the USA in the comparisons. The majority of textbook analysis research focused on China, followed by studies comparing Japan and South Korea to the USA. In total, 17 textbook analysis studies comparing China to other education systems were identified in our data source. Singapore and Taiwan were analyzed relatively less frequently. Only five textbook analysis studies were conducted based on Singapore, and four studies involved Taiwan.

Table 1.2 presents a frequency count of the type of analysis (content, problem, or both), grade level, and topic. Of the 31 articles surveyed, six strictly analyzed the content of the textbooks. An article was counted as strictly content analysis if the

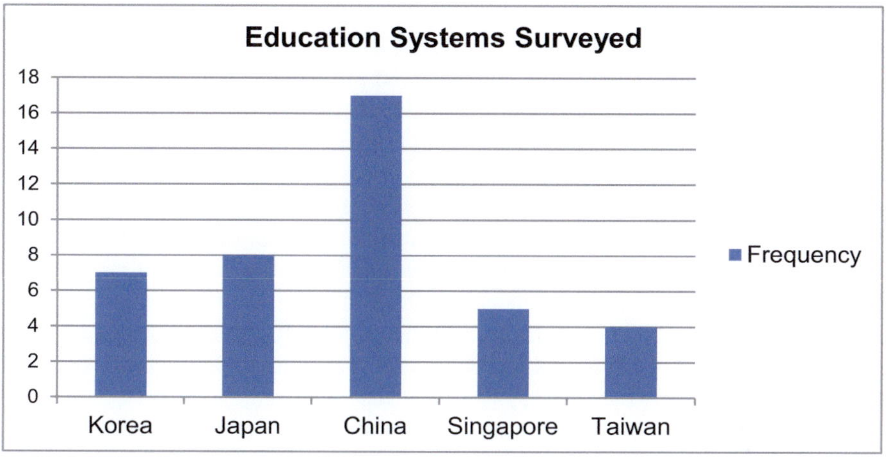

Fig. 1.1 Frequency of textbook analysis research by education system

Table 1.2 Frequency of research articles with respect to analysis, grade level, and topics

Themes	Frequency
Analysis foci	
• Content	6
• Problem	11
• Both	14
Grade level	
• 1	8
• 2	6
• 3	7
• 4	7
• 5	8
• 6	10
• 7	11
• 8	10
• 9	4
• 10	5
• 11	2
• 12	1
Topic	
• Whole numbers	5
• Fractions	8
• Integers	3
• Algebra	4
• Probability	2
• Geometry	2
• Others (e.g., average, percent)	7

researchers looked at the overall structure of the book, which includes factors such as the grade placement of topics, size of the books and number of pages, methods of introducing topics and developing concepts and procedures, the use of technology, and the number of worked examples. Eleven of the articles strictly analyzed the problems and tasks within the student textbooks or supplemental materials (e.g., student workbooks, review materials, and assessments). These articles addressed factors such as tasks that require an answer only vs. those that also require explanation, frequency of tasks involving real-world applications, and cognitive demand and expectations required in solving problems. The 14 remaining articles analyzed both content and problem tasks.

With respect to grade level, the most frequently analyzed grade levels were the elementary grades 1–5, followed by the middle grades, especially grades 6, 7, and 8. High school levels (grades 9–12) were the least analyzed. Note that occasionally topics do not appear at the same grade level in the US and Asian education systems. This represents a particular challenge for those engaged in textbook analysis if grade bands are used, such as elementary, middle, and high school grades. Thus, we presented the frequency of the topic analyzed by grade. Because research articles that analyzed more than one grade level were counted once for each grade level covered, the total frequency counts for grade level is higher than the 31 total articles surveyed.

A wide range of topics was discovered in our survey. The most widely analyzed topic was fractions. This includes the introduction of fractions as well as operations with fractions. A total of eight articles had some fractional component to them (Alajmi, 2012; Charalambous et al., 2010; Li et al., 2009; Son, 2012; Son & Senk, 2010; Sun, 2011; Sun & Kulm, 2010; Yang, Reys, & Wu, 2010). Son and Senk (2010), for example, examined how fraction multiplication and division were introduced, and what types of problems were used to facilitate the development of procedural and conceptual understanding in textbooks in the USA and South Korea. Five surveyed articles analyzed tasks and the placement of either whole number operations (Fuson et al., 1988; Watanabe, 2003) or introductions to whole number operations (Boonlerts & Inprasitha, 2013; Kang, 2014; Xin, Liu, & Zheng, 2011). For example, Watanabe (2003) compared and contrasted the number of lessons, problem situations, and types of representations used in the initial treatment of multiplication of whole numbers in the USA and Japan. Four studies focused on algebraic thinking, including the distributive property, the equals sign, and algebraic problems (Ding & Li, 2010; Hong & Choi, 2014; Li, 2007; Li, 2007). For example, Ding and Li (2010) compared elementary textbooks in the USA and China when analyzing instances of the distributive property. Other topics include probability, average, the use of justifications, and explanations of work.

After establishing the type of analysis conducted, grade level, and topics, each article's framework and findings were explored in greater detail. The following sections detail the common themes discovered in our survey, broken down into content analysis and problem analysis.

Table 1.3 Frequency of textbook analysis research by sub-dimensions of content analysis

Content analysis	Frequency
Topic placement	10
Size/length of book	5
Repetition of content	2
Development of concepts and procedures	10
Others (e.g., the use of technology)	6

Content Analysis: Common Themes

Content analysis includes comparing content coverage, topic placement, textbook size, allocation of time, methods of introducing topics and developing concepts and procedures, repetition of content, the use of technology, and the characteristics of worked examples. A total of 20 articles fell into the category of content analysis. Of these, six strictly utilized content analysis and the remaining studies analyzed both content and problems. Table 1.3 presents the frequency of research articles addressing each of the subcomponents in content analysis. The most prevalent themes for content analysis were *topic placement* and *development of concepts and procedures.*

Topic Placement

Topic placement refers to the grade or chapter of the textbook where the content appeared. Nine surveyed articles reported on topic placement (Boonlerts & Inprasitha, 2013; Cai, Lo, & Watanabe, 2002; Choi & Park, 2013; Fuson et al., 1988; Hong & Choi, 2014; Kang, 2014; Li et al., 2009; Son, 2012; Son & Senk, 2010; Yang et al., 2010). Of the nine articles focusing on topic placement, eight reported that mathematical ideas tended to occur earlier in Asian education systems than in the USA (Cai et al., 2002; Choi & Park, 2013; Fuson et al., 1988; Hong & Choi, 2014; Li et al., 2009; Son & Senk, 2010). For instance, in their analysis of fifth- and sixth-grade textbooks Son and Senk (2010) found that multiplication of fractions appears earlier in Korean texts than in a standards-based US textbook (*Everyday Mathematics*). Some multiplication and division of fractions topics that appear in Korean textbooks did not appear at all in *Everyday Mathematics*. Similar results were found in Li et al.'s (2009) analysis of fraction division. They found that US textbooks vary the introduction of fraction division anywhere between grades six and eight. However, all the covered Chinese and Japanese textbooks introduce fraction division in sixth grade. Yang et al. (2010) also found similar results for computation of fractions in Singapore, Taiwan, and the USA. In their study, US students in the sixth grade were only expected to compare fractions, while in Taiwan sixth-grade students were expected to subtract proper fractions. Fifth-grade students in Singapore were required to subtract mixed numbers. In their analysis of computing averages, Cai et al. (2002) similarly found that China,

Japan, and Taiwan introduce the topic as early as fourth grade, but averages do not appear until fifth grade in the USA.

Continuing the theme of the earlier introduction of content in Asian education systems, in their analysis of the addition and subtraction of whole numbers, Fuson et al. (1988) found uniformity of grade level placement in Japan and China, but not in the USA. Topics seem to appear earlier and disappear earlier in Asian texts than in US texts. Similarly, Hong and Choi (2014) found that topics relating to quadratic equations in grades 9, 10, and 11 are introduced earlier in Korean textbooks than in standards-based US textbooks, and some topics appearing in Korean textbooks did not appear at all in the standards-based US textbooks.

One study reported that geometry topics occur during the same grades in both Korea and the USA. However, this study looked at the intended curriculum of both countries, rather than at actual textbooks (Choi & Park, 2013). The final study that analyzed topic placement for the introduction and development of multiplication found that Singaporean textbooks begin teaching multiplication in first grade, while Japanese and Thai textbooks begin in grade 2 (Boonlerts & Inprasitha, 2013). However, this study did not compare its findings with US textbooks.

Introduction and Development of Mathematical Concepts and Procedures

Introduction and development of mathematical concepts and procedures was another prevalent theme in our survey. We found ten articles that analyzed this concept, reporting on methods of topic introduction, development of concepts and procedures, and how topics related to other content areas (Cai et al., 2002; Cheng & Wang, 2012; Han, Rosli, Capraro, & Capraro, 2011; Kim, 2012; Li, Ding, Capraro, & Capraro, 2008; Li et al., 2009; Son & Senk, 2010; Sun, 2011; Sun & Kulm, 2010; Watanabe, 2003). Of the ten, three articles analyzed how fractions and fraction operations are introduced and developed in US and Chinese series (Son & Senk, 2010; Sun, 2011; Sun & Kulm, 2010). Sun and Kulm (2010) analyzed one Chinese textbook series and one standards-based US textbook series (*Connected Mathematics*) with respect to fraction concepts. They found that the US textbooks focused on measurement and part-whole sub-constructs, while Chinese texts emphasized fractions as division and fractions as part-whole. According to Sun and Kulm, 55.6% of the content in the standards-based US textbook series involved the use of fraction strips as a measurement tool and 30.5% of the content related to the part-whole sub-construct in a real-world context. In contrast, 66.7% of the content in the Chinese textbook series focused on fractions as division, while 27.8% of the content was devoted to the part-whole sub-construct with an emphasis on equal sharing.

Sun (2011) also found variations in developing fraction division between four US and three Chinese series. According to Sun (2011), the Chinese textbook solidifies new concepts through abbreviated problem sets with conceptual connections, while the US series first utilizes repetition for retention of the fraction division procedure and then develops fluency. For example, Chinese texts treated fraction division as the inverse computation of fraction multiplication, with further explanation on how these two computations are conceptually related. In contrast, US textbooks used the idea of the reciprocal and the "flip-and-multiply" algorithm. Although the US textbooks also used verbal explanations and illustrations, they did not explain why the procedure works.

Continuing the exploration of developing mathematical concepts and procedures in US and Chinese texts, Cheng and Wang (2012) analyzed one standards-based (*Investigations*) and one traditional US series (*Mathematics*), along with two Chinese textbooks, for the development of number sense. They found that textbooks in both education systems stress the counting property of number sense. However, US traditional and reformed textbooks pay much more attention to number sense properties, such as various ways of counting, number meaning and representation, place value and base-ten concepts, and different number composition, compared to the Chinese traditional and reformed textbooks.

Further analysis of fractional concepts was performed by Li et al. (2009) on US, Chinese, and Japanese textbook series. The concept of fraction division was analyzed in three standards-based (*Mathematics in Context, Connected Mathematics*, and *MathScape*) and one traditional US textbook (*Glencoe*), alongside three Chinese and three Japanese series. Like Sun (2011), Li et al. (2009) found that all the books from Japan and China treated fraction division as an inverse computation of fraction multiplication, with further explanation of how these two computations are conceptually related. For example, Japanese textbooks introduced "$\frac{5}{8} \div \frac{1}{3}$" through a context-based problem and then introduced at least two ways of solving the problem: one with the anticipated computation of fraction division (i.e., the inverse computation of fraction multiplication, $\frac{5}{8} \div \frac{1}{3} = \frac{5}{8} \times \frac{3}{1} = \frac{15}{8}$) and the other with an alternative solution using multiplication (i.e., $\frac{5}{8} \div \frac{1}{3} = \left(\frac{5}{8} \div 1\right) \times 3 = \frac{5}{8 \times 1} \times 3 = \frac{5 \times 3}{8 \times 1}$). However, US textbooks emphasized how the division computation can be explained in a way similar to the division of whole numbers. For example, US textbooks tended to explain the meaning of division using "partitive" or "measurement" interpretations, with whole number division word problems. According to Li et al. (2009), Chinese texts tended to provide verbal explanations, numerical computations, and pictorial representations in connection with fraction division computation. In contrast, in the US textbooks verbal explanations and numerical expressions were only used to complement the process of fraction division. Japanese texts used some explanations, but were the first to use line segment representations to show relationships.

Analyzing the concept of averaging in US, Chinese, Japanese, and Taiwanese textbook series, Cai et al. (2002) found that the US textbook series focused on averages as measures of central tendency where Asian series focused on the meaning of average. Additionally, being able to solve complex problems is an explicit standard for China and Taiwan. All series used the process of evening out to find an average, but the Asian series used it as a model to mediate learning of the concept while the US series used it as representative of a data set. Watanabe (2003) also analyzed US and Japanese texts for representations of the multiplication of whole numbers. He found that all series use array situations and equal sets. Also, both the Japanese and the US textbook series provided in-depth discussions of the rationale for specific instructional decisions. However, the Japanese series did not emphasize the multiplier/multiplicand distinction and focused on only one property instead of multiple properties/strategies.

The introduction and development of mathematical concepts and procedures also varied between US and Korean textbook series. One standards-based US textbook series (*Everyday Mathematics*) and three Korean textbooks were analyzed by Son and Senk (2010). Looking at fraction multiplication and division, Son and Senk found that while the US textbook series introduced fraction multiplication as "part of a fractional part," the Korean textbooks introduced it as repeated addition and expanded this meaning from "part of whole units" to "part of a fractional part." Han et al. (2011), who analyzed two Korean, two Malaysian, and four US traditional textbooks for probability, also reported variations in terms of definitions of probability, and noted that only one textbook included experimental probability.

Textbook Size and Allocation of Content

Textbook size and *allocation of content* was the third most-researched theme in our survey. Textbook size refers to the total number of pages in the textbook, the overall dimensions of the textbook, and the number of pages per chapter. Allocation of content refers to the number of pages spent on a particular mathematical concept. Five articles fell under this theme (Alajmi, 2012; Choi & Park, 2013; Li, 2007; Saminy & Liu, 1997; Yan & Lianghuo, 2006). All five studies reported very similar findings. When comparing US and Korean textbooks, Choi and Park (2013) reported that US textbooks had more textbook pages and more overall chapters than Korean textbooks. The US textbooks also tended to be physically larger and have a significantly greater number of pages than Japanese texts (Alajmi, 2012; Saminy & Liu, 1997). However, while Saminy and Liu (1997) reported that Japan had a larger number of chapters in each textbook than the USA, Alajmi (2012) reported later that the US series had a greater number of chapters than Japanese textbooks. Finally, Li (2007) reported that US textbooks are longer than those in Hong Kong, China, and Singapore.

Allocation of time related to specific content areas tended to be greater in the US than in Asian education systems. Choi and Park (2013) found that more pages of the Korean textbooks were allocated to geometry, but the number of chapters devoted to geometry was similar in the USA and Korea. In terms of fractions, US textbooks had a greater number of chapters on fractions and a greater percentage of pages with fractions, as well as a greater number of fraction lessons than the textbooks in Japan (Alajmi, 2012). Similarly, Yan and Lianghuo (2006) found that the overall number of problems in the US textbook they analyzed was greater than the overall number of problems in the Chinese texts. However, they reported that the number of problems in a single section was nearly the same for textbooks in both countries. Yan and Lianghuo (2006) did not explain how each section can have similar amounts of problems when the overall number is higher in US textbooks. They also found that the ratio of exercise problems to text problems was higher in the USA than in China. Exercise problems are tasks which students are to complete on their own, whereas text problems are designed to include teacher intervention during the lesson. Other dimensions of content analysis, including repetition of content, the use of technology, and the use of worked examples, were not prevalent in our survey.

Summary and Implications of Content Analysis

Our survey of textbook analysis studies emphasizing content analysis suggests some variations and commonality in terms of learning opportunities presented in the US and Asian mathematics textbooks. First, in both the early grades and the later grades, mathematical topics tend to appear earlier in Asian education systems than they do in the USA. Textbooks in the USA tend to be physically larger as well as to contain more pages than Asian textbooks. Individual textbooks vary in size, but it is commonly reported that US textbooks spend more time on specific content areas, as well as on revisiting previously taught material. With regard to textbook exercises, the USA tends to have a larger ratio of exercise (practice) problems to text problems (in-class activities) than Asian countries. Furthermore, there are some commonalities but variations in how certain mathematical ideas are introduced and developed among the surveyed education systems.

Problem Analysis: Common Themes

Problem analysis entails classifying textbook exercises and problems by various kinds of schemes, as shown in Table 1.4. In problem analysis, problems can be defined as tasks appearing in textbooks, teacher manuals, and supplemental materials which are done during the instructional lesson or as independent practice by the students. Although a mathematical task can be a set of problems with a particular goal, we used *problems* and *tasks* interchangeably. Thus, problem analysis includes mathematical task analysis. Our survey of the problem analysis articles was based on the dimensions of problem requirements set forth by Li (2000). We categorized findings based on the following dimensions: *mathematical features* (the number of steps required in solving problems), *contextual features* (whether problems are presented in purely mathematical contexts, real-world contexts, or with illustrations), *response types* (type of answers required), *cognitive demand* (Stein et al., 1996), *cognitive expectation* (Son & Senk, 2010), *depth of knowledge* (Webb, 1999), and *the relevance of non-textual elements* like photos, pictorial illustrations, visual representations, and pictures (Kim, 2012). Table 1.4 gives a frequency count of peer-reviewed articles addressing these components. A total of 25 articles fell into the problem analysis group. Of these, 11 were strictly problem analysis and the remaining studies analyzed both content and problems.

Table 1.4 Frequency of the sub-dimensions of problem analysis

Problem analysis	Frequency
Mathematical features	
Single/multiple computational steps	4
Contextual features	
Purely mathematical/real world or illustrative	12
Response type	
Answer only/numerical expression required/explanation required	6
Cognitive demand	
High (doing mathematics/procedures with connections) Low (memorization; procedures without connections)	1
Cognitive expectation	
Conceptual knowledge, procedural knowledge, problem-solving, representation, and mathematical reasoning	10
Depth of knowledge:	1
Level 1; Level 2; Level 3; Level 4	
Relevance of non-textual elements	5

In the USA, there are two major textbook formats with differing pedagogical approaches, referred to here as traditional (commercial) and reform-oriented (standards-based) curricular materials. Standards-based materials are those that adopt the recommendations of the NCTM (1989, 2000) with the support from National Science Foundation, i.e., to include a classroom pedagogy that fosters the understanding of discrete concepts through communication and problem-solving (Senk & Thompson, 2003). Traditional textbooks tend to utilize direct instructional methods and reinforce concepts through individual practice (Senk & Thompson, 2003). While many traditional textbooks cite the content recommendations of the NCTM, ideological and political disputes have allowed them to retain their traditional pedagogy (Schoenfeld, 2004). In the USA, the choice of a mathematics textbook often occurs at the school level, and school districts have had a choice between traditional and reform curriculum materials (Reys, Reys, Lapan, & Holliday, 2003). Both types can be described as highly utilized across the USA. However, standards-based textbooks have been reported to have a higher level of conceptual questions than traditional textbooks (Son & Senk, 2010). This is important to recognize when interpreting the results of problem analysis. Table 1.4 shows that the most prevalent themes for problem analysis are *contextual features* and the depth and breadth of *cognitive requirements (cognitive expectation, cognitive demand,* and *depth of knowledge).*

Mathematical Features

Mathematical features involve the number of steps required to answer a problem: tasks are classified as a single computational procedure or multiple computational procedures (Li, 2000). Four surveyed articles reported on mathematical features of the problems, with mixed results (Li, 2000; Li et al., 2009; Son & Senk, 2010; Yan & Lianghuo, 2006). When comparing the multiplication of fractions in US standards-based and Korean textbook series, Son and Senk (2010) found that 17% of the tasks in the Korean series required a student to use multiple steps to solve a problem, while only 2% of the US textbooks required the use of multiple steps. When analyzing five US and four Chinese textbooks, Li (2000) found that in both US and Chinese textbook series, the majority of the problems (80% in both countries) required a single step only. Unlike Li (2000), Yan and Lianghuo (2006) did find differences between US standards-based and Chinese texts, without looking at a specific content area. They found that in the US textbook over 63% of the tasks required only a single computational step, while 52% of the tasks in the Chinese textbook required only a single computational step. Supporting these results, Li et al. (2009) found that in both standards-based and traditional US textbooks, fraction division problems mainly required a single computation step, while Chinese and Japanese texts included many more multistep problems.

Contextual Features

The most prevalent theme for problem analysis was *contextual features*. This term refers to the setting of the task, which involves whether a problem that is presented with illustrations including representations and/or real-life contexts or presented purely mathematically (Li, 2000). Twelve of the articles we surveyed were counted as addressing this aspect (Alajmi, 2012; Choi & Park, 2013; Han et al., 2011; Hong & Choi, 2014; Kang, 2014; Li, 2000, 2007; Son, 2012; Son & Senk, 2010; Sun & Kulm, 2010; Yan & Lianghuo, 2006; Yang et al., 2010). Kang (2014) analyzed one traditional (*Harcourt*) and one standards-based US textbook series (*Investigations*), along with one Korean textbook series. In his analysis of addition and subtraction in first grade, Kang found that the traditional US textbook series lacked attention to real-life contexts, with a whopping 91% of the tasks being purely mathematical. The Korean textbook contained 68% purely mathematical questions, with the US standards-based textbook having the fewest purely mathematical questions at 63%. For fourth-grade materials, Kang reported similar findings. Han et al. (2011) analyzed US and Korean textbooks for probability. Four US traditional textbook series (*Glencoe, Saxon, McDougal Littell,* and *Prentice Hall*) and two Korean textbooks were found to focus on routine, closed-ended, non-contextual problems. Choi and Park (2013) found similar results when looking at geometry tasks. They found that while the standards-based US textbook (*Connected Mathematics 2*) typically introduced geometry using real-life applications, only a small portion of real-life tasks were included in Korean texts.

When comparing US and Chinese textbooks, the results of our survey indicate that both US and Chinese texts contain a majority of purely mathematical tasks. Li (2000) found that textbooks in both countries (5 in the USA and 4 in China) had a majority of tasks that were purely mathematical and only required a single computational step for integer addition and subtraction. Similarly, Yan and Lianghuo (2006) found that a standards-based US textbook developed by the University of Chicago School Mathematics Project and a Chinese textbook published by the People's Education Press contained a majority of purely mathematical tasks. However, the US textbook contained more application problems than the Chinese series. Sun and Kulm (2010) found similar results when analyzing the learning of fractions. Sun and Kulm noted that the US standards-based textbook also used more real-world representations (51%) than the Chinese textbook (11%).

In comparing US, Chinese, and Singaporean textbooks, Li (2007) found that all the textbooks were dominated by purely mathematical contexts. However, the five US textbooks analyzed had an average of 15% real-world problems with only 6% of the Asian tasks being real-world examples. Li reports that 62% of the US and 90% of the Asian series required the use of routine procedures. When comparing US, Taiwanese, and Singaporean textbooks for the development of fractions, Yang et al. (2010) found that the standards-based US series had more real-world problems than the Taiwanese and Singaporean series. Over 95% of the tasks in the US series were coded as real-world, while only 48% and 55% were coded as real-world in Taiwanese and Singaporean textbooks, respectively.

Response Types

Response types, the third most-researched theme in problem analysis, examined tasks that require students either to provide an answer only or to explain or justify their reasoning (Li, 2000). Six surveyed articles reported on the type of student responses (Hong & Choi, 2014; Li, 2007, 2014; Son & Senk, 2010; Sun & Kulm, 2010; Xin et al., 2011). Findings were mixed when comparing response types in US and Chinese texts (Li, 2007, 2014; Sun & Kulm, 2010; Xin et al., 2011). When looking at algebraic problems, Li (2007) found that out of five US curriculum series, one Chinese textbook, and one textbook from Singapore, US texts put more emphasis on explanation. Approximately 6.9% of the problems in US textbooks required an explanation, followed by China at 2.9% and finally Singapore at 0.1%. Later, Li (2014) found that out of five US textbooks and four Chinese textbooks, none of the Chinese textbook problems required an explanation when analyzing problems that immediately followed the introduction of integer addition and subtraction. However, 19% of the problems in US textbooks required an explanation. Unlike Li (2007, 2014), when analyzing the development of fraction concepts Sun and Kulm (2010) found that the Chinese series used more questions that required an explanation or justification than the US standards-based textbook series did. Furthermore, Xin et al. (2011) analyzed four traditional and two standards-based US series, along with one series from China. They found that the US traditional books had the fewest problems that required an explanation, followed by the Chinese books, whereas the standards-based series had the most problems that required an explanation.

Comparing multiplication and division of fractions in US and Korean textbooks, Son and Senk (2010) found that Korean texts had a greater number of problems that required an explanation than the US standards-based textbook (*Everyday Mathematics*) had. However, different tendencies were reported by Hong and Choi (2014). In their analysis of quadratic equations in three US standards-based textbooks and two Korean textbooks, Hong and Choi found more than 30% of the problems in the US series required explanations while only 15.7% of problems in the Korean books required an explanation. This tendency suggests that based on the content area, grade level, and the type of textbook analyzed in each country, the findings of textbook analysis of response types can be quite varied.

Cognitive Demand

Table 1.5 presents similarities and differences among cognitive demand, cognitive expectation, and depth of knowledge. As mentioned earlier, according to Stein et al. (1996), cognitive demand refers to the kind and level of thinking required when students are working on mathematical problems and tasks. According to Stein, Grover, and Henningsen, different mathematical problems place differing cognitive demands on students and can be categorized into two kinds—(1) low cognitive demand tasks (i.e., " procedures without connections" and "memorization") and (2) high cognitive demand tasks (i.e., "doing mathematics" and "procedures with connections"). In a similar vein, Webb (1999, 2002) developed the Depth-of-Knowledge framework that measures (more) specific cognitive demands of mathematical problems. The framework, which features three cognitive complexity levels (i.e., low, moderate, and high) and four levels of knowledge depth (Level 1: Recall/Reproduce; Level 2: Basic application of skill/concept; Level 3: Strategic

Table 1.5 Framework used for analysis of mathematical problems in textbooks

Focus questions	Aspects investigated
1. What is expected in terms of the depth or level of cognitive demand?	• Cognitive demand – High level – Low level
2. What is expected in terms of the breadth of cognitive complexity?	• Cognitive expectation – Conceptual knowledge – Procedural knowledge – Mathematical reasoning – Representation – Problem-solving
3. What is expected in terms of the depth of cognitive complexity?	• Depth of knowledge (DOK) – Level 1: Recall/reproduce – Level 2: Skill/concept – Level 3: Strategic thinking – Level 4: Extended thinking

thinking; and Level 4: Extended thinking), references the complexity of mental processing that must occur to answer a question or perform a task. While cognitive demand and depth of knowledge explain the depth of cognitive complexity of the mathematical problems presented in textbooks, cognitive expectation, as operationalized by Son and Senk (2010), can capture the problems' breadth of cognitive complexity by examining their cognitive expectations, i.e., knowledge and process required for students to solve mathematical problems and tasks.

Only one article was classified as addressing cognitive demand based on Stein, Grover, and Henningsen's framework (Hong & Choi, 2014). In their analysis of quadratic equations in three US standards-based textbooks and two Korean textbooks, Hong and Choi (2014) found that US standards-based textbooks included a higher percentage of problems with higher level cognitive demands than Korean textbooks. However, the majority of problems in both US standards-based and Korean textbook series require lower level cognitive demand: around 87% of the problems in the US standards-based text, and around 94% of the problems in the Korean textbooks.

Cognitive Expectations

Cognitive expectations, the second most-researched theme in problem analysis, refer to the type of mathematical knowledge or processes students should acquire and use when solving mathematical problems, and includes conceptual knowledge, procedural knowledge, problem-solving, representation, and mathematical reasoning. This measure is a composite of the work of Li (2002), Kilpatrick et al. (2001), and NCTM (2000). Referring to NCTM (2000), Son and Senk combined Li's (2002) four components of problems' cognitive expectations with Kilpatrick et al.'s (2001) five interrelated components of mathematical proficiency. Son and Senk then operationalized the cognitive expectation of tasks as the kind of knowledge and processes required when students are working on mathematical problems. Ten articles were counted as addressing cognitive expectation (Ding & Li, 2010; Kang, 2014; Li, 2000, 2007; Li et al., 2008; Son & Senk, 2010; Xin, 2007; Xin et al., 2011; Yan & Lianghuo, 2006; Yang et al., 2010). Note that although these articles did not directly refer to either Son and Senk (2010) or Li (2000), to some extent they looked at the *kind* of knowledge and processes required when students work on mathematical problems and tasks, by categorizing mathematical problems as either concept-based tasks or procedural tasks.

Our survey of articles on cognitive expectations yielded varied results based on the textbooks analyzed and the countries compared. Overall, the articles that compared Korean and US textbooks reported that US standards-based textbooks had a heavier focus on conceptual problems than Korean texts (Kang, 2014; Son & Senk, 2010). While Kang (2014) compared a traditional US textbook, US standards-based textbooks, and Korean textbooks, Son and Senk (2010) compared standards-based textbooks to Korean textbooks. Kang (2014) found that the Korean textbooks fell in between the traditional and standards-based books when it came to a focus on conceptual understanding. The standards-based textbook had the highest number of problems involving conceptual understanding while the traditional US textbook had the least. In a similar vein, Son and Senk reported that while the US series introduced conceptual understanding before algorithms, Korean texts developed conceptual understanding and procedural fluency simultaneously.

The majority of articles which analyzed the cognitive expectation of problems compared US and Chinese textbooks. However, their results differ from the Korean comparisons. Three articles found that US and Chinese texts are similar in their distribution of procedural and conceptual problems, regardless of the type of US book analyzed—Li (2000), Yan and Lianghuo (2006), and Xin (2007). Li (2000) found that the majority of the problems in both US and Chinese textbooks required a single computational procedure. Li analyzed five US textbooks and four Chinese texts, none of which were described as standards-based or traditional. Yan and Lianghuo (2006) found similar results without focusing on a single concept for analysis. Yan and Lianghuo looked at one standards-based US series and one Chinese textbook and found that both books contained a majority of routine, procedure-based problems. However, the Chinese texts had more problems that required multiple steps (problems that cannot be solved using one direct operation). Xin (2007) analyzed word problem-solving tasks, and again found that US traditional textbooks and Chinese textbooks had a similar distribution of word problems.

Alternatively, three other surveyed articles reported Chinese textbooks having higher numbers of conceptually based problems than US textbooks (Ding & Li, 2010; Li et al., 2009; Xin et al., 2011). When analyzing problems in multiplication and division, Xin et al. (2011) found that one Chinese series required a higher level of conceptual understanding than the four standards-based and two traditional US textbook series. When analyzing the distributive property, Ding and Li (2010) found that the Chinese textbook they analyzed aimed at conceptual understanding and utilized a heavy amount of word problems. The two US textbook series analyzed were not defined as being standards-based or traditional, but the authors found that the distributive property was mostly used for computation, rather than for conceptual understanding.

A comparison of US, Chinese, and Singaporean texts revealed that Singaporean and Chinese texts expected more from students when they were analyzing traditional algebraic problems than did the five US texts (Li, 2007). However, only 62% of the US textbooks and 90% of the Asian textbooks were based on routine procedures, indicating that US texts have a higher number of conceptually based problems. Finally, Yang et al. (2010) found that US standards-based textbooks emphasize conceptual knowledge while Singaporean and Taiwanese texts were more focused on procedural knowledge.

Depth of Knowledge

Only one article was found to use Webb's depth of knowledge framework. Son (2012) analyzed one standards-based US textbook (*Everyday Mathematics*) and two Korean textbook series on the topic of fraction addition and subtraction. In particular, to characterize the reform efforts in South Korea, she compared the quality of the mathematical problems in the reformed version of the Korean textbooks to the previous version. When comparing the two Korean series, the reformed version of the Korean textbooks provided better opportunities for students to learn fraction addition and subtraction than its previous version, in terms of the depth of cognitive complexity. However, the selected standards-based US series provided a more balanced level of depth of knowledge than the revised version of the Korean reform textbooks, by providing more opportunities for students to use strategic thinking and extended thinking.

Relevance of Non-textual Elements to Concepts or Problems

Existing research has emphasized the importance of both visual representations and pictorial representations in teaching and learning, because they provide students with concrete and concise images of related concepts (NCTM, 2000). *Non-textual elements* refer to the context in which a mathematical concept or problem is visually presented, whether in the form of drawings, illustrations, pictures, or mathematical representations such as mathematical diagrams. While contextual features only capture the presence of *non-textual elements* in textbooks, *relevance of non-textual elements* provides additional information on how non-textual elements are related to mathematical problems or concepts.

Four of the articles we surveyed focused on the relevance of *non-textual elements* (Kim, 2012; Li, 2007; Li et al., 2009; Saminy & Liu, 1997). Saminy and Liu (1997) found that an American textbook used 60 picture sets within 13 pages for the concept of subtraction, while a Japanese book used only 28 picture sets within 8 pages. Thus, the American text in their study had 4–5 pictures per page while the Japanese one had only 3–4. In addition, Saminy and Liu reported that in the American text, some pictures might be visually entertaining to students, but they did not always relate to the concept the textbooks intended to teach. This is because one important criterion in the section of pictures in the American textbook was the book's themes, e.g., "Tumble through the Jungle" or "Fun through the Seasons." For example, in the unit on subtraction, two pictures of forests (one in spring and one in winter) were presented to teach the subtraction facts of eight, but they were not relevant to the concept of subtraction. In contrast, in the Japanese text, mathematical concepts are the primary organizing criterion for the selection of pictures. Similarly, Li (2007), who analyzed algebra content in mathematics textbooks from four education systems, reported that the US algebra textbooks had illustrations and figures on nearly every page while the textbooks from Hong Kong, China, and Singapore were basically black-and-white with only a few illustrations. Li et al. (2009) focused on the role and relevance of pictorial representations in illustrating mathematical concepts in three countries' textbooks. They found that in US and Chinese textbooks, pictorial representations played a major role in illustrating the solution to a fraction division problem, and followed up with explanations and numerical representations. In contrast, Japanese textbooks mainly relied on numerical expressions to explain the fraction division algorithm.

Similar findings were observed by Kim (2012), who reported a significant difference across topics and textbooks in regard to students' learning opportunities though non-textual elements. Two standards-based (*Connected Mathematics, MathThematics*) and one traditional US textbook series (*Holt Middle School Mathematics*), along with three Korean textbook series, were analyzed. In her analysis of non-textual elements in mathematics textbooks, Kim developed a conceptual framework which includes the following four aspects: *accuracy* (i.e., how non-textual elements represent concepts and ideas correctly in mathematical ways), *connectivity* (i.e., how closely non-textual elements are related to the mathematical content), *contextuality* (i.e., whether mathematical problems are presented in realistic contexts), and *conciseness* (i.e., how a non-textual element is concise and neat in presenting a concept or problem without any unnecessary or distracting factors). Looking at angle, slope, and prime factorization, Kim found that overall, non-textual elements were accurate, well connected, and concise in both countries. However, mathematical representations tended to be used more often than pictorial representations in both US and Korean textbook series, and the pictorial representations were relatively weaker in terms of accuracy, connectivity, and conciseness compared with mathematical representations.

Summary and Implications of Problem Analysis

Many articles reported on the contextual features (pure math vs. illustrative) of textbook tasks. The majority of tasks in both US and Asian textbooks are purely mathematical. Results were mixed when comparing response types in US and Asian textbooks. Some studies showed that US textbooks had a greater number of tasks that required students to explain or justify their reasoning, compared to Chinese texts. However, other studies reported contradictory findings. The same contradictions are found when comparing US and Korean texts. These conflicting results can be attributed to the types of US textbooks, the content area, and the grade levels (e.g., elementary vs. secondary) analyzed. Similarly, with regard to cognitive requirements (i.e., cognitive demand, cognitive expectation, and depth of knowledge), results are mixed. Some studies suggested that the US and Chinese textbooks have similarly cognitively demanding tasks, while others report that Chinese books had more tasks that require high cognitive demand. Considering the type of US textbooks analyzed did not resolve this conflict. However, there were clearer results when comparing Korea to other Asian countries. The US textbooks tended to have a greater number of cognitively demanding tasks than textbooks in Korea (at the secondary level), Taiwan, and Singapore, regardless of the type of US textbook series analyzed. Non-textual elements (e.g., drawings, pictures) tended to be used more frequently in US texts than in Asian countries. However, they were not always relevant to the mathematical content in US textbooks.

Summary and Limitations

Our survey of 31 textbook analysis studies has shed some light on the relationship of textbook characteristics among the USA and five high-achieving Asian education systems. It appears that most studies utilizing textbook analysis have consistently revealed, to a greater or lesser degree, the inadequacy of textbooks in presenting mathematics content, topics, and problem-solving. Remarkable differences were found in textbooks from different series and particularly from different education systems. These results indicate both challenges and a need for researchers, curriculum developers, policymakers, and schoolteachers to conduct further research and action. More specifically, by breaking down our analysis into the structure of the content and problems in those textbooks, we were able to determine common themes throughout the articles we surveyed. Additionally, we were able to determine the most widely analyzed education systems, grades, and topics. We found that textbooks in China were compared most frequently with US series. Taiwanese textbooks were least often compared to their US counterparts. The majority of articles focused on elementary grades (1–5), with high school grades (9–12) being the least analyzed. Conceptual components of fraction development and operations with fractions were the most analyzed topic in our survey.

Based on comparative content analysis of textbooks in American and Asian education systems, several conclusions can be drawn. In both the early grades and the later grades, mathematical topics appear earlier in Asian education systems than they do in the USA. Additionally, textbooks in the USA tend to review material more frequently than in Asian education systems. While this was not a prevalent finding among the majority of articles, it is important to consider for textbook publishers and curriculum developers. Textbooks in the USA also tend to be physically larger and contain more pages than Asian textbooks. Results on the number and length of chapters vary, but the US books tend to have more chapters and more pages per chapter on specific content areas. With regard to textbook exercises, the USA tends to have a greater number of problem tasks as well as exercise-to-text problems than Asian countries. Non-textual elements are more colorful and are also used more frequently in US texts than in Asian series (Li, 2007; Mayer, Sims, & Tajika, 1995). However, non-textual elements are not always relevant, or well connected to the math content (Carter et al., 1997).

Topic introduction and development tend to be different in US textbooks than has been reported for Asian textbooks. Based on these findings, we attempted to identify textbook signatures that might represent the features of each education system's textbooks (see Table 1.6). Researchers might examine the hypothesis that different textbooks offer different learning opportunities to students and consequently contribute to differences in student achievement. Specifically, based on the results of this study, we hypothesize the following characteristics of textbooks that may partly account for the superior achievement of Asian students to US students on international assessments.

Table 1.6 Textbook signatures in the US and five high-achieving Asian education systems

Textbook	Content analysis	Problem analysis
US Traditional	1. Larger than Asian series 2. More pages than Asian series 3. Later presentation of topics than Asian series 4. Greater number of exercises than Asian series 5. Greater number of illustrations than Asian series and illustrations that relate less closely to concepts	1. Greater number of tasks requiring a single step than Asian series 2. Fewer number of real-world tasks than Asian series and US standards-based series 3. Heavier focus on conceptually based problems than Korean textbooks, but fewer than standards-based US textbooks. For US vs. Chinese books, equal focus on concepts, though some studies say China has more
US Standards-Based	1. Larger than Asian series 2. More pages than Asian series 3. Later presentation of topics than Asian series 4. Greater number of exercises than Asian series 5. Greater number of non-textual elements than Asian series and non-textual elements that relate less closely to concepts	1. Greater number of tasks requiring a single step than Asian series 2. Greater number of real-world tasks than Asian series and US traditional series 3. Heavier focus on conceptually based problems than Korean series. For US vs. Chinese books, equal focus on concepts, though some studies say China has more. 4. More hands-on activities than Chinese books 5. Connects concepts to other mathematical content 6. More use of pictorial representations
Korea	1. Topics occur earlier than US standards and traditional 2. Texts have less pages and less chapters than US standards and traditional 3. More pages devoted to geometry than US standards-based	1. Required greater use of multiple steps in dealing with fractions than US standards-based 2. Amount of purely mathematical tasks fell in between US standards-based and US traditional 3. Conflicting findings regarding the requirement of an explanation when comparing US texts 4. Fewer conceptually based problems than US standards and traditional books 5. Less non-textual elements than the USA, but were more connected to content

(continued)

Table 1.6 (continued)

Textbook	Content analysis	Problem analysis
China	1. Topics occur earlier than US standards-based and traditional series 2. Textbooks are shorter than US standards-based and traditional series 3. Smaller number of overall problems than US standards-based series 4. Fewer exercise-to-text problems than US standards-based series 5. Fewer illustrations than US series	1. Greater number of tasks that require multiple steps than US standards-based and traditional series 2. Fewer number of real-world problems than US standards-based series 3. Conflicting findings regarding the requirement for an explanation when compared to US texts 4. Similar focus on procedural and conceptual tasks when compared to US traditional and standards-based series, but conflicting results find that China has more conceptual tasks
Japan	1. Topics occur earlier than US standards-based and traditional series 2. Texts are smaller, with fewer pages than US standards-based and traditional series 3. Fewer pages, chapters, and lessons regarding fractions than US traditional series 4. Fewer illustrations than US traditional series; non-textual elements do not always relate to concepts	1. Greater number of tasks that require multiple steps than US standards-based and traditional series 2. Fewer number of real-world tasks than US traditional texts
Taiwan	1. Topics occur earlier than US standards-based and traditional series	1. Fewer real-world problems than US standards-based texts 2. More focused on procedural questions than US series
Singapore	1. Textbooks are shorter than US standards-based and traditional 2. Fewer illustrations than the USA	1. Fewer real-world problems than US standards-based texts 2. Fewer problems that require an explanation vs. US or Chinese series 3. More focused on procedural questions than US series

First, decreasing the number of superfluous and irrelevant illustrations would reduce the number of distractions students encounter. Textbook publishers, especially those in the USA, might increase the amount of time dedicated to the depth of knowledge needed for specific content areas. Second, changes to the actual problem tasks students are exposed to can be updated as well. Tasks' levels of cognitive demand and their context (real world vs. pure math) can play a significant role in student achievement on tests comparing US and Asian education systems. However, since results on response types varied dramatically based on education systems and type of textbooks analyzed, we cannot determine that response type is a contributing factor to the assessment gaps.

Finally, when analyzing different representations of mathematical concepts, it appears that US textbooks seem to use multiple representations, ones not necessarily directly connected to the mathematical concepts, and they revisit concepts in order to solidify mathematical fluency. Asian textbooks tend to be succinct in their representations of mathematical concepts. Given the results of varied representations, it may be beneficial to consider limiting the amount of variation before moving on to alternate representations.

Implications and Future Directions

No matter how large a gap might exist between the intended and potentially implemented curriculum, teachers who provide additional instruction for their students can narrow this gap (Cai & Wang, 2010; Cai et al., 2014). In our collection of textbook analysis studies, many studies only looked at the intended curriculum or potentially implemented curriculum, and not the implemented curriculum. Only seven studies indicated this aspect as one of the limitations of their respective studies. As mentioned previously, we do not support the direct connection between textbook analysis and student achievement because there are many other factors to be considered. Textbook analysis might be able to predict, but can never conclude with confidence, the actual classroom use of texts. We thus highlight the important role of teachers and suggest that teachers need to be aware of what is intended and what is presented in textbooks for teaching mathematics, and then work toward helping students make sense of mathematics.

This chapter summarized the findings from prior research on textbook analysis with respect to content and problem analysis, and raised questions and issues for mathematics education researchers to further advance the research on mathematics textbook analysis. First, we think that it is necessary for researchers to establish a more solid fundamental conceptualization and theoretical underpinning for analytical frameworks. Unfortunately, our survey of common themes for textbook analysis faced an added layer of difficulty due to the lack of a common analytical framework. Many researchers used their own frameworks to suit the needs of the particular content they were examining. It was thus difficult to compare their findings directly, so we identified common themes based on findings, rather than on a given analytical framework.

To address these problems, future researchers might use a framework that encompasses wider educational and social contexts. Few textbook analyses focus on gender, ethnicity, and equity issues in exploring students' learning opportunities presented in textbooks. Moreover, only two studies looked at the language used in the books—Herbel-Eisenmann (2007) and O'Keeffe and O'Donoghue (2015). Only looking at the tasks themselves is not enough: the author's voice can have an impact on how students see those tasks. Furthermore, we observed some methodological issues in our collection of textbook analysis studies. Patton (2002) stated that validity and reliability are two factors which any researcher should be concerned about while designing a study, analyzing results, and judging the quality of the study. Given that different researchers use different frameworks, it is important to establish clear research questions and valid analysis methods, including coding systems. However, clearly defined research questions were lacking in several studies (e.g., Ding & Li, 2010). Eight studies did not clearly articulate their framework with examples of tasks using the coding method (e.g., Sun & Kulm, 2010). Of the 31 articles reviewed, only 16 studies clearly stated some type of inter-rater reliability along with percentages of agreement of the inter-rater reliability.

Indeed, as Fan, Zhu, and Miao (2013) pointed out, there is a strong need for more confirmatory research about the relationship between textbooks and students' learning outcomes. As reported earlier, the results are mixed when comparing learning opportunities presented in US and Asian textbooks and even within the same education system, so the research evidence for a positive correlation between textbooks and students' learning outcomes is weak and inconclusive (Fan et al., 2013). This is because the results are often based on a comparison of selected textbooks, investigating the differences between textbooks in different countries and comparing grade levels in these countries. In these studies, the issues of whether the selected textbooks are a good representation of all the available textbooks, and whether the selected framework and methodology are solid, were often ignored or taken for granted. Similar to our perspective and findings, Cai and Cirillo (2014) also call for more careful attentions to the variety of the methods used in textbook analysis studies in order to move the field forward in understanding the potential role of the curriculum on students' learning. Future researchers should therefore consider the following questions:

- How many textbooks should we analyze?
- Which textbook(s) should we analyze?
- What text in the textbook(s) series should we analyze?
- How much of that text should we analyze?
- How should we analyze it?
- What research questions should guide our analysis?
- What framework should we use?
- How can a researcher persuade his or her audiences that the research findings of textbook analysis are relevant and worthwhile?

References

Bos, K., & Kuiper, W. (1999). Modelling TIMSS data in a European comparative perspective: Exploring influencing factors on achievement in mathematics in Grade 8. *Educational Research and Evaluation, 5*(5), 157–179.

Cai, J. (2010). Evaluation of mathematics education programs. In E. Baker, B. McGraw, & P. Peterson (Eds.), *International encyclopedia of education* (Vol. 2, pp. 653–659). Oxford: Elsevier.

Cai, J., & Cirillo, M. (2014). What do we know about reasoning and proving? Opportunities and missed opportunities from curriculum analyses. *International Journal of Educational Research, 64*, 132–140.

Cai, J., Ding, M., & Wang, T. (2014). How do exemplary Chinese and U.S. mathematics teachers view instructional coherence? *Educational Studies in Mathematics, 85*(2), 265–280.

Cai, J., & Howson, A. G. (2013). Toward an international mathematics curriculum. In M. A. Clements, A. Bishop, C. Keitel, J. Kilpatrick, & K. S. F. Leung (Eds.), *Third international handbook of mathematics education research* (pp. 949–978). New York: Springer.

Cai, J., Mok, I., Reedy, V., & Stacey, K. (2016). *International comparative studies in mathematics: Lessons for improving students' learning*. New York, NY: Springer.

Cai, J., Ni, Y., & Lester, F. K. (2011). Curricular effect on the teaching and learning of mathematics: Findings from two longitudinal studies in China and the United States. *International Journal of Educational Research, 50*(2), 63–64.

Cai, J., & Wang, T. (2010). Conceptions of effective mathematics teaching within a cultural context: Perspectives of teachers from China and the United States. *Journal of Mathematics Teacher Education, 13*, 265–287.

Cai, J., Wang, N., Moyer, J. C., Wang, C., & Nie, B. (2011). Longitudinal investigation of the curriculum effect: An analysis of student learning outcomes from the LieCal Project. *International Journal of Educational Research, 50*(2), 117–136.

Carter, J., Li, Y., & Ferrucci, B. J. (1997). A comparison of how textbooks present integer addition and subtraction in PRC and USA. *The Mathematics Educator, 2*(2), 197–209.

Charalambous, C., Delaney, S., Hsu, H., & Mesa, V. (2010). A comparative analysis of the addition and subtraction of fractions in textbooks from three countries. *Mathematical Thinking and Learning, 12*, 117–151.

Creemers, B. P. M. (1994). *The effective classroom*. London: Cassell.

Fan, L., Zhu, Y., & Miao, Z. (2013). Textbook research in mathematics education: Development status and directions. *ZDM-International Journal of Mathematics Education, 45*(2), 633–645.

Freeman, D. T., & Porter, A. C. (1989). Do textbooks dictate the content of mathematics instruction in elementary schools? *American Educational Research Journal, 26*(3), 343–356.

Fuson, K. C., Stigler, J. W., & Bartsch, K. (1988). Grade placement of addition and subtraction topics in Japan, Mainland China, the Soviet Union, Taiwan, and the United States. *Journal for Research in Mathematics Education, 19*, 449–456.

Herbel-Eisenmann, B. A. (2007). From intended curriculum to written curriculum: Examining the "voice" of a mathematics textbook. *Journal for Research in Mathematics Education, 38*(4), 344–369.

Kilpatrick, J., Swafford, J., & Findell, B. (Eds.). (2001). *Adding it up: Helping children learn mathematics*. Washington, DC: National Academy Press.

Kim, R. Y. (2012). The quality of non-textual elements in mathematics textbooks: An exploratory comparison between South Korea and the United States. *ZDM-International Journal of Mathematics Education, 44*(2), 175–187.

Kupari, P. (2006). Student and school factors affecting Finnish mathematics achievement: Results from TIMSS 1999 data. *Contexts of learning mathematics and science, Lessons learned from TIMSS*, 127–140.

Li, Y. (2002). A comparison of problems that follow selected content presentations in American and Chinese mathematics textbooks. *Journal for Research in Mathematics Education, 31*(2), 234–241.

Li, Y. (2007). Curriculum and culture: An exploratory examination of mathematics curriculum materials in their system and cultural contexts. *The Mathematics Educator, 10*(1), 21–38.

Li, Y., Chen, X., & An, S. (2009). Conceptualizing and organizing content for teaching and learning in selected Chinese, Japanese and US mathematics textbooks: The case of fraction division. *ZDM-International Journal of Mathematics Education, 41*(6), 809–826.

Li, Y. (2014). Learning about and improving teacher preparation for teaching mathematics from an international perspective. In S. Blömeke, F.-J. Hsieh, G. Kaiser & W. H. Schmidt (Eds.). International perspectives on teacher knowledge, beliefs and opportunities to learn. (pp. 49–57). Dordrecht: Springer.

Mesa, V. (2004). Characterizing practices associated with functions in middle school textbooks: An empirical approach. *Educational Studies in Mathematics, 56*, 255–286.

Mullis, I. V. S., Martin, M. O., Foy, P., & Arora, A. (2012). *TIMSS 2011 international results in mathematics*. Chestnut Hill, MA: TIMSS & PIRLS International Study Center, Boston College.

National Council of Teachers of Mathematics. (1989). *Curriculum and evaluation standards.* Reston, VA: Author.

National Council of Teachers of Mathematics. (2000). *Principles and standards for school mathematics.* Reston, VA: Author.

Newton, D. P., & Newton, L. D. (2007). Could elementary mathematics textbooks help give attention to reasons in the classroom? *Educational Studies in Mathematics, 64*(1), 69–84.

O'Keeffe, L., & O'Donoghue, J. (2015). A role for language analysis in mathematics textbook analysis. *International Journal of Science and Mathematics Education, 13*(3), 605–630.

Papanastasiou, C. (2008). A residual analysis of effective schools and effective teaching in mathematics. *Studies in Educational Evaluation, 34*, 24–30.

Patton, M. Q. (2002). *Qualitative evaluation and research methods* (3rd ed.). Thousand Oaks, CA: Sage Publications, Inc..

Remillard, J. T. (2005). Examining key concepts in research on teachers' use of mathematics curricula. *Review of Educational Research, 75*(2), 211–246.

Reys, R., Reys, B., Lapan, R., & Holliday, G. (2003). Assessing the impact of Standards-based middle grades mathematics curriculum materials on student achievement. *The Journal for Research in Mathematics Education, 34*(1), 74–95.

Schmidt, W. H., McKnight, C., & Raizen, S. (1997). *A splintered vision: An investigation of U.S. science and mathematics education*. Dordrecht: Kluwer.

Schmidt, W. H., McKnight, C. C., Houang, R. T., Wang, H. A., Wiley, D. E., Cogan, L. S., et al. (2001). *Why schools matter: A cross-national comparison of curriculum and learning*. San Francisco: Jossey-Bass.

Schmidt, W. H., McKnight, C. C., Cogan, L. S., Jakwerth, P. M., & Houang, R. T. (1999). *Facing the Consequences: Using TIMSS for a Closer Look at US Mathematics and Science Education.* Dordrecht, The Netherlands: Kluwer.

Schoenfeld, A. H. (2004). The math wars. *Educational Policy, 18*(1), 253–286.

Senk, S., & Thompson, D. (2003). *Standards-based school mathematics curricula: What are they? What do students learn?* New Jersey: Lawrence Erlbaum Associates.

Silver, E. (2009). Cross-national comparisons of mathematics curriculum materials: What might we learn? *ZDM-The International Journal on Mathematics Education, 41*, 827–832.

Son, J., & Kim, O. (2015). Teachers' selection and enactment of mathematical problems from textbooks. *Mathematics Education Research Journal, 27*(4), 491–518.

Son, J., & Senk, S. (2010). How reform curricula in the USA and Korea present multiplication and division of fractions. *Educational Studies in Mathematics, 74*(2), 117–142.

Son, J., & Hu, Q. (2016). The initial treatment of the concept of function in selected secondary mathematics textbooks in the US and China. *International Journal of Mathematical Education in Science and Technology, 47*(4), 503–530.

Stein, M. K., Grover, B. W., & Henningsen, M. A. (1996). Building student capacity for mathematical thinking and reasoning: An analysis of mathematical tasks used in reform classrooms. *American Educational Research Journal, 33*(2), 455–488.

Valverde, G. A., Bianchi, L. J., Wolfe, R. G., Schmidt, W. H., & Houang, R. T. (2002). *According to the book: Using TIMSS to investigate the translation of policy into practice through the world of textbooks*. Dordrecht: Kluwer.

Wagemaker, H. (2003). IEA: International studies, impact and transition. Retrieved September 24, 2015 from http://www.iea.nl/fileadmin/user_upload/IRC2004/Wagemaker.pdf.

Watanabe, T. (2003). Teaching multiplication: An analysis of elementary school mathematics teachers' manuals from Japan and the United States. *The Elementary School Journal, 104*(2), 111–125.

Webb, N. L. (1999). Research monograph No. 18. Alignment of science and mathematics standards and assessments in four states. Washington, DC: Council of Chief State School Officers.

Webb, N. L. (2002). *An Analysis of the alignment between mathematics standards and assessments for three states*. Paper presented at the American Educational Research Association Annual Meeting. New Orleans, LA.

Appendix: Surveyed Textbook Analysis Articles

Alajmi, A. H. (2012). How do elementary textbooks address fractions? A review of mathematics textbooks in the USA, Japan, and Kuwait. *Educational Studies in Mathematics, 79*(2), 239–261.

Boonlerts, S., & Inprasitha, M. (2013). The textbook analysis on multiplication: The case of Japan, Singapore and Thailand. *Creative Education, 4*(4), 259.

Cai, J., Lo, J. J., & Watanabe, T. (2002). Intended treatments of arithmetic average in US and Asian school mathematics textbooks. *School Science and Mathematics, 102*(8), 391–404.

Carter, J., Li, Y., & Ferrucci, B. J. (1997). A comparison of how textbooks present integer addition and subtraction in PRC and USA. *The Mathematics Educator*, *2*(2), 197–209.

Charalambous, C. Y., Delaney, S., Hsu, H. Y., & Mesa, V. (2010). A comparative analysis of the addition and subtraction of fractions in textbooks from three countries. *Mathematical Thinking and Learning, 12*(2), 117–151.

Cheng, Q., & Wang, J. (2012). Curriculum opportunities for number sense development: A comparison of first-grade textbooks in China and the United States. *International Journal for Mathematics Teaching and Learning*.

Choi, K. M., & Park, H. J. (2013). A comparative analysis of geometry education on curriculum standards, textbook structure, and textbook items between the US and Korea. *Eurasia Journal of Mathematics, Science & Technology Education, 9*(4), 379–391.

Ding, M., & Li, X. (2010). A comparative analysis of the distributive property in US and Chinese elementary mathematics textbooks. *Cognition and Instruction, 28*(2), 146–180.

Fuson, K. C., Stigler, J. W., & Bartsch, K. (1988). Grade placement of addition and subtraction topics in Japan, mainland China, the Soviet Union, Taiwan, and the United States. *Journal for Research in Mathematics Education, 19*(5), 449–456.

Han, S., Rosli, R., Capraro, R. M., & Capraro, M. M. (2011). The textbook analysis on probability: The case of Korea, Malaysia and U.S. textbooks. *Journal of the Korean Society of Mathematical Education Series D, 15*(2), 127–140.

Hong, D. S., & Choi, K. M. (2014). A comparison of Korean and American secondary school textbooks: The case of quadratic equations. *Educational Studies in Mathematics, 85*(2), 241–263.
Kang, H. J. (2014). A cross-national comparative study of first- and fourth-grade math textbooks between Korea and the United States. *Curriculum and Teaching Dialogue, 16*(1–2), S91.
Kim, R. Y. (2012). The quality of non-textual elements in mathematics textbooks: An exploratory comparison between South Korea and the United States. ZDM Mathematics Education, 44(2), 175–187.
Li, X., Ding, M., Capraro, M. M., & Capraro, R. M. (2008). Sources of differences in children's understandings of mathematical equality: Comparative analysis of teacher guides and student texts in China and the United States. *Cognition and Instruction, 26*(2), 195–217.
Li, Y. (2000). A comparison of problems that follow selected content presentations in American and Chinese mathematics textbooks. *Journal for Research in Mathematics Education, 31*(2), 234–241.
Li, Y. (2007a). Curriculum & cognition: A study on math problems. *Academic Exchange Quarterly, 11*(2), 86.
Li, Y. (2007b). Curriculum and culture: An exploratory examination of mathematics curriculum materials in their system and cultural contexts. *The Mathematics Educator, 10*(1), 21–38.
Li, Y., Chen, X., & An, S. (2009). Conceptualizing and organizing content for teaching and learning in selected Chinese, Japanese and US mathematics textbooks: The case of fraction division. *ZDM Mathematics Education, 41*(6), 809–826.
Mayer, R. E., Sims, V., & Tajika, H. (1995). Brief note: A comparison of how textbooks teach mathematical problem solving in Japan and the United States. *American Educational Research Journal, 32*(2), 443–460.
Saminy, K. K., & Liu, J. (1997). A comparative study of selected United States and Japanese first-grade mathematics textbooks. *Focus on Learning Problems in Mathematics, 19*(2), 1–13.
Son, J. W. (2012). A cross-national comparison of reform curricula in Korea and the US in terms of cognitive complexity: The case of fraction addition and subtraction. *ZDM-International Journal of Mathematics Education, 44*(2), 161–174.
Son, J. W., & Senk, S. L. (2010). How reform curricula in the USA and Korea present multiplication and division of fractions. *Educational Studies in Mathematics, 74*(2), 117–142.
Sun, X. (2011). "Variation problems" and their roles in the topic of fraction division in Chinese mathematics textbook examples. *Educational Studies in Mathematics, 76*(1), 65–85.
Sun, Y., & Kulm, G. (2010). A comparison of US and Chinese middle grades textbook development of fraction concepts. *The Mathematics Educator, 12*(2), 51–62.
Watanabe, T. (2003). Teaching multiplication: An analysis of elementary school mathematics teachers' manuals from Japan and the United States. *The Elementary School Journal, 104*(2), 111–125.
Xin, Y. P. (2007). Word problem solving tasks in textbooks and their relation to student performance. *The Journal of Educational Research, 100*(6), 347–360.
Xin, Y. P., Liu, J., & Zheng, X. (2011). A cross-cultural lesson comparison on teaching the connection between multiplication and division. *School Science and Mathematics, 111*(7), 354–367.
Yan, Z., & Lianghuo, F. (2006). Focus on the representation of problem types in intended curriculum: A comparison of selected mathematics textbooks from Mainland China and the United States. *International Journal of Science and Mathematics Education, 4*(4), 609–626.
Yang, D. C., Reys, R. E., & Wu, L. L. (2010). Comparing the development of fractions in the fifth- and sixth-graders' textbooks of Singapore, Taiwan, and the USA. *School Science and Mathematics, 110*(3), 118–127.

Chapter 2
Intended Treatment of Fractions and Fraction Operations in Mathematics Curricula from Japan, Korea, and Taiwan

Tad Watanabe, Jane-Jane Lo, and Ji-Won Son

Abstract In spite of extensive research efforts, teaching and learning fractions remain challenging throughout the world. Although students' mathematics learning is influenced by many factors, one important factor is the learning opportunities afforded by their textbooks. Therefore, we examined how textbooks from Japan, Korea, and Taiwan—three high-achieving countries prominent in comparative studies—introduced and developed fraction concepts and fraction arithmetic. We used the content analysis method (National Research Council, On evaluating curricular effectiveness: Judging the quality of K-12 mathematics evaluations, 2004) to analyze the problems presented in the textbooks. Our analysis revealed that there were many similarities among the textbooks from these three countries, including the overall flow of the topics related to fraction concepts and fraction arithmetic. However, significant differences included how various fraction subconstructs were integrated in the textbooks and how fraction multiplication and division were discussed. These similarities and differences among high-achieving countries suggest fruitful directions for future research in the area of fraction teaching and learning.

Keywords Curriculum analysis • Textbook analysis • Fractions • Elementary school education • Number concepts and operations

---

T. Watanabe (✉)
Department of Mathematics, Kennesaw State University, Kennesaw, GA, USA
e-mail: twatanab@kennesaw.edu

J.-J. Lo
Department of Mathematics, Western Michigan University, Kalamazoo, MI, USA
e-mail: jane-jane.lo@wmich.edu

J.-W. Son
Department of Learning and Instruction, University at Buffalo—The State University of New York, Buffalo, NY, USA
e-mail: jiwonson@buffalo.edu

J.-W. Son et al. (eds.), *What Matters? Research Trends in International Comparative Studies in Mathematics Education*, Research in Mathematics Education, DOI 10.1007/978-3-319-51187-0_2

Introduction

The teaching and learning of fractions have long attracted the attention of mathematics education researchers (National Research Council [NRC], 2004). However, in spite of almost a half-century of research, these tasks continue to challenge mathematics teachers and students throughout the world. It is generally agreed that developing a deep understanding of fractions is critical for students' success in more advanced mathematics. The National Mathematics Advisory Panel (2008) in the United States, for example, listed fractions as one of the foundational topics for algebra.

While these challenges are widespread, cross-national comparison studies suggest that both teachers and students from East Asian countries seem to possess a deeper understanding of fractions than their counterparts in the United States (Mullis, Martin, & Foy, 2008; Son & Senk, 2010). For example, Mullis, Martin, and Foy (2008) noted that students from Hong Kong, Japan, Korea, Singapore, and Taiwan generally outperformed students from the United States. Table 2.1 shows some of the released Grade 8 mathematics problems from TIMSS 2011, for which significantly higher percentages of students from Japan, Korea, and Taiwan answered correctly compared to students from the United States. Comparing Chinese and US elementary school teachers, Ma (1999) noted that Chinese teachers possessed the deep understanding of elementary school mathematics, including division of fractions, necessary to teach it effectively. Similarly, Lo and Luo (2012) showed that Taiwanese prospective elementary school teachers understand division of fractions more deeply than their US counterparts.

Although a variety of factors influence student achievement, these performance differences might be attributed to variations in mathematical curricula (Reys, Reys, & Chávez, 2004). Textbooks are generally considered the bridge between the intended curriculum and the implemented curriculum. As Kilpatrick, Swafford, and Findell (2001) pointed out, "what is actually taught in classrooms is strongly influenced by the available textbooks" (p. 36). Moreover, while the Chinese teachers in Ma's (1999) study gained their deep understanding of elementary mathematics, at least partly from studying their textbooks, Ball (1996) questioned whether US textbooks are written with teachers' learning in mind. Thus, examining the content of textbooks as a possible contributing factor to achievement gaps seems fruitful, and a growing number of cross-national researches analyzing the content of textbooks have been conducted in recent years. Some of those studies have examined the treatment of specific ideas related to fractions. For example, Charalambous, Delaney, Hsu, and Mesa (2010) examined the treatment of addition and subtraction of fractions in textbooks from Cyprus, Ireland, and Taiwan. Li, Chen, and An (2009) examined how selected textbooks from China, Japan, and the United States discussed division of fractions. Son and Senk (2010) also investigated the treatment of multiplication and division of fractions in textbooks from Korea and the United States.

Table 2.1 Student performance on selected TIMSS 2011 Grade 8 problems, by country (Mullis, Martin, Foy, & Arora, 2012)

	Student percent correct				
Item number/problem statement	JPN (%)	KOR (%)	TAI (%)	US (%)	Int'l Avg. (%)
M032064: Ann and Jenny divide 560 zeds between them. If Jenny gets $\frac{3}{8}$ of the money, how many zeds will Ann get?	45	67	60	25	27
M032094: $\frac{4}{100} + \frac{3}{1000}$	77	89	85	63	62
M032662: 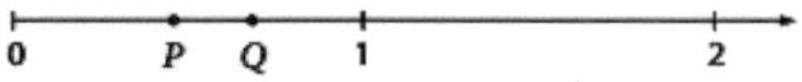 *P* and *Q* represent two fractions on the number line above. $P \times Q = N$. Which of these shows the location of *N* on the number line? 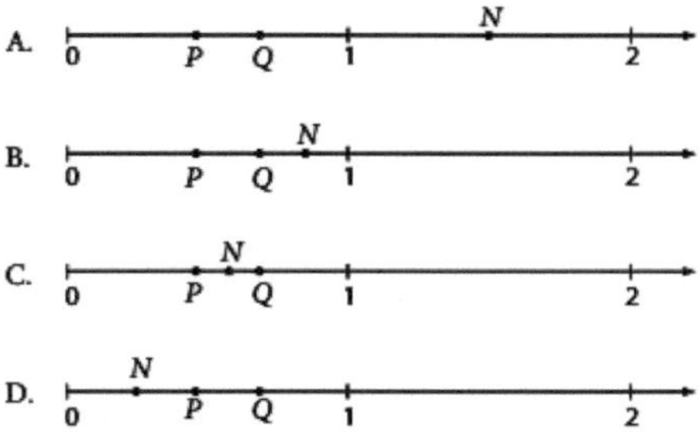 	43	44	53	22	23
M052228: Which shows a correct method for finding $\frac{1}{3} - \frac{1}{4}$? A. $\frac{1-1}{4-3}$ B. $\frac{1}{4-3}$ C. $\frac{3-4}{3\times4}$ D. $\frac{4-3}{3\times4}$	65	86	82	29	37

The purpose of the current study is to add to the growing knowledge base on the content of textbooks from high-achieving East Asian countries. In particular, we hope to deepen our knowledge of how textbooks from Japan, Korea, and Taiwan introduce and develop the mathematically challenging idea of fractions.

Theoretical Perspectives

Textbook Analysis

Textbook analysis—in particular, cross-national textbook analysis—is a relatively new field of inquiry. Some of the existing research has investigated the overall structures of textbooks, often focusing on what mathematics is taught at what grade level (e.g., Schmidt, McKnight, Valverde, Houang, & Wiley, 1997), while other studies examined the treatment of a particular mathematical topic (e.g., Cai, Lo, & Watanabe, 2002; Son & Senk, 2010) or mathematical process (e.g., Fan & Zhu, 2007; Mayer, Sims, & Tajika, 1995). Charalambous et al. (2010) referred to the former approach as horizontal analysis and to the latter as vertical analysis, while Li et al. (2009) called the former type "macroanalysis" and the latter "microanalysis."

Although cross-national horizontal, or macro, analyses of textbooks give us a general sense of what topics are discussed in what grade level across different educational systems, they do not reveal much about the actual learning opportunities offered by different textbooks. On the other hand, because vertical, or micro, analyses of textbooks focus on a single mathematical topic, they can reveal different approaches taken by different textbooks. However, such an analysis does not reveal what influences other topics might have on the treatment of a particular topic. Furthermore, because mathematics consists of many interrelated "topics," it may be difficult to identify the boundaries of a single topic. For example, if we were to examine the treatment of a division algorithm, would we need to examine how division is introduced? What about the treatment of a multiplication algorithm or algorithms? Thus, some researchers chose to examine textbooks by integrating both horizontal (or macro) and vertical (or micro) analysis (e.g., Charalambous et al., 2010; Li et al., 2009).

Selecting which textbooks to include in a cross-national study is also an important consideration. Some researchers selected textbooks based on the characteristics of the education systems. For example, both Boonlerts and Inprasitha (2013) and Charalambous et al. (2010) selected their textbooks from countries with centralized education systems and national curriculum standards. Other studies consider the achievements of the targeted students, either explicitly or implicitly. Those studies will often include textbooks from high-achieving Asian countries and other countries of interest to the researchers. For example, Li et al. (2009) examined textbooks from China, Japan, and the United States, while Boonlerts and Inprasitha (2013) examined textbooks from Japan, Singapore, and Thailand. In many cases, the authors' familiarity with the textbook's language appears to play a role in the selection of textbooks.

Fractions

Teaching and learning fractions have been recognized as problematic for quite some time. Research has revealed a variety of misconceptions children possess about fractions. For example, some students do not appear to understand fractions as numbers or quantities, as the following excerpt from Simon (2002) shows:

> In a fourth-grade class, I asked the students to use a blue rubber band on their geoboards to make a square of a designated size, and then to put a red rubber band around one half of the square. Most of the students divided the square into two congruent rectangles. However, Mary, cut the square on the diagonal, making two congruent right triangles. The students were unanimous in asserting that both fit with my request that they show halves of the square. Further, they were able to justify that assertion.
>
> I then asked the question, "Is Joe's half larger; is Mary's half larger, or are they the same size?" Approximately a third of the class chose each option. In the subsequent discussion, students defended their answers. However, few students changed their answers as a result of the arguments offered.
>
> (Simon, 2002, p. 992)

Another common misconception occurs when students misapply their understanding of whole numbers to fractions. Thus, some students conclude that $\frac{1}{3}$ is greater than $\frac{1}{2}$ because 3 is greater than 2. Larson (1980) noted that many students had difficulty locating fractions on number lines, and Greer (1987) reported on difficulties students had in selecting the appropriate operation to solve word problems. The fact that fractions comprise a multifaceted construct has been identified as contributing to these complexities (Lamon, 2007). Kieren (1976) articulated that fractions consist of five subconstructs—part-whole, measure, quotient, operator, and ratio. Table 2.2 provides a simple summary of these five subconstructs.

The goal of fraction instruction is to help students "recognize nuances in meaning; to associate each meaning with appropriate situations and operations; and, in general, to develop insight, comfort, and flexibility in dealing with the rational numbers" (Lamon, 2007, p. 636). Unfortunately, fraction instruction in the United States rarely extends beyond the part-whole meaning of fractions, despite the consensus that focusing solely on the part-whole meaning of fractions is limiting (e.g., Lamon, 2007).

Table 2.2 Interpretations of $\frac{3}{4}$ according to the five subconstructs (Lamon, 2007)

Part-whole	3 parts out of 4 equal parts of a unit
Measure	3 pieces of $\frac{1}{4}$-units, for example, the distance of $3\frac{1}{4}$-units on a number line
Operator	$\frac{3}{4}$ of something; $\frac{3}{4}$ is a rule that tells how to operate on a unit
Quotient	3 divided by 4
Ratio	3 of A are compared to 4 of B in a multiplicative sense

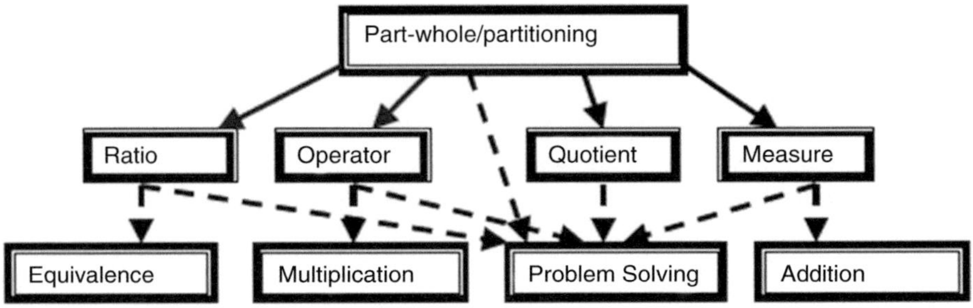

Fig. 2.1 Five subconstructs of fractions and their relationships (Behr et al., 1983)

Behr, Lesh, Post, and Silver (1983) further developed Kieren's (1976) ideas and proposed a theoretical model linking the different interpretations of fractions to the basic operations of fractions, as shown in Fig. 2.1.

According to Behr et al., the part-whole subconstruct of rational numbers is fundamental for developing understanding of the four subordinate constructs of fractions. Moreover, the operator and measure subconstructs are helpful for developing understanding of the multiplication and addition of fractions, respectively. However, there are many unanswered questions about how to incorporate these subconstructs in a mathematics curriculum. For example, is it better for students to be exposed to all five subconstructs early, or is it better to focus on one (beyond part-whole)? If it is better to focus on one, which? Do students need to understand all five subconstructs before algebra? These are some of the outstanding questions that demand mathematics education researchers' attention (Lamon, 2007).

Mack (1990, 1995) examined how educators might be able to take advantage of children's informal understanding of fractions in the formal study of fractions. Her studies suggest that instruction starting with partitioning of a whole might be effective. Pothier and Sawada (1983, 1989) also show that there is a pattern in young children's development of partitioning strategies and their justifications for equality of parts. Armstrong and Larson (1995) asked students in middle grades (Grades 4, 6, and 8) to compare areas of rectangles and triangles embedded in other geometric figures. They found that more students used justifications based on part-whole relationships or partitioning as they became more familiar with fractions. These studies suggest the foundational nature of partitioning activities in the early instruction of fractions.

Steffe, Olive, Tzur, and their colleagues have embarked upon an ambitious multipart study to articulate children's construction of fraction understanding (e.g., Olive, 1999; Steffe, 2002; Tzur, 1999, 2004). Their studies showed that children's whole number concepts did not interfere with their conceptualization of fractions (Olive, 1999; Steffe, 2002; Tzur, 1999, 2004). In fact, the types of units and operations children construct in their whole number sequences can support their development of fraction schemes. However, these studies suggest that teaching which supports students' development of fraction understanding requires a more coherent approach, not only toward fractions but also toward other related ideas, such as multiplication and division of whole numbers. For example, multiplication must go far beyond repeated addition: it must be understood as a way to quantify something when it is composed of several copies of identical size. Such an understanding of multiplication can help students view fraction $\frac{m}{n}$ as m times $\frac{1}{n}$ instead of "m out of n," which does not necessarily signify a quantity. Thompson and Saldanha (2003) noted that "we rarely observe textbooks or teachers discussing the difference between thinking of $\frac{3}{5}$ as 'three out of five' and thinking of it as '3 one fifths'" (p. 107).

Because fractions themselves are multifaceted constructs, students, and often teachers, may have difficulty with fraction arithmetic. As a result, a large number of studies have been conducted to examine students' understanding of fractions, including some cross-national examinations of textbooks (e.g., Charalambous et al., 2010; Li et al., 2009; Son & Senk, 2010). Fraction division in particular has attracted the attention of many researchers as it is recognized as one of the most challenging mathematics topics in the middle grades. Too often, fraction instruction focuses on the invert-and-multiply algorithm of division. However, a major difficulty for students is knowing when division is the appropriate calculation (e.g., Greer, 1987; Siegler & Lortie-Forgues, 2015), in part due to the difficulty of interpreting fraction division. While partitive division—that is, the divisor being the number of equal groups and the quotient being the group size—is more common with whole number division (e.g., Fischbein, Deri, Nello & Marino, 1985), it is easier to interpret fraction division with quotitive division than with partitive division. Some researchers (e.g., Zambat, 2015) recommend students first learn fraction division in quotitive situations, leading to the common denominator algorithm instead of the invert-and-multiply algorithm. On the other hand, many of the Chinese teachers Ma (1999) interviewed were able to give both partitive and quotitive problem situations for fraction division problems. This may suggest important differences in approaches to fraction multiplication and division in East Asia.

In the United States, the Common Core State Standards for Mathematics (CCSSM, Common Core State Standard Initiatives, 2010) suggests a progression of fraction arithmetic. The CCSSM approaches addition and subtraction of fractions by utilizing the idea of non-unit fractions as collections of unit fractions, which seems to be consistent with the idea of Steffe and his colleagues. With multiplication, the CCSSM first focuses on multiplication of fractions by whole numbers in Grade 4, and then multiplication of fractions by fractions in Grade 5. The CCSSM approaches division of fractions by first exploring division of unit fractions by whole numbers and whole numbers by unit fractions in Grade 5, and then fractions divided by fractions in Grade 6, leading to the invert-and-multiply algorithm.

Research Questions

In spite of the multitude of research studies and recommendations described above, fraction teaching and learning remain challenging, particularly in the United States. Our study focuses on textbooks from three high-achieving East Asian countries: Japan, Korea, and Taiwan. By examining their textbooks, we hope to gain some insights into how we might support both teachers and students as they tackle this mathematically challenging topic. Specifically, we examine the following questions:

1. What are the similarities and differences in the intended learning progressions of fraction concept development among the three high-achieving Asian curricula? In particular, what are the similarities and differences with respect to (1) the sequence of topics and (2) the integration of fraction subconstructs?
2. What are the similarities and differences in the treatment of the four arithmetic operations with fractions among the three high-achieving Asian curricula? In particular, what are the similarities and differences with respect to (1) the types of problem situations utilized in discussing each operation, (2) the intended computational algorithms, if any, and (3) the use of visual representations (set, line, or area)?

Because our study focuses on the single topic of fractions, it principally involves a *micro* analysis of the textbooks. However, because the topic is broader than addition/subtraction of fractions (Charalambous et al., 2010) or division of fractions (Li et al., 2009), our study also shares some characteristics of *macro* analysis. The scope of the analysis is still limited to topics directly related to fractions.

Methodology

NRC (2004) argues that content analysis should be about a specific standard and comparison curricula should be selected judiciously. For a cross-national study, there is no common standard on which to focus. Instead, we chose to use a specific mathematical topic, fractions, and examine how the selected textbooks introduce and develop the ideas of fractions and fraction operations. We focused on Japan, Korea, and Taiwan for three reasons. First, they are high-achieving countries in various international achievement studies. Second, their educational systems are similar—centralized with national curriculum standards published by the respective Ministries of Education. Finally, all three curricula complete the discussion of fractions within elementary school (i.e., Grades 1–6). The background of the research team members, who are natives of the three countries, was also a factor.

The textbooks selected (see Table 2.3) were in alignment with the national curriculum standards at the time of the study—the 2008 standards for Japan,

Table 2.3 Textbooks analyzed in this study

Country	Textbook series
Japan	Fujii, T. & Iitaka, S. (2011). *Atarashii Sansuu*. Tokyo: Tokyo Shoseki Co. Ltd.
Korea	Korean Ministry of Education and Human Resources Development. (2014). *Mathematics* (Grades 3–4). Seoul: DaeHan Printing and Publishing Co., Ltd. Korean Ministry of Education and Human Resources Development. (2015). *Mathematics* (Grades 5–6). Seoul: DaeHan Printing and Publishing Co., Ltd.
Taiwan	Kang Hsuan Educational Publishing Group. (2012). *Kang Hsuan elementary school mathematics textbooks*. (4A) Tainan, Taiwan: Author. Kang Hsuan Educational Publishing Group. (2013). *Kang Hsuan elementary school mathematics textbooks*. (3A, 4B, 5A, 6A) Tainan, Taiwan: Author. Kang Hsuan Educational Publishing Group. (2014). *Kang Hsuan elementary school mathematics textbooks*. (3B, 5B, 6B) Tainan, Taiwan: Author.

2013 for Korea, and 2008 (for Grades 1–3) and 2003 (for Grades 4–6) for Taiwan. The Japanese textbook series is commercially published by Tokyo Shoseki, one of six textbook series approved by the Ministry of Education. It is the most widely used elementary mathematics textbook (Naigaikyouiku, 2010). The Korean textbook series examined is the only elementary mathematics textbook series in Korea and is written by the Ministry of Education. The Taiwanese textbook series is one of the four commercially published textbooks in Taiwan. It is one of the two most widely used textbook series (S. Law, personal communication, July 13, 2015). The textbooks were analyzed in their original languages by researchers who are native speakers—the first author analyzed the Japanese textbooks, the second author the Taiwanese textbooks, and the third author the Korean textbooks. Because some aspects of our analysis—for example, word problem contexts and fraction subconstructs—are not visually verifiable, we used the English translation of the Japanese series (Fujii & Iitaka, 2012) to calibrate our analysis. The researchers independently analyzed segments of the translated Japanese textbook on those aspects, and then compared analyses. Whenever a discrepancy in the analyses occurred, the particular instance was discussed until a consensus was reached.

The study reported in this chapter is a content analysis of three Asian textbook series. The analysis took place in three stages. First, we identified the sequence and the grade placement of the major topics related to fractions in each textbook. Then, we analyzed each textbook's treatment of addition and subtraction, focusing on problem types, the use of diagrams, and the target algorithms, if any. For problem types, we first determined the frequencies of word problems, calculation exercises, and others. For word problems, we examined the addition/subtraction problem situations using the Cognitively Guided Instruction framework, which categorizes addition and subtraction word problems based on the four problem situations—join, separate, part-part-whole, and compare—and the unknown quantity in the situation (Carpenter, Fennema, & Romberg, 1992). Table 2.4 summarizes the 11 addition and subtraction word problem types according to this framework. Finally, we examined the treatment of multiplication and division in these textbooks, again focusing on problem types, the use of diagrams, and the target algorithms, if any.

Table 2.4 Problem types based on the CGI framework (Carpenter et al., 1992)

Problem type	Unknown factors		
Join (add to)	**(Result Unknown)** Connie had 5 marbles. Juan gave her 8 more marbles. How many marbles does Connie have altogether?	**(Change Unknown)** Connie had 5 marbles. How many marbles does she need to have 13 marbles altogether?	**(Start Unknown)** Connie had some marbles. Juan gave her 5 more. Now she has 13 marbles. How many marbles did Connie have to start with?
Separate (take from)	**(Result Unknown)** Connie had 13 marbles. She gave 5 to Juan. How many marbles does Connie have left?	**(Change Unknown)** Connie had 13 marbles. She gave some to Juan. Now she has 5 marbles left. How many marbles did Connie give to Juan?	**(Start Unknown)** Connie had some marbles. She gave 5 to Juan. Now she has 8 marbles left. How many marbles did Connie have to start with?
Part-Part-Whole (put together/ take apart)	**(Whole Unknown)** Connie has 5 red marbles and 8 blue marbles. How many marbles does she have altogether?		**(Part Unknown)** Connie has 13 marbles: 5 are red, and the rest are blue. How many blue marbles does Connie have?
Compare	**(Difference Unknown)** Connie has 13 marbles. Juan has 5 marbles. How many more marbles does Connie have than Juan?	**(Larger Unknown)** Juan has 5 marbles. Connie has 8 more than Juan. How many marbles does Connie have?	**(Smaller Unknown)** Connie has 13 marbles. She has 5 more marbles than Juan. How many marbles does Juan have?

Findings

Overall, the treatment of fractions in the three textbook series is more similar than different. However, there are some significant differences in the way some fraction topics are discussed in these textbooks. In the following sections, we will share the findings in accordance with the two research questions.

Intended Learning Progression

Table 2.5 summarizes the grade placements of the major fraction topics in the textbooks from each country. Clearly, some topics, like addition and subtraction of fractions, are discussed in multiple grade levels. However, by simply examining the grade level in which each topic is introduced, we found that all three textbooks introduce these topics in an identical order. Likewise, all three textbook series emphasize the idea that a non-unit fraction is made up of unit fractions. For

Table 2.5 Grade placements of major fraction topics in the textbooks from Japan, Korea, and Taiwan

	Japan	Korea	Taiwan
Fractions as equal shares	2	3	3
Fraction as number	3/4	3/4	3/4
Comparison	3/4/5	3/4/5	3/4
Addition/subtraction	3/4/5	3/4/5	3/4/5
Equivalent fractions	4/5	4/5	4
Fractions as quotients	5	4	4
Multiplication	5/6	5	4/5
Division	5/6	5/6	6

example, Fig. 2.2(a) shows how the Japanese textbook uses this idea to deal with fractions greater than 1 (Problem 2). Question (2) in Fig. 2.2(b) shows the Taiwanese textbook asking students "How many $\frac{1}{10}$ pieces are needed to make up a $\frac{7}{10}$ piece?" As we will see later, all three textbook series make use of this way of looking at fractions as they discuss addition and subtraction of fractions.

A few differences do occur in the overall flow of the curricula. First, in the textbooks from Japan and Korea, the idea of equivalent fractions is first introduced in Grade 4, but the formulas to create equivalent fractions, i.e., $\frac{a}{b} = \frac{a \times k}{b \times k}$ and $\frac{a}{b} = \frac{a \div k}{b \div k}$ (a, b, and k are nonzero whole numbers), are not discussed until Grade 5. However, the Taiwanese textbook develops this formula in Grade 4. Another difference is the grade placement of the quotient meaning of fractions; that is, $\frac{a}{b} = a \div b$ (a and b are whole numbers, $b \neq 0$). Both the Korean and the Taiwanese textbooks introduce this idea in Grade 4, but the Japanese textbook introduces it in Grade 5.

Whereas the treatments of addition and subtraction in all three textbooks are very similar, the treatments of multiplication and division illustrate significant differences among the three textbooks. (We will discuss the similarities and the differences of the actual treatments in greater detail later.) The Taiwanese textbook first introduces fraction multiplication in Grade 4. Both the Japanese and the Korean textbooks introduce multiplication and division of fractions in Grade 5, while the Taiwanese textbook does not introduce division of fractions until it completes the discussion of multiplication of fractions. Although multiplication and division are both introduced in Grade 5 in the Japanese and the Korean textbooks, the Korean textbook completes the discussion of multiplication in Grade 5 while the Japanese textbook extends the discussion of both operations into Grade 6.

Integration of Fraction Subconstructs

Table 2.6 summarizes which fraction subconstructs are discussed in the three Asian textbooks at different grade levels. Once again, the integration of various subconstructs among the three textbook series is more similar than different. All three textbook series integrate the five subconstructs into their discussions of fractions. Additionally, the part-whole and measure subconstructs clearly play a

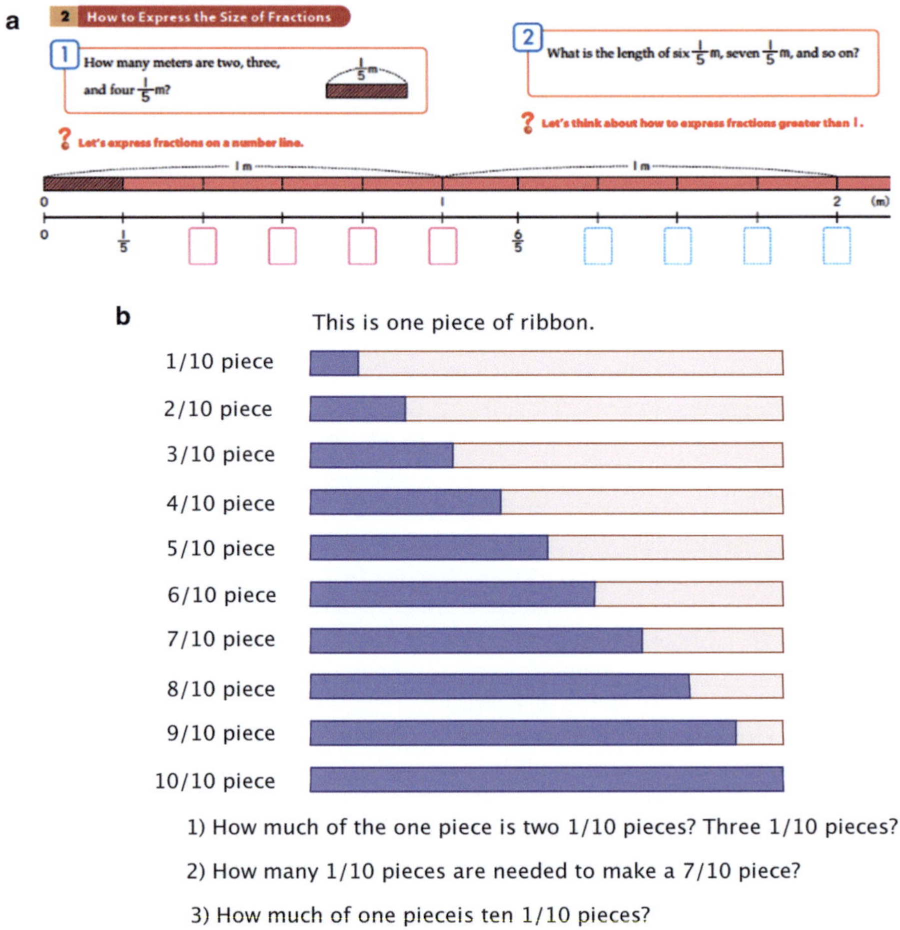

Fig. 2.2 (**a**) The idea that non-unit fractions are made up of unit fractions is emphasized in these Japanese Grade 3 problems (Fujii & Iitaka, 2012, Grade 3, pp. B48–B49). (Although the analysis was conducted using the original textbook in Japanese, we use images from the English translation so that we will not have to provide the translation separately.) (**b**) Problems from Taiwanese Grade 3 textbook (translated from Kang Hsuan Educational Publishing Group, 2014, Grade 3B, p. 35)

central role in the initial instruction on fractions in all three series. The quotient subconstruct is introduced, as expected, with the quotient meaning of fractions—during Grade 4 in Korea and Taiwan, and during Grade 5 in Japan. In all three series, the ratio subconstruct was introduced last.

Although the measure subconstruct seems to play a central role in all three textbook series in early grades, the operator subconstruct begins to play a more important role in the Korean and the Taiwanese textbooks than in the Japanese textbook when they discuss multiplication by fractions, that is, when the multiplier becomes a fraction. For example, Fig. 2.3 shows a problem from the Taiwanese

Table 2.6 Fraction subconstructs appearing in the three Asian textbooks

Grade	Japan	Korea	Taiwan
2	**Part-whole**		
3	Part-whole **Measure**	**Part-whole** **Measure** **Operator**	**Part-whole** **Measure**
4	Part-whole Measure	Part-whole Measure **Quotient**	Part-whole Measure **Quotient**
5	Part-whole Measure **Quotient**	Part-whole Measure Quotient Operator	Part-whole Measure Quotient **Operator**
6	Part-whole Measure **Operator** **Ratio**	Part-whole Measure Operator **Ratio**	Part-whole Measure Quotient **Ratio**

Note: Bold-faced letters indicate the first time the particular subconstruct is discussed in the textbook series

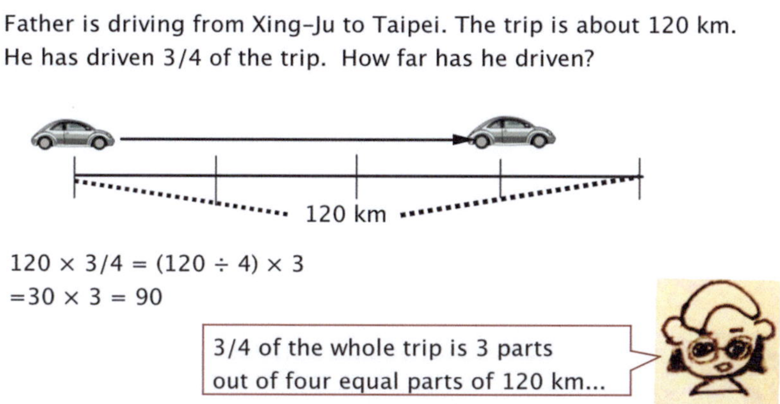

Fig. 2.3 A multiplication problem from the Taiwanese textbook (translated from Kang Hsuan Educational Publishing Group, 2014, Grade 5B, p. 7)

series. Note that in this problem $\frac{3}{4}$ is the multiplier, and it is given as an operator fraction.

In contrast, Fig. 2.4 shows an introductory word problem found in the Japanese Grade 6 unit on multiplication of fractions. In this case, $\frac{2}{3}$ is the multiplier and represents a measured quantity, not an operator fraction.

The Korean textbook actually incorporates the operator subconstruct much earlier than either the Japanese or the Taiwanese series. Figure 2.5 shows a Grade 3 problem from the Korean series. Although this problem can be interpreted as a

Fig. 2.4 An introductory problem in the unit on multiplication of fractions (Fujii & Iitaka, 2012, Grade 6, p. A 23)

Fig. 2.5 Translated from Korean Ministry of Education and Human Resources Development, 2014, Grade 3-B student book, p. 109

multiplication problem it appears in the introductory unit, in which the focus is helping students understand the meaning of fractions.

Find out how many are in $\frac{3}{4}$ of the set if 8 apples are a whole set.

Addition and Subtraction

The ways in which addition and subtraction are introduced and developed in all three series are quite similar. Addition and subtraction of fractions are first introduced in word problems. Table 2.7 summarizes the addition and subtraction situations used in the three textbook series. Although the Taiwanese series includes all but two of the possible addition and subtraction situations, all three series generally use simpler situations.

Figure 2.6 shows an introductory problem, of a part-part-whole whole unknown type, from the English translation of the Japanese series.

As noted earlier, all three textbook series emphasize the idea of a non-unit fraction being made up of unit fractions. In discussions regarding how to add or subtract fractions with like denominators, they all make use of this unitary perspective. Thus, $\frac{7}{10} - \frac{3}{10}$ can be thought as taking away 3 $\frac{1}{10}$-units from 7 $\frac{1}{10}$-units.

Table 2.7 Word problem situations found in the three Asian textbook series

Join result unknown *JKT*	Join change unknown *T*	Join start unknown
Separate result unknown *JKT*	Separate change unknown *T*	Separate start unknown *T*
Part-part-whole whole unknown *JKT*	Part-part-whole part unknown *T*	
Compare difference unknown *JKT*	Compare smaller unknown	Compare larger unknown *T*

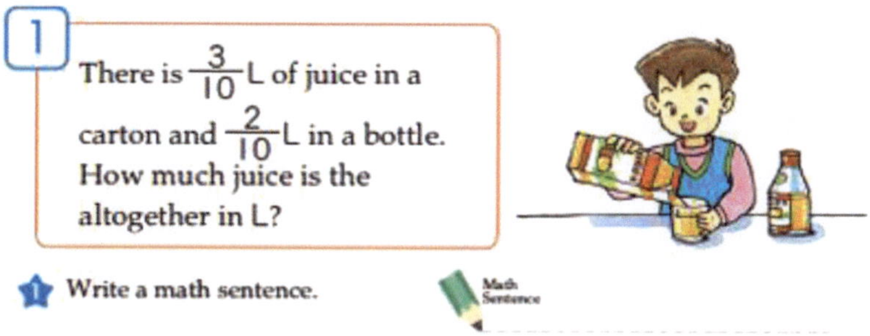

Fig. 2.6 An introductory problem addition of fractions from the Japanese series (Fujii & Iitaka, 2012, Grade 3, p. B51)

Therefore, the difference is (7–3) $\frac{1}{10}$-units, or 4 $\frac{1}{10}$-units, i.e., $\frac{4}{10}$. Figure 2.7 shows this approach as it appears in the Taiwanese series.

Although all three textbook series use word problems to introduce the addition and subtraction of fractions, they all seem to focus on helping students develop computational mastery once the reasoning behind the calculation is established. As a result, about $\frac{3}{4}$ of the problems found in the units on addition and subtraction are purely calculation exercises.

The three Asian textbook series incorporate a variety of visual representations to support students' reasoning with addition and subtraction. Figure 2.8(a) shows how the Korean textbook uses an area model to represent $1\frac{1}{5}+2\frac{2}{5}$, while Fig. 2.8 (b) shows a linear model found in the Taiwanese series.

Multiplication and Division

Unlike addition and subtraction, more significant differences exist among the three East Asian textbook series in how they present multiplication and division of fractions. Overall, the Korean and the Taiwanese series' treatments of multiplication and division are similar, while the Japanese series incorporates some unique approaches. One common aspect among the three series is that they all discuss multiplication and division by whole numbers separately from multiplication and

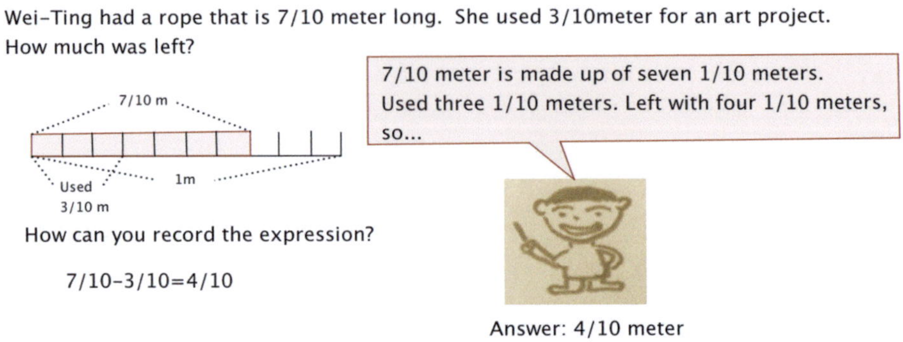

Fig. 2.7 This example from the Taiwanese series shows the typical approach, found in all three Asian series, to thinking about subtraction of fractions with like denominators (translated from Kang Hsuan Educational Publishing Group, 2014, Grade 3B, p. 20)

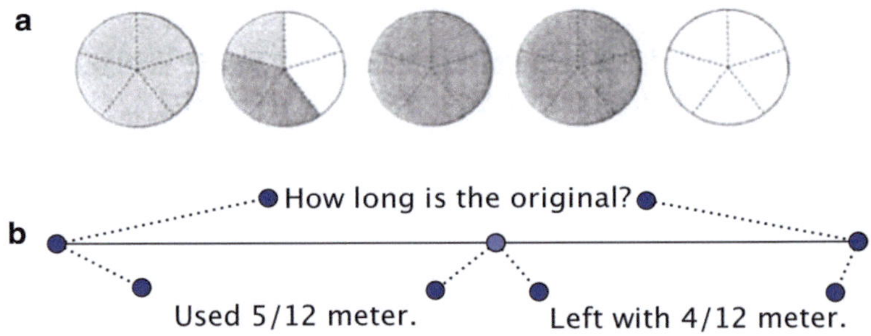

Fig. 2.8 (**a**) An area model from the Korean textbook (translated from Korean Ministry of Education and Human Resources Development, 2014, Grade 4B, p. 83). (**b**) A linear model showing $\frac{5}{12}+\frac{4}{12}$ in the Taiwanese series (translated from Kang Hsuan Educational Publishing Group, 2014, Grade 3B, p. 42)

division by fractions. Thus, the discussion of multiplication of fractions begins with situations where there are whole-number groups of fractional quantities (e.g., $3 \times 2/5$)—occurring in Grade 4 for the Taiwanese series and in Grade 5 for the Japanese and the Korean series. Furthermore, all three series continue to use the idea that non-unit fractions are made up of unit fractions to help students make sense of the process of multiplying fractions by whole numbers. Figure 2.9 shows an example from the Taiwanese series showing $\frac{2}{10}$ multiplied by 5.[1]

[1] In all three Asian textbook series, a multiplication equation is written in the form (multiplicand) $\times$ (multiplier) $=$ (product), or (group size) $\times$ (number of groups) $=$ (product). In this chapter, we adopt the convention that seems to be more common in English-speaking countries, (multiplier) $\times$ (multiplicand) $=$ (product). However, we keep the Asian notation in figures or quotes taken directly from the textbooks.

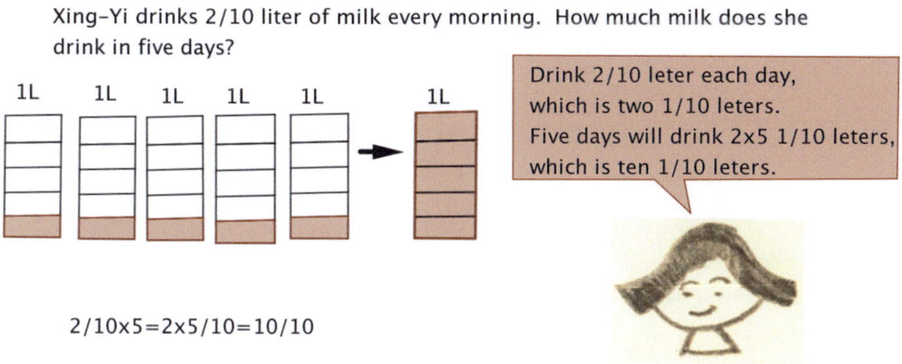

Fig. 2.9 An example of conceptualizing fraction multiplication through unit fraction (translated from Kang Hsuan Educational Publishing Group, 2012, Grade 4A, p. 102)

After this initial discussion of multiplication of fractions by whole numbers, the approach of the Japanese series diverges from both the Korean and the Taiwanese series. First, as noted earlier, both the Korean and the Taiwanese series wait to discuss division of fractions until they complete their discussions of multiplication of fractions, including multiplication by fractions. However, the Japanese series discusses dividing fractions by whole numbers in Grade 5, before discussing multiplication by fractions. Figure 2.10 shows the initial problem from the Japanese series that discusses division of a fraction by a whole number.

In addition to the difference in the overall sequencing of multiplication and division, there are differences in the sequences of topics related to multiplication of fractions among the three Asian textbook series. Table 2.8 summarizes the sequence of topics related to multiplication.

Sequence similarities exist between the Korean and the Taiwanese textbook series. However, unit fractions seem to play a foundational role in the Taiwanese series. Thus, as the series discusses multiplication of whole numbers by fractions, multiplication of fractions by whole numbers, and multiplication of two fractions, it starts with unit fractions. However, regarding multiplying whole numbers by fractions and fractions by whole numbers, the Korean series treats unit fractions as a special case of proper fractions. Thus, W × UF and UF × W appear in the exercise sets after the textbook discusses W × PF and PF × W, respectively. In the case of multiplying two fractions, however, both the Korean and the Taiwanese series start with the UF × UF situation.

The Japanese series also considers unit fractions as a special case of proper fractions. Thus, W × UF is found in the exercise set after it discusses W × PF, like in the Korean series. However, the Japanese series keeps the same perspective when it discusses multiplication of two fractions. Moreover, in their discussion of multiplying by fractions, the textbook authors seem to consider whole numbers as a special case of fractions. Thus, the series begins the discussion of fraction multipliers with a situation that involves multiplication of two proper fractions, e.g., $\frac{2}{3} \times \frac{4}{5}$, instead of P × W like the Korean series or UF × W like the Taiwanese series.

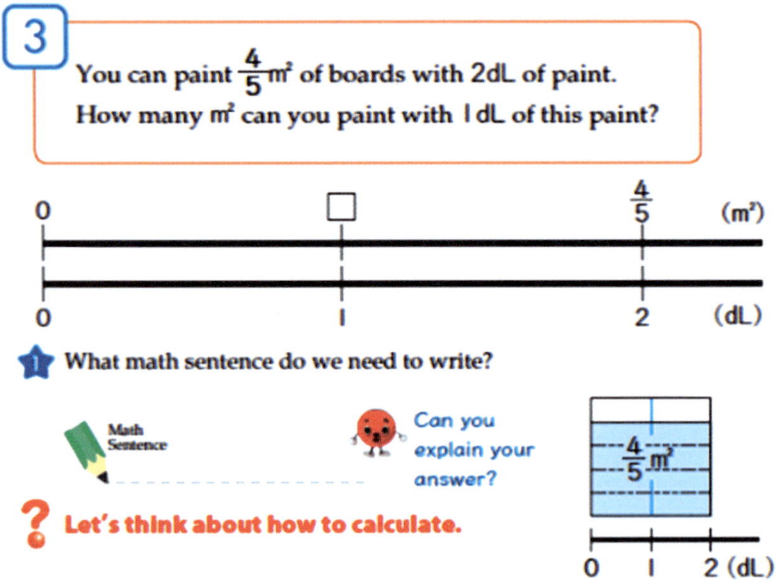

Fig. 2.10 The Japanese textbook discusses dividing fractions by whole numbers before discussing multiplication by fractions (Fujii & Iitaka, 2012, Grade 5, p. B91)

Table 2.8 Sequence of multiplication-related topics in the three Asian textbook series

Japan	Korea	Taiwan
Whole number multiplier		
W × PF[a]	W × PF[a]	W × UF
	W × M	W × PF
		W × M
Fraction multiplier		
P × P	P × W[b]	UF × W
P × W	M × W	P × W
P × M[c]	UF × UF	M × W
	P × P	UF × UF
	M × M	P × P
		P × M & M × P
		M × M

W: whole numbers; UF: unit fractions; PF: proper fractions; M: mixed fractions

[a]W × UF is included in the exercise set after this topic is discussed

[b]U × W is included in the exercise set after this topic is discussed

[c]M × M is included in the exercise set after this topic is discussed

Additional variations in the treatment of division of fractions occur among the three textbook series. Both the Japanese and the Korean textbooks discuss dividing fractions by whole numbers in Grade 5, but the Taiwanese textbook does not discuss this as a separate topic. In fact, the Taiwanese series only has one problem that considers dividing a fraction by a whole number, and it appears near the end of the discussion of division of fractions. Although the Japanese and the Korean series discuss division of fractions by whole numbers as a separate topic in Grade 5, there are some significant differences between these two series. In the Japanese series, division of fractions by whole numbers is treated immediately after multiplication of fractions by whole numbers and before the discussion of multiplication by fractions, a Grade 6 topic. In contrast, the Korean series discusses division of fractions by whole numbers after the completion of the discussion of multiplication by fractions. A major goal in the Japanese series is to develop the algorithm $\frac{a}{b} \div n = \frac{a}{b \times n}$. On the other hand, the Korean series tries to lay the foundation for the invert and multiply algorithm by helping students understand that division by a whole number is the same as multiplying by the unit fraction which is the reciprocal of the divisor.

All three series discuss division of fractions by fractions in Grade 6. While the Japanese series begins with a word problem that is solved by $\frac{2}{5} \div \frac{3}{4}$, both the Korean and the Taiwanese textbooks start with word problems that involve dividing a fraction by a unit fraction with a common denominator: $\frac{5}{6} \div \frac{1}{6}$ for the Korean series and $\frac{8}{9} \div \frac{1}{9}$ for the Taiwanese. While the word problem for the Japanese series is a partitive division problem, both the Korean and the Taiwanese series use quotitive division problems. These two series follow up the initial problems with division problems where the numerator of the dividend is not divisible by the numerator of the divisor, which the Taiwanese series calls "fraction division with remainder." For example, in the Korean series, students are asked to find how many $\frac{2}{6}$ *m* are in $\frac{5}{6}$ *m*. The textbook provides a bar diagram showing $\frac{5}{6}$ *m* and then asks how many 2 *m* are in 5 *m*, accompanied by a bar diagram showing 5 *m* (see Fig. 2.11). Then, by comparing these two situations, the series develops the common denominator algorithm for division of fractions.

In both series, the invert-and-multiply algorithm is discussed only after the common denominator algorithm is established. For example, in the Taiwanese series, students are given a quotitive word problem that can be solved by $\frac{13}{12} \div \frac{5}{12}$. The textbook then illustrates the calculation process to show how the quotient may be found by multiplying the dividend by the reciprocal of the divisor:

$$\frac{7}{8} \div \frac{3}{5} = \frac{7 \times 5}{8 \times 5} \div \frac{3 \times 8}{5 \times 8} = (7 \times 5) \div (8 \times 3) = \frac{7}{8} \times \frac{5}{3} = \frac{35}{24}.$$

Similarly, the Korean textbook series addresses how the common denominator method can be connected to the invert-and-multiply method with the problem $\frac{3}{4} \div \frac{2}{5}$, as follows:

Activity 1. Figure out how to calculate $\frac{5}{6} \div \frac{2}{6}$.

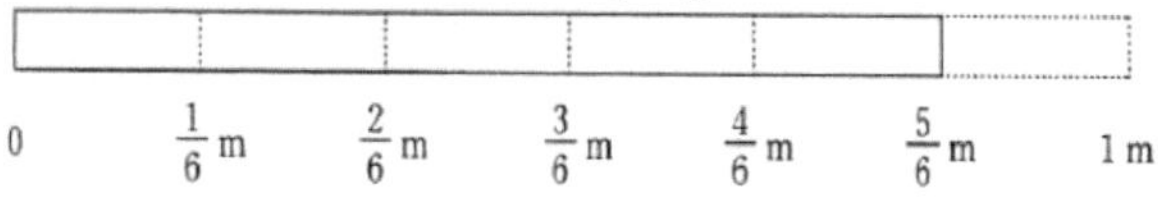

- Cut $\frac{5}{6}$ into $\frac{2}{6}$.
- When $\frac{5}{6}$ is divided by $\frac{2}{6}$, there are 2 pieces of $\frac{2}{6}$ and half of $\frac{2}{6}$.

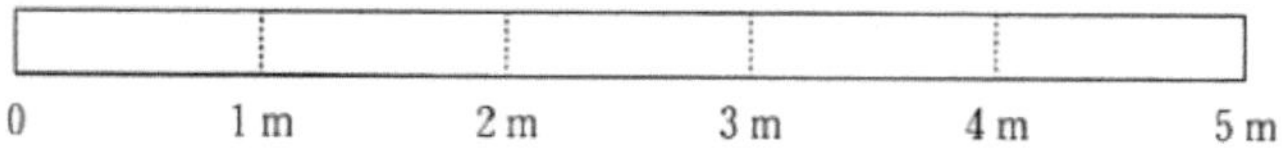

- Cut 5 m into 2 m.
- When 5 m is divided by 2m, there are 2 pieces of 2m and half of 2m.
- Is the quotient of $5 \div 2$ the same as that of $\frac{5}{6} \div \frac{2}{6}$?
- Why do you think so?
- Construct an expression to calculate $\frac{5}{6} \div \frac{2}{6}$.

$$\frac{5}{6} \div \frac{2}{6} = [\ ] \div [\ ]$$

Fig. 2.11 A fraction division problem from the Korean textbook that requires students to use the common denominator algorithm for dividing fractions by comparing $\frac{5}{6} \div \frac{2}{6}$ and $5 \div 2$ (translated from Korean Ministry of Education and Human Resources Development, 2015, Grade 6–1, p. 2)

$$\frac{3}{4} \div \frac{2}{5} = \frac{3 \times 5}{4 \times 5} \div \frac{2 \times 4}{5 \times 4} = (3 \times 5) \div (2 \times 4) = \frac{3 \times 5}{2 \times 4}.$$

Because $\frac{3 \times 5}{2 \times 4} = \frac{3}{4} \times \frac{5}{2}, \frac{3}{4} \div \frac{2}{5} = \frac{3}{4} \times \frac{5}{2}$.

Note that these textbooks implicitly apply the commutative property at different steps. As stated above, while both the Korean and the Taiwanese textbook series introduce division by fractions using quotitive word problems, the Japanese series introduces division by fractions with a partitive word problem. Figure 2.12 shows the opening problem in the unit of division by fractions.

Not only is the problem situation partitive, the calculation involves dividing by a fraction less than 1, which has been shown to be challenging (Greer, 1987). Thus, the Japanese series' initial emphasis is helping students understand why this problem can be solved by $\frac{2}{5} \div \frac{3}{4}$. The textbook includes explanations by two hypothetical students. One student uses the generalized equation [Amount painted] $\div$ [Amounts of paint used (dL)] $=$ [Area we can paint with 1 dL], derived by thinking about whole-number divisors. The other student uses the double-number line representation to argue that $\frac{2}{5}$ is obtained by multiplying the missing quantity by $\frac{3}{4}$. Then, by using the relationship between multiplication and division,

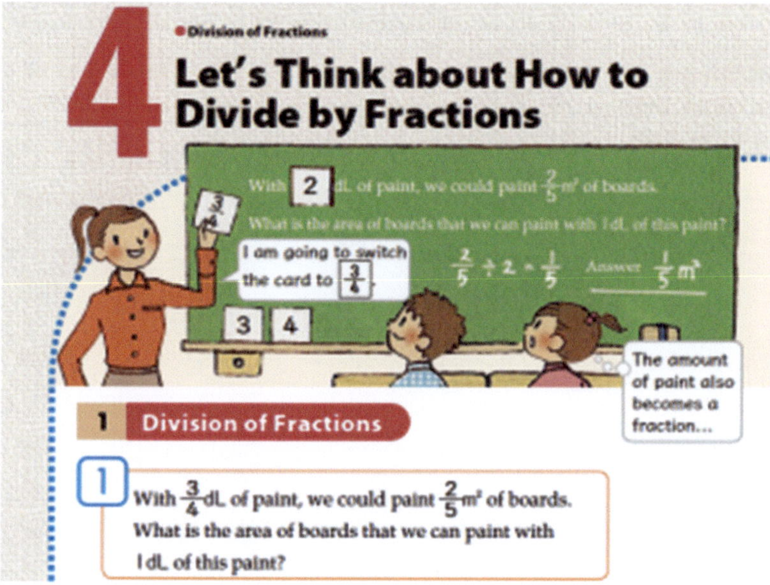

Fig. 2.12 The Japanese series introduces division by fractions using a partitive division situation (Fujii & Iitaka, 2012, Grade 6, p. A34)

the student justifies that the operation to find the missing quantity is $\frac{2}{5} \div \frac{3}{4}$ (see Fig. 2.13).

The use of double-number line diagrams to represent the relationships in a given problem situation is another unique feature of the Japanese textbook series. The diagram is used in all introductory problems as the authors discuss multiplying fractions by whole numbers, dividing fractions by whole numbers, multiplying fractions by fractions, and dividing fractions by fractions. In each instance, the double-number line diagram is used to justify the calculation needed to find the missing quantity. The Japanese series uses a different diagram to consider ways of actually carrying out the calculation. While the Korean and the Taiwanese textbooks use different diagrams to support students' reasoning with multiplication and division of fractions—area diagrams for multiplication and bar diagrams for division—the Japanese series uses a diagram that combines the area model of fractions with a number line (see Fig. 2.14).

Discussion and Implications

The findings discussed above clearly show that there are many similarities among the three Asian textbook series' initial treatment of fractions. In particular, all three series make heavy use of the measure subconstruct and the idea that non-unit fractions are collections of unit fractions. The three series approach addition,

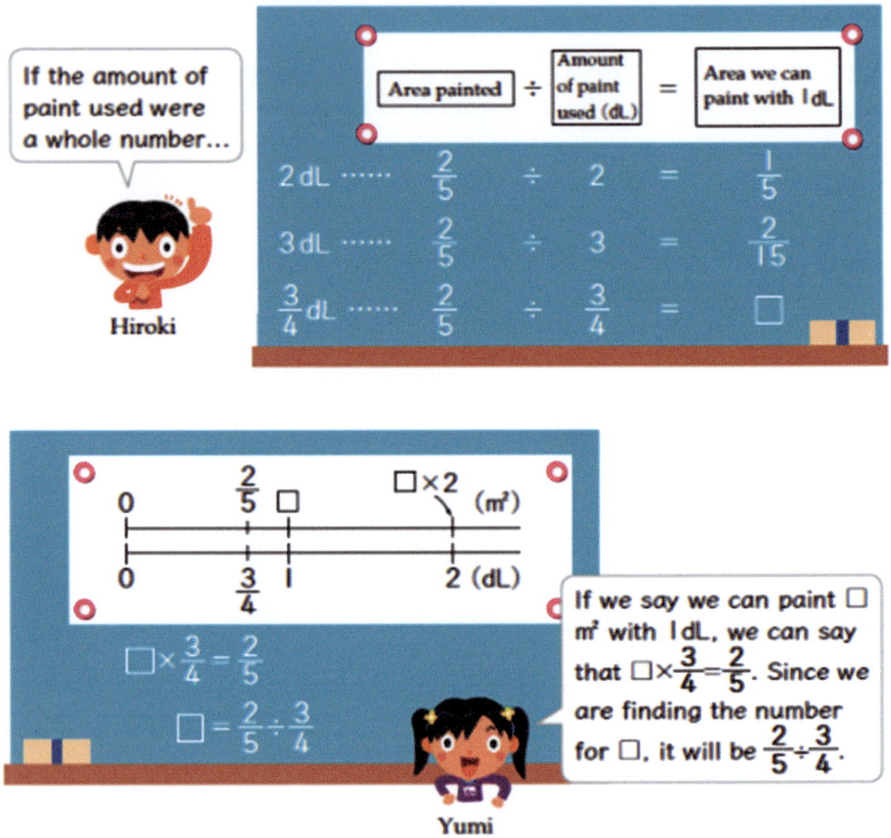

Fig. 2.13 Two ways the Japanese series justifies that the opening problem can be solved by $\frac{2}{5} \div \frac{3}{4}$ (Fujii & Iitaka, 2012, Grade 6, p. A35)

subtraction, and multiplication of fractions by whole numbers, i.e., whole-number groups of fractional quantities, using these tools. Their approach is consistent with Behr et al.'s (1983) hypothesis that the measure subconstruct supports students' development of addition and subtraction with fractions. As stated earlier, Thompson and Saldanha (2003) noted that thinking about non-unit fractions as collections of unit fractions is rare in US textbooks. However, this approach is emphasized in the CCSSM, and our findings support the CCSSM authors' claim that they have used high-achieving Asian curriculum materials as benchmarks.

In regard to addition/subtraction word problem situations, the three Asian curricula generally include simpler situations. It is as though the authors of these curricula attempt to develop the understanding that the operation necessary to answer a problem is determined by the situation and the missing quantity, not by the type of numbers. Once that understanding is achieved, they can then focus on helping students develop ways of calculating sums and differences in a meaningful manner.

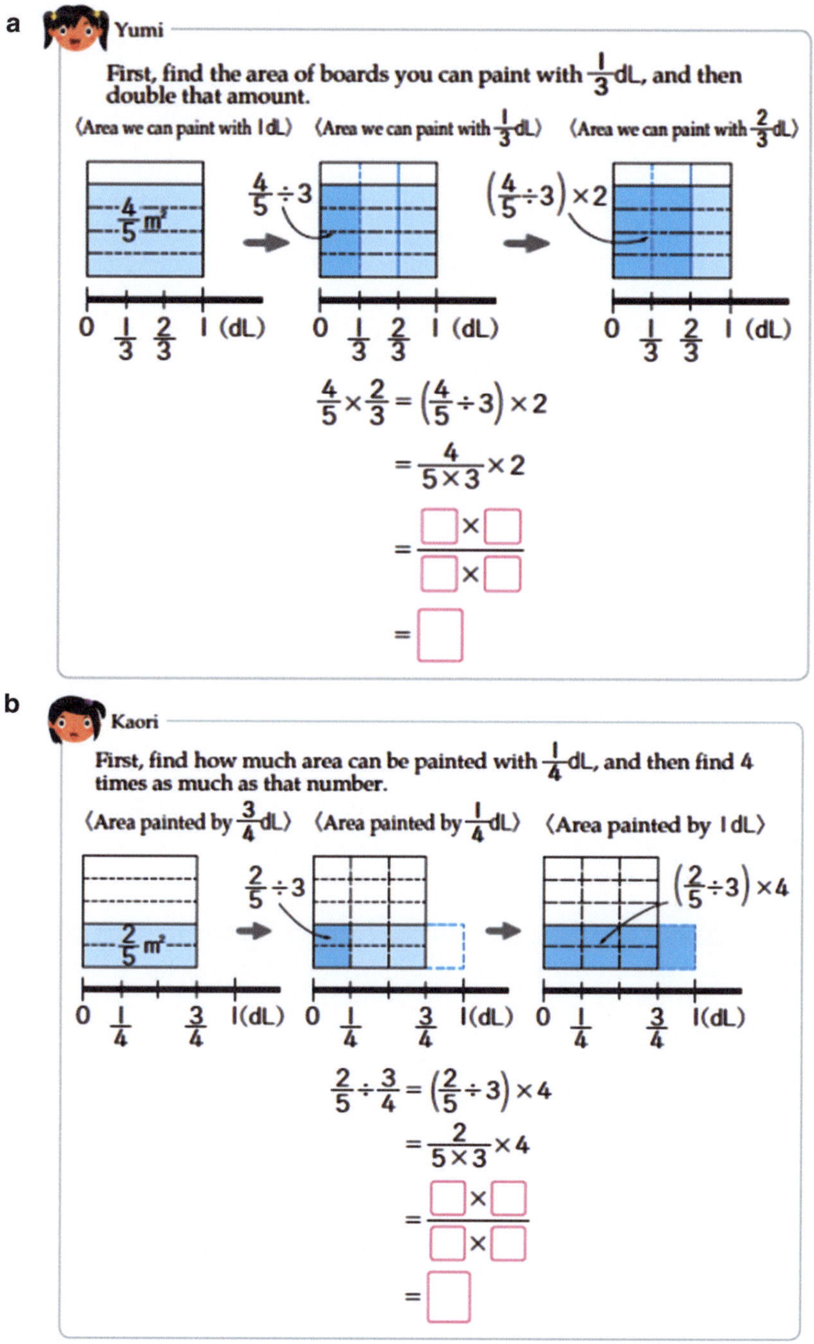

Fig. 2.14 The Japanese series uses a combination of the area model and the number line to illustrate the process of multiplying two fractions (**a**) and dividing a fraction by another fraction (**b**) (Fujii & Iitaka, 2012, Grade 6, p. A 25 & p. A36)

Although the three Asian textbook series' approaches to fractions support some aspects of the model proposed by Behr et al. (1983), they raise questions about other aspects. For example, according to Behr et al., the ratio subconstruct is helpful for developing the idea of equivalence. However, none of the three series incorporate the ratio subconstruct before they discuss equivalent fractions. Of course, this does not mean that the ratio subconstruct is not useful for developing the idea of equivalence. The three Asian textbook series simply show that there are other approaches to helping students develop the idea of equivalent fractions. The Asian models also suggest that the operator subconstruct is helpful for supporting students' development of multiplication. Indeed, both the Korean and the Taiwanese series utilize the operator construct to discuss multiplication by fractions by considering multiplication as an operation to find the fractional amount of the given quantity. However, the Japanese series approaches multiplication by fractions differently. While one justification for multiplication as the appropriate operation uses the idea of multiplicative comparison, the fractions in the problem situations are measured quantities. Further examination of the role the operator subconstruct may play in supporting students' development of multiplication is needed.

Lamon (2007) noted that "Is it better to teach one rational number subconstruct or all five?" and "If one, which should it be?" are two of the remaining researchable questions. As noted already, the three textbook series in the current study do not discuss the ratio subconstruct until the end of the fraction instruction in elementary schools. However, the Korean series seems to introduce the remaining four subconstructs intentionally early, while the Japanese series takes the most deliberate approach. Moreover, although the operator subconstruct plays a key role in the discussion of multiplication by fractions in both the Korean and the Taiwanese series, it is not quite clear what advantages the Korean textbook affords by introducing the subconstruct sooner than the Taiwanese series does. Thus, the current study offers mixed answers to these questions.

Because of the similar approaches taken by these three textbook series to the initial instruction of fractions, the differences in the way multiplication and division are treated are rather surprising. Overall, the approaches in the Korean and the Taiwanese series appear to be similar to US textbook series that are aligned with the CCSSM (Son, Lo, & Watanabe, 2015). However, the Japanese approach is intriguing for a couple of reasons. First, as noted in our findings, the measure subconstruct is a major emphasis of the early fraction instruction in all three Asian textbook series. However, in the Korean and the Taiwanese series, the measure subconstruct does not play a significant role in later instruction. On the other hand, in the Japanese approach, the idea of non-unit fractions being composed of unit fractions plays an important role in explaining the process of multiplication and division (see Fig. 2.14). In the CCSSM, 5.NF.4.a states that "Interpret the product $(a/b) \times q$ as a parts of a partition of q into b equal parts; equivalently, as the result of a sequence of operations $a \times q \div b$.[2]" In the Japanese series, the quotient $q \div b$ is explicitly

[2]In order to match the verbal description, this expression should really be written as $a \times (q \div b)$.

interpreted as the amount corresponding to the unit fraction $\frac{1}{b}$. Moreover, because students must be able to divide fractions by whole numbers if q is a fraction, the Japanese series discusses division of fractions by whole numbers prior to multiplication by fractions.

Another interesting aspect of the Japanese approach is its consistency in problem context and visual representations across multiplication and division. As noted above, the Japanese series uses the same problem context to introduce multiplying and dividing fractions by whole numbers and multiplying and dividing by fractions. The series also uses the same representations—(1) double-number line diagrams to represent the relationships among the quantities in the problem situations, and (2) the combined area model and number line to illustrate the process of calculation. Although they discuss multiple ways to find the results of calculations, one approach involves the same reasoning process—first finding the amount corresponding to the unit fraction of the multiplier or the divisor, and then multiplying the result. These consistencies seem to emphasize the connection between multiplication and division operations, an important mathematical implication of the invert-and-multiply algorithm.

Limitations and Future Research

Because of the connoisseurial nature of textbook analyses, the NRC (2004) recommends that such a study make explicit the identity of those who conduct the analysis. The three researchers who conducted this study are natives of the three Asian countries whose textbooks were examined. As a result, they are fluent in the respective languages. They all received their doctorates in mathematics education from US institutions: Florida State University for the first two authors, and Michigan State University for the third author. Each has experience in content analysis of textbooks (e.g., Cai, Lo, & Watanabe, 2002; Son & Senk, 2010; Watanabe, 2003). The first two authors have also examined teaching and learning of fractions with children (e.g., Lo & Watanabe, 1996). Although none of the researchers are professional mathematicians, the first two authors hold master's degrees in mathematics. Thus, the researchers are well qualified to engage in this study. One limitation of the study, though, is that there is only one native speaker of each of the three Asian languages.

Another limitation of the study is that, for Japan and Taiwan, we examined only one of each country's existing elementary mathematics textbook series. Although each series is the most widely used series in its home country, there are other series. While past studies seem to suggest that textbooks from Japan are very similar (e.g., Li et al., 2009), nevertheless the differences in the way multiplication and division are treated in the three series make us wonder if there are some within-nation differences. Since each country has national standards, the grade placement of a particular topic should be the same in different textbooks. However, how topics within a grade level are ordered and developed can vary. Fujii (personal communication, 2010) noted that if Japanese mathematics education research has not reached a consensus on the teaching and learning of a particular mathematical idea, the ways different Japanese textbook series treat the topic can be different.

In this study, building on the existing studies, we intentionally expanded the scope of our analysis to the treatment of fractions from its introduction to its conclusion at the end of elementary school. We did so in part because we felt the way a particular idea is discussed is influenced by earlier discussion on related topics. Our findings show clear benefits of this expansion. For example, we see that the Korean textbook lays the foundation for multiplication by fractions by introducing the operator subconstruct of fractions in the introductory stage. We also see how the Japanese series utilizes a consistent approach to discuss both multiplication and division by fractions. However, our findings also suggest that it may be important to analyze how other related topics are treated in these textbooks. For example, how are decimal numbers introduced and developed? What are similarities and differences in the ways multiplication and division of decimal numbers are discussed? What about ratios and proportions? Are the ways ratios and proportions are discussed influenced by the ways fractions are treated in the textbooks? Further textbook analyses are definitely needed.

As we noted earlier, the difference in the way multiplication and division of fractions are treated in the three textbook series was a surprise for us. It will be interesting to see how the different emphases these textbook series place may impact students' understanding of fraction multiplication and division in particular. For example, how do Japanese students use visual representations in determining the appropriate operation for a given problem? Would they use a double-number line diagram, as emphasized by the textbook series?

Finally, it should be once again noted that textbooks are only an approximation of the implemented curriculum. They may reflect the image of the ideal implemented curriculum envisioned by the authors. However, it is obvious that teachers may use the same textbook and teach the same lesson very differently. For example, each of the three Asian textbook series includes a number of worked-out examples. However, how these examples are treated in actual classrooms can vary drastically. Some teachers may simply explain an example and assign students the exercise set that follows it. Other teachers may have the students actually tackle the problem on their own and use the worked-out solution only as one of the anticipated solutions by students. Clearly, those classrooms would be experiencing different implemented curricula. Thus, we need to be cautious how we interpret the results from this and other textbook content analyses.

References

Armstrong, B., & Larson, C. (1995). Students' use of part-whole and direct comparison strategies for comparing partitioned rectangles. *Journal for Research in Mathematics Education, 26*, 2–19.

Ball, D. L. (1996). Connecting to mathematics as a part of teaching to learn. In D. Schifter (Ed.), *What's happening in math class? Reconstructing professional identities* (Vol. 2, pp. 26–45). New York: Teachers College Press.

Behr, M., Lesh, R., Post, T., & Silver, E. (1983). Rational number concepts. In R. Lesh & M. Landau (Eds.), *Acquisition of mathematics concepts and processes* (pp. 91–125). New York: Academic Press.

Boonlerts, S., & Inprasitha, M. (2013). The textbook analysis on multiplication: The case of Japan, Singapore and Thailand. *Creative Education, 4*, 259–262.

Cai, J., Lo, J., & Watanabe, T. (2002). Intended treatment of arithmetic average in US and Asian school mathematics textbooks. *School Science and Mathematics, 102*, 391–404.

Carpenter, T. P., Fennema, E., & Romberg, T. A. (1992). Toward a unified discipline of scientific inquiry. In T. P. Carpenter, E. Fennema, & T. A. Romberg (Eds.), *Rational numbers: An integration of research* (pp. 1–11). NJ: Lawrence Erlbaum Associates.

Charalambous, Y. C., Delaney, S., Hsu, H., & Mesa, V. (2010). A comparative analysis of the addition and subtraction of fractions in textbooks from three countries. *Mathematical Thinking and Learning, 12*, 117–151.

Common Core State Standard Initiatives (CCSSI) (2010). Common Core State Standards for Mathematics. Washington, DC: National Governors Association Center for Best Practices and the Council of Chief State School Officers. http://www.corestandards.org/assets/CCSSI_Math Standards.pdf.

Fan, L., & Zhu, Y. (2007). Representation of problem-solving procedures: A comparative look at China, Singapore, and US mathematics textbooks. *Educational Studies in Mathematics, 66*, 61–75.

Fischbein, E., Deri, M., Nello, M. S., & Marino, M. S. (1985). The role of implicit models in solving verbal problems in multiplication and division. *Journal for Research in Mathematics Education, 16*, 3–17.

Fujii, T., & Iitaka, S. (2011). *Atarashii sansuu*. Tokyo: Tokyo Shoseki Co. Ltd..

Fujii, T., & Iitaka, S. (2012). *Mathematics international*. Tokyo: Tokyo Shoseki Co. Ltd..

Greer, B. (1987). Non-conservation of multiplication and division involving decimals. *Journal for Research in Mathematics Education, 18*, 37–45.

Kang Hsuan Educational Publishing Group. (2012). *Kang Hsuan elementary school mathematics textbooks*. (4A) Tainan, Taiwan: Author.

Kang Hsuan Educational Publishing Group. (2013). *Kang Hsuan elementary school mathematics textbooks*. (3A, 4B, 5A, 6A) Tainan, Taiwan: Author.

Kang Hsuan Educational Publishing Group. (2014). *Kang Hsuan elementary school mathematics textbooks*. (3B, 5B, 6B) Tainan, Taiwan: Author.

Kieren, T. E. (1976). On the mathematical, cognitive, and instructional foundations of rational numbers. In R. Lesh (Ed.), *Number and measurement: Papers from a research workshop* (pp. 101–144). Columbus, OH: ERIC/SMEAC.

Kilpatrick, J., Swafford, J., & Findell, B. (2001). *Adding it up: Helping children learn mathematics*. Washington, DC: National Academy Press.

Korean Ministry of Education and Human Resources Development (2014). *Mathematics*. (Grades 3–4) Seoul: DaeHan Printing and Publishing Co., Ltd.

Korean Ministry of Education and Human Resources Development (2015). *Mathematics*. (Grade 5–6) Seoul: DaeHan Printing and Publishing Co., Ltd.

Lamon, S. J. (2007). Rational numbers and proportional reasoning: Toward a theoretical framework for research. In F. K. Lester Jr. (Ed.), *Second handbook of research on mathematics teaching and learning, National Council of Teachers of Mathematics* (pp. 629–668). Charlotte, NC: Information Age Publishing.

Larson, C. N. (1980). Locating proper fractions on number lines: Effect of length and equivalence. *School Science and Mathematics, 80*, 423–428.

Li, Y., Chen, X., & An, S. (2009). Conceptualizing and organizing content for teaching and learning in selected Chinese, Japanese and US mathematics textbooks: The case of fraction division. *ZDM Mathematics Education, 41*, 809–826.

Lo, J., & Luo, F. (2012). Prospective elementary teachers' knowledge of fraction division. *Journal of Mathematics Teacher Education, 15*, 481–500.

Lo, J., & Watanabe, T. (1996). Developing ratio and proportion schemes: A story of a fifth grader. *Journal for Research in Mathematics Education, 28*, 216–236.

Ma, L. (1999). *Knowing and teaching elementary mathematics: Teachers' understanding of fundamental mathematics in China and the United States*. Mahwah, NJ: Lawrence Erlbaum Associates.

Mack, N. K. (1990). Learning fractions with understanding: Building on informal knowledge. *Journal for Research in Mathematics Education, 21*, 16–32.

Mack, N. K. (1995). Confounding whole-number and fraction concepts when building on informal knowledge. *Journal for Research in Mathematics Education, 26*, 422–441.

Mayer, R. E., Sims, V., & Tajika, H. (1995). A comparison of how textbooks teach mathematical problem solving in Japan and the United States. *American Educational Research Journal, 32*, 443–460.

Mullis, I. V. S., Martin, M. O., & Foy, P. (2008). *TIMSS 2007 International Mathematics Report: Findings from IEA's Trends in International Mathematics and Science Study at the Fourth and Eighth Grades*. Chestnut Hill, MA: Boston College.

Mullis, I. V. S., Martin, M. O., Foy, P., & Arora, A. (2012). *TIMSS 2011 international results in mathematics*. Chestnut Hill, MA: TIMSS & PIRLS International Study Center, Boston College.

Naigaikyouiku. (2010). *2011 nendo shougakko kyoukasho saitaku joukyou: Monkashou matome (Elementary school textbook market share for the 2011 school year: Summary by the Ministry of Education)*. Tokyo, Japan: Jijitsushinsha.

National Mathematics Advisory Panel. (2008). *Foundations for success: The final report of the National Mathematics Advisory Panel*. Washington, DC: US Department of Education.

National Research Council. (2004). On evaluating curricular effectiveness: Judging the quality of K-12 mathematics evaluations. Committee for a Review of the Evaluation Data on the Effectiveness of NSF-Supported and Commercially Generated Mathematics Curriculum Materials. In J. Confrey & V. Stohl (Eds.), *Mathematical Sciences Education Board, Center for Education, Division of Behavioral and Social Sciences and Education*. Washington, DC: The National Academies Press.

Olive, J. (1999). From fractions to rational numbers of arithmetic: A reorganization hypothesis. *Mathematical Thinking and Learning, 1*, 279–314.

Pothier, Y., & Sawada, D. (1983). Partitioning: The emergence of rational number ideas in young children. *Journal for Research in Mathematics Education, 14*, 307–317.

Pothier, Y., & Sawada, D. (1989). Children's interpretation of equality in early fraction activities. *Focus on Learning Problems in Mathematics, 11*(3), 27–38.

Reys, B. J., Reys, R. E., & Chávez, O. (2004). Why mathematics textbooks matter. *Educational Leadership, 61*(5), 61–66.

Schmidt, W. H., McKnight, C. C., Valverde, G., Houang, R. T., & Wiley, D. E. (1997). *Many visions, many aims: A cross-national investigation of curricular intentions in school mathematics*. Dordrecht, The Netherlands: Kluwer.

Siegler, R. S., & Lortie-Forgues, H. (2015). Conceptual knowledge of fraction arithmetic. *Journal of Educational Psychology, 107*(3), 909–918.

Simon, M. A. (2002). Focusing on key developmental understandings in mathematics. In D. S. Mewborn et al. (Eds.), *Proceedings of the Twenty-Fourth Annual Meeting of the North*

American Chapter of the International Group for the Psychology of Mathematics Education (PME-NA), Volume 2 (pp. 991–998). Athens, Georgia: PME-NA.

Son, J., Lo, J., & Watanabe, T. (2015). Intended treatments of fractions, fraction addition and subtraction in mathematics curriculum from Japan, Korea, Taiwan, and US. In T. G. Bartell et al. (Eds.), *Proceedings of the 37th annual meeting of the North American Chapter of the International Group for the Psychology of Mathematics Education (pp. 96–103).* East Lansing, MI: Michigan State University.

Son, J., & Senk, S. (2010). How reform curricula in the USA and Korea present multiplication and division of fractions. *Educational Studies in Mathematics, 74*(2), 117–142.

Steffe, L. P. (2002). A new hypothesis concerning children's fractional knowledge. *The Journal of Mathematical Behavior, 20*, 267–307.

Thompson, P. W., & Saldanha, L. A. (2003). Fractions and multiplicative reasoning. In J. Kilpatrick, W. G. Martin, & D. Schifter (Eds.), *A research companion to Principles and Standards in School Mathematics* (pp. 95–113). Reston, VA: National Council of Teachers of Mathematics.

Tzur, R. (1999). An integrated study of children's construction of improper fractions and the teacher's role in promoting that learning. *Journal for Research in Mathematics Education, 30* (4), 390–416.

Tzur, R. (2004). Teacher and students' joint production of a reversible fraction conception. *The Journal of Mathematical Behavior, 23*, 93–114.

Watanabe, T. (2003). Teaching multiplication: An analysis of elementary school mathematics teachers' manuals from Japan and the United States. *The Elementary School Journal, 104*, 111–125.

Zambat, I. O. (2015). An alternative route to teaching fraction division: Abstraction of common denominator algorithm. *International Electronic Journal of Elementary Education, 7*(3), 399–422.

Chapter 3
Comparing the Difficulty Level of Junior Secondary School Mathematics Textbooks in Five Nations

Yiming Cao, Libao Wu, and Lianchun Dong

Abstract This study examines the difficulty level of mathematics textbooks in junior secondary schools in China, the USA, South Korea, Singapore, and Japan. The analysis uses a novel framework which focuses on content breadth, content depth, difficulty level of worked examples, and difficulty level of exercises. Based on the analysis of five selected topics—numbers and calculation, equations, triangles, solids, and statistics—China has the most difficult textbooks, followed by South Korea and Singapore. The US difficulty levels are quite similar to those in Japan, which has the easiest textbooks. In addition, the selected Chinese mathematics textbook seems to value "shapes and geometry" most significantly among the five nations, whereas the selected US textbook seems to emphasize "numbers and calculation." Furthermore, it is found that the selected Japanese mathematics textbook series involves the least amount of mathematics, and also shows a lack of statistics.

Keywords Junior secondary school • Mathematics • Textbook • Difficulty level • Comparative study

The development and study of textbooks has been a significant aspect of academic research in mathematics curriculum and instruction (Kilpatrick, Swafford, & Findell, 2001; Wagemaker, 2003). The quality of mathematics textbooks could

Y. Cao (✉)
School of Mathematical Sciences, Beijing Normal University, Beijing, China
e-mail: caoym@bnu.edu.cn

L. Wu
Teacher Education College, Tianjin Normal University, Tianjin, China
e-mail: wlb0723@163.com

L. Dong
Melbourne Graduate School of Education, University of Melbourne, Melbourne, VIC, Australia
e-mail: donglianchun1986@163.com

J.-W. Son et al. (eds.), *What Matters? Research Trends in International Comparative Studies in Mathematics Education*, Research in Mathematics Education, DOI 10.1007/978-3-319-51187-0_3

influence the quality of classroom mathematics teaching, and it is also an important factor in determining the extent to which mathematics curriculum reform could be implemented (Cai, Mok, Reedy, & Stacey, 2016; Schmidt et al., 2001; Son & Senk, 2010; Wu & Cao, 2014). Thus, the investigation of textbooks is of great interest for school mathematics teachers and principals, especially at the junior secondary school level (Wu & Cao, 2014).

This study is part of a key Chinese government-funded project entitled "A comparative study of the difficulty level of junior school mathematics textbooks in ten nations" (2010–2015). This part of the project evaluates the difficulty level of mathematics textbooks in junior secondary schools (years 6–8 or 7–9) in China, the USA, South Korea, Singapore, and Japan. We first constructed a mathematical model to examine the degree of difficulty of junior secondary school mathematics textbooks based on a literature review, interviews with mathematics educators, and follow-up exploratory factor analysis. The model focuses on four aspects: content breadth, content depth, difficulty level of worked examples, and difficulty level of exercises. Using this model, the degree of difficulty of five sets of recommended junior secondary school mathematics textbooks (one from each nation listed above) was analyzed.

A cross-national exploration of school mathematics textbooks could contribute to a better understanding of school mathematics textbooks in each nation (Silver, 2009; Son & Diletti, 2017). For example, this study could provide implications for further development and revision of school textbooks in each nation. In addition, the findings of this study would also provide important references for policy makers, mathematics curriculum designers, and mathematics education researchers, helping them reflect on the school mathematics textbooks in their own nations and abroad. Furthermore, this study could also allow mathematics teachers to understand in more depth the mathematics curriculum documents in different cultures, and thereby draw lessons from other nations' documents to improve their own classroom teaching.

Selected Nations and Textbooks

Nations

This study selected five sets of recommended junior secondary school mathematics textbooks (in years 7, 8, and 9 or years 6, 7, and 8) from China, Japan, South Korea, Singapore, and the USA. Although these five nations vary in terms of economy, culture, technology, and education, their mathematics education systems have a substantial influence worldwide, and the USA, China, and Japan are ranked in the world's top ten economies. In addition, students from the Asian nations consistently showed strong performance in mathematics on large-scale international assessments, such as the Trends in International Mathematics and Science Study

(TIMSS) and the Program for International Student Assessment (PISA). For example, the fourth graders and eighth graders from Singapore were ranked third in TIMSS 2007 and second in mathematics in PISA 2009. And in 2012, for a second time, Chinese students' performance topped the list of PISA results in mathematics, science, and literacy (Mullis, Martin, Foy, & Arora, 2012).

Year Level

The school systems in these five nations have distinct differences, but in Asian countries years 7, 8, and 9 are typically covered in the junior secondary school level. China, Japan, and Korea follow this pattern, while Singapore junior secondary school includes 4 or 5 years, including years 7–9. By contrast, in the USA years 6–8 are considered as junior secondary school level. In our analysis, we focused on textbook analysis in years 7–9 in the four Asian countries and in years 6–8 in the USA because they are comparable grades.

Textbooks

Textbooks in this study only refer to the student books which are written based on curriculum standards, systematically reflecting the content of the subject, and exclude the teaching workbooks, teacher's manual, teaching reference books, educational software, and other teaching materials. All five nations have more than one type of mathematics textbook. In this study, the selected textbooks are either developed (or recommended) by the national education department or recognized as being influential in their own countries. Detailed information about the selected textbooks is listed in Table 3.1.

Table 3.1 Details of the selected mathematics textbooks

Nation	Title	Publisher	Publish time	Abbreviation
China	Mathematics	People's Education Press	2008	PEPM
USA	IMPACT Mathematics	McGraw-Hill	2009	IM
Singapore	Math Insights Secondary Normal	Pearson	2011	SM
South Korea	Mathematics	Visang Education Inc	2009	HM
Japan	Mathematics	Dainippon Tosho Publishing Co., Ltd	2010	JM

Research Design and Results

Analyzed Topics

In this study, five mathematics topics were chosen for content analysis based on recommendations from the National Council of Teachers of Mathematics [NCTM] (NCTM, 2000) and the Chinese Ministry of Education (2011). These five topics are "numbers and calculation," "equations," "triangles," "solids," and "statistics." Note that although the Japanese textbook does not cover statistics, all five topics are common to the other selected textbooks.

For each textbook, the total number of pages counts material from the first page of the first chapter to the last page of the last chapter, and excludes prefaces, tables of contents, appendices, and references. The proportions of these five topics in each textbook, calculated as the quotient of each topic's pages divided by the total pages in the textbooks, are listed in Table 3.2 (Wu, 2013a, 2013b).

Table 3.2 shows that the selected US textbook series tends to have a greater number of pages on the topic of numbers and calculation, whereas the Chinese mathematics textbook significantly values the topic of triangles (e.g., shapes and geometry) compared to the other books. The Japanese mathematics textbook lacks a section on statistics.

Modeling Textbook Difficulty

Understanding the degree of difficulty of secondary school textbooks internationally can significantly help a country develop its own primary or secondary school textbooks. However, only a small number of research studies have tried to investigate the degree of difficulty of secondary school mathematics textbooks (Bao, 2002, 2009). In mathematics education research, Nohara (2001) first proposed a model for the overall difficulty of mathematical problems, in a report submitted to the US National Center for Education Statistics. This model includes four factors: (1) the percentage of "scalability issues" (the so-called scalability problem refers to problems that ask students to draw their own conclusions and explain the process of problem-solving); (2) the percentage of problems with "real background"; (3) the

Table 3.2 The proportions of the five topics in each textbook

Nation	Numbers and calculation	Equations	Triangles	Solids	Statistics	Total
USA	0.3360	0.1102	0.0278	0.0310	0.0508	0.5554
China	0.1056	0.1182	0.1444	0.0513	0.0638	0.4603
Singapore	0.2148	0.0874	0.0813	0.0510	0.1141	0.5486
South Korea	0.1481	0.0717	0.0669	0.0406	0.0550	0.3823
Japan	0.1090	0.1106	0.0593	0.0497	–	0.3286

percentage of problems with "operations" excluding the problems in the area of "amount"; and (4) the percentage of problems of "multistep reasoning" (p. 14). This framework was developed to assess the difficulty level of mathematics test items, but Bao (2002) further developed and applied it to examine the difficulty level of mathematics textbooks. Bao suggested that factors such as "background," "calculating," "amount of knowledge," "reasoning," and "exploring" would jointly influence the difficulty of a mathematics problem. (p. 5) Based on these five factors, he developed a model to measure the comprehensive difficulty of a mathematics problem and compared the difficulty of intended mathematics curricula, implemented mathematics curricula, and enacted mathematics curricula in China and the United Kingdom.

In addition, Shi, Kong, and Li (2005) proposed the use of knowledge points (e.g., mathematical concepts, formulas, properties, and propositions) when examining the content breadth and content depth of mathematics textbooks. They argued that the degree of difficulty of a curriculum or textbook is influenced by three basic elements: the depth of the curriculum (i.e., the depth of thinking required by the curricular content), the width of the curriculum (i.e., the scope and width of the curricular content, which is quantified by the amount of content), and time (i.e., the time needed to complete the curricular content, which can be quantified by lesson hours). According to Shi et al., most students are able to understand the curricular content as long as enough time is provided. From this point of view, they established a model to measure the degree of difficulty of a mathematics curriculum: $N = \alpha\frac{S}{T} + (1 - \alpha)\frac{S}{T}$ where N refers to the degree of difficulty of a mathematics curriculum, G refers to the width of the curriculum (i.e., content breadth), S refers to the depth of the curriculum (i.e., content depth), and T is the time needed to complete the curricular content. In this model, S/T is the comparable depth, G/T is the comparable width, and $\alpha (0 < \alpha < 1)$ is the weight coefficient, reflecting the weighting degree of the comparable depth (S/T) or the comparable width G/T, which was 0.5 in their study.

In this study, on the basis of the models by Bao (2002) and Shi et al. (2005) reviewed above, a model developed by Cao and Wu (2015) was used to measure the difficulty level (N) of the aforementioned five mathematics topics in the selected mathematics textbooks. The difficulty level of each mathematics topic and the overall difficulty level of the textbooks for the five selected topics were calculated by looking at the following four aspects: content breadth (G), content depth (S), difficulty level of the worked examples (L), and difficulty level of the exercises (X). After constructing a mathematical model to examine the degree of difficulty of junior secondary school mathematics textbooks based on our literature review, we conducted interviews with mathematics educators, and used follow-up exploratory factor analysis to validate the model. Finally, after discussing our findings with experts involved in the project, the following four-factor model was chosen for the study:

$$N = f(G, S, L, X)$$

$$G_i = A_1 G_{i1} + A_2 G_{i2} + A_3 G_{i3} + A_4 G_{i4} + A_5 G_{i5}$$

$$S_i = \frac{1{*}A + 2{*}B + 3{*}C}{A + B + C} \quad (i = 1, 2, 3, 4, 5)$$

$$LT_{ij} = \alpha * LYQ_{ij} + \beta * LZS_{ij} + \gamma * LBJ_{ij} \, (i = 1, 2, 3, 4, 5; j = 1, 2, 3, 4, 5);$$

$$XT_{ij} = \alpha{*}XYQ_{ij} + \beta{*}XZS_{ij} + \gamma{*}XBJ_{ij} \, (i = 1, 2, 3, 4, 5; j = 1, 2, 3, 4, 5)$$

In this model, N refers to the degree of difficulty of junior secondary mathematics textbooks, which is represented as a function of the content breadth (G), content depth (S), difficulty level of the worked examples (L), and difficulty level of the exercises (X). G refers to the width of content (the amount of knowledge), which is quantified by the number and scope of mathematics topics covered in textbooks. S refers to the depth of content, which is quantified by the total number of mathematical concepts and propositions that are demonstrated by direct description (A), by analogy and induction (B), and by deduction (C) in the selected mathematics topics. L refers to the degree of difficulty of the worked examples, which is quantified by the level of cognitive requirements (YQ), the number of mathematics knowledge points included (ZS), and the level of contexts (BJ). X refers to the degree of difficulty of the exercises, and is quantified by the level of cognitive requirements (YQ) and the number of mathematics knowledge points included (ZS). In this model, i refers to the number of nations and ranges from 1 (USA) to 5 (Japan); j refers to the specific mathematical topic and ranges from 1 (numbers and calculation) to 5 (statistics). More details could be found in Table 3.3.

We first identified content breadth, content depth, difficulty level of the worked examples, and difficulty level of the exercises with respect to each topic, and then calculated the overall content breadth and depth and the overall difficulty levels of worked examples and exercises for the five mathematics topics in each nation's textbook. We determined the weight coefficients, and thereby to determine the weighting degree of content breadth (a), content depth (b), difficulty level of

Table 3.3 The proportion of the five topics

		Topic				
No. (i)	Nation	Numbers and calculation (A_1)	Equations (A_2)	Triangles (A_3)	Solids (A_4)	Statistics (A_5)
1	USA	0.6045	0.1983	0.05	0.0558	0.0914
2	China	0.2185	0.2446	0.2988	0.1061	0.132
3	Singapore	0.3915	0.1593	0.1482	0.093	0.208
4	South Korea	0.3874	0.1875	0.175	0.1062	0.1439
5	Japan	0.3317	0.3366	0.1805	0.1512	–

worked examples (c), and difficulty level of exercises (d), based on the results of interviews with mathematics educators. Then, the overall difficulty level of each textbook for the five selected topics was calculated, based on the following formula: $N_{ij} = \alpha * G_{ij} + b * S_{ij} + c * L_{ij} + d * X_{ij} \, (i = 1, 2, 3, 4, 5; j = 1, 2, 3, 4, 5)$.

In the following section, we describe our model in detail and present the corresponding findings of this study. As mentioned previously, there are quantitative studies on the difficulty of mathematics courses or mathematical exercises, but no quantitative studies on the difficulty of mathematics textbooks or comparative studies on the difficulty of mathematics textbooks across nations (Bao, 2002). This study first constructed a model of the degree of difficulty of junior secondary school mathematics textbooks, and then used it to examine the degree of difficulty of five sets of junior secondary school mathematics textbooks in the selected nations.

Content Breadth

Content breadth in this chapter refers to the number of mathematics knowledge points covered in the five mathematics topics, i.e., numbers and calculation, equations, triangles, solids, and statistics. A knowledge point could be a mathematical concept, a mathematical formula, or a mathematical property covered within the mathematics topics in the textbooks. The content breadth of the mathematics textbook in the ith country is calculated by the formula $G_i = A_1G_{i1} + A_2G_{i2} + A_3G_{i3} + A_4G_{i4} + A_5G_{i5}$. In this model, A_1, A_2, A_3, A_4, and A_5 represent the proportion taken up by each of the five mathematics topics in the selected textbooks (see Table 3.3), and G_{ij} refers to the relative content breadth of the five mathematics topics for the ith nation's textbook, where j refers to the value of mathematical topics, ranging from 1 (numbers and calculation) to 5 (statistics). G_{ij} was calculated by the formula $G_{ij} = \frac{a_j}{b} (i = 1, 2, 3, 4, 5; j = 1, 2, 3, 4, 5)$, where i refers to the country number and j refers to the topic number. In this formula, a_j represents the number of mathematics knowledge points covered within the five mathematics topics and b refers to the highest number of mathematics knowledge points in a topic area across the five nations. The number of mathematics knowledge points was determined based on its overall frequency, regardless of repetitiveness across grades. For example, if the same concept/formula/property is discussed in more than one grade level, it was counted based on the number of appearances.

More specifically, the calculation of the content breadth of five mathematics topics in the selected US textbooks is demonstrated below, taking the example of the mathematics topic "numbers and calculation" (coded 1) in the USA (coded 1).

1. The number of mathematics knowledge points covered within the mathematics topic "numbers and calculation" for US textbooks was counted and labelled as a_{11}, equations as a_{12}, triangles as a_{13}, solids as a_{14}, and statistics as a_{15}.
2. The largest number of $a_j (j = 1, 2, 3, 4, 5)$ was labelled as b and refers to the largest number of mathematics knowledge points among the five topics. Here,

Table 3.4 Relative content breadth and the overall content breadth of each nation's textbook

Nation	Relative content breadth by topic					Overall breadth	Ranking
	Topic						
	Number/ calculation	Equations	Triangles	Solids	Statistics		
USA	0.9394	0.7037	0.3030	0.6429	1	0.8498	2
China	0.7879	1	1	0.7857	0.5909	0.8769	1
Singapore	0.8182	0.4074	0.6667	1	0.6818	0.7188	4
South Korea	1	0.4815	0.7576	0.7143	0.5455	0.7646	3
Japan	0.2727	0.3704	0.2121	0.9286	–	0.3938	5

let's suppose that the mathematics topic "numbers and calculation" has the largest number of mathematics knowledge points.

3. The relative content breadth of the mathematics topic "numbers and calculation" and other five topics for the US book was calculated by the formula $G_{ij} = \frac{a_j}{b}\,(i = 1,2,3,4,5; j = 1,2,3,4,5)$.
4. The content breadth of the mathematics textbook in the USA is calculated by the formula $G_1 = A_1\frac{a_{11}}{b} + A_2\frac{a_{12}}{b} + A_3\frac{a_{13}}{b} + A_4\frac{a_{14}}{b} + A_5\frac{a_{15}}{b}$, where $A_1 = 0.6045$, $A_2 = 0.1983$, $A_3 = 0.05$, $A_4 = 0.0558$, and $A_5 = 0.0914$ (drawn from Table 3.3).

The results of these calculations, the relative content breadth of each mathematics topic, and the overall content breadth of each textbook series are listed in Table 3.4.

Content Depth

Content depth concerns the depth of thinking required by mathematical concepts and propositions. There are three ways of demonstrating the depth of mathematical concepts and propositions, namely by direct description, by analogy and induction, and by deduction. These three ways of demonstration were separately given the values of 1 (direct description), 2 (analogy and induction), and 3 (deduction). The following formula calculates the depth of each nation's mathematics knowledge point:

$$S_i = \frac{1{*}A + 2{*}B + 3{*}C}{A + B + C},$$

where A is the total number of mathematical concepts and propositions that are demonstrated by *direct description*, B is the total number of mathematics knowledge points demonstrated by *analogy* and *induction*, and C is the total number of mathematics knowledge points demonstrated by *deduction* within the ith nation's mathematics knowledge point. After calculating the depth of all the mathematics knowledge points included in each mathematics topic, the depth of this

Table 3.5 Relative depth of content by topic and overall depth of the textbooks in five nations

Nation	Relative depth of content by topic					Overall depth	Ranking
	Topic						
	Numbers and calculation	Equations	Triangles	Solids	Statistics		
USA	0.6724	0.7538	0.7176	0.509	1	0.7116	5
China	0.8322	0.8972	1	0.5719	0.6667	0.8488	3
Singapore	0.8408	0.8658	0.9332	0.509	0.5777	0.7729	4
South Korea	0.9665	0.7262	0.9145	1	0.6932	0.8766	1
Japan	1	1	0.5465	0.5509	–	0.8502	2

mathematics topic could be obtained by $S = \frac{\sum_{j=1}^{n} S_{ij}}{n_j} (i = 1, 2, 3, 4, 5; j = 1, 2, 3, 4, 5)$, where n is the highest value of the mathematics knowledge points within one mathematics topic across the five nations. S_{ij} is the depth of the ith nation's mathematics knowledge point, which is quantified by the total number of mathematical concepts and propositions that are demonstrated by direct description (A), by analogy and induction (B), and by deduction (C) in the five mathematics topics. The values for the relative depth of one mathematics topic in all five nations were then standardized by dividing these values by the highest value in the five nations. Table 3.5 presents the relative depth of content by topic and the depth of the textbook across the five nations.

Difficulty Level of the Worked Examples and Exercises

The difficulty level of the worked examples and exercises involves three aspects, namely the level of cognitive requirements, the number of mathematics knowledge points included, and the level of contexts (Wu, 2013a, 2013b; Wu, Wang, & Cao, 2014). This is quantified by the level of cognitive requirements (YQ) (i.e., the number of worked examples or exercises requiring the imitation, comprehension, application, and investigation level of cognitive requirements), the number of mathematics knowledge points included (ZS) (i.e., the number of worked examples or exercises that require one, two, three, and four mathematics knowledge points), and the level of contexts (BJ) (i.e., the number of worked examples or exercises having no context (or personal contexts, or public life contexts, or scientific contexts)).

The level of cognitive requirements (YQ).[1] In the mathematics curriculum document released by the Chinese Ministry of Education (2011), there are four

[1]The abbreviations for different terms in this chapter are based on the corresponding Chinese terms. For example, "cognitive requirements" in Chinese is "Ren Zhi Yao Qiu," so we took YQ, the initials of the last two words, as the abbreviation.

verbs used to describe learning objectives: “know,” “understand,” “master,” and “apply.” The corresponding levels of cognitive requirements are “knowing,” “understanding,” “application,” and “investigation” which are separately given values of 1, 2, 3, and 4. The cognitive requirement level of the worked examples and exercises is separately labelled as $LYQ_{ij}, XYQ_{ij}(i = 1,2,3,4,5; j = 1,2,3,4,5)$ and obtained by the formulas below:

$$LYQ_{ij} = \frac{1{*}A + 2{*}B + 3{*}C + 4{*}D}{A + B + C + D}(i = 1,2,3,4,5; j = 1,2,3,4,5);$$

$$XYQ_{ij} = \frac{1{*}A + 2{*}B + 3{*}C + 4{*}D}{A + B + C + D}(i = 1,2,3,4,5; j = 1,2,3,4,5)$$

where A is the number of the worked examples (or exercises) that have the imitation level of cognitive requirements, B is for comprehension, C is for application, and D is for investigation.

The number of mathematics knowledge points included (ZS). The identification of the mathematics knowledge points is the same as the process for determining the content breadth. There are four values for the number of the mathematics knowledge points: 1, 2, 3, and 4, which means that there are one, two, three, or four mathematics knowledge points involved in one worked example or exercise. For those worked examples and exercises where there are more than four mathematics knowledge points, the value is set to 4. The calculation formula is below:

$$LZS_{ij} = \frac{1{*}A + 2{*}B + 3{*}C + 4{*}D}{A + B + C + D}(i = 1,2,3,4,5; j = 1,2,3,4,5);$$

$$XZS_{ij} = \frac{1{*}A + 2{*}B + 3{*}C + 4{*}D}{A + B + C + D}(i = 1,2,3,4,5; j = 1,2,3,4,5)$$

where A (or B, or C, or D) is the number of worked examples (or exercises) which involve 1 (or 2, or 3, or 4) mathematics knowledge points within each mathematics topic in the nation’s textbook.

The level of contexts (BJ). There are four levels in terms of the contexts of the worked examples and exercises: no context, personal contexts, public life contexts, and scientific contexts. These four levels are given values of 1, 2, 3, and 4. The level of context for the worked examples and exercises in the jth mathematics topic is labelled as LBJ_{ij}, $XBJ_{ij}(i = 1,2,3,4,5; j = 1,2,3,4,5)$. The calculation formula is below:

$$LBJ_{ij} = \frac{1{*}A + 2{*}B + 3{*}C + 4{*}D}{A + B + C + D}(i = 1,2,3,4,5; j = 1,2,3,4,5);$$

$$XBJ_{ij} = \frac{1{*}A + 2{*}B + 3{*}C + 4{*}D}{A + B + C + D}(i = 1,2,3,4,5; j = 1,2,3,4,5)$$

Table 3.6 The difficulty level of the worked examples in five nations

Nation	Relative difficulty level by topic					Overall difficulty level	Ranking
	Topic						
	Numbers and calculation	Equations	Triangles	Solids	Statistics		
USA	0.623	0.6952	0.6294	0.8131	0.8208	0.6663	5
China	0.7099	0.8353	0.907	0.6636	0.9611	0.8277	4
Singapore	0.8963	1	0.8513	0.7039	0.7297	0.8536	2
South Korea	1	0.8155	1	1	1	0.9654	1
Japan	0.8015	0.9091	0.6958	0.9354	–	0.8389	3

where A (or B, or C, or D) is the number of worked examples (or exercises) having no context (or personal contexts, or public life contexts, or scientific contexts) in the ith mathematics knowledge point.

Overall difficulty level of the worked examples and exercises. For each textbook, the formulas to calculate the difficulty level of the worked examples and exercises are below:

$$LT_{ij} = \alpha^*LYQ_{ij} + \beta^*LZS_{ij} + \gamma^*LBJ_{ij}(i = 1,2,3,4,5; j = 1,2,3,4,5)$$

$$XT_{ij} = \alpha^*XYQ_{ij} + \beta^*XZS_{ij} + \gamma^*XBJ_{ij}(i = 1,2,3,4,5; j = 1,2,3,4,5)$$

where α, β, γ are the weights of the level of cognitive requirements, the number of mathematics knowledge points included, and the level of context. The values are separately 0.38, 0.36, and 0.26. The results are then standardized in a similar way as when calculating the content depth. The detailed results are below, in Tables 3.6 and 3.7.

Overall difficulty level of mathematics textbooks in five mathematics topics. In order to calculate the overall difficulty level of each topic, and that of each nation's mathematics textbooks in the five mathematics topics, the following models are used:

$$N_j a * G_j + b * S_j + c * L_j + d * X_j,$$

$$N = A_1N_1 + A_2N_2 + A_3N_3 + A_4N_4 + A_5N_5$$

For the overall difficulty level (N) of the jth mathematics topic, content breadth (G), content depth (S), difficulty level of the worked examples (L), and difficulty level of the exercises (X) in jth mathematics topic are summed, where a, b, c, d mean the weights of content breadth, content depth, difficulty level of worked examples, and difficulty level of exercises.

Table 3.7 The difficulty level of the exercises in five nations

Nation	Relative difficulty level by topic					Overall difficulty level	Ranking
	Topic						
	Numbers and calculation	Equations	Triangles	Solids	Statistics	Difficulty level	Ranking
USA	0.6604	0.6324	0.6897	0.5251	1	0.6798	5
China	0.8932	1	1	0.6771	0.9673	0.9381	1
Singapore	1	0.8773	0.8535	0.761	0.7993	0.8948	2
South Korea	0.8819	0.7951	0.8817	0.6709	0.8973	0.8454	4
Japan	0.964	0.7731	0.8274	1	–	0.8805	3

Table 3.8 Survey results of weights for mathematics textbooks' difficulty levels

	Content breadth	Content depth	Worked examples	Exercise
Researchers and textbook authors	0.273	0.320	0.195	0.221
Teachers	0.293	0.269	0.215	0.224

To get the information about the weights of the above four factors (a, b, c, d) (i.e., the relative impact of the four factors on the difficulty level), we surveyed a group of 23 mathematics education researchers and textbook authors from Beijing City, Tianjin City, Jilin Province, and Sichuan Province in China, and a group of 161 experienced junior secondary mathematics teachers from Beijing City, Shandong Province, and Sichuan Province in China. The results are listed in Table 3.8. Then, by employing the analytic hierarchy process (Zhou & Wang, 2013), the final values of weights were obtained to be used in the model, which are 0.28, 0.30, 0.20, and 0.22, respectively, for content breadth, content depth, difficulty level of worked examples, and difficulty level of exercises.

Similar approaches were used to obtain the weights of the four factors for the difficulty level of mathematics textbooks across the five nations. In the next section, we present the values and rankings of the difficulty level for the five mathematics topics in the selected mathematics textbooks.

Interpretation and Discussions

Tables 3.9 and 3.10 present the weighted difficulty levels of the five mathematics topics and the overall weighted difficulty level of the five nations' textbooks. Weighted mean values of all five areas (four areas in the case of Japan, since the selected textbook does not cover statistics) were calculated to get the values of the difficulty level of the whole textbook.

Table 3.9 Values and rankings of the difficulty level for the five mathematics topics

	Topic									
	Numbers and calculation		Equations		Triangles		Solids		Statistics	
Nation	Value	Ranking	*V*	*R*	*V*	*R*	*V*	*R*	*V*	*R*
USA	0.7346	5	0.7013	4	0.5777	4	0.6109	5	0.9642	1
Chins	0.8088	3	0.9362	1	0.9814	1	0.6733	4	0.7705	2
Singapore	0.8806	2	0.7668	2	0.8247	3	0.7409	3	0.686	4
South Korea	0.9640	1	0.6907	5	0.8805	2	0.8476	1	0.7581	3
Japan	0.7487	4	0.7556	3	0.5445	5	0.8324	2	–	–

Note: V refers to value and *R* stands for ranking.

Table 3.10 The overall difficulty level of the select textbooks in five nations

	Nation				
Difficulty level	China	South Korea	Singapore	USA	Japan
Overall in four dimensions	0.8721	0.8561	0.8007	0.7343	0.7268
Rankings	1	2	3	4	5
Content breadth	0.8769	0.7646	0.7188	0.8498	0.3938
Rankings	1	3	4	2	5
Content depth	0.8488	0.8766	0.7729	0.7116	0.8502
Rankings	3	1	4	5	2
Difficulty of exercise	0.8277	0.9654	0.8536	0.6663	0.8389
Rankings	4	1	2	5	3
Difficulty of worked examples	0.9381	0.8454	0.8948	0.6798	0.8805
Rankings	1	4	2	5	3

The Overall Difficulty Level of the Five Textbooks

Table 3.10 shows that the selected Chinese textbook is ranked No. 1 with an overall difficulty level of 0.8721 across five mathematics topics. Out of the five mathematics topics, the Chinese textbook has the highest level of difficulty in "equations" and "triangles," and the second-highest level of difficulty in "statistics" (see Table 3.9). In addition, compared with other textbooks, the worked examples in Chinese textbooks are particularly difficult (Son & Hu, 2015).

The South Korean textbook is the second most difficult, with a difficulty level of 0.8561. Out of the five mathematics topics, "numbers and calculation" and "solids" in the Korean textbook are ranked the highest in terms of difficulty level, and "triangles" is ranked second highest. Roughly speaking, although there are slight differences, the Chinese and South Korean textbooks are quite similar in terms of the difficulty level.

The selected Singaporean mathematics textbook's difficulty level is 0.8007, and ranked third among the five nations. Compared with the other nations' textbooks, two out of five mathematics topics in Singapore are the second most difficult.

Another two topics are the third most difficult, and the last one is ranked fourth. As a whole, with respect to the difficulty level, the Singaporean mathematics textbook stands between the Chinese and South Korean textbooks.

The selected US textbook series is ranked fourth in terms of the overall difficulty level in the five mathematics topics. In particular, the US textbook is ranked second and fifth in content breadth and content depth, respectively, which reflects its wide but shallow coverage. Moreover, the worked examples and exercises in the selected US textbook series were found to be the least difficult.

The least difficult textbook is the selected Japanese textbook series. Although there is no statistics content in this textbook, as a whole it covers fewer mathematics topics than any other nation's textbook series. In particular, compared with other textbooks, the content breadth of the Japanese textbook is the lowest, revealing that it seems to be more focused. The difficulty level of the worked examples and exercises is ranked third. It is important to note that the selected Japanese textbook series had a market share of slightly under 7% in 2012.

Content Breadth

The selected Chinese textbook covers the largest amount of mathematics. In terms of content breadth, the Chinese book's coverage of "equations" and "triangles" is ranked first. For example, it has the most systematic approach to "equations," and this topic also covers a significantly larger amount of mathematics than any other nation's textbook. The contents in "equations" include the concept of an equation, setting up equations, equations' properties, linear equations, solving linear equations, applications of linear equations, systems of linear equations with two variables, solving systems of linear equations with two variables, method of elimination, method of substitution, application of (systems of) linear equations with two variables, systems of linear equations with three variables, fractional equations (only involving the equations that could be converted to linear equations), solving fractional equations, quadratic equations, solving quadratic equations by completing the squares, solving quadratic equations by formulas (such as taking out the common factor, differences of squares formula, and the perfect square formula), solving quadratic equations by factorization, relationships between the coefficients of a quadratic equation and its solutions, applications of quadratic equations, systems of linear and quadratic equations, linear equations and linear functions, linear functions and systems of linear equations, and quadratic functions and quadratic equations.

The selected US textbook is ranked second in terms of content breadth. The content breadth rankings of the US textbook for "numbers and calculation," "equations," "triangles," "solids," and "statistics" are, respectively, 2nd, 2nd, 4th, 5th, and 1st (see Table 3.9). The content breadth of the overall textbook is 0.8498, which ranks second among the five nations. In particular, the US textbook covers the largest amount of statistics among the five selected textbooks. Although the value for its content breadth of "solids" is smaller than all the other four nations, the differences are not significant. In addition, the proportion of "solids" is relatively smaller, so it exerts relatively slight influence on the content breadth of the textbook. Despite the low ranking in terms of "triangles" and "solids," the content breadth in "calculation" and "statistics" is ranked in the top 3. To some extent, this might reflect that US educators and curriculum developers more strongly emphasize the role of algebra and statistics in primary mathematics learning (Cao & Wang, 2007).

The selected South Korean textbook is ranked third, and the Singaporean book is ranked fourth. The content breadth of South Korea's textbook is 0.7646, third among the five nations. But it is worth mentioning that its value for content breadth is the highest for "numbers and calculation" and the second highest for "triangles." The Singaporean textbook is ranked fourth with a content breadth value of 0.7188.

The Japanese textbook covers the least amount of mathematics. The outstanding characteristic of the selected Japanese textbooks is its narrow focus. For one thing, there is no "statistics" section in the junior secondary-level textbook. For another, the content breadth values for all the other topics are quite low, except for "solids," for which the content breadth is ranked second.

Content Depth

The mathematics textbook from South Korea is ranked first overall, with a content depth value of 0.8766. Within the topics "triangles" and "solids," most of the mathematical concepts and propositions are demonstrated by deduction, whereas for "numbers and calculation" and "equations," most of the concepts are demonstrated by induction and fewer are demonstrated by direct description. Japan and China follow South Korea in terms of content depth, respectively, ranked second and third. Overall, the differences between the Japanese and Chinese textbooks are slight. For the topic "triangles," the Chinese textbook is much deeper than the Japanese textbook, while for the topics "numbers and calculation" and "equations," the Japanese textbook is found to be much deeper. Fourth in the ranking is Singapore, whose textbook content depth is 0.7729. The US textbook ranks last in terms of content depth, with the value of 0.7116. In the US textbook series, most of the mathematics concepts and propositions are demonstrated by direct description.

The Difficulty Level of Worked Examples and Exercises

The worked examples and exercises in the Chinese textbook are the most difficult (0.9381), whereas those in the US textbook are the least difficult. The selected textbooks in Singapore and Japan are ranked second and third, respectively, with values of 0.8948 and 0.8805. South Korea is fourth in terms of the difficulty level of worked examples and exercises (0.8454). These findings reflect that in cultural settings that have been influenced by Confucius, the difficulty level of worked examples and exercises is uniformly high. In particular, the level of cognitive requirements is relatively high. But when examining the number of exercises in the textbooks, the highest in the ranking list is the US textbook, which includes 4192 exercise tasks. This number is far larger than that of the Chinese textbook, where there are only 670 exercise tasks (Wu, Song, & Yang, 2013; Yang & Wu, 2014). The textbook from South Korea ranks highest in the difficulty level of worked examples in all five topics, except for "equations." There are slight differences among the textbooks in Singapore, Japan, and China, which are ranked second, third, and fourth. The worked examples are the least difficult in the US textbook.

Conclusions and Implications

The findings in this chapter are based on analysis of five selected mathematics topics in five sets of textbooks from five nations. Thus we must warn against overgeneralization of our findings. We acknowledge that the selected textbook series may not be the only representative of the textbooks in use in the five nations. In addition, due to differences among the countries' school systems, we analyzed different grade bands. However, despite these limitations, the comparison of the degree of difficulty of the five selected textbooks allows us to make the following conclusions.

Regional Features May Emerge in Terms of the Textbooks' Difficulty Levels

We suggest that there exist regional features when examining the difficulty levels of mathematics textbooks across the five nations. The top three countries in the ranking list are South Korea, China, and Singapore, all of which have been significantly influenced by Confucian values. The only exception is the Japanese textbook, which includes relatively less mathematics content and a relatively lower difficulty level. The US textbook series mainly emphasizes the "numbers and calculation" topic, for which the number of exercise tasks takes up half of the total exercise tasks across all five mathematics topics.

Teaching Materials Are Changing into Learning Materials

The worldwide trend is that the functions of textbooks are changing from assisting teaching to assisting learning (Mullis et al., 2012). This is also reflected in all five nations' textbooks. In all five textbooks, most of the mathematical concepts and propositions are demonstrated by direct descriptions, and the learners' capabilities, interests, and needs are taken into consideration when designing the worked examples and exercises (see Tables 3.7 and 3.8). Meanwhile, these textbooks prioritize the learners' development and thus support their mathematics learning. For example, the Singaporean textbook emphasizes the construction of context and backgrounds for mathematics learning (Yang, Reys, & Wu, 2010), and tasks pertaining to mathematics investigation exist in every unit of the mathematics textbook in South Korea (Son & Senk, 2010).

Practical Applications of Mathematics Are Emphasized

In all five textbooks, the connections between mathematics and real life or other school subjects are explicitly emphasized. These textbooks focus significantly on developing students' abilities to solve real-life tasks by using mathematics. New mathematics topics tend to be introduced with real-life problems so as to arouse students' motivations to move on. In particular, the US textbook emphasizes mathematics investigation and provides students with contexts that are connected to students' everyday life, local culture, and history (Son & Senk, 2010). Further, there are particular chapters focusing on application tasks in the US textbook. For the exercise tasks, the Chinese textbook also includes a particular module requiring students to consider the applications of mathematics and solve application tasks (Li, 2000). The emphasis on application is also very evident in the other textbooks, which include application tasks related to culture, business, and finance.

The Integration of Information Technology into Mathematics Is Evident

The development of information technology has brought dramatic changes into mathematics research, teaching, and learning. In almost every nation, the mathematics curriculum encourages the integration of information technology into mathematics teaching and learning (Guo & Cao, 2012). This is also reflected in all five nations' textbooks, as students are encouraged to use graphing calculators, computers, and software to assist mathematics teaching and learning activities. For example, in South Korea's mathematics textbook, there are particular sections discussing how to use calculators or computers to solve mathematical tasks or to investigate mathematical knowledge.

Although only one type of mathematics textbook was selected from each nation, it is noteworthy to point out that a variety of mathematics textbooks are being used in each nation. For many nations, there is no national mathematics textbook. Instead, many types of mathematics textbooks are available for teachers and schools to choose. For example, there are 14 types of mathematics textbooks at the junior secondary level in South Korea. Therefore, the findings of this study should be carefully interpreted. Future work can learn much more from different nations, and thus provide more implications to improve mathematics curriculum and textbook development.

References

Bao, J. (2002). *A comparative study of difficulty level of mathematics textbooks in junior secondary level in China and UK* (In Chinese). Unpublished doctoral dissertation, East China Normal University, Shanghai.

Bao, J. (2009). A comparative study of mathematics textbooks' cognitive requirements in China and US—An example of quadratic equations and functions (In Chinese). *Journal of Mathematics Education, 18*(4), 57–60.

Cao, Y., & Wang, Z. (2007). A study of teaching algebraic thinking as key principle of mathematics (In Chinese). *Journal of Mathematics Education, 16*(1), 8–11.

Cao, Y., & Wu, L. (2015). An international comparison of mathematics textbooks in junior secondary level (In Chinese). *Journal of Mathematics Education, 24*(4), 3–7.

Cai, J., Mok, I. A., Reddy, V., & Stacey, K. (2016). International Comparative Studies in Mathematics: Lessons for Improving Students' Learning. In *International Comparative Studies in Mathematics* (pp. 1–36). Springer International Publishing

Guo, K., & Cao, Y. (2012). An international comparison on the use of information technology in mathematics curriculum (In Chinese). *China Educational Technology, 2012*(7), 108–113.

Kilpatrick, J., Swafford, J., & Findell, B. (Eds.). (2001). *Adding it up: Helping children learn mathematics.* Washington, DC: National Academy Press.

Li, Y. (2000). A comparison of problems that follow selected content presentations in American and Chinese mathematics textbooks. *Journal for Research in Mathematics Education, 31*(2), 234–241.

Ministry of Education (China). (2011). *Curriculum Standards for Year 1–9 School Mathematics (in Chinese).* Beijing, P.R. China: Beijing Normal University Publishing Group.

Mullis, I. V. S., Martin, M. O., Foy, P., & Arora, A. (2012). *TIMSS 2011 international results in mathematics.* Chestnut Hill, MA: TIMSS & PIRLS International Study Center, Boston College.

National Council of Teachers of Mathematics. (2000). *Principles and standards for school mathematics.* Reston, VA: National Council of Teachers of Mathematics.

Nohara, D. (2001). *A Comparison of the National Assessment of Educational Progress (NAEP), the Third International Mathematics and Science Study-Repeat (TIMSS-R), and the Programme for International Student Assessment (PISA) (NCES 2001–07).* U.S. Department of Education. Washington, DC: National Center for Education Statistics Working Paper.

Schmidt, W. H., McKnight, C. C., Houang, R. T., Wang, H. A., Wiley, D. E., Cogan, L. S., et al. (2001). *Why schools matter: A cross-national comparison of curriculum and learning.* San Francisco: Jossey-Bass.

Shi, N., Kong, F., & Li, S. (2005). A model of degree of difficulty of course: Contrast of difficulty degree of geometry course in China's compulsory education (In Chinese). *Journal of Northeast Normal University (Philosophy and Social Sciences), 6*, 151–155.

Silver, E. (2009). Cross–national comparisons of mathematics curriculum materials: What might we learn? *ZDM–The International Journal on Mathematics Education, 41*, 827–382.

Son, J., & Diletti, J. (2017). Chapter 1: What can we learn from textbook analysis? In J. Son, T. Watanabe, & J. J. Lo (Eds.), *What matters?: Research trends in international comparative studies in mathematics education.* New York: Springer.

Son, J., & Senk, S. (2010). How reform curricula in the USA and Korea present multiplication and division of fractions. *Educational Studies in Mathematics, 74*(2), 117–142.

Son, J., & Hu, Q. (2015). A comparative study of the initial treatment of the concept of function in selected secondary mathematics textbooks in the US and China. *International Journal of Mathematical Education in Science and Technology*. doi: 10.1080/0020739X.2015.1088084.

Wagemaker, H. (2003). *IEA: International studies, impact and transition*. Retrieved from http://www.iea.nl/fileadmin/user_upload/IRC2004/Wagemaker.pdf.

Wu, L. (2013a). An international comparison of mathematics curriculum in junior secondary level (In Chinese). *Journal of Educational Studies, 9*(2), 29–36.

Wu, L. (2013b). A comparative study of worked examples in mathematics textbooks in China and Australia (In Chinese). *Journal of Mathematics Education, 22*(2), 58–61.

Wu, L., & Cao, Y. (2014). Strategies to analyze mathematics textbooks in secondary level (In Chinese). *ZhongGuo Jiao Yu XueKan, 1*, 60–64.

Wu, L., Wang, J., & Cao, Y. (2014). An international comparison of mathematics textbooks in junior secondary level (In Chinese). *Curriculum, Teaching Materials and Teaching Methods, 34*(2), 112–117.

Wu, L., Song, W., & Yang, F. (2013). The US IM mathematics textbook's characteristics and implication for China (In Chinese). *Primary and Secondary Schooling Abroad, 2003*(8), 60–64.

Yang, D. C., Reys, R. E., & Wu, L. L. (2010). Comparing the development of fractions in the fifth- and sixth-graders' textbooks of Singapore, Taiwan, and the USA. *School Science and Mathematics, 110*(3), 118–127.

Yang, F., & Wu, L. (2014). A comparative study of worked examples in mathematics textbooks in China and US (In Chinese). *Journal of Mathematics Education, 23*(5), 56–59.

Zhou, J., & Wang, Y. (2013). A comparative study of the difficulty level of mathematics textbooks in China and UK (In Chinese). *Journal of Mathematics Education, 22*(3), 18–23.

Chapter 4
Uncovering the Label "Asian" in International Comparative Studies of Mathematics Education

Yoshinori Shimizu

Abstract In mathematics education scholarship, the results of international comparative studies tend to be reported with dichotomies: "high-performing" versus "low-performing," "teacher-centered" versus "student-centered," and even "Eastern" versus "Western." The label "Asian" is quite often used in such contexts to contrast differences in reported research findings between Asian countries or regions and other areas. This chapter problematizes such dichotomies and discusses possibilities of going beyond them, by evaluating similarities and differences in educational practices in "similar" Asian countries based on data from the Learner's Perspective Study. A lesson event where teachers sum up during mathematics lessons is similarly used to illustrate similarities and differences in classroom practices. While recognizing that international comparative studies on classroom teaching provide researchers and policy makers opportunities for understanding their own implicit theories about how teachers teach and how children learn mathematics in their local contexts, this chapter emphasizes the importance of the different cultural assumptions underpinning teaching and learning in the international debates on mathematics education.

Keywords International comparisons • Dichotomy • Cultural activity • East Asia • LPS

Introduction

Mathematics education research in the last decade has included more international endeavors than ever before (Dindyal, 2014; Shimizu & Kaur, 2013; Singh & Ellerton, 2013). International comparative studies have started to recognize the need to focus more on diverse voices and perspectives among members of local

Y. Shimizu (✉)
College of Education, University of Tsukuba, Tsukuba, Japan
e-mail: yshimizu@human.tsukuba.ac.jp

J.-W. Son et al. (eds.), *What Matters? Research Trends in International Comparative Studies in Mathematics Education*, Research in Mathematics Education, DOI 10.1007/978-3-319-51187-0_4

communities (Clarke, 2003). As the globalization and internationalization of research activities have continued to expand, the field of mathematics education research has clearly shown a diversification of perspectives on teaching and learning in classrooms embedded in local contexts (Shimizu & Williams, 2013).

Prior to this recognition of the importance of local contexts, much has been documented about the high mathematics performance of students from Asian countries or regions on international assessments such as the Trends in International Mathematics and Science Studies (TIMSS) and the OECD-Programme for International Student Assessment (PISA). The recent "Teacher Education and Development Study: Learning to Teach Mathematics" (TEDS-M) has also shown some characteristic outcomes related to Asian teachers' content knowledge and pedagogical knowledge. Further, a growing body of research on Lesson Study (which originated in Japan) has been conducted, with a focus on the form and function of professional learning around the world. The umbrella term "Asian" in these contexts functions as a label which is used to highlight characteristics of educational practices found in the geographical region and to contrast them with the countries beyond Asia, but occasionally this term betrays an ignorance of significant differences within Asian education.

One of the major challenges confronting the international mathematics education community is how best to learn from each other's classroom practices. Keitel and Kilpatrick (1999) questioned the assumptions on which international comparative studies of school mathematics had been predicated. They questioned, in particular, the treatment of the mathematics curriculum as unproblematic and the associated assumption that a single test could give comparative measures of curriculum effects across countries. Even within regions with the "same" cultural traditions and where mathematics curricula and classroom practices may have commonalities, there should still be significant opportunities to learn from colleagues' classroom practices.

The results of international comparative studies tend to be reported with typical dichotomies such as "high-performing" versus "low-performing," "teacher-centered" versus "student-centered," or even "Eastern" versus "Western." This chapter problematizes such dichotomies and labels to discuss the possibilities of going beyond them, by searching for similarities and differences in educational practices in "similar" cultural contexts in East Asian countries. Drawing on data from the Learner's Perspective Study (Clarke, Keitel, & Shimizu, 2006), an analysis of the lesson event where teachers sum up during mathematics lessons in Tokyo and Shanghai illustrates some possibilities of identifying similarities and differences in classroom practices within Asia. The form and function of this lesson event are discussed in relation to the Japanese pedagogical term "Matome," which means summing things up, and the cultural values attached to it.

Dichotomies Found in International Comparative Studies

Research in mathematics education that crosses national boundaries provides new insights into the development and improvement of the teaching and learning of mathematics. In particular, cross-national comparisons lead researchers to more explicit understandings of their own implicit theories about how teachers teach and how children learn mathematics in their local contexts, as well as what is going on in school mathematics in other countries. In searching for the identity of mathematics education in East Asia, Leung (2001) tries to describe its distinctive features by focusing on key differences between the East Asian and the Western traditions in mathematics education. Leung expresses these differences using six dichotomies: "product (content) versus process," "rote learning versus meaningful learning," "studying hard versus pleasurable learning," "intrinsic versus extrinsic motivation," "whole-class teaching versus individualized learning," and "competence of teachers: subject matter versus pedagogy" (p. 35). For "product (content) versus process," for example, Leung (2001) describes East Asian mathematics classroom as emphasizing mathematics content and procedures or skills while putting basic knowledge and basic skills in the foreground, whereas Western education in recent decades tends to focus more on the process of doing mathematics.

The dichotomy "East versus West" has also been foregrounded by international benchmark testing, and has led to a qualitative focus on learning in different geographical regions. Accumulated research over the past few decades has contributed to our understanding of similarities and differences in mathematics teaching and learning between East Asia and the West (e.g., Leung, Graf, & Lopez–Real, 2006) or between Eastern and Western cultures (Cai, Perry, & Wong, 2007). The discussion document for the ICMI study argued that "those based in East Asia and the West seem particularly promising for comparison"(Leung et al., 2006, p. 2). In this study a comparison was made between "Chinese/Confucian tradition on one side, and the Greek/Latin/Christian tradition on the other" (Leung et al., 2006, p. 4).

Juxtaposing the two different cultures indicates that the researchers wanted to examine teaching and learning in each cultural context by contrasting differences between them. The labels "East/Eastern" and "West/Western," however, could be problematic in several ways. First, the terms East and West literally mean geographical areas but not cultural regions. Needless to say, there are significant diversities in ethnicity, tools, and habits that are tied to the corresponding cultures. Further, Cobb and Hodge (2011) argue that there are two different views of culture in the mathematics education literature on the issue of equity, and that both are relevant to the goal of ensuring that all students have access to significant mathematical ideas. "In one view," they note, "culture is treated as a characteristic of readily identified and thus circumscribable communities, whereas in the other view it is treated as a set of locally instantiated practices that are dynamic and improvisational" (p. 179). With the second view, in particular, it is problematic to specify different cultures based on geographical areas.

Second, it is possible to oversimplify and misread the cultural influence on students' learning within each cultural tradition by using the same label for different communities. For example, there are studies that suggest that early child education in Japan diverges from the Confucian approach in "East Asia" (Lewis, 1995). Also, in a special journal issue on exemplary mathematics instruction and its development in selected education systems in East Asia, the authors documented a variety of approaches to accomplishing quality mathematics instruction in these different systems in East Asia (Li & Shimizu, 2009). Thus, any framework for differentiating cultural traditions runs the risk of oversimplifying the cultural interplay. In particular, there is a need to question whether polarizing descriptors such as "East" and "West," or "Asian" and "European," are maximally useful. We need more useful ways to examine differences, and similarities as well, for the purposes of learning from each other and identifying ways to optimize learner practices.

The countries in East Asia in the Confucian Heritage Culture certainly share commonalities, and mathematics classroom practices in this region exhibit similarities in various aspects of teaching and learning (Leung, Park, Shimizu, & Xu, 2015). However, educational systems are embedded in their respective societies with particular cultural and historical backgrounds, and educational practices in classrooms have been shaped by their own policies and faced with various context-specific issues. When we look into mathematics classrooms in local contexts in different countries, even within East Asia, we immediately realize the diversity of practices in teaching and learning. Teachers in different countries or regions behave differently when teaching the same mathematical content, and consequently students in each country learn the topic differently.

Beyond Dichotomies: Finding Differences in Similarities and Identifying Similarities in Differences

The mathematics education research community has recognized that mathematics classrooms need to be considered as cultural and social environments in which individuals participate, and that teaching and learning activities taking place in the environments should be studied with this in mind (e.g., Cobb & Hodge, 2011). Teachers' actions that appear normal in a classroom reflect the social values, norms, or traditions that are prevalent outside it. These realizations have led to studies of differences between teaching behaviors and learning outcomes in different countries. Thus, similarities and differences have been explored in topics such as exemplary mathematics classroom instruction (Li & Shimizu, 2009) and teachers' perspectives on effective mathematics teaching (Cai et al., 2007).

One of the major reasons for studying aspects of teaching and learning in classrooms across cultures is that teaching is by nature a cultural activity (Stigler & Hiebert, 1999), one which takes place in particular cultural and social environments. Because cultural activities vary little within a given community or society, they are often transparent and unnoticed. Cross-cultural comparison is a powerful approach to uncover unnoticed but ubiquitous practices, inviting examination of the things "taken for granted" in our teaching, as well as suggesting new approaches that have not evolved in our own societies (Stigler, Gallimore, & Hiebert, 2000). International comparative classroom research is viewed as the exploration of similarity and difference in order to expand our understanding of what is possible in mathematics classrooms, through consideration of what constitutes "good practice" in culturally diverse settings.

In sum, given the growing globalization and internationalization of educational research, and since the education community has given higher priority to international research in the last decade, it is timely to examine the insights from comparative analyses of aspects of teaching and learning of mathematics that are situated in different cultures as well as in similar cultures. The contrasts and unexpected similarities offered by cross-cultural studies reveal and challenge existing assumptions and theories (Clarke, 2003). Specifically, this chapter examines the possibilities of going beyond the dichotomies found in international comparative studies by looking at similarities and differences in teaching and learning in classrooms in Tokyo and Shanghai.

Analyzing Lesson Events Cross-nationally: The Learner's Perspective Study

Cross-national studies provide an opportunity to not only understand better what is going on in classrooms in each educational system, but also question our implicit assumptions about the nature of teaching and learning. The Learner's Perspective Study (LPS) is an international project investigating the practices and learning outcomes of mathematics classrooms in 14 countries and regions, and aims to uncover such implicit assumptions (Clarke, Emanuelsson, Jablonka, & Mok, 2006; Clarke et al., 2006; Leung, Park, Holton, & Clarke, 2014; Shimizu, Kaur, Huang, & Clarke, 2010). Drawing on LPS data, an analysis of the same lesson event (the teacher summing up during a lesson) in eighth-grade mathematics classrooms in Shanghai and Tokyo is provided below.

Data collection in the LPS used a three-camera approach (teacher camera, student camera, and whole-class camera) that included the onsite mixing of the teacher and student camera images into a picture-in-picture video recording, which was then used in post-lesson interviews to stimulate participant-reconstructive accounts of classroom events (Clarke, 2006). These data were collected for sequences of at least ten consecutive mathematics lessons occurring in the "well-taught" eighth-grade classrooms of teachers in 14 countries/regions, including Japan and Shanghai/China.

Each participating country/region used the same research design to collect videotaped classroom data, and to conduct post-lesson video-stimulated interviews with at least 20 students in each of the three participating eighth-grade classrooms. The three mathematics teachers in each country were identified for their locally defined teaching competence and for their placement in demographically diverse government schools in major urban settings. The local criteria for teacher selection, which was made by each local research group, included such things as status within the profession, respect of peers or the school community, or visibility in presenting at teacher conferences or contributing to teacher professional development programs.

In LPS, eighth-grade mathematics classrooms in Australia, Germany, Hong Kong, Japan, China (Macao and Shanghai), and the USA are analyzed with a focus on lesson events (Clarke et al., 2006). A lesson event is conceived as an event type sharing certain features common across the classrooms of the different countries studied. Lesson events included beginning the lesson, learning tasks, guided development, setting the task, instruction between desks, and summing up. Each of these lesson events could be found in some form in the classroom data from all of the countries studied. In each classroom, both within a culture and between cultures, there were idiosyncratic features that distinguished each teacher's enactment of each lesson event, particularly with regard to the function of the particular event. At the same time, common features could be identified in the enactment of lesson events across the entire international data set, and across the data set specific to each country.

Shimizu (2006) analyzed a particular lesson event, "summing up." This event was observed in the LPS classrooms in different countries, and featured specific observable behaviors (Shimizu, 2006). On the one hand, the "summing up" lesson event is observable internationally with some variations. On the other hand, there is a specific Japanese pedagogical term, "Matome," used to describe a teacher's summing up a lesson in the classroom. In Japanese, Matome refers to an event in which the teacher talks with the whole class to highlight and summarize the main point of the lesson. This summary briefly reviews what the students have discussed in the lesson, and highlights what they have learned in a whole-class setting. In reporting on his research of Matome as a classroom phenomenon, Shimizu (2006) stated: "For the Japanese teachers, the event Matome appeared to have the following principal functions: (i) highlighting and summarising the main points of the lesson, (ii) promoting students' reflection on what they have done, (iii) setting the context for introducing a new mathematical concept or term based on previous experiences, and (iv) making connections between the current topic and previous ones" (Shimizu, 2006, p. 141). The following section compares examples of teachers' summing up in Tokyo and Shanghai LPS classrooms, to examine similarities and differences between them. In these examples, the teachers taught topics in algebra, simultaneous linear equations with two variables, and linear function, all in eighth-grade mathematics classrooms.

Matome in Tokyo LPS Classrooms

Typically in the final phase of the lesson, though sometimes in the middle of it, the teacher will review with the students what the classroom engaged in, and emphasize the main point of the lesson. The teacher may ask a few students to tell to the whole class what he or she learned in the lesson. He or she may also write the main points or key mathematical terms on the chalkboard, sometimes using colored chalk, and then refer to the corresponding page of the textbook. In some cases, the teacher then announces practice exercises to apply what has been just highlighted.

The transcript below (Example 1) shows an example from one of the three Japanese LPS classrooms in Tokyo. In this lesson, the students were learning to solve simultaneous linear equations, and the teacher summarized and highlighted what they had done in the form of general comments. These comments were made in the final minutes of the lesson. He noted that the class had done "something extremely important" (00:43:08:15), emphasizing that the students "would be able to solve tons of equations" (00:43:22:13) and they "should be able to solve everything" (00:43:35:18). Also, he encouraged the students to "check the calculation when you need to." At the end of the lesson, after some discussions of two alternative ways of checking the solution to simultaneous linear equations, the teacher again emphasized that what they had done was extremely important. He then asked the students to jot things they have learned down in their notebooks. In this case, the teacher appeared to promote students' reflection on what they had done and on the importance of checking the results. The teacher also pointed out the part of blackboard on which an important idea was described.

Example 1: *J3-L03, (00:43:08:15 to 00:43:35:18).*

00:43:08:15T: Yes, um, today, we will end here but we did something extremely important today. Um, it will have to be next week, solving the equation from KINO's question will have to be next week.

00:43:22:13T: But if we finish up to here, I think you'll be able to solve tons of equations. Check the calculation when you need to and I'll ask you sometimes. I'll ask you to show me how much you can do, ok?

The lesson event in Example 2 took place in the middle of the third lesson. This example shows that Matome can take place not only at the end of the lesson but also in the middle of the lesson, to pull together students' activities from multiple lessons.

Example 2: *J1-L03 (00:24:43:06 to 00:27:17:26).*

00:24:47:00T: Because they are all shown as a linear equation, they are called linear functions. Note that somewhere in your notebook.

00:25:05:12T: And please look at this type of equations, in which B is zero.

00:25:18:28T: They are in the same group, but are linear functions. What did we call this kind of equations in seventh grade?

00:25:33:27T: Huh? Do you all remember that? How did you call them? How did you describe the relation between x and y?

00:25:48:28S: X is directly proportional to y.

00:25:49:23T: Yes, the proportion.

00:25:51:07T: When we learned this in seventh grade, we said that they are proportional.

00:25:54:15T: But actually you already understood it as a kind of linear functions; the only difference is if it has B or not.

00:26:06:07T: You can see this in the textbook. Please open it and see it yourself. It's on page fifty-seven. [Writing on the blackboard]

00:26:41:10T: Okay, look at the tenth line, no I mean the ninth line. Just what we talked about. It's summarized there.

00:26:57:13T: Draw an underline from the ninth line.

00:27:07:05T: We can see what just we talked about in words.

00:27:08:25T: Given two variables x and y, if y can be expressed with the linear equation of x, then we call it a linear function.

00:27:17:26T: And a linear function is expressed as y = x + b. Okay?

In Example 2, the teacher introduced the term "linear function" as a formal mathematical term, by reflecting on several examples of linear functions that appeared as a result of previous activities in the classroom: "these are called linear functions. Because they are all shown as a linear equation, they are called linear functions" (00:24:43:06 to 00:24:47:00). Then she asked the students to make a note of this fact. In this case, she tried to sum up what the students had worked on in three consecutive lessons. She next tried to make a connection between the concept of linear functions and the concept of direct proportionality, a special case of a linear function which her students had learned in the previous year: "And please look at this type of equations, in which B is zero. They are in the same group, but linear functions. What did we call this kind of equations in seventh grade?" (00:25:05:12–00:25:18:28).

After introducing the formal term, she asked the students to "*Draw an underline from the ninth line,*" and then wrote the point on the chalkboard using yellow chalk. Finally, she repeated the main point by reading the corresponding page in the textbook: "We can see what just we talked about in words; having two variables x and y, if y can be expressed with the linear equation of x, then we call it a linear function. And a linear function is expressed as y = x + b. Okay?" (00:27:07:0–00:27:17:26). As Example 2 illustrates, Matome includes the teacher's effort to make connections among lessons to reflect on what the students have been doing. The event serves to set the stage for introducing a new mathematical term, in this case based on examples of linear functions examined in the activities across three lessons, including the current one.

The Japanese lesson pattern identified by the TIMSS 1995 Video Study included "highlighting and summarizing the main point" as the final segment (Stigler & Hiebert, 1999, pp. 79–80). The above example, however, shows that Matome can also take place in the middle of the lesson, to pull together the students' activities in multiple lessons. Both these sample lesson events reveal teachers' intentions of both reflecting on what students have learned and making connections among mathematical concepts in the classroom.

Summing Up in Shanghai LPS Classrooms

The teachers in the three Shanghai LPS classrooms often highlighted and summarized their lessons, mostly at the end of lessons. They began to sum up by reviewing what the class had done in the lesson and then emphasized the main points, as illustrated by the examples below:

SH1-L06, 00:41:48:00 So today we have talked about using the method of substitution to solve a linear equation in two unknowns.

SH1-L06, 00:43:17:11: These are the points which we should pay attention to when we want to solve linear equations in two unknowns.

SH2-L03, 00:41:49:27 to 00:42:16:16: Good. Okay, today we've talked about some concepts of system of linear equations in two unknowns. System of linear equations in two unknowns, its solutions and how to solve the system, we've talked about one of the way to solve the system, basically it is to change two unknowns into one unknown. Today, we've used method of substitution, and we will talk about the other methods later on.

These teachers interacted with their students while summing up. They also summarized the main points using an overhead projector or the blackboard. A teacher in the school SH3, for example, used slides or the blackboard to sum up in 12 out of 15 lessons.

The excerpt from the SH2-L03 transcript later shows that the teacher summed up the lesson by recalling the topics and by asking the students some questions, to encourage students to find the answers by themselves in the textbook. For example, the teacher asked, "I want to ask, how many equations are needed as minimal, in order to solve the system?" (00:42:24:17). In response to a student answer of "two," he then asked another question: "Two, we can find out the solutions only when there are two independent equations. Tell me, how many solutions are there for an equation?" (00:42:35:05). The teacher thus summed up the lesson by asking his students questions to check their understanding. In general, teachers' behaviors during summing up, and the students' behaviors in response to them, were very similar in Shanghai and Japanese classrooms. It is noteworthy, however, that in all the examples examined, the teachers in Shanghai highlighted and summarized the lesson at the end of lessons, and that the focus of summing up was mainly on the mathematical content taught in each lesson.

Looking at Values Attached to Teachers' Behavior

There are both similarities and differences in teachers' behaviors during summing up events. While teachers from both countries highlighted and summarized the main points of the lesson, the Japanese teacher summed things up even in the middle of the lesson, while the Shanghai teachers mainly focused on mathematical content taught in their lessons. Japanese mathematics teachers often organize an entire lesson around the multiple solutions to a single problem, in a whole-class instruction mode. Since the teachers emphasize finding alternative ways to solve a problem, Japanese classes often consider several strategies. It would be natural, then, for the classes to discuss problem-solving strategies from various viewpoints, such as mathematical correctness, brevity, and efficiency. A teaching style with an emphasis on finding many ways to solve a problem naturally invites certain summarizing behaviors. If the whole-class discussion reaches a point of thinking retrospectively about what they have considered, even in the middle of the lesson, a teacher may have Matome.

There seem to be supporting conditions and shared beliefs among the Japanese teachers that justify often having Matome at the end of the lessons or at the end of subunits. Every lesson has an opening, a core, and a closing. This is particularly the case for Japanese lessons, which begin and end with the students bowing. Teachers regard their lessons as dramas, which have a beginning and lead to a climax. In fact, one of the characteristics of Japanese teachers' lesson planning is the deliberate structuring of the lesson around a climax, "Yamaba" in Japanese (Shimizu, 2006, p. 143). Most teachers think that a lesson should have a highlight. The essential point is that Japanese mathematics teachers have access to a sophisticated and coherent vocabulary that allows them to discuss the components of the mathematics lesson, reflect on their teaching, and offer and receive advice. This structure provides a powerful tool for preservice and in-service teacher education. These pedagogical terms are learned by teachers through participation in Lesson Study, which is a Japanese approach to improving teaching and learning mathematics through a particular form of activities by a group of teachers, including planning, implementing, and discussing actual lessons (Shimizu, 1999). It is important to note that these pedagogical terms are used in the discourse of particular contexts embedded in a whole system, to describe a particular style of teaching. Structured problem-solving is often used to describe the system, with an emphasis on students' thinking on problem posed. Japanese mathematics teachers often organize an entire lesson by posing just a few problems, and focus on students' various solutions to them. Educating teachers about lesson plans includes making sure that they understand key pedagogical terms.

Stigler and Perry (1988) found significant reflectivity and coherence in Japanese mathematics classrooms. The meaning they attached to the coherence is similar to that used in the literature on story comprehension. Stigler and Perry (1988) noted:

> A well-formed story, which also is the most easily comprehended, consists of a protagonist, a set of goals, and a sequence of events that are causally related to each other and to the eventual realization of the protagonist's goals. An ill-formed story, by contrast, might consist of a simple list of events strung together by phrases such as "and then . . .," but with no explicit reference to the relations among events . . .. The analogy between a story and a mathematics classroom is not perfect, but it is close enough to be useful for thinking about the process by which children might construct meaning from their experience in mathematics class. A mathematics class, like a story, consists of sequences of events related to each other and, hopefully, to the goals of lesson. (p. 215)

The lesson event Matome appeared to promote reflection by the Japanese teacher and students, which is consistent with Stigler and Perry's (1988) attention to reflectivity. They pointed out that the Japanese teachers stress the process by which a problem is worked and exhort students to carry out procedures patiently, with care and precision. The event type seems to rest on a tacit set of core beliefs about what should be valued and esteemed in the classroom. As Lewis noted, within Japanese schools, as within the larger Japanese culture, Hansei—self-critical reflection—is emphasized and esteemed (Lewis, 1995). The practice of teaching is thus closely related with values attached to the importance of reflection.

Concluding Remarks

International comparisons of mathematics classrooms offer insights into the novel, interesting, and adaptable practices employed in other school systems, and into the unquestioned routines in our own school systems. These studies provide researchers and policy makers opportunities for understanding their own implicit theories about how teachers teach and how children learn mathematics in their local contexts. Namely, by comparing the teaching and learning processes in different countries, the uniqueness of one's own practices appears. In learning from another country or region, we should consider not just others' educational practices but also the cultural values behind those practices. It is crucial to take into account the different cultural assumptions underpinning teaching and learning. Comparative studies in mathematics education of classrooms from eastern and western cultures, as well as those comparing similar cultures, are pertinent to the understanding of the why, what, and how of mathematics teaching and learning in these cultures.

References

Cai, J., Perry, R., & Wong, N. Y. (2007). What is effective mathematics teaching? A dialogue between East and West. *ZDM—The International Journal on Mathematics Education, 39*(4), 311–318.

Clarke, D. J. (2003). International comparative studies in mathematics education. In A. J. Bishop, M. A. Clements, C. Keitel, J. Kilpatrick, & F. K. S. Leung (Eds.), *Second international handbook of mathematics education* (pp. 145–186). Dordrecht: Kluwer.

Clarke, D. (2006). The LPS research design. In D. Clarke, C. Keitel, & Y. Shimizu (Eds.), *Mathematics classrooms in twelve countries: The insider's perspective* (pp. 15–36). Rotterdam: Sense Publishers.

Clarke, D., Emanuelsson, J., Jablonka, E., & Mok, I. A. C. (Eds.). (2006a). *Making connections: Comparing mathematics classrooms around the world.* Rotterdam: Sense Publishers.

Clarke, D., Keitel, C., & Shimizu, Y. (Eds.). (2006b). *Mathematics classrooms in twelve countries: The insider's perspective.* Rotterdam: Sense Publishers.

Cobb, P., & Hodge, L. L. (2011). Culture, identity, and equity in the mathematics classroom. In E. Yackel, K. Gravemeijer, & A. Sfard (Eds.), *A journey in mathematics education research: Insights from the work of Paul Cobb* (pp. 179–195). New York, NY: Springer.

Dindyal, J. (2014). International comparative studies in mathematics: An overview. In S. Lerman (Ed.), *Encyclopedia of mathematics education* (pp. 320–325). New York: Springer.

Keitel, C., & Kilpatrick, J. (1999). The rationality and irrationality of international comparative studies. In G. Kaiser, E. Luna, & I. Huntley (Eds.), *International comparisons in mathematics education* (pp. 241–256). London: Falmer Press.

Leung, F. K. S. (2001). In search of an East Asian identity in mathematics education. *Educational Studies in Mathematics, 47*(1), 35–51.

Leung, F. K. S., Graf, K., & Lopez–Real, R. (Eds.). (2006). *Mathematics education in different cultural traditions: A comparative study of East Asian and the West.* New York, NY: Springer.

Leung, F. K. S., Park, K., Holton, D., & Clarke, D. (Eds.). (2014). *Algebra teaching around the world.* Rotterdam: Sense Publishers.

Leung, F. K. S., Park, K., Shimizu, Y., & Xu, B. (2015). Mathematics education in East Asia. In S. J. Cho (Ed.), *The Proceedings of the 12th International Congress on Mathematical Education: Intellectual and Attitudinal Challenges* (pp. 123–143). New York, NY: Springer.

Lewis, C. (1995). *Educating heart and mind: Reflections on Japanese preschool and elementary education.* New York, NY: Cambridge University Press.

Li, Y., & Shimizu, Y. (2009). Exemplary mathematics instruction and its development in East Asia. *ZDM—The International Journal on Mathematics Education, 41*(3), 257–262.

Shimizu, Y. (1999). Aspects of mathematics teacher education in Japan: Focusing on teachers' role. *Journal of Mathematics Teacher Education, 2*(1), 107–116.

Shimizu, Y. (2006). How do you conclude today's lesson?: The form and functions of "matome" in mathematics lessons. In D. Clarke, J. Emanuelsson, E. Jablonka, & I. A. C. Mok (Eds.), *Making connections: Comparing mathematics classrooms around the world* (pp. 127–145). Rotterdam: Sense Publishers.

Shimizu, Y., & Kaur, B. (Eds.). (2013). Learning from similarities and difference: Cross–national studies on aspects of teaching and learning mathematics. *ZDM—The International Journal on Mathematics Education, 45*(1), 1–5.

Shimizu, Y., Kaur, B., Huang, R., & Clarke, D. (2010). The role of mathematical tasks in different cultures. In Y. Shimizu, B. Kaur, R. Huang, & D. Clarke (Eds.), *Mathematical tasks in classrooms around the world* (pp. 1–14). Rotterdam: Sense Publishers.

Shimizu, Y., & Williams, G. (2013). Studying learners in intercultural contexts. In M. A. (Ken) Clements, A. Bishop, C. Keitel, J. Kilpatrick, & F. K. S. Leung (Eds.), *Third international handbook of mathematics education* (pp. 145–167). New York, NY: Springer.

Singh, P., & Ellerton, N. F. (2013). International collaborative studies in mathematics education. In M. A. (Ken) Clements, A. Bishop, C. Keitel, J. Kilpatrick, & F. K. S. Leung (Eds.), *Third international handbook of mathematics education* (pp. 827–860). New York, NY: Springer.

Stigler, J. W., Gallimore, R., & Hiebert, J. (2000). Using video surveys to compare classrooms and teaching across cultures: Examples and lessons from TIMSS Video Studies. *Educational Psychologist, 35*(2), 87–100.

Stigler, J. W., & Hiebert, J. (1999). *The teaching gap: Best ideas from the world's teachers for improving education in the classroom.* New York, NY: Free Press.

Stigler, J. W., & Perry, M. (1988). Cross cultural studies of mathematics teaching and learning: Recent findings and new directions. In D. A. Grouws, T. J. Cooney, & D. Jones (Eds.), *Perspectives on research on effective mathematics teaching* (pp. 194–223). Reston, VA: NCTM.

Chapter 5
Achievement Gaps in Mathematics and Opportunities to Learn: Insights from PISA 2012

Yan Zhu

Abstract According to large-scale comparisons, East Asian students have consistently outperformed students from other nations in mathematics. However, despite the extensive research on these students' cognitive skills, their noncognitive attributes have been the focus of research far less frequently. This study compares East Asian and US students' attainments in both cognitive and noncognitive aspects via a secondary analysis of the Program for International Student Assessment (PISA) 2012. It explores the between-system gaps from the perspective of opportunity to learn, and discusses the implications of between-system similarities and differences.

Keywords Cognitive achievement • Noncognitive achievement • Opportunity to learn • Mathematics • East Asia

East Asian students consistently outperform their Western competitors in large-scale international comparisons (e.g., TIMSS and PISA) of mathematics achievement, a fact that often catches the attention of researchers, educators, policy makers, and the general public in recent years. For instance, in the USA, the achievement gap between East Asian students and American students has been widely cited in education policy documents at the local, state, and national levels. In fact, these extraordinary academic accomplishments have not only impressed many other nations, but also increasingly prompted reflections or criticisms about education in these academically strong nations (Zhao et al., 2010).

As Porter (2014) argued, in the last 40 years more attention has been given to achievement gaps than to any other topic in education. One important reason for this attention is that people generally believe that weaker academic skills bode poorly for a student's prospects in the global economy, as today's youths need to compete with their international counterparts for employment opportunities.

Y. Zhu (✉)
East China Normal University, Shanghai, China
e-mail: yzhu@kcx.ecnu.edu.cn

J.-W. Son et al. (eds.), *What Matters? Research Trends in International Comparative Studies in Mathematics Education*, Research in Mathematics Education, DOI 10.1007/978-3-319-51187-0_5

Consequently, the achievement gaps reported in international assessments often serve as barometers of economic vitality. Hanushek, Jamison, Jamison, and Wößmann (2008) even used achievement data from 50 nations to predict the average annual growth rates in per capita gross domestic product (GDP) in those nations. They found that an increase of 50 points in achievement could boost a nation's annual economic growth rate by 0.63% points. According to them, if the USA could reach the 1990 goal set by President Bush to become No. 1 in mathematics and science by 2000, its GDP in 2015 would be 4.5% higher than it would be without any achievement gains. This 4.5% increment is equivalent to what the USA spends yearly on K-12 education (Czehut, 2012).

A "New" Look at School Achievement

Earlier studies of school achievement often focused on cognitive aspects, such as ability, IQ (intelligence quotient), and other measures of innate aptitude. One consequence is that though Trends in International Mathematics and Science Study (TIMSS) and Program for International Student Assessment (PISA) are both designed to serve multiple purposes, much public attention has focused on test scores and country rankings, which in some sense has turned such studies into an academic Olympics. Such a single-minded focus often leads to an overinterpretation of the meaning and importance of these studies. However, researchers have found that IQ can only explain about 25% of the variance in achievement levels (e.g., Jensen, 1998; Neisser et al., 1996). Moreover, achievement test scores can only predict a small fraction of the variance in learners' later life success (e.g., Kautz, Heckman, Diris, ter Weel, & Borghans, 2014). It has been suggested that other domains, such as individuals' affective and motivational characteristics, also serve as important factors in school achievement (Mo, Singh, & Chang, 2012). Moreover, a growing body of empirical research shows that noncognitive skills actually have stronger predictive power than cognitive ones in relation to learners' life outcomes.

Though noncognitive achievement is still largely ignored in evaluations of schools and interventions, the existing research has revealed some insightful results. For instance, based on TIMSS 2003 data, Leung (2006) found that East Asian students' high test scores in mathematics did not seem to be accompanied by correspondingly positive attitudes toward mathematics and mathematics learning. In particular, students from Japan, Chinese Taipei, and Korea were the least likely to think that mathematics was important. Lee's (2009) analysis of PISA 2003 data revealed some contradictory results between students' cognitive performance and noncognitive performance: while students from East Asia and Europe dominated the PISA top 10 on mathematics achievement, they tended to lie at opposite ends of the table on noncognitive variables.

What Drives East Asian Students' Cognitive Success in Mathematics?

While East Asian students' cognitive advantage in mathematics is obvious, the reasons for it are far less clear. In searching for the keys to East Asia's success, researchers have proposed quite a number of possible explanations, many of which are not related to ability. Among these efforts, Stevenson and Stigler's *The Learning Gap* (1992) is the touchstone study: a cited reference search on ISI Web of Science shows that their work has been referenced in over 400 published studies (Czehut, 2012). Stevenson and Stigler tended to explain the achievement gaps between East Asia and the USA from two separate perspectives: cultural beliefs and educational structures. Though it seems that the former may influence and further shape the latter, the relationship between the two aspects needs further clarification.

Many scholars have tried to attribute the differences in classroom practices and achievements, particular those between East Asia and the West, to cultural factors (e.g., Watkins & Biggs, 1996; Wong, 1998). For instance, Leung (2006) examined the relationship between a number of characteristics of the commonly shared Confucian culture and high student achievement in mathematics in East Asia. He claimed that while there were indeed different cultural values pertinent to education that could explain the differences in mathematics education between East Asia and the West, it is hard to prove that cultural differences actually caused the differences in student achievement. He argued that this may be partially due to the absence of clues from variables at other levels. In fact, in his earlier work, Leung (2001) tried to identify the differences in mathematics education between East Asia and the West in terms of six dichotomies, including product (content) vs. process, rote learning vs. meaningful learning, studying hard vs. pleasurable learning, extrinsic vs. intrinsic motivations, whole-class teaching vs. individualized learning, and competence of teachers: subject matter vs. pedagogy. According to him, these distinctive features were expressions of distinctive underlying cultural values, and it is important for educators to understand that different practices are based on deep-rooted cultural values and paradigms.

Besides cultural influences, researchers have also explored many other aspects of mathematics education in relation to achievement gaps between East Asia and the West. For instance, Askew, Hodgen, Hossain, and Bretscher (2010) identified a list of themes as potential reasons for high performance in mathematics learning, using research evidence from countries with high attainment. These themes are further classified into three broader types: cultural influences (e.g., wider learning goals within the context of mathematics education in Japan, the Asian emphasis on effort), contextual influences (e.g., parents' expectations of their children's mathematical attainment, widespread shadow education), and pedagogical influences (e.g., mathematically informed procedural teaching, well-constructed textbooks, teachers' mathematical subject knowledge). Though the sheer number of possible explanations suggest that teasing out factors driving East Asian educational success is not an easy task, this does not stop authoritative figures such as Andreas Schleicher (head of PISA) from making strong suggestions about "[w]hat Asian schools can teach the rest of the world"[1] (Jerrim, 2014).

Investigating Achievement Gaps from the Perspective of Opportunity to Learn

Among factors which may influence students' cognitive attainment, whether or not students have had opportunities to learn the assessed topics is an important concern. Regarding the subject of mathematics, Grouws and Cebulla (2000) claimed that the extent of students' opportunity to learn mathematics content bears directly and decisively on their achievement. In general, Herman, Klein, and Abedi (2000) suggested that opportunity to learn (OTL) can serve as an indicator for progress, verify that students from diverse backgrounds have had the same level of opportunity to meet expected standards, and provide feedback to schools. More specifically, Porter (1993) proposed three possible uses of OTL standards: a basis for school-by-school accountability, an indicator system, and a clearer vision for challenging curriculum and pedagogy.

Since OTL was first studied in the First International Mathematics Study (FIMS) (Husen, 1967), its definition has evolved over time. In particular, the notion of OTL has expanded from a limited focus on instructional time and content coverage in the 1960s to curriculum in the mid-1980s, and to a broader set of teacher qualifications, access to resources, funding, teaching methods, etc. in the 1990s and 2000s (Kennedy, 2011). It is generally found that there is a positive correlation between OTL and student achievement scores, with high levels of OTL associated with high achievement (e.g., Collie-Patterson, 2000; Gau, 1997; Wang & Goldschmidt, 1999; Wiley & Yoon, 1995).

However, as Pianta, Jay, Renate, and Fred (2007) argued, research attention to the effects of OTL on student achievement has been far from sufficient. Similarly, Reeves, Carnoy, and Addy (2013) claimed that while it appears logical that low student achievement in many developing countries could be due to students' relatively rare opportunities to learn the skills needed for academic success, surprisingly little empirical evidence was available to support the hypothesized explanation. In particular, little has been done to link variation in OTL to variation in student learning outcomes. In fact, researchers' interest in OTL as a potential factor facilitating teaching and learning did not begin until the 1990s (e.g., Muthen et al., 1995; Wiley & Yoon, 1995). Overall, international comparative studies seem to pay more attention to OTL to investigate the cross-national achievement gaps. It is believed that differences revealed in international comparisons could be explained by differential OTL levels across schools and countries (e.g., Schmidt, Houang, &

[1] http://edition.cnn.com/2013/12/03/opinion/education-rankings-commentary-schleicher/.

Cogan, 2002; Schmidt, Schmidt, & McKnight, 1995; Webster, Young, & Fisher, 1999; Wiley & Yoon, 1995). In other words, the investigation of OTL in the context of differences between educational systems could provide insights into the importance and limits of OTL as an explanation of student learning.

Research Questions and Purposes

The purpose of this study is to generate a more comprehensive comparison between East Asian systems and the USA in terms of students' cognitive and noncognitive attainments in mathematics, as well as the opportunities to learn those students received in their respective systems. In particular, five broad research questions are addressed via a secondary analysis of PISA 2012 data:

1. Do 15-year-old students from East Asian systems and the USA perform differently in mathematics literacy assessment?
2. Are 15-year-old students from East Asia and the USA different in their attitudes towards mathematics?
3. Do East Asian and US systems offer 15-year-olds different opportunities to learn?
4. Are the relations between students' academic achievement and their opportunities to learn different between East Asian systems and the USA?
5. Are the relations between students' attitudes towards mathematics and their opportunities to learn different between East Asian systems and the USA?

Research Methods

Data Sources

Data for this study were taken from PISA 2012, which were retrieved from the official PISA website. Besides the USA, the top seven East Asian systems were selected for study: those from Shanghai-China, Singapore, Hong Kong-China, Chinese Taipei, Korea, Macao-China, and Japan. Both the final sampling weights and the replicate weights for students were used in this study to make the sample reflective of the corresponding populations.

Measures

Academic achievement in mathematics. To assess 15-year-olds' literacy in reading, mathematics, and science, the PISA 2012 randomly assigned each student one of

22 rotated booklets containing questions about one or more of the three testing domains; all the booklets included mathematics material. In this sense, each student only attended to a part of the assessment item pool, so the raw test scores are incomparable across students. To resolve the problem, the PISA used a multiple imputation approach to estimate the unobservable latent achievement for all students. As a result, sets of plausible values were produced for each student for their overall mathematics performance as well as domain-specific performance, and all these values were used in this study to calculate the corresponding parameter estimates.

Attitudes towards mathematics. In order to measure students' attitudes towards school subjects, a range of indicators were set in the PISA 2012 student questionnaire. In this study, four of them focusing on mathematics learning were selected, containing ten sets of items covering *motivation*, *self-beliefs*, *dispositions*, and *participation in mathematics activities* (for details, see the *PISA 2012 Technical Report*).

Opportunity to learn. In PISA 2012, opportunity to learn (OTL) includes three indices related to student-perceived experiences and familiarity with mathematics tasks, student-perceived teaching practices, and student-perceived teaching quality (for details see *PISA 2012 Technical Report*). All three sets of items are included in this study.

Data Process and Analysis

To explore the differences in students' academic achievement, attitudes towards mathematics, and opportunities to learn between East Asian systems and the USA, descriptive analyses were first carried out on all the indices by system, followed by ranking the corresponding means in the entire international league table. Based on the system mean scores, a series of cluster analyses were performed by indices as well as by dimensions, so as to classify the eight systems into groups in accordance to their similarities. Furthermore, correlations between the two types of achievements and those between achievements and different OTL values were also investigated.

Comparisons of Students' Mathematics Achievement

PISA 2012 assessed students' mathematics achievement in both cognitive and noncognitive domains. In particular, students' cognitive achievement was distinguished by three interrelated aspects: mathematical content, mathematical processes, and mathematical contexts. The mathematical content covers four overarching areas, including "change and relationships," "space and shape," "quantity," and "uncertainty and data," and they are secondarily in relation to curricular strands. The mathematical processes are defined in terms of three categories: "formulating situations mathematically"; "employing mathematical concepts, facts, procedures, and reasoning"; and "interpreting, applying, and evaluating mathematical outcomes". These processes describe what individuals do to connect the context of a problem with the mathematics and thus solve the problem. The mathematical context is defined as the aspect of an individual's world in which the problems are placed, and features four categories: personal, occupational, societal, and scientific types.

In terms of students' overall performance, Shanghai-China and the USA are at the 2 ends of the league table of the 8 systems, and are ranked 1st and 36th, respectively, among the 65 participating systems. The other six East-Asian systems are ranked from second to seventh (see Table 5.1). Further analysis shows that the differences among Hong Kong-China, Chinese Taipei, and Korea, as well as those between Macao-China and Japan, are not statistically significant. Similar patterns were also observed when comparing students' performance on each mathematics subscale.

Among the four mathematical content areas, *space and shape* is the strongest one for the East Asian systems but the weakest for the USA, while *uncertainty and data* shows the opposite pattern. Another inconsistency between the East Asian systems and the USA is that the Asian students tended to have better knowledge on *change and relationships* than *quantity*, which is reversed for the American students. Moreover, it can be seen that the students in Shanghai-China illustrate the largest performance variation across the four content areas ($\Delta_{max} = 58$ points), while those in Hong Kong-China show a more balanced pattern ($\Delta_{max} = 14$ points).

Regarding the three types of mathematical processes, it is interesting that *interpreting, applying, and evaluating mathematical outcomes* is the strongest process type for the USA but the weakest for the East Asian systems, while *formulating situations mathematically* displays an opposite pattern. Similar to the findings on content knowledge, Shanghai-China students again showed the largest performance variation across the different processes ($\Delta_{max} = 46$ points), which is more than three times of the US variation ($\Delta_{max} = 14$ points).

To assess students' noncognitive achievement in mathematics, their attitudes towards the subject were measured. Ten indices, using 67 items, were constructed, including motivation (INTMAT, INSTMOT, FAILMAT), self-beliefs (MATHEFF, ANXMAT, SCMAT), dispositions (MATINTFC, SUBNORM), and participation in mathematics activities (MATBEH, MATWKETH). Table 5.2 lists the international rankings for all eight systems. It can be seen that the pattern is very much different from that in students' cognitive achievement; that is, while the seven East Asian systems took up the first seven places in the mathematics test, their mean scores on some of the attitude-related noncognitive indices were below the Organization for Economic Co-operation and Development (OECD) averages. In particular, five out of the seven East Asian systems had about half or more of those indices.

Table 5.2 reveals that both Japanese and Korean students have strongly negative attitudes towards mathematics. It can be seen that on majority of the indices, these

Table 5.1 Students' mathematics performance among eight education systems: overall and subscales

		Mathematical contents				Mathematical processes		
	Overall	[1]	[2]	[3]	[4]	[1]	[2]	[3]
Shanghai-China	613 [1]	624 [1]	649 [1]	591 [1]	592 [1]	624 [1]	613 [1]	579 [1]
Singapore	573 [2]	580 [2]	580 [3]	569 [2]	559 [2]	582 [2]	574 [2]	555 [2]
Hong Kong-China	561 [3]	564 [3]	567 [5]	566 [3]	553 [3]	568 [4]	558 [3]	551 [3]
Chinese Taipei	560 [4]	561 [4]	592 [2]	543 [4]	549 [4]	578 [3]	549 [5]	549 [4]
Korea	554 [5]	559 [5]	573 [4]	537 [6]	538 [5]	562 [5]	553 [4]	540 [6]
Macao-China	538 [6]	542 [6]	558 [6]	530 [9]	525 [9]	545 [7]	536 [7]	530 [8]
Japan	536 [7]	542 [7]	558 [7]	518 [14]	528 [7]	554 [6]	530 [8]	531 [7]
USA	481 [36]	488 [29]	463 [39]	478 [40]	488 [29]	476 [37]	480 [38]	490 [33]

Note. (a) Mathematical content includes [1] change and relationships, [2] space and shape, [3] quantity, and [4] uncertainty and data. (b) Mathematical processes include [1] formulating situations mathematically; [2] employing mathematical concepts, facts, procedures, and reasoning; and [3] interpreting, applying, and evaluating mathematical outcomes. (c) Numbers along with the mean scores represent rankings in the league table of 65 systems

Table 5.2 The international rankings of indices on students' attitudes towards mathematics by systems

	Motivation			Self-beliefs			Disposition		Participation	
	INTMAT	INSTMOT	FAILMAT	MATHEFF	ANXMAT	SCMAT	MATINTFC	SUBNORM	MATBEH	MATWKETH
QCN	17	39	3	1	*25*	*40*	27	36	15	12
SGP	4	9	4	3	*37*	9	23	8	18	10
HKG	21	*52*	5	8	*33*	*55*	*61*	*40*	34	*44*
TAP	40	*58*	6	9	*51*	*64*	*57*	*54*	40	*60*
KOR	*58*	*62*	9	*61*	*50*	*63*	*59*	*51*	32	*65*
MAC	32	*55*	20	11	*40*	*60*	55	*60*	29	36
JAP	*60*	*64*	1	*63*	*54*	*65*	*33*	*65*	*60*	*57*
USA	39	31	8	14	12	6	30	25	*55*	16

Note. The rankings *italics* indicate that the mean scores of the corresponding indices are below the OECD averages

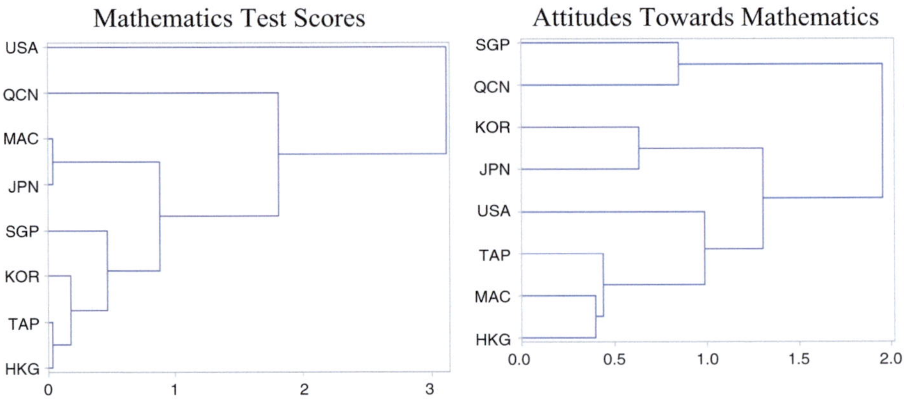

Fig. 5.1 Cluster analysis of students' mathematics performance (based on PV1MATH to PV5MATH) and attitudes towards mathematics among seven East Asian education systems and the USA

two systems were at the bottom of the international table (Japan: 9 vs. Korea: 8). Only on *attributions to failure in mathematics* did the two systems, along with the other systems, provide highly positive views. Though the USA has only one index with a mean score below the OECD average, a cluster analysis reveals that the USA shares more commonalities with Chinese Taipei, Macau-China, and Hong Kong-China (see Fig. 5.1), whereas Shanghai-China vs. Singapore and Korea vs. Japan are more similar to each other. On the other hand, it is obvious that the seven East Asian systems are more similar to each other than to the USA in terms of students' cognitive achievement, which is not the case when students' noncognitive achievement is concerned.

Comparisons of Opportunity to Learn

It is believed that the quantity and quality of educational resources will have no effect on learning if students are not exposed to them. Correspondingly, OTL is a way to measure and report whether students and teachers have access to the different ingredients that make up quality education. The more OTL ingredients that are present to students, the more opportunities for students to benefit from a high-quality education. The PISA 2012 constructs three OTL indices, which are related to content (EXAPPLM, EXPUREM, FAMCON, FAMCONC), teaching quality (TEACHSUP, COGACT, MTSUP, CLSMAN, DISCLIMA), and teaching practices (TCHBEHTD, TCHBEHFA, TCHBEHSO).

OTL: Content

In the PISA 2012, content-related OTL is defined as coverage of problem types and content categories (OECD, 2012, technical report). In particular, two problem types are classified: pure tasks vs. applied tasks. The data show that while on pure tasks all the systems but Chinese Taipei have mean scores above the OECD average, on applied mathematics tasks only Korea, Singapore, and Shanghai-China do so. On both task types, Korea provides the highest exposure levels (pure: 0.43 vs. applied: 0.40), followed by Singapore (pure: 0.33 vs. applied: 0.31). On applied tasks, Japan (−0.18) and Hong Kong-China (−0.14) provide the lowest levels of exposure among the eight systems. The USA's exposure levels are more similar to Chinese Taipei and Macao-China on the applied type and to Shanghai-China on the pure type. It is interesting to observe that Shanghai-China is the only system among the eight where the students have a higher level of exposure to applied mathematics tasks than pure mathematics tasks ($\Delta = 0.12$).[2] The largest exposure difference between the two types of mathematics tasks is observed in Japan ($\Delta = 0.38$) followed by Macao-China ($\Delta = 0.32$), while Singapore ($\Delta = 0.02$) and Korea ($\Delta = 0.03$) provide nearly identical exposure levels on the two types of mathematics tasks.

In terms of familiarity with a range of mathematical concepts, Shanghai-China shows an outstandingly high level (1.2), which is nearly twice the level for the next highest system (Macao-China: 0.7). On this scale, Japan and the USA present the lowest levels (0.3). To avoid overclaiming, one additional index was constructed to adjust the scale indicating familiarity with mathematical concepts for single detection. Again, Shanghai-China's mean score is about 1 *SD* higher than the OECD average (1.1), and Korea becomes another system having mean score over 1 *SD* on this complementary scale (1.3). With such an adjustment, the USA's and Singapore's mean scores are just slightly higher than the OECD average, which presents the bottom level for the eight systems.

OTL: Teaching Quality

In terms of teaching quality, the USA shares more commonality with both Shanghai-China and Singapore. In particular, the three systems provide the highest level on three teaching-quality indices (TEACHSUP, COGACT, and CLSMAN). In particular, on *cognitive activation* the mean scores from all the other five systems are below the OECD average. On this scale, Japan and Korea's mean scores are more than 0.5 *SD* lower than the OECD average, and the difference between the

[2]All the scales in PISA 2012 questionnaires are scaled to have an OECD average of 0 and a standard deviation of 1. Correspondingly, the between-scale comparisons involve the relative differences with the OECD average level as a reference in terms of standard deviation.

USA and Korea is greater than 1 *SD*. Moreover, the two East Asian systems also presented the lowest mean scores on two scales related to teacher support. On *mathematics teacher support* (MTSUP), Korea and Japan are the only systems among the eight having below-OECD-average mean scores. In contrast, Singapore and Shanghai-China's mean scores on this scale are about 0.5 *SD* higher than the OECD average.

Although the USA has a similarly high level of *classroom management* to Shanghai-China and Singapore, its *disciplinary climate* level is incomparable with the two systems, particularly with Shanghai-China ($\Delta = 0.5$). It is somewhat unexpected that Chinese Taipei's mean score on this scale is slightly lower than the OECD average (−0.01) and the USA's mean score is only slightly higher than Chinese Taipei ($\Delta < 0.1$). While Japan's *classroom management* level is only about 0.1 *SD* higher than the OECD average, its *disciplinary climate* level is the highest among the eight systems, nearly 0.7 *SD* higher than the OECD average. Though on the *classroom management* scale three East Asian systems (Korea, Hong Kong-China, and Chinese Taipei) are at below-OECD-average levels, both Hong Kong-China and Korea's *disciplinary climate* levels are not that low. In particular, on this scale, Hong Kong-China presents the third highest level among the eight systems.

OTL: Teaching Practices

Regarding teaching practices, the USA provides the highest level of *student orientation* as well as *formative assessment* activities, where most East Asian systems provide a below-OECD-average level. On *formative assessment*, Shanghai-China and Singapore are the only two East Asian systems having levels above the OECD average. Both Korea and Japan provide a particularly low level on this scale, which is nearly 1 *SD* below the US level. On *student orientation*, both Macao-China and Singapore have mean scores above the OECD average but not for the other systems in the same region. Hong Kong-China provides the lowest level on this scale, which is close to 0.7 *SD* away from the US level.

It is interesting to observe that the USA provides a high level not only on *student orientation* teaching but also on *teacher-directed instruction*, which, in fact, is only lower than Shanghai-China. In contrast, on this scale, the mean scores of all the East Asian' systems but Shanghai-China and Singapore are below the OECD average. Moreover, it is found that Shanghai-China has the largest variance between the two types of teaching practices ($\Delta = 0.7$), while the level differences between the two scales in all the other East Asian systems and the USA are no more than 0.2 *SD*. In fact, the USA maintains nearly the same level on all three types of teaching practices, and greater variances can be observed with the East Asian systems.

Figure 5.2 reveals that the USA shares more commonalities with Singapore on all three OTL dimensions. For both teaching quality and teaching practices, Shanghai-China's characteristics are more similar to the USA and Singapore than

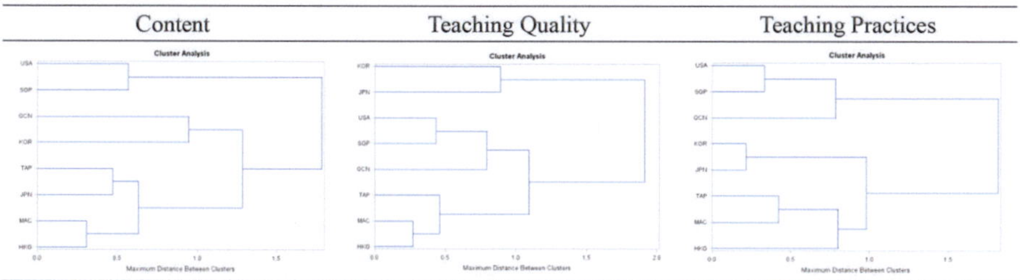

Fig. 5.2 Cluster analysis of students' OTL by dimensions among seven East Asian education systems and the USA

to the other five East Asian systems. However, this is not the case for learning contents, where all the East Asian systems but Singapore are more similar to each other. Moreover, it can be observed that in terms of teaching, Japan and Korea are always clustered together.

Comparisons of Relations Between Test Scores and Attitudes

Both cognitive and noncognitive achievements are assessed in PISA 2012. Among the various noncognitive indices, students' self-beliefs show the strongest correlations with test scores. In particular, the magnitudes of the correlations[3] between students' mathematics self-efficacy and their mathematics scores in all the eight systems are fairly large, and students' mathematics self-concepts have medium correlations with test scores. Correlations related to students' mathematics anxiety levels show greater inconsistencies across systems. The corresponding correlations in both Korea and Japan ($|r| < 0.2$) are much weaker than those in all the other systems, and the strongest correlation was observed in the USA.

Compared to students' self-beliefs, the correlations related to the indices in the other three noncognitive aspects show more inconsistencies across the eight systems. For instance, Chinese Taipei is the only system where all three motivation indices consistently have medium correlations with students' test scores. In contrast, in Singapore students' *intrinsic* and *instrumental motivation to learn mathematics* only have negligible correlations with their test scores ($|r| < 0.1$). In both Japan and Korea, while both *intrinsic* and *instrumental motivation to learn mathematics* show medium correlations with test scores, only the correlations related to *attributions to failure in mathematics* surpass the threshold. In fact, on the latter index, all but Chinese Taipei present a small correlation with test scores.

For the two indices related to dispositions towards mathematics learning, negligible correlations with test scores are again found in Singapore. Similar results are

[3]Correlation coefficients of 0.1, 0.3, and 0.5 are defined as small, medium, and large, respectively.

also observed in Macao-China and the USA. In fact, it is generally found that the two indices have small or negligible correlations with test scores in all the eight systems with two exceptions. That is, students' *mathematics intentions* in Chinese Taipei and *subjective norms in mathematics* in Korea show medium correlations with test scores.

Another consistency between Chinese Taipei and Korea is found in the correlations related to students' participation in mathematics activities. In both systems, students' *mathematics work ethic* and *mathematics behavior* show medium correlations with test scores. In contrast, negligible correlations are observed in Singapore and the USA on the two indices. In the remaining systems, the corresponding correlations are all small, except that *mathematics behavior* shows a medium correlation with test scores in Shanghai-China.

Comparisons of Correlations Between Opportunity to Learn and Academic Achievement

PISA 2012 assesses students' opportunity to learn via three aspects: learning content, teaching practices, and teaching quality. As expected, content-related indices show stronger correlations with test scores than teaching-related ones. The strongest correlations are observed with students' *familiarity with mathematical concepts*, ranging from 0.3 in Shanghai-China to 0.6 in Korea. Compared to *experience with applied mathematics tasks at school*, students' *experience with pure mathematics tasks at school* has a stronger correlation with their PISA math test scores. In all but Shanghai-China ($r = 0.04$) and Macao-China ($r = 0.13$), the magnitude of the correlation is about medium (i.e., $r > 0.3$). In fact, students' *experience with applied mathematics tasks at school* in both Shanghai-China and Macao-China also shows weak correlations with their test scores ($|r| < 0.1$), and similar results are observed in Hong Kong-China and Singapore as well. In contrast, it seems that such experience has stronger correlations with students' test scores in Chinese Taipei, Japan, and Korea, and the corresponding strengths are approaching medium.

Classroom instruction can be broadly differentiated between teacher directed and student oriented. In all eight systems, *teacher-directed instruction* has a negligible correlation with students' test scores, while *student orientation* shows medium correlations in Chinese Taipei and the USA. The weakest correlation related to this index is found in Japan ($r = -0.1$). Though there are some inconsistencies in terms of the correlation strengths, all eight systems show that the more *student orientation* is practiced in classes, the lower test scores students tend to achieve. Similar patterns are also observed with the usage of *formative assessment*, with the corresponding magnitudes either negligible or small. The indices related to teaching quality have an overall weak correlation with students' test scores. Comparatively, *disciplinary climate* has the strongest correlation among the five relevant indices. In particular, this index appears to be more influential in Shanghai-China and Singapore than in the other systems ($r > .30$).

Comparisons of Correlations Between Opportunity to Learn and Attitudes Towards Mathematics

Opportunity to Learn and Motivation

Compared to teaching-related OTL, content-related OTL shows stronger correlations with students' motivation to learn mathematics. In particular, students' *familiarity with mathematical concepts* has medium or nearly medium correlations with *intrinsic* and *instrumental motivation to learn* in all but Macao-China, Shanghai-China, and Singapore. It is interesting to find that all four indices of content-related OTL have medium or nearly medium correlations with students' intrinsic and instrumental motivation in Korea. In a majority of the systems, students' *experience with applied mathematics tasks* tends to have slightly stronger correlations with their motivation to learn mathematics than *experience with pure mathematics tasks*. However, content-related OTL only has a negligible or small correlation with students' *attribution to failure in mathematics*.

Among the five teaching quality-related indices, *math teaching*, *cognitive activation*, and *teacher support* show medium or nearly medium correlations with students' *intrinsic* and *instrumental motivation*. In most cases, both *classroom management* and *disciplinary climate* have small correlations with students' motivation with the exception of Shanghai-China, where *classroom management* shows medium correlations. While *teacher-directed instruction* shows medium or nearly medium correlations with students' *intrinsic* and *instrumental motivation*, *student orientation* has negligible or small correlations in all the eight systems. More variances across systems are observed on the correlations related to *formative assessment*. In particular, medium and nearly medium correlations are found in Chinese Taipei, Shanghai-China, and the USA, but correlations are small in the remaining systems. Both teaching-related types of OTL have overall small correlations with students' *attributions to failure in mathematics*.

Opportunity to Learn and Self-beliefs

Content-related OTL has strong correlations with students' mathematics self-efficacy. In particular, students' *familiarity with mathematical concepts* shows large or near correlations in all eight systems. Slightly weaker correlations are observed with students' *mathematics self-concept*, which in all but Japan and Shanghai-China are medium or nearly medium. In Singapore and the USA, this index also has medium correlations with students' *anxiety*. Moreover, students' experiences with *applied mathematics tasks* and *pure mathematics tasks* illustrate medium correlations with students' *mathematics self-efficacy* in Chinese Taipei, Japan, Korea, and the USA. The two indices exhibit negligible or small correlations with students' *mathematics anxiety* and *mathematics self-concept* in all the systems with the exception of Korea, where *experience with pure mathematics tasks* shows a medium correlation with students' *mathematics self-concept*. The correlations with teaching-related OTL in all eight systems are either negligible or small.

Opportunity to Learn and Disposition

The correlations between students' *mathematics intention* and all three types of OTL are either negligible or small. Different from the findings with students' motivation and self-beliefs, teaching-related OTL shows stronger correlations with students' *subjective norms in mathematics* than content-related OTL. In fact, in all eight systems, the correlations between students' *subjective norms in mathematics* and all four content-related OTL indices are generally about small or negligible. The only exception is found in Korea, where students' *familiarity with mathematical concepts* has a medium correlation with *subjective norms in mathematics*.

It is observed that *cognitive activation*, *teacher support*, *teacher-directed instruction*, and *formative assessment* have medium or nearly medium correlations with students' *subjective norms in mathematics* in all eight systems. The corresponding correlations related to *disciplinary climate* and *student orientation* are the weakest in their respective dimensions. Moreover, Shanghai-China is the only system where *classroom management* shows a medium correlation with students' *subjective norms in mathematics*.

Opportunity to Learn and Participation

Content-related OTL is found to have an overall stronger correlation with students' *mathematics behavior* than *mathematics work ethic*, while the opposite relation is observed with teaching-related OTL. Compared to the findings related to the other three types of noncognitive achievements, more inconsistencies across systems arise with students' participation in mathematics activities. For instance, while all the content-related OTL indices have medium or large correlations both with students' *mathematics work ethic* and *mathematics behavior* in Korea, the corresponding correlations in Macao-China, Singapore, and the USA are only negligible or small. In the other four systems, *familiarity with mathematical concepts* shows medium correlations with both participation-related indices. In addition, Japan is another system where students' *experience with applied mathematics tasks* has a medium correlation with *mathematics behavior*.

Among the five indices related to teaching quality OTL, *disciplinary climate* shows the weakest correlation with students' *mathematics work ethic* and Shanghai-China is the only system where the corresponding correlation is about medium. More consistencies are found with the other four indices, and the relevant correlations in all the systems are about medium or nearly medium. In contrast, on *mathematics behavior* only *cognitive activation* shows a medium correlation in Hong Kong-China and Shanghai-China, and in the other six systems the correlations are about small.

Similarly, stronger correlations with teaching practice OTL are also found with *mathematics work ethic*. In particular, *teacher-directed instruction* in all eight systems shows medium or nearly medium correlations with *mathematics work ethic* but small correlations with *mathematics behavior*. Practice on *formative assessment* shows medium or nearly medium correlations with *mathematics work ethic* in Chinese Taipei, Shanghai-China, Singapore, and the USA, and with *mathematics behavior* in Shanghai-China and Singapore. In this aspect, *student orientation* shows small or even negligible correlations with the two participation indices in all the systems.

Conclusions and Discussions

East Asian students have consistently outperformed their counterparts in the West in the large-scale international comparisons of mathematics achievement (e.g., TIMSS and PISA). In the most recent PISA, East Asian systems filled the top seven spots in the international league table. However, such an extraordinary accomplishment is only in terms of students' academic scores. In fact, recent research has shown that students' school achievement should be more than cognitive performance, and that other domains actually have greater influences on students' later life success. With a secondary analysis of PISA 2012 mathematics data, this study provides a more comprehensive comparison of students' mathematics achievement, including both cognitive (assessment scores) and noncognitive (attitudes) aspects, opportunity to learn, as well as their relationships in the top seven East Asian systems and the USA.

The analysis showed that East Asian students topped the international league table in terms of not only their overall performance but also their performance on all the subscales. However, it also demonstrated that in some top-performing systems, the knowledge/skill development across contents and procedures is unbalanced, most obviously in Shanghai-China, while the variances for the USA students are much smaller. Despite their extraordinary test scores, students from some East Asian systems had extremely negative attitudes towards mathematics learning. This is most severe in Japan and Korea, where on a majority of the indices their students were ranked at the bottom of the international league table. Unlike the findings on cognitive performance, more attitudinal differences than similarities are found within the seven East Asian systems. Interestingly, the US students' attitudes towards mathematics are closer to those from Chinese Taipei, Macao-China, and Hong Kong-China.

The comparison of opportunity to learn (content) shows that students from the USA have a similar level of experience with applied mathematics tasks compared to those from Chinese Taipei and Macao-China, while on pure mathematics tasks American students' experiences are closer to those from Shanghai-China. In terms of students' familiarity with mathematical concepts, it is no surprise to see that students from Shanghai-China showed an outstandingly high level, more than twice that of students from the USA.

In general, the USA shared more commonality with Shanghai-China and Singapore on teaching-related OTL. However, a large difference was found in the disciplinary climate levels, though the three systems present similarly high levels of classroom management. An opposite pattern was observed in Japan, which presents a low level of classroom management but the highest level in disciplinary climate. Regarding teaching practices, the USA provides a high level not only on *student orientation* teaching but also on *teacher-directed instruction*, which is not observed with East Asian systems. Indeed, most of them provide below-OECD-average levels on both types of teaching practices.

Among the four attitude-related indices, students' self-beliefs show the strongest correlation with their test scores, and the corresponding magnitudes in all eight systems are consistently large. Students' anxiety levels show a much weaker correlation with their test scores. More inconsistencies across systems were found on the correlations related to the other three noncognitive aspects. Interestingly, greater inconsistencies were more often observed within East Asian systems rather than between East Asian systems and the USA. This trend has been repeatedly observed when analyzing correlations with three OTL components.

References

Askew, M., Hodgen, J., Hossain, S., & Bretscher, N. (2010). *Values and variables: Mathematics education in high-performing countries*. London: Nuffield Foundation.

Collie-Patterson, J. M. (2000, November). *The effects of four selected components of opportunity to learn mathematics achievement of Grade 12 students in New Providence, Bahamas*. Paper presented at the annual meeting of the Mid-South Educational Research Association, Bowling Green, KY.

Czehut, K. (2012). *The achievement gap, revisited: An empirical assessment of what we can learn from East Asian education*. Doctoral dissertation, Harvard University.

Gau, S.-J. (1997, March). *The distribution and the effects of opportunity to learn on mathematics achievement*. Paper presented at the annual meeting of the American Educational Research Association, Chicago, IL.

Grouws, D. A., & Cebulla, K. J. (2000). *Improving student achievement in mathematics. Part 1: Research findings*. Geneva: International Academy of Education.

Hanushek, E. A., Jamison, D. T., Jamison, E. A., & Wößmann, L. (2008). Education and economic growth: It's not just going to school but learning something while there that matters. *Education Next, 8*(2), 62–70.

Herman, J. L., Klein, D. C. D., & Abedi, J. (2000). Assessing students' opportunity to learn: Teacher and student perspectives. *Educational Measurement: Issues and Practice, 19*(4), 16–24.

Husen, T. (Ed.). (1967). *International study of achievement in mathematics: A comparison of twelve systems*. New York: Wiley.

Jensen, A. R. (1998). The suppressed relationship between IQ and the reaction time slope parameter of the hick function. *Intelligence, 26*(1), 43–52.

Jerrim, J. (2014). *Why do East Asian children perform so well in PISA? An investigation of Western-born children of East Asian descent* (Working Paper No. 14–16). Department of Quantitative Social Science, Institute of Education, University of London.

Kautz, T., Heckman, J. J., Diris, R., ter Weel, B., & Borghans, L. (2014). *Fostering and measuring skills: Improving cognitive and non-cognitive skills to promote lifetime success*. Paris: OECD.

Kennedy, J. J. (2011, April). Opportunity to learn: Understanding why the achievement gap exists. *EDUC* 872. Retrieved from http://docslide.net/documents/opportunity-to-learn-understanding-why-the-achievement-gap-exists.html.

Lee, J. (2009). Universals and specifics of math self-concept, math self-efficacy, and math anxiety across 41 PISA 2003 participating countries. *Learning and Individual Differences, 19*, 355–365.

Leung, F. K. S. (2001). In search of an East Asian identity in mathematics education. *Educational Studies in Mathematics, 47*, 35–51.

Leung, F. K. S. (2006). Mathematics education in East Asia and the West: Does culture matter? In F. K. S. Leung, D. Graft, & F. Lopez-Real (Eds.), *Mathematics education is different cultural traditions—A comparative study of East Asia and the West: The 13th ICMI study* (pp. 21–46). New York, NY: Springer.

Mo, Y., Singh, K., & Chang, M. (2012). Opportunity to learn and student engagement: A HLM study on eighth grade science achievement. *Educational Research for Policy and Practice, 12* (1), 3–19.

Muthen, B., Huang, L.-C., Jo, B., Khoo, S.-T., Goff, G. N., Novak, J. R., & Shih, J. C. (1995). Opportunity-to-learn effects on achievement: Analytical aspects. *Educational Evaluation and Policy Analysis, 17*(3), 371–403.

Neisser, U., Boodoo, G., Bouchard, T. J., Boykin, A. W., Brody, N., Ceci, S. J., et al. (1996). Intelligence: Knowns and unknowns. *American Psychologist, 51*, 77–101.

OECD. (2012). *PISA 2012 technical report*. Paris: Author Retrieved from http://www.oecd.org/pisa/pisaproducts/PISA-2012-technical-report-final.pdf.

Pianta, R. C., Jay, B., Renate, H., & Fred, M. (2007). Teaching: Opportunities to learn in America's elementary classrooms. *Science, 315*(5820), 1795–1796.

Porter, A. (1993). Opportunity to learn. Brief No. 7. *Access to Education, 7*, 2–6.

Porter, A. (2014). *Rethinking the achievement gap*. Retrieved August 3, 2015, from http://www.gse.upenn.edu/node/269.

Reeves, C., Carnoy, M., & Addy, N. (2013). Comparing opportunity to learn and student achievement gains in southern African primary schools: A new approach. *International Journal of Educational Development, 33*, 426–435.

Schmidt, W. H., Houang, R., & Cogan, L. (2002). A coherent curriculum: The case of mathematics. American Educator, 26(2), 10–26, 47–48.

Stevenson, H. W., & Stigler, J. W. (1992). *The learning gap: Why our schools are failing and what we can learn from Japanese and Chinese education*. New York: Summit Books.

Wang, J., & Goldschmidt, P. (1999). Opportunity to learn, language proficiency, and immigrant status effects on mathematics achievement. *Journal of Educational Research, 93*(2), 101–111.

Watkins, D. A., & Biggs, J. B. (1996). *The Chinese learner: Cultural, psychological and contextual influences*. Hong Kong: Comparative Education Research Center, The University of Hong Kong/Australian Council for Educational Research.

Webster, B. J., Young, D. J., & Fisher, D. L. (1999, April). *Gender and socioeconomic equity in mathematics and science education: A comparative study*. Paper presented at the annual meeting of the American Educational Research Association, Montreal, Quebec, Canada.

Wiley, D. E., & Yoon, B. (1995). Teacher reports on opportunity to learn: Analyses of the 1993 California Learning Assessment System (CLAS). *Educational Evaluation and Policy Analyis, 17*(3), 355–370.

Wong, N. Y. (1998). Mathematics curriculum reform: The Hong Kong experience. In H. S. Park, Y. H. Choe, H. Shin, & S. H. Kim (Eds.), *Proceedings of the ICMI-East Asia regional conference on mathematic education* (Vol. 2, pp. 89–102). Seoul: Korea Society of Mathematical Education.

Zhao, Y., Lei, J., Li, G., He, M., Okano, K., Megahed, N., et al. (Eds.). (2010). *Handbook of Asian education: A cultural perspective*. New York: Routledge.

Chapter 6
Toward Understanding the Influence of Curriculum Resources on Students' Mathematics Learning: Cross-National Perspectives on What Matters Where, When, and for Whom

Edward A. Silver

Abstract Over several decades, international surveys of students' mathematics achievement, such as TIMSS and PISA, have reported the generally superior performance of students in East Asian locales. Each of the chapters in this section of the book makes a contribution to the growing evidence base regarding mathematics education in East Asia. They do this either by examining selected aspects of mathematics curriculum in East Asian locales or by investigating characteristics of students and teachers in these countries that interact with curriculum as it is implemented in classrooms. Collectively they add nuance to our understanding of the superior performance of East Asian students on international assessments even as they challenge us to resist overly simplistic characterizations and overgeneralized attributions that may not apply in all East Asian settings.

Keywords Mathematics • Learning • Curriculum • Textbooks • Cross-national

Despite variations in language, culture, and customs across the globe, in virtually every country, at every grade level, and on every day during the academic year, a student attending school is likely to receive mathematics instruction. Moreover, among all the school subjects that a student may study, the learning of mathematics is most likely to be contingent upon what happens in school. Whereas students may have many opportunities to sharpen their literacy skills or to learn history or science independent of classroom instruction, it is generally acknowledged that learning mathematics depends to a great extent on the quantity and quality of classroom instruction and the completion of associated assignments. Thus in most countries across the globe there is great interest on the part of educators, public policy

E.A. Silver (✉)
University of Michigan, Ann Arbor, MI, USA
e-mail: easiler@umich.edu

J.-W. Son et al. (eds.), *What Matters? Research Trends in International Comparative Studies in Mathematics Education*, Research in Mathematics Education, DOI 10.1007/978-3-319-51187-0_6

professionals, and the general public in determining the extent to which mathematics is being learned in school.

Large-scale international surveys of the mathematics achievement of students around the world, such as TIMSS and PISA, have reported considerable variation in performance across participating countries. One consistently observed phenomenon, however, has been the generally superior performance of students in East Asian locales. For example, Hong Kong, Japan, Korea, Shanghai (representing China), Singapore, and Taiwan (usually called Chinese Taipei in the assessments) have been among the highest performers on recent international assessments. In response to this observation, researchers have engaged in a multi-decade search for factors that explain the superior performance of East Asian students. Though those who attempt to compare education in different countries face a daunting task, aptly characterized by Torsten Husén (1983) as "comparing the incomparable" (p. 455), the search has been vigorous.

Many factors have been examined, with attention to affective as well as cognitive factors and to cultural influences as well as the technical details of schooling. Some investigations, patterned after the pioneering The International Mathematics and Science Study (TIMSS) which featured analysis of videotaped classroom instruction from three participating education systems (Germany, Japan, and the United States) (Kawanaka, Stigler, & Hiebert, 1999), have examined samples of classroom instruction in East Asia (e.g., Li & Huang, 2013; Li & Shimizu, 2009). But curriculum has been the dominant focus of attention when investigating factors contributing to East Asian student performance on mathematics assessments.

Derived from the Latin word *currere*—meaning "to run"—the word curriculum referred initially to the course that runners followed in a competition. Today, it is a word with multiple meanings, including the sequence of courses that a student may complete, the topics that are contained in a given grade, or the specific expectations regarding content, skills, competencies, and habits of mind that are deemed necessary for educational or societal reasons. These varied meanings all focus on the content of the curriculum, as it appears in textbooks or official documents such as syllabi, frameworks, and catalogues. However, there are other distinctions that may be even more critical for our understanding of how curriculum might play a role in student achievement.

More than two decades ago, Travers and Westbury (1989) suggested an important distinction between curriculum as planned and curriculum as enacted. In the Second International Mathematics Study (SIMS), they contrasted the intended curriculum (as represented in official documents and textbooks) with the implemented curriculum (measured through questionnaires given to teachers). These were, in turn, related to the attained curriculum, as reflected in students' performance on SIMS test items. Distinguishing among these three different "versions" of curriculum draws attention to the variation that was found in SIMS and in many other studies that compare the topics that appear in official curricula to those taught and emphasized in classrooms, and then to the proficiency obtained by the students taught in those classrooms.

The tripartite distinction offered by Travers and Westbury suggests both the promise and peril of pinning one's hopes for understanding critical aspects of mathematics teaching and learning in a country to an analysis of intended curriculum. Though official curriculum documents and widely used textbooks serve as blueprints for school mathematics, and they can serve as tools to influence classroom instruction, they do not completely determine the content and conduct of classroom instruction (i.e., the implemented curriculum). Nevertheless, the long-standing research focus on curriculum as a strategic site for investigating the basis for East Asian students' superior performance in mathematics has been quite reasonable.

Within the realm of mathematics classroom instruction, textbooks and other curriculum resources play an essential role. They vary greatly across countries and grade levels, both in how they are organized and in how they provide instruction to students and support to teachers. Yet their centrality to classroom instruction in almost all compulsory school settings has been demonstrated through the rigorous analysis of empirical evidence, and verified by the informal observation of everyday experience in mathematics classrooms across the globe. Mathematics classroom instruction is generally organized around and delivered through the mathematical tasks, activities, and problems found in curriculum resources, especially (though not exclusively) textbooks. Textbooks and other curriculum resources thus play a central role in both shaping students' opportunities to learn and mediating between the intended and implemented curriculum, that is, between the formal statements of expectations regarding what should be taught or learned in school and what actually happens in classrooms as a consequence of the interactions of teachers, students, and academic content.

Each of the chapters in this section of the book makes a contribution to the growing evidence base regarding mathematics education in East Asia. They do this either by examining selected aspects of mathematics curriculum in East Asian locales or by investigating characteristics of students and teachers in these countries that interact with curriculum as it is implemented in classrooms. Collectively, they add nuance to our understanding of the superior performance of East Asian students on international assessments, even as they challenge us to resist overly simplistic characterizations and overgeneralized attributions that may not apply in all East Asian settings.

In the opening chapter, Son and Diletti provide a useful review of prior cross-national research that has focused on mathematics textbooks. They summarize the methods and findings of 31 different studies, using a framework they propose to distinguish among analytic foci in the various investigations. Applying this framework to the set of studies examined, they draw attention to variations in the extent of attention to grade levels and topics within textbooks, as well as variations in the frequency with which different countries were included in the studies.

Their analysis reveals that there has been more attention paid to fractions and whole numbers than to algebra, geometry, and probability; more attention to middle grades (or lower secondary grades) than to upper secondary grades; and more attention to China than to Singapore or Taiwan. These data-based observations should be useful to researchers wishing to synthesize findings across studies, as well as to those wishing to aim new investigations at less frequently researched targets. Also likely to be useful is the detailed analysis the authors provide of studies that have examined in detail the cognitive demand, cognitive expectations, and depth of knowledge characteristics for tasks found in textbooks in different countries.

Regrettably the authors have some discouraging news regarding the overall quality of the research they reviewed, as they identified methodological weaknesses in at least half of the studies. Clearly, we have much work to do as a research community to improve the quality of research design, research conduct, and research reporting in this domain. Fortunately, some of the other chapters in this section may offer possible assistance in this regard.

Watanabe, Lo, and Son compare the treatment of fractions and fraction operations in mathematics textbooks in Japan, Korea, and Taiwan. In their content analysis they employ two established and validated frameworks on children's understanding of mathematical ideas; one pertains to children's understanding of arithmetic operations as expressed in story problems with whole numbers, and the other pertains to identified fraction subconstructs. Though the Watanabe et al. study was limited to an examination of a single textbook from each of their three countries, it was quite comprehensive in another sense: examining the treatment of fractions in the textbooks across a span of 4 or 5 years. In this way, the authors were able to identify similarities and variations in the frequency and ordering of fraction subconstructs in the textbooks, and variations in the ways those subconstructs interacted with the treatment of fraction operations in the texts. For example, their analysis reveals that the operator interpretation of fractions was prevalent in the Korean textbook across the grade span, whereas it appeared only in grade 6 (the highest grade examined) in the Japanese textbook. This finding stands alongside another major finding regarding the intercountry difference: the Japanese textbook treated multiplication and division differently than the Taiwanese and Korean texts.

Variation in textbook treatments of selected mathematics topics across countries was also evident in the study reported by Cao, Wu, and Dong. They analyze textbooks used at the junior secondary level in five systems: China, Korea, Singapore, Japan, and the United States. As with the Watanabe et al. study, Cao et al. analyze only one textbook from each country, but they looked across multiple years, in this case a span of three grades. In their analysis of overall textbook difficulty Cao et al. focus on two major dimensions: content coverage (including both breadth and depth) and content difficulty (including both student exercises and textbook worked-out examples).

The reported analysis contains some familiar results, such as the finding that US textbooks tended to rank lower in difficulty than those from the other countries, and that the Chinese textbook was rated as the most difficult overall. But there are some surprises as well. For example, the Japanese textbook ranked lower than the US textbook in overall difficulty, though it ranked higher on the difficulty ratings of exercises and worked examples. Though the authors chose to aggregate across different rating dimensions to produce an overall difficulty rating, I think the disaggregated ratings they present could be even more useful to other researchers.

In the two textbook comparison studies summarized above, a common limitation is the inclusion of only a single textbook from each country. This highlights one of the challenges in cross-national comparative research: it is expensive and complicated to obtain adequate samples that allow conclusions that can be generalized at the country level. One way for individual investigators and others with fairly limited financial and human resources to overcome this challenge is to make use of data collected as part of large-scale, comprehensive cross-national research endeavors, such as TIMSS, PISA, or the Learner's Perspective Study (LPS). The remaining chapters in this section provide examples of how researchers used data from such large-scale studies as the basis for their investigations, which provide further insights into the factors that may influence the performance of East Asian students and the intercountry variations that may lurk within the designation *East Asian*.

In his chapter Shimizu uses LPS data (Clarke, Keitel, & Shimizu, 2006) to examine similarities and differences in a particular aspect of mathematics classroom teaching practice that he calls "summing up." He contrasts samples of mathematics teaching in Shanghai and Japan. Shimizu argues that the data reveal how this aspect of teaching practice differs in classrooms in the two countries, and he attributes the differences to variations in culture and tradition. In so doing his argument resonates with the findings of the textbook comparison studies summarized above, by pointing to variations across East Asian countries that are often masked in our quest for simplistic explanations of observed international test performance. Variation among the high-performing countries of East Asia was also amply evident in the final chapter in this section. Zhu uses PISA data to examine the relationship of cognitive and noncognitive factors to the oft-researched notion of "opportunity to learn" in the US and the seven East Asian systems that were the top scorers in PISA 2012.

Researchers can find in these chapters examples of potentially productive ways to advance scholarship in this domain, including structured literature reviews, empirical investigations using data specifically collected for the reported study, and secondary analyses of data collected in large-scale, multipurpose ventures. The authors of the chapters likewise raise important issues regarding theoretical framing, research methods, and interpretation of findings in cross-national studies, and these discussions should be useful to other investigators interested in exploring this territory.

Those with a more practical orientation toward improving education in lower performing countries like the United States can also extract value from these chapters. However, anyone seeking a recipe for the "magic sauce" that accounts for the superior performance of East Asian students on international mathematics assessments is likely to be disappointed after reading the five chapters in this section of the book. Perhaps the ingredients will be revealed elsewhere in the volume, but perhaps not. My reading of these chapters suggests that a *simple* explanation for why East Asian students do so well in international assessments is unlikely to be found. Not only is the explanation likely to be quite complex, but also these papers amply demonstrate that East Asia is not a unitary educational entity any more than the United States is one.

Notwithstanding the caution offered above, we can learn much by examining features and factors that vary across countries. As Shimizu reminds us in his chapter, those who seek to improve mathematics teaching and learning in the United States and other lower performing countries would be wise to recognize the value in not only recognizing the differences across education systems, but also trying to understand how it is that different educational practices arise and flourish within countries. Rather than searching for the secret sauce in another country's cuisine, or borrowing one ingredient from each of several different cuisines to obtain a mixture that is not appealing to our national palate, we might be wise to pursue a more holistic approach to improving our local cuisine. Toward that goal, these chapters give us much to contemplate, and even more to chew on!

References

Clarke, D., Keitel, C., & Shimizu, Y. (2006). *Mathematics classrooms in twelve countries: The insider's perspective* (Vol. 1). Rotterdam: Sense Publishers.

Husén, T. (1983). Are standards in U.S. schools really lagging behind those in other countries? *Phi Delta Kappan, 64*, 455–461.

Kawanaka, T., Stigler, J. W., & Hiebert, J. (1999). Studying mathematics classrooms in Germany, Japan, and the United States: Lessons from the TIMSS videotape study. In G. Kaiser, E. Luna, & I. Huntley (Eds.), *International comparisons in mathematics education* (pp. 86–103). London: Falmer Press.

Li, Y., & Huang, R. (2013). *How Chinese teach mathematics and improve teaching*. New York: Routledge.

Li, Y., & Shimizu, Y. (2009). Exemplary mathematics instruction and its development in selected education systems in East Asia. *ZDM—The International Journal on Mathematics Education, 41*(3), 257–262.

Travers, K. J., & Westbury, I. (1989). *The IEA study of mathematics I: Analysis of mathematics curricula*. New York: Pergamon Press.

Part II
Research on Institutional System of Mathematics Teacher Education

Chapter 7
Knowledge Expectations Matter: Mathematics Teacher Preparation Programs in South Korea and the United States

Rae Young Kim and Seung Hwan Ham

Abstract In this chapter, we discuss comparative knowledge expectations for mathematics teachers in their respective sociohistorical contexts. Specifically, this study examines the official educational aims and curricula of 49 mathematics teacher preparation programs in South Korea and the United States, where substantial differences have been observed in both student achievement and teacher knowledge. Overall, our findings suggest that transnational commonalities and national differences simultaneously affect social expectations for teacher knowledge. We argue that attending to both *culturally contextualized* and *semantically decontextualized* dimensions offers a more balanced comparative perspective from which we can better evaluate the current status of teacher education. Constructive international dialogue can be facilitated by such a balanced perspective, and may further enrich teacher education without ignoring either profound differences in sociohistorical contexts or important commonalities in epistemic models of teacher education across countries.

Keywords Teacher education/development • Social expectations for teacher knowledge • Mathematics education • Sociohistorical contexts • Transnational discourses on education

---

This chapter is a revised version of an earlier publication in *Journal of Teacher Education, 62*(1), 48–61.

R.Y. Kim (✉)
Ewha Womans University, Seoul, South Korea
e-mail: kimrae@ewha.ac.kr

S.H. Ham
Hanyang University, Seoul, South Korea

J.-W. Son et al. (eds.), *What Matters? Research Trends in International Comparative Studies in Mathematics Education*, Research in Mathematics Education, DOI 10.1007/978-3-319-51187-0_7

Introduction

It has been well documented that the quality of mathematics education for schoolchildren is never independent of the quality of the teaching force. Given the central position of mathematics as a core subject in the school curriculum, researchers and educational reformers around the world have been paying sustained attention to issues around and strategies for improving the quality of mathematics teachers. In this context, this study explores social expectations around adequate knowledge for prospective secondary mathematics teachers in two different countries—South Korea and the United States. By social expectations for teacher knowledge, we mean discursive practices institutionalized around what teachers should be taught to teach, according to the collective image or epistemic model of "the ideal teacher" in a given society (Oser, 1994; Tatto, 2008). Analyzing the educational aims and curricula of teacher preparation programs in the two countries, we shed light on the possibility that mathematics teacher knowledge may be influenced not only by social expectations in particular sociohistorical contexts but also by evolving transnational discourses on education. Although many scholars have called attention to the importance of mathematics teacher knowledge in order to enhance student achievement, relatively little effort has been made to examine what kinds of and how much knowledge prospective mathematics teachers are expected to have in different national contexts. This study aims to deepen our understanding of teacher knowledge from a comparative perspective. By reviewing prior research and analyzing some comparative data, this study hopes to make a unique contribution to the current debates on teacher education reforms in both countries, and it considers implications for both policy and practice.

In the United States, results from international assessments of student performance such as the *Trends in International Mathematics and Science Study* (TIMSS) and the *Programme for International Student Assessment* (PISA) have prompted significant policy efforts to improve student achievement. In these circumstances, considerable scholarly attention has been given to teacher knowledge as a key factor in improving student achievement (Hiebert, Gallimore, & Stigler, 2002; Hill, Rowan, & Ball, 2005; Kennedy, Ahn, & Choi, 2008; Wayne & Youngs, 2003). In fact, the findings from the cross-national study *Mathematics Teaching in the 21st Century* (MT21) reveal that there may be a "preparation gap" across countries (Schmidt et al., 2007). Its report emphasizes that prospective secondary mathematics teachers in the United States are not as well prepared as those in other high-achieving countries such as South Korea and Taiwan.

Thus, we further investigate teacher preparation programs in South Korea and the United States, where substantial differences have been observed in both student achievement and teacher knowledge (Blömeke & Paine, 2008; Schmidt et al., 2007). Indeed, US-based scholars have asked how South Korea has achieved excellence in teacher quality as well as considerable equity in students' learning opportunities (e.g., Akiba, LeTendre, & Scribner, 2007; Kang & Hong, 2008). In addition, in South Korea a strong sense of the need for reforming teacher education is shared by many educational researchers and policy makers, whose debates often refer to US teacher education to explore some alternative possibilities (e.g., Jang, 2001; Shin, 2009). Despite the complex mixture of similarities and differences across the spectrum of policy issues concerning teacher education reform in the two countries, we believe that a closer look at such differences and similarities, in relation to societal contexts both within and beyond national borders, will help educational researchers and policy makers explore possibilities to improve teacher education in both countries.

Previous comparative studies have often failed to capture the analytical significance of understanding the mixture of national patterns and transnational commonalities in teacher knowledge (see Jeynes, 2008; Wang & Lin, 2005). Focusing only on national differences, we could miss transnational influences on educational structural convergence throughout the world. If we consider only transnational similarities, we might likewise underestimate the importance of local meanings of and variations in education ácross different societal contexts. By considering the synthesis of competing frameworks for understanding teacher knowledge from international comparative perspectives, this study attempts to make a unique contribution to the literature on teacher knowledge.

In particular, this study addresses two research questions: (1) To what extent are the educational aims of mathematics teacher preparation programs in South Korea and the United States contingent upon (or independent of) each country's particular sociohistorical context? (2) To what extent are the curricular structures of mathematics teacher preparation programs different (or similar) between the two countries? Inasmuch as we explore knowledge expectations for teachers rather than the actual knowledge they possess, we do not intend to make claims about the relationship between teachers' knowledge and students' learning of mathematics. Rather, this exploratory study is intended to provoke further inquiry into the complexities involved in understanding multiple layers of societal contexts, within which social discourses on teacher knowledge are constituted, challenged, and modified both nationally and internationally.

In the following sections, we first present some theoretical perspectives that serve as conceptual frameworks for this study. Both comparative perspectives on and pedagogical conceptualizations of teacher knowledge are briefly reviewed and assessed. Secondly, the sociohistorical evolution of teacher education is examined for both countries, especially with regard to social expectations for teacher knowledge in teacher preparation programs. We distinguish social expectations for teacher knowledge from the actual knowledge possessed by individual teachers. This study focuses on the former, as it comprises an important aspect of teacher knowledge that has been largely ignored in research on teacher education. As part of this investigation, we examine both the curricular structure of mathematics teacher preparation programs and their educational aims, in order to see how the social expectations for mathematics teacher knowledge mirror social epistemologies both within and across national borders. Finally, we consider some implications from our findings in light of current policy debates on teacher education reform in each country.

Conceptual Perspectives

Comparative Perspectives: A Conceptual Reflection

Despite the abundance of comparative studies, recent scholarly debates on teacher knowledge have reached seemingly contrasting conclusions about whether such knowledge is similar or different across countries. Some scholars argue that there are different national patterns in teacher knowledge, since it is contextualized within the distinctive sociohistorical conditions of individual national societies (Alexander, 2000; Anderson-Levitt, 2002). Other groups of researchers understand teacher knowledge to be largely shaped by the cultural dynamics in the wider environment beyond national borders, emphasizing that what teachers know and how they behave are considerably similar across different countries despite different sociohistorical settings (LeTendre, Baker, Akiba, Goesling, & Wiseman, 2001). We, however, argue that these seemingly different views about teacher knowledge are not at odds as they seem to be. Rather, they describe two different dimensions of teacher knowledge: teacher knowledge reflects not only nationally (or locally) constructed cultural scripts (Fernandez & Yoshida, 2004; Paine, 1990; Stigler & Hiebert, 1999) but also transnationally shared meanings of and beliefs about teaching and learning (Baker & LeTendre, 2005; Desimone, Smith, Baker, & Ueno, 2005). We conceptualize the former as the *culturally contextualized* dimension of teacher knowledge and the latter as the *semantically decontextualized* dimension. We assume that some global structural models of education may exist (Meyer & Ramirez, 2000), but we also suppose that such global models are continuously recontextualized by local agency, and accordingly that they may have different meanings across societal contexts (Phillips, 2004; Schriewer, 2003).

We see teacher knowledge as semantically decontextualized because most pedagogical and mathematical theories purport to be neutral in meaning as they evolve from scientific elaboration. We emphasize that it is the "semantic fields or zones of meaning" (Berger & Luckmann, 1967, p. 41) that are often decontextualized because of the "scientization" of educational discourses around the world (Drori, Meyer, Ramirez, & Schofer, 2003; Ham & Cha, 2009). By investigating both of these dimensions of secondary mathematics teacher knowledge in South Korea and the United States, we are able to devise a more comprehensive analytical framework for comparing teacher knowledge at an international level.

Perspectives on Knowledge Bases for Teachers

Although there have been many competing debates about teacher knowledge with regard to its nature and typology, it is widely believed that teachers should have knowledge in multiple areas in order to teach mathematics effectively (Ball,

Thames, & Phelps, 2008; Shulman, 1987; Wilson, Floden, & Ferrini-Mundy, 2001). Teacher preparation programs, as integral agents in producing a teaching force, are expected to provide prospective teachers with various opportunities to acquire such knowledge through their curricula (Kennedy et al., 2008; Tatto et al., 2008). Shulman (1987), for example, proposed that teachers should have many different kinds of knowledge for teaching, including "subject matter content knowledge," "pedagogical content knowledge," "general pedagogical knowledge," "curricular knowledge," "knowledge of learners and their characteristics," "knowledge of educational contexts," and "knowledge of educational ends, purposes, and values, and their philosophical and historical grounds." According to his conceptualization, subject matter content knowledge is defined as knowing not only the accepted truth in a domain but also why it is and how it is warranted, while pedagogical knowledge consists of "broad principles and strategies of classroom management and organization that appear to transcend subject matter" (Shulman, 1987, p. 8). Also, he delineated pedagogical content knowledge as a kind of the "subject matter knowledge *for teaching*" (Shulman, 1986, p. 9), including knowledge about the most useful methods of representing ideas; the most convincing ways of giving examples, providing explanations, and showing demonstrations; and the most appropriate strategies for modifying a task to accommodate different cognitive levels of students.

Newly developed ideas about cognition and learning have added another important framework for understanding teacher knowledge. In particular, sociocultural perspectives view knowledge as *situated*, *distributed*, and *social* (Putnam & Borko, 2000). Whereas traditional cognitive perspectives focus on the individual since they view learning as the acquisition of knowledge and skills, sociocultural perspectives understand knowledge as involving distributed and shared cognitions that are the products of interactions among people over time. Such views focus on the processes through which participants interact with one another and the ways in which people acquire knowledge by using tools and participating in activities within certain contexts. According to this theoretical perspective, preservice teachers can learn, for example, by situating experiences in K-12 classrooms as authentic contexts. Such situated experiences can provide preservice teachers with coordinated opportunities to make sense of theoretical ideas learned through formal coursework, within the concrete contexts of actual practice.

In the next sections, we explore how such knowledge expectations for teachers are embedded in larger sociohistorical contexts around the profession of secondary mathematics teaching and teacher education in South Korea and the United States. By doing so, we can better understand how models of teacher knowledge and teacher education have evolved within each country, in conjunction with social meanings of the profession of teaching that have been constructed at both national and international levels.

Societal Expectations for Teachers: The Two Countries

Institutional Arrangements for Teacher Preparation and Certification

In order to contextualize knowledge expectations for secondary mathematics teachers better within the societal contexts of each country, it is necessary to compare their institutional arrangements for teacher preparation and certification. First of all, compared to US teacher preparation programs, South Korean programs are relatively uniform because of the country's centralized system of teacher education and certification. In most cases, those who want to acquire a secondary teacher certificate in South Korea first need to gain admission to a teacher preparation program upon entrance to a university. Although there are some alternative routes available at either the undergraduate or the graduate level, such routes are highly competitive. Once students have completed all the requirements in a teacher preparation program, they must pass the national teacher employment test to teach in public secondary schools with civil servant status. This test is administered annually by the Korea Institute for Curriculum and Evaluation, a government-funded educational research institution (Kang & Hong, 2008). Upon passing this test, teachers gain tenure to teach in public secondary schools. Usually, this test is extremely competitive (Kwon, 2004). However, such competition is not surprising, considering that the teaching profession has long been highly appreciated in South Korea (Sorensen, 1994).

In contrast, the routes to become teachers vary significantly in the United States, depending on the policies of individual states. Although there are some common requirements for prospective teachers such as completing the requirements for a teacher preparation program, a bachelor's degree, and a teaching certificate, there are considerable variations in teacher preparation and certification across states (Hess & Petrilli, 2006). For example, there are variations across states with regard to entry standards for undergraduate mathematics education programs. In addition, various alternative teacher certification programs have been created in response to teacher shortages (Feistritzer, 2008). Despite such variations across the United States, it is important to note that national accreditation organizations, including the National Council for the Accreditation of Teacher Education (NCATE), provide standards for teacher certification and preparation programs, so that such variations can be bounded by professional guidelines (NCATE, 2008). Further, 41 states have employed teacher certification tests as a criterion for admission to teacher preparation programs, to certify the completion of teacher training, to control the quality of teachers, and/or to manage the supply of new teachers (National Center for Education Statistics, 2009). These teacher certification tests are primarily designed to measure the basic skills, content knowledge, and pedagogical knowledge of teacher candidates, i.e., to determine that they have a minimum amount of knowledge for teaching (Mitchell, Robinson, Plake, & Knowles, 2001).

Core Knowledge in Teacher Preparation

Despite the rather different institutional arrangements described above, teacher preparation programs in South Korea and the United States both require candidates to complete courses *n* subject area content, education theory, and pedagogy, and to gain student teaching experience (Ingersoll, 2007). Although the details of the curricular contents in teacher preparation programs vary within each country, general curricular structures are largely similar between the two countries. For example, in the United States, Darling-Hammond and Bransford (2005) delineated the core curriculum of teacher education: learners and learning in social contexts, curriculum and subject matter, and teaching with a vision and professional practice. Similarly, many Korean scholars see the core curriculum for teacher preparation programs as being composed of educational foundations (e.g., historical and philosophical approaches to education, as well as psychological and sociological bases for education), subject matter studies, professional studies including pedagogical (content) knowledge courses (e.g., mathematics education theories and mathematics teaching methods), and teaching practice (Jang, 2001). Although different teacher preparation programs have somewhat different categorizations and descriptions of their curricula, there is a general consensus on the appropriate curricular structure within and across both countries.

Knowledge Expectations for Teachers in Sociohistorical Contexts

Along with structural similarities in teacher education curricula between the two countries, unique sociohistorical contexts are also central to understanding differences in social expectations for teachers and teacher education, as well as differences in the goals of education in general. Considering that education is deeply rooted in social and cultural assumptions and historical situations wherein the ideas of the "ideal person" have evolved (Cummings, 2003), it is important to understand the contexts in which South Korea and the United States have developed distinct social meanings of curricula and educational aims in their teacher preparation programs.

In the early nineteenth century in America, teaching could be done in any place (e.g., homes, town offices, and churches), by anyone (e.g., parents, town officers, preachers, and adults in the neighborhood) (Labaree, 2008). Although qualifications for teachers varied, the knowledge expectation for teachers was usually not very high. The assumption of teacher qualification was that teachers needed to reach the knowledge level at which they would be teaching, and the most important characteristic of teachers was the ability to keep order among students (Sedlak, 1989). As the common school systems were developed in the middle of the nineteenth century, formal preparation for teachers also appeared in the form of summer teacher institutes, which provided lectures and on-the-job training over a period ranging from 1 to 8 weeks (Labaree, 2008).

The common school movement triggered an increased demand for more teachers and, at the same time, for higher qualifications. Normal schools were established to train teachers, and they had a mixture of liberal arts courses, subject matter courses, and professional courses. Initial models of mathematics teacher education programs were developed in the late nineteenth century, and they offered preparation in mathematics content, specialized training in mathematics pedagogy, social sciences, and practice teaching (Donoghue, 2006). Although such components of mathematics teacher preparation programs have been widely used and remain even today, the topics and emphases within each component have shifted over time. In particular, the leaders of normal schools eventually chose "relevance over rigor" (Labaree, 2008) because emphasizing relevance was considered an easier way to meet the increasing demands for teachers than preserving academic rigor. This tension between relevance and rigor still exists in US debates regarding contemporary teacher education (Fraser, 2007).

Along with this history of teacher education, teachers in the United States have been viewed as "special but shadowed" (Lortie, 2002). That is, although teachers have been seen as performing a "special" mission in society, they have often been "shadowed" by other more favored positions requiring higher levels of expertise. This general situation for teachers applies to mathematics teachers as well. Many talented people who have majored in mathematics are likely not to enter the profession of teaching in the United States. Although some of them get teaching positions in schools, most of them choose to enter the profession of teaching not because they see teaching as a truly desirable occupation for themselves, but rather because they consider teaching as a reasonable or temporary alternative (Clewell & Forcier, 2001).

Unlike the situation in the United States, the knowledge expectation for teachers in South Korea has been very high as education has long been a top priority in the country. Education was highly valued during Korea's Chosŏn Dynasty (1392–1897), when access to education was limited to the hereditary ruling class. Although there was no official qualification system for teachers, many prominent scholars built schools and taught history, philosophy, and poetry to their students. At the local level, those who were considered the most intelligent and best moral exemplars became teachers. *Gunsabu-ilche*, a popular Korean traditional adage meaning "Kings, teachers, and parents deserve equal respect," shows how much Korean society has appreciated the role of teachers.

When Korea gained independence from Japan in 1945, many normal schools were established to meet the increasing demands for teachers, and tertiary-level teacher preparation institutions replaced normal schools, especially during the 1960s (Woo & Ahn, 2006). As the Korean economy has grown rapidly and the society has changed dramatically since the end of the Korean War, a high level of education has come to be viewed as the most valuable capital and the surest route to success and upward social mobility (Bae, 1991; Robinson, 1994; Sorensen, 1994). Often called Korea's "education fever" (Seth, 2002), the South Korean people's enthusiasm for learning has sustained and reinforced the traditional image of teachers who deserve deference and respect. In this sociohistorical context, teachers have been expected to possess a high level of knowledge for teaching as professionals, and to have respectable moral character (Kang & Hong, 2008; Sorensen, 1994). Thus, an applicant for admission to a college of education must satisfy multiple criteria such as a high-school record of academic excellence, a competitive score on the college scholastic aptitude test, and acceptable scores on both the aptitude test for teaching and a separate interview named the "humaneness test" (Kwon, 2004). In fact, a recent report from the Korea Labor Institute showed that the top 10% of high-school graduates enter teacher education programs (Lee et al., 2005).

Leung (2001) claimed that such values regarding education specifically influence mathematics education. Since mathematics is one of the core subjects influencing a student's opportunity to go to college, mathematics teachers' rigorous content knowledge is also seen as crucial to student achievement. Despite the fact that Western theories dominate mathematics teacher education in many East Asian countries, there have been distinctive social expectations for mathematics teachers in those countries. Kim (2002) asserts that mathematics teachers in South Korea are expected to be mathematicians who have deep and broad knowledge of advanced mathematics in addition to school mathematics. That is, teachers are expected to be scholars before being facilitators of learning (Leung, 2001).

Comparing Knowledge Expectations for Teachers: Inside Teacher Preparation

Data and Methods

In order to compare knowledge expectations for mathematics teachers between the United States and South Korea, we selected a sample of teacher preparation institutions in each country: 28 institutions were sampled in the United States and 21 in South Korea. These samples, however, were not nationally representative. Selecting a nationally representative sample of teacher preparation institutions would itself be a challenging task, especially in the United States, where there are over 1300 teacher preparation institutions with varying characteristics (Schmidt et al., 2007). Our alternative approach used purposive sampling of some teacher education institutions that have been nationally accredited in each country. We can plausibly assume that teacher preparation programs that are guided by the standards of a national level professional association, or of their national government, follow guidelines or criteria that embody certain models of teacher education that are regarded as appropriate in society. Thus, despite the limited generalizability of the data, this purposive sampling strategy helps us to see the features that are widely assumed to be appropriate in teacher preparation within each country. For South

Korea, all the teacher preparation programs we selected had been guided and approved by the Ministry of Education. For the United States, we collected data among teacher preparation programs accredited by NCATE, which serves as a professional accrediting body for teacher preparation and provides guidelines for teacher education programs.

The sample included universities whose prestige and geographical locations varied. Among the 21 South Korean universities sampled, 8 were located in Seoul, the capital city, and the rest were located across all 13 of the provinces/metropolitan cities of South Korea, reflecting the ratio of the total number of Seoul-based mathematics teacher preparation programs to the total number of such programs in non-Seoul regions. Our 28 universities sampled in the United States were located across 22 states. In both countries, we focused on undergraduate-level teacher preparation programs with a single major in secondary mathematics education; we excluded graduate-level teacher preparation programs from this analysis because such programs often focus on "professional" training rather than providing the full range of courses classified by our analytical categories of teacher knowledge.

Based on these teacher education institutions sampled from both countries, we first identified some key ideas about teacher knowledge that were embedded in the official educational aims of individual teacher preparation programs in South Korea and the United States. Considering that the educational aims of teacher preparation programs are shaped largely by social values and philosophies about education, we can plausibly assume that they reflect not only the purposes and visions for educating future teachers, but also social and cultural beliefs about the meaning of teaching and learning. The documents used as textual data for this portion of the analysis included official documents of teacher education institutions, especially the program introduction section of teacher education program/department handbooks or equivalent information on official websites. We did not assume that these texts were isolated or detached from historical contingency, but rather treated them as parts of "social events" shaped by social "orders of discourse" (Fairclough, 2003).

Identifying key embedded ideas was, in fact, a highly interpretive task. The interpretive nature of the task, however, does not necessarily mean that the key ideas we identified were arbitrary. Among many different ideas that were initially identified from the educational aims, only those that had a clear connection to knowledge expectations for prospective teachers were finally selected. Table 7.1 shows our selection of key embedded ideas, along with some quotes from our data. If a given key idea was found in the educational aims of a given teacher preparation institution, "1" was assigned to the institution; otherwise, "0" was assigned. The inter-rater agreement was measured by Cohen's kappa, and its coefficient ranged from .796 to .831 across different key ideas, indicating a very high degree of agreement between the two raters. One external bilingual rater was asked to code the data where there was a disagreement between the two internal raters, so that we could use this external rater's score for our final coding.

Table 7.1 Key embedded ideas about teacher knowledge

Ideas about teacher knowledge	Key quotes
Wide range of disciplinary knowledge	A wide variety of general education courses; interdisciplinary nature of educational inquiry; comprehensive studies in general education
Deep content knowledge	Rigorous studies in mathematics; professional expertise in the content area; profound understanding of mathematical principles and concepts
Knowledge about instructional methods	Effective instructional methods; a variety of strategies to enhance student learning; creativity in teaching; appropriate assessment and evaluation
Knowledge about how to deal with student diversity/equity issues	Understanding of human diversity; diverse learners; cultural and linguistic diversity; students' special needs; promoting both equity and excellence
Knowledge gained from situated experiences	Field experiences; student teaching; practicum; internship in schools; clinical experiences

Next, we analyzed curricular structures of mathematics teacher preparation programs in each country. Relying on conventional conceptualizations of categories of teacher knowledge, we classified the courses provided by secondary mathematics teacher preparation programs into five different types of knowledge: content knowledge (CK), pedagogical content knowledge (PCK), pedagogical knowledge (PK), general knowledge (GK), and field experience (FLD). Table 7.2 shows our conceptual categories of teacher knowledge and some examples of related courses. In our definition, CK courses include pure and applied mathematics courses that do not directly focus on school mathematics. PCK courses are those that focus on school mathematics or on methods of teaching school mathematics. PK courses refer to general education theories or educational foundations courses. GK courses include many other courses such as university-wide requirements and electives that are not directly related to either mathematics or education. Finally, we define FLD courses as practice-based classes in classroom settings. We separate them from other categories whose courses are taught on campus. According to this classification, we examined the curricular emphasis given to each type of knowledge, by assessing the credit hours allocated to each corresponding type of course as a percentage of the total minimum credit requirements of a given program.[1] This strategy allowed us to see the relative curricular emphasis devoted to each type of course across programs, thereby making the data comparable between the two countries.

Of course, it should be noted that official educational aims and official course titles/descriptions of teacher preparation programs are often "loosely coupled" (Weick, 1976) with teacher educators' actual expectations and practices. However, much of the sociology of education literature suggests that, precisely because of

[1]Since US students generally enter a teacher preparation program in their second year of college, we excluded their first-year courses from our analysis where appropriate.

Table 7.2 Conceptual categories of teacher knowledge and related courses

Types of knowledge	Types of courses	Examples of courses
CK	Mathematics courses	Algebra; geometry; calculus; topology; complex analysis; real analysis; discrete mathematics; number theory; probability and statistics; differential equations
PCK	Mathematics education courses; mathematics teaching methods courses	Secondary mathematics curriculum and instruction; secondary mathematics methods; teaching methods in algebra; foundations of mathematics education
PK	General education courses; educational theory courses	Introduction to education; educational psychology; educational sociology; educational history; educational administration; classroom management; counseling; general teaching methods
GK	Other courses, e.g., university-wide requirements, electives	Modern history; contemporary literature; modern foreign languages; social psychology; environmental issues; introduction to political science; science, technology, and society; family and society
FLD	Student teaching; internship; clinical experiences	Field experience in mathematics education; student teaching; internship

Note: *CK* content knowledge, *PCK* pedagogical content knowledge, *PK* pedagogical knowledge, *GK* general knowledge, *FLD* field experience

such loose coupling, formal structures of educational organizations can easily conform to epistemic models constituted and rationalized in broad societal contexts, both within and beyond national borders (Meyer & Rowan, 1983; Wiseman & Baker, 2006). Since this study aims to examine social expectations for teacher knowledge rather than the actual knowledge possessed by individual teachers, we believe that formal curricular structures of teacher preparation programs deserve systematic comparative analysis as social artifacts.

Educational Aims of Teacher Preparation Programs

The findings from this analysis suggest that transnational commonalities and national differences exist simultaneously, and examining both is necessary in order to understand teacher knowledge better. Figure 7.1 shows the percentages of secondary teacher preparation programs whose official educational aims mention different key ideas about knowledge expectations for prospective teachers across the set of institutions in each country. On the one hand, it seems that there is no doubt about the importance of mathematics teacher knowledge in both countries, in terms of discursive practices expressed in the official educational aims of secondary mathematics teacher preparation institutions. At least one-fifth of the institutions in

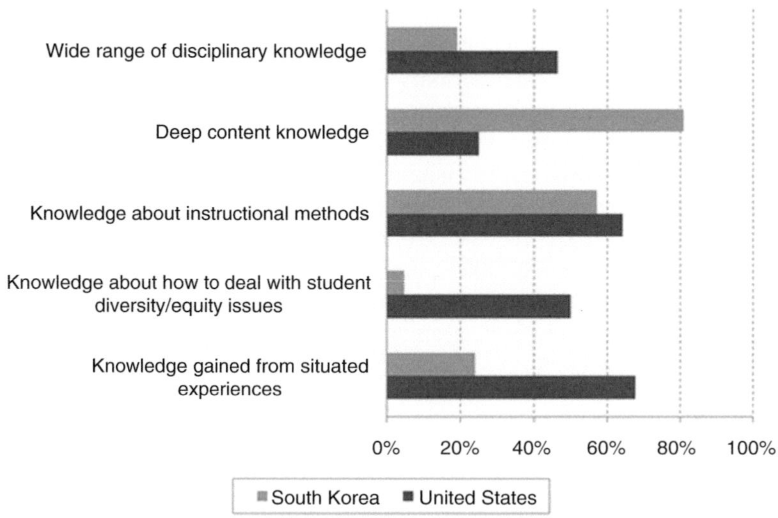

Fig. 7.1 Percentages of secondary teacher preparation programs whose official educational aims mention different key ideas about teacher knowledge

both countries appear to have educational aims that refer to the importance of each of the following: a wide range of disciplinary knowledge, deep content knowledge, knowledge about instructional methods, and knowledge gained from situated experiences. In particular, a striking agreement exists between the two countries in terms of the teacher preparation institutions' mention of the importance of teachers' knowledge about instructional methods: about three-fifths of the institutions in both countries mention it in their official educational aims.

On the other hand, the data presented in Fig. 7.1 also suggest the possibility that the models of teacher knowledge expressed in the official educational aims of secondary teacher preparation programs may have been shaped by their cultural assumptions, historical contexts, and social epistemologies of education. Among the US programs in our data, one-quarter of them mentioned the importance of deep content knowledge in their official education aims, but the proportion was more than three times as high for the South Korean programs. One possible explanation for this contrast may be that South Korean programs tend to assume that prospective teachers' strong content knowledge can be applied in practice. This assumption appears to bolster the societal expectations for "scholar teachers" (Leung, 2001) who have rigorous knowledge in advanced mathematics. At the same time, more than three-fifths of the US teacher preparation programs in our dataset mentioned the significance of the knowledge preservice teachers can gain from situated experiences. This appears congruent with their assumption that prospective teachers need to be provided with sufficient opportunities to learn in school settings where they can make connections between theory and practice. This contrasts with South Korea, where less than one-quarter of the programs mentioned something about situated experiences.

With regard to prospective teachers' knowledge about how to deal with student diversity/equity issues, half of the US programs considered it important in their official educational aims, whereas only one South Korean program specifically mentioned it. This clear difference may be in part due to different degrees of societal diversity between the two countries. For example, in terms of the ethno-linguistic fractionalization index, which ranges theoretically from zero for no fractionalization to one for perfect fractionalization, the United States is .58, indicating a high degree of diversity, while South Korea is less than .01, indicating extreme homogeneity (Annett, 2001). Another reason for such contrasting degrees of attention to student diversity may be related to the different degrees of institutionalization of multicultural education discourses between the two countries (Cha, Dawson, & Ham, 2010). Indeed, the academic and policy attention regarding multicultural issues in education is a fairly recent phenomenon in South Korea.

Curriculum Models for Teacher Education

Table 7.3 shows the percentages of credit requirements allocated to different types of courses in secondary mathematics teacher preparation programs. Two similar patterns are evident between South Korea and the United States. First, secondary mathematics teacher preparation programs in both countries require all the five types of courses: GK, CK, PCK, PK, and FLD. Almost nine out of ten US programs, and virtually all South Korean programs, provide preservice teachers with opportunities to learn all five different types of knowledge within their formal curriculum. Second, on average GK and CK, the two biggest parts, account for about two-thirds of the curricular requirements in both countries, and PCK, PK, and FLD, which all relate to pedagogy-related knowledge, comprise the remaining one-third of the curricular requirements. Such similar curricular structures between the two countries demonstrate a certain degree of consensus on the ratio of general and non-pedagogy knowledge to pedagogy-related knowledge around which curricular

Table 7.3 Percentages of credit requirements allocated to different types of teacher knowledge

	South Korea ($n = 21$)		United States ($n = 28$)			
	Mean	*SD*	Mean	*SD*	t^a	p-Value (two tailed)
GK	23.03	9.13	39.38	15.81	-4.55^a	<.001
CK	45.29	10.60	26.33	10.36	6.26^a	<.001
PCK	16.35	6.04	8.90	6.65	4.03^a	<.001
PK	12.94	3.87	12.66	10.10	$.14^a$	.893
FLD	2.40	1.84	12.73	6.91	-7.56^a	<.001

Note: The sum of the means of all five curricular components equals 100 within each country. Independent-sample t-tests were conducted using either Student's t-test or Welch's t-test, depending on the statistical significance of Levene's test for equality of variances; Student's t-test was used unless Levene's test was significant at the $p < .05$ level

[a]Indicates a Welch's t-test statistic

requirements are organized in secondary mathematics teacher preparation programs. These patterns suggest that although the details of curricular contents in teacher preparation programs may vary within each country, the general curricular structures in both countries are largely in accordance with what international research and policy recommends. Indeed, such similar curricular structures are not surprising, considering that "most teacher education programs [around the world] offer some combination of coursework in subject matter, teaching methods and materials, child/adolescent development, and other education courses such as psychology, history, and philosophy of education, along with practice teaching" (OECD, 2005, p. 107).

Along with these similarities, some national patterns are also evident. As also shown in Table 7.3, while curricular requirements are focused most heavily on CK in South Korean programs, GK tends to be given more emphasis than CK in US programs. In terms of credit requirements, 45% of credits are allocated to CK courses in South Korea, which is the largest proportion of the five types of courses in the country. In the United States, however, the credit requirement for CK courses is only 26%, which is lower than the credit requirement for GK courses by 13%. This clear contrast between the two countries is congruent with evidence from previous research, which found that mathematics teacher knowledge in the United States is seen as developing from exposure to a wide range of academic disciplines (Roth, 1999), whereas mathematics teacher preparation programs in South Korea usually locate the centrality of teacher knowledge in academic expertise in mathematics (Kwon, 2004; Pang, 2003). Similarly, Table 7.4 shows that only slightly more than one-third of US curricular requirements pertain to content-related components (CK+PCK). This proportion of credit requirements is significantly lower than in South Korea, where more than three-fifths of the curricular requirements are for content-related components.

This pattern is also consistent with the ideas embedded in the official educational aims analyzed in Fig. 7.1. Prospective teachers' knowledge gained from situated experiences is emphasized considerably more in the United States than in South Korea. As Table 7.3 also shows, the percentage of credit requirements allocated to field-based experiences is less than 3% in South Korea while it is almost 13% in the United States. Despite the larger standard deviation among US programs than among the South Korean ones, the difference in curricular emphasis given to field-based experiences between the two countries is statistically significant. This

Table 7.4 Percentages of credit requirements allocated to content-related (CK + PCK) and pedagogy-related (PK + PCK + FLD) components

	South Korea ($n = 21$)		United States ($n = 28$)			
	Mean	*SD*	Mean	*SD*	t	p-Value (two tailed)
Content	61.64	10.08	35.23	11.42	8.42	<.001
Pedagogy	31.69	7.05	34.29	12.93	−.83	.409

Note: For each t-test, equality of variances between the two groups was assumed because Levene's test was statistically insignificant

significant difference is quite suggestive, considering that curricular emphasis given to pedagogy-related components (PK + PCK + FLD) is not very different between the two countries. Although the percentage of credit requirements allocated to pedagogy-related components is slightly higher in the United States than in South Korea by 2.6%, the *t*-test result in Table 7.4 indicates that this difference is not statistically significant. That is, despite the fact that the South Korean and US programs place a similar amount of curricular emphasis on pedagogy-related components as a whole, the South Korean programs tend to place most of their curricular emphasis on on-campus coursework rather than on field-based experiences, among pedagogy-related components. These findings partly account for why mathematics teacher education in South Korea is often criticized for its weak emphasis on pedagogy-related practical knowledge (i.e., math teachers as math experts rather than practitioners) (Pang, 2003), while mathematics teachers' weak CK has long been criticized in the United States (i.e., math teachers as practitioners rather than math experts) (Ball, Lubienski, & Mewborn, 2001; Wise, 1999).

Implications for Policy and Practice

Scholars from both countries have reached much agreement on the requirements for becoming a teacher in terms of teacher qualifications, the structure and content of undergraduate teacher education programs, and the need to strengthen teacher knowledge. In both countries, a person who wants to be a secondary mathematics teacher must complete an approved program, earn acceptable grades, pass required tests, and complete some student teaching. He or she must take courses not only in mathematical subject matter but also in educational theory and pedagogy; in addition, student teaching experiences are required. The data presented here show that there are common qualification systems and considerable agreement on the integral components of teacher preparation programs that shape preservice teachers' knowledge for teaching in both countries. In terms of the knowledge expectations expressed in official educational aims of teacher preparation programs, prospective teachers in both countries are expected to have competence in both theory and practice as fundamental bases for teaching. As many studies suggest, it seems that there are some common international models of teacher quality and teacher qualifications that have been rationalized through professional discourses at an international level (Akiba et al., 2007; Cha, 2002). Such similarities can lead to constructive international dialogue on mathematics teacher preparation, because similarities allow us to find some common grounds from which we can see differences more clearly without creating "the other" (Abu-Lughod, 1991).

Still, we should not underestimate the profound peculiarities that have been shaped within the sociohistorical contexts of each country. As discussed earlier, there are differences between South Korea and the United States in curricular emphasis, as well as in many aspects of the educational aims of teacher preparation programs. Despite variations within each country, it appears that there are some general national patterns of how to organize teacher preparation. For example, the widely acknowledged assumption that the United States is a pluralist and democratic society has led teacher education programs to emphasize diversity and equity issues to integrate people from different cultural backgrounds. A wide range of general knowledge and field-based knowledge have become crucial for teachers to deal with such issues in the classroom. Existing research about preparing teachers for dealing with diversity in educational settings has emphasized the importance of teacher candidates' multicultural awareness and their broader perspectives on diversity and equity issues in different social contexts (Banks, 2007; Zeichner et al., 1998).

The situation is rather different in South Korea, where the nation has less cultural diversity and where societal expectations for teachers lead to the assumption that teachers should have rigorous content knowledge to improve students' cognitive achievement. Teachers are expected to be "scholar teachers" (Leung, 2001) with high academic competence in mathematics. Considering the sociohistorical context of South Korea, where education has long been considered the only available avenue toward upper social mobility and success in society (Bae, 1991; Robinson, 1994; Sorensen, 1994), such a high expectation is not surprising. This is especially true from the perspective of most parents in South Korea, who believe that the future success or failure of their children depends heavily on teachers and their teaching.

Given such differences, many scholars from South Korea and the United States have given different recommendations for teacher preparation programs with regard to prospective teachers' adequate knowledge. Scholarly and policy debates in the United States have emphasized resolving inadequate subject matter preparation of preservice teachers in teacher preparation programs (Ball et al., 2001; Ma, 1999; Schmidt et al., 2007). A general consensus appears to be emerging on the need to ensure the depth and breadth of prospective mathematics teachers' subject matter knowledge, because many future teachers have been ill prepared for teaching mathematics for understanding (Ball, 1990). Ma (1999) suggested that teachers' content knowledge should be fostered in the United States because "low-quality school mathematics education and low-quality teacher knowledge of school mathematics reinforce each other" (p. 145). As many studies have stressed, policy in the United States should take seriously the strong relationship between teachers' subject matter preparation and teaching performance (Goulding, Rowland, & Barber, 2002; McEwan & Bull, 1991; Wilson et al., 2001).

In the meantime, many Korean scholars have recently drawn attention to the need of preservice teachers for field-based experience, which would allow them to have "knowledge in practice" (Cochran-Smith & Lytle, 1999) and to connect practice with the theory that they learn in teacher preparation programs (Jang, 2001; Jin, Kwak, & Jo, 2007; Shin & Lee, 2004). Preservice teachers in South Korea are currently not provided with enough opportunities to learn from practice through teacher preparation programs. In this situation, prospective teachers often have difficulty connecting college mathematics (i.e., what they have learned in teacher preparation programs) with school mathematics (i.e., what they will teach on the job) because the theory-based collegiate mathematics emphasized in South Korea's teacher preparation programs cannot be used as it is in their future teaching. Similarly, preservice teachers in South Korea are provided with insufficient opportunities to learn the "mathematics of mathematics teachers" (Park, 1998) or to experience the actual challenges which they might encounter in the classroom when they begin teaching. The situated cognition view of teacher knowledge has received increased attention, with many mathematics teacher educators in South Korea claiming that prospective teachers need more systematic preparation for teaching in schools, including more clinical experiences and opportunities to learn useful skills and methods in connection with both theory and practice in education (Kim, 2002; Shin & Lee, 2004).

Conclusion

Scrutinizing both national differences and transnational commonalities in knowledge expectations for mathematics teachers in South Korea and the United States, we have argued that understanding both is useful in trying to capture the whole "glo*c*al" (Robertson, 1995) picture of culturally contextualized and semantically decontextualized dimensions of teacher knowledge. We use the term glo*c*al, combining the global and the local, in order to emphasize that they frequently intersect rather than constituting a polarized dichotomy. Much of the comparative research on teacher knowledge seems to revolve around whether the fundamental character of teacher knowledge is based on national patterns or transnational commonalities. We, however, argue that these seemingly different characters of teacher knowledge are not so contradictory. Rather, they describe different dimensions of the same phenomenon; our data suggest that societal expectations for teacher knowledge reflect a set of both nationally constructed cultural scripts and transnationally shared assumptions and beliefs about teaching and learning.

Attending to both culturally contextualized and semantically decontextualized dimensions of teacher knowledge allows us to have a more balanced comparative perspective, from which we can better assess current conditions of teacher education and better inform policy makers on how to improve teacher education. Indeed, international comparisons can help us move "beyond the familiar ... to see how easily we fall into the trap of thinking only in locally bounded ways that restrict the development of our theories and the reform of our practices" (Blömeke & Paine, 2008). Cross-national differences in models of teacher knowledge and teacher education can be the intellectual reservoir of many different alternatives in both academic debates and policy discourses. At the same time, we have to look carefully at transnational similarities as well. This is because policy and practice would hardly be transferable to another national context without considerable similarities; differences can be given meaning only when they are grounded in some shared similarities. Educational policy makers can benefit from such a balanced comparative perspective, as it allows for constructive international dialogue that may help them determine how to further enrich teacher education programs without ignoring either profound differences in sociohistorical contexts or important commonalities in epistemic models of teacher education across countries.

References

Abu-Lughod, L. (1991). Writing against culture. In R. Fox (Ed.), *Recapturing anthropology: Working in the present* (pp. 137–162). Santa Fe, NM: School of American Research Press.

Akiba, M., LeTendre, G. K., & Scribner, J. P. (2007). Teacher quality, opportunity gap, and national achievement in 46 countries. *Educational Researcher, 36*(7), 369–387.

Alexander, R. J. (2000). *Culture and pedagogy: International comparisons in primary education*. Malden, MA: Blackwell.

Anderson-Levitt, K. M. (2002). *Teaching cultures: Knowledge for teaching first grade in France and the United States*. Cresskill, NJ: Hampton.

Annett, A. (2001). Social fractionalization, political instability, and the size of government. *IMF Staff Papers, 48*(3), 561–592.

Bae, C.-K. (1991). Education: Top reason behind rapid growth. *Koreana, 5*(2), 56–68.

Baker, D. P., & LeTendre, G. K. (2005). *National differences, global similarities: World culture and the future of schooling*. Stanford, CA: Stanford University Press.

Ball, D. L. (1990). The mathematical understandings that prospective teachers bring to teacher education. *Elementary School Journal, 90*(4), 449–466.

Ball, D. L., Lubienski, S., & Mewborn, D. (2001). Research on teaching mathematics: The unsolved problem of teachers' mathematics knowledge. In V. Richardson (Ed.), *Handbook of research on teaching* (4th ed., pp. 433–456). New York, NY: Macmillan.

Ball, D. L., Thames, M. H., & Phelps, G. (2008). Content knowledge for teaching: What makes it special? *Journal of Teacher Education, 59*(5), 389–407.

Banks, J. A. (2007). *Educating citizens in a multicultural society* (2nd ed.). New York, NY: Teachers College Press.

Berger, P. L., & Luckmann, T. (1967). *The social construction of reality: A treatise in the sociology of knowledge*. New York, NY: Anchor Books.

Blömeke, S., & Paine, L. W. (2008). Getting the fish out of the water: Considering benefits and problems of doing research on teacher education at an international level. *Teaching and Teacher Education, 24*(8), 2027–2037.

Cha, Y.-K. (2002). Chodeung gyosa gyoyukgwajeong-ui jedo-jeok giwon mit segye-jeok hwaksan-e daehan bigyo yeongu [Institutional origins and the globalization of the primary teacher education program: A cross-national study]. *Korean Journal of Comparative Education, 12*(1), 113–128.

Cha, Y.-K., Dawson, W. P., & Ham, S.-H. (2010). *Global civil society and the cross-national adoption of multicultural education policies*. Paper presented at the XIV World Congress of the World Council of Comparative Education Societies, Istanbul, Turkey.

Clewell, B. C., & Forcier, L. B. (2001). Increasing the number of mathematics and science teachers: A review of teacher recruitment programs. *Teaching and Change, 8*(4), 331–361.

Cochran-Smith, M., & Lytle, S. L. (1999). Relationships of knowledge and practice: Teacher learning in communities. *Review of Research in Education, 24*, 249–305.

Cummings, W. K. (2003). *The institutions of education*. Oxford: Symposium Books.

Darling-Hammond, L., & Bransford, J. (2005). *Preparing teachers for a changing world: What teachers should learn and be able to do*. San Francisco, CA: Jossey-Bass.

Desimone, L. M., Smith, T., Baker, D. P., & Ueno, K. (2005). Assessing barriers to the reform of U.S. mathematics instruction from an international perspective. *American Educational Research Journal, 42*(3), 501–535.

Donoghue, E. F. (2006). The education of mathematics teachers in the United States: David Eugene Smith, early twentieth-century pioneer. *Paedagogica Historica, 42*(4), 559–573.

Drori, G. S., Meyer, J. W., Ramirez, F. O., & Schofer, E. (Eds.). (2003). *Science in the modern world polity: Institutionalization and globalization*. Stanford, CA: Stanford University Press.

Fairclough, N. (2003). *Analyzing discourse: Textual analysis for social research*. London: Routledge.

Feistritzer, C. E. (2008). Introduction and overview. In C. E. Feistritzer (Ed.), *Building a quality teaching force: Lessons learned from alternate routes* (pp. 1–9). Upper Saddle River, NJ: Pearson Education.

Fernandez, C., & Yoshida, M. (2004). *Lesson study: A Japanese approach to improving mathematics teaching and learning*. Mahwah, NJ: Lawrence Erlbaum Associates.

Fraser, J. W. (2007). *Preparing America's teachers: A history*. New York, NY: Teachers College Press.

Goulding, M., Rowland, T., & Barber, P. (2002). Does it matter? Primary teacher trainees' subject knowledge in mathematics. *British Educational Research Journal, 28*(5), 689–704.

Ham, S.-H., & Cha, Y.-K. (2009). Positioning education in the information society: The transnational diffusion of the information and communication technology curriculum. *Comparative Education Review, 53*(4), 535–557.

Hess, F. M., & Petrilli, M. J. (2006). *No child left behind: Primer*. New York: Peter Lang.

Hiebert, J., Gallimore, R., & Stigler, J. W. (2002). A knowledge base for the teaching profession: What would it look like and how can we get one? *Educational Researcher, 31*(5), 3–15.

Hill, H. C., Rowan, B., & Ball, D. L. (2005). Effects of teachers' mathematical knowledge for teaching on student achievement. *American Educational Research Journal, 42*(2), 371–406.

Ingersoll, R. M. (Ed.). (2007). *A comparative study of teacher preparation and qualifications in six nations*. Philadelphia: Consortium for Policy Research in Education.

Jang, U.-H. (2001). Gyosa gyoyuk program gyoyukgwajeong-ui munjegeom-gwa gaeseonbangahn [Problems in curricula for teacher education programs and suggestions for improvement]. *Educational Research, 21*(1), 65–79.

Jeynes, W. (2008). What we should and should not learn from the Japanese and other East Asian education systems. *Educational Policy, 22*(6), 854–874.

Jin, S., Kwak, D.-J., & Jo, D.-J. (2007). Hyeonjang jungsim gyosagyoyuk yeongu: Yeongguk Oxford daehakgyo-wa Canada Queens daehakgyo-ui sarye-reul jungsim-euro [A study on field-based teacher education programs: Reflective practices at Oxford University, U.K., and Queens University, Canada]. *Korean Journal of Comparative Education, 17*(3), 241–268.

Kang, N.-H., & Hong, M. (2008). Achieving excellence in teacher workforce and equity in learning opportunities in South Korea. *Educational Researcher, 37*(4), 200–207.

Kennedy, M. M., Ahn, S., & Choi, J. (2008). The value added by teacher education. In M. - Cochran-Smith, S. Feiman-Nemser, & J. McIntyre (Eds.), *Handbook of research on teacher education: Enduring issues in changing contexts* (3rd ed., pp. 1249–1273). Mahwah, NJ: Lawrence Erlbaum Associates.

Kim, Y.-K. (2002). Suhakgyosa yangseong-eul wihan gyoyuk-ui isang-kwa hyeonsil [Educational ideals and realities and mathematics teacher education]. *Journal of Korean Society of Mathematics Education Series E: Communications of Mathematical Education, 14*, 453–468.

Kwon, O.-N. (2004). *Mathematics teacher education in Korea*. Paper presented at the International Congress on Mathematical Education (ICME-10), Copenhagen, Denmark.

Labaree, D. (2008). An uneasy relationship: The history of teacher education in the university. In M. Cochran-Smith, S. Feiman-Nemser, & J. McIntyre (Eds.), *Handbook of research on teacher education: Enduring issues in changing contexts* (3rd ed., pp. 290–306). Washington, DC: Association of Teacher Educators.

Lee, B.-H., Kim, J.-S., Ahn, J.-Y., Jung, J. H., Nam, K. G., Rhyu, J. S., et al. (2005). *Education and the labor market in Korea*. Seoul: Korea Labor Institute. Retrieved August 15, 2009, from http://kli.re.kr/.

LeTendre, G. K., Baker, D. P., Akiba, M., Goesling, B., & Wiseman, A. W. (2001). Teachers' work: Institutional isomorphism and cultural variation in the U.S., Germany, and Japan. *Educational Researcher, 30*(6), 3–15.
Leung, F. K. S. (2001). In search of an East Asian identity in mathematics education. *Educational Studies in Mathematics, 47*(1), 35–51.
Lortie, D. C. (2002). *Schoolteacher: A sociological study* (2nd ed.). Chicago, IL: University of Chicago Press.
Ma, L. (1999). *Knowing and teaching elementary mathematics: Teachers' understanding of fundamental mathematics in China and the United States*. Mahwah, NJ: Lawrence Erlbaum Associates.
McEwan, H., & Bull, B. (1991). The pedagogic nature of subject matter knowledge. *American Educational Research Journal, 28*(2), 316–334.
Meyer, J. W., & Ramirez, F. O. (2000). The world institutionalization of education. In J. Schriewer (Ed.), *Discourse formation in comparative education* (pp. 111–131). Frankfurt am Main: Peter Lang.
Meyer, J. W., & Rowan, B. (1983). The structure of educational organizations. In J. W. Meyer & W. R. Scott (Eds.), *Organizational environments: Ritual and rationality* (pp. 71–97). Beverly Hills, CA: Sage.
Mitchell, K. J., Robinson, D. Z., Plake, B. S., & Knowles, K. T. (2001). *Testing teacher candidates: The role of licensure tests in improving teacher quality*. Washington, DC: National Academies Press.
National Center for Education Statistics. (2009). *State education reforms*. Retrieved August 15, 2009, from http://nces.ed.gov/programs/statereform/tab3_1.asp.
National Council for Accreditation of Teacher Education [NCATE]. (2008). *Professional standards for the accreditation of teacher preparation institutions*. Washington, DC: NCATE.
Organization for Economic Co-operation and Development [OECD]. (2005). *Teachers matter: Attracting, developing, and retaining effective teachers*. Paris, France: OECD.
Oser, F. K. (1994). Moral perspectives on teaching. *Review of Research in Education, 20*(1), 57–127.
Paine, L. (1990). Teacher as virtuoso: A Chinese model for teaching. *Teachers College Record, 92* (1), 49–81.
Pang, J. S. (2003). Suhakgyosa hakseup-kwa jeonmunseong sinjang-e kwanhan sogo [Mathematics teacher learning and professional development]. *Journal of Educational Research in Mathematics, 13*(2), 143–157.
Park, H.-S. (1998). *Suhakgyosa-ui suhak* [*Mathematics for mathematics teachers*]. Paper presented at the ICMI-East Asia Regional Conference on Mathematics Education 1, Cheongwon, South Korea.
Phillips, D. (2004). Toward a theory of policy attraction in education. In G. Steiner-Khamsi (Ed.), *The global politics of educational borrowing and lending* (pp. 54–67). New York, NY: Teachers College Press.
Putnam, R. T., & Borko, H. (2000). What do new views of knowledge and thinking have to say about research on teacher learning? *Educational Researcher, 29*(1), 4–15.
Robertson, R. (1995). Glocalization: Time-space and homogeneity-heterogeneity. In M. Featherstone, S. Lash, & R. Robertson (Eds.), *Global modernities* (pp. 23–44). London: Sage.
Robinson, J. (1994). Social status and academic success in South Korea. *Comparative Education Review, 38*(4), 506–530.
Roth, R. A. (1999). University as context for teacher development. In R. A. Roth (Ed.), *The role of the University in the preparation of teachers* (pp. 180–195). Philadelphia, PA: Falmer Press.
Schmidt, W. H., Tatto, M. T., Bankov, K., Blömeke, S., Cedillo, T., Cogan, L., et al. (2007). *The preparation gap: Teacher education for middle school mathematics in six countries (MT21 report)*. East Lansing, MI: Center for Research in Mathematics and Science Education/ Michigan State University.

Schriewer, J. (2003). Globalization in education: Process and discourse. *Policy Futures in Education, 1*(2), 271–283.
Sedlak, M. W. (1989). Let us go and buy a schoolmaster. In D. Warren (Ed.), *American teachers: Histories of a profession at work* (pp. 257–290). New York, NY: Macmillan.
Seth, M. J. (2002). *Education fever: Society, politics, and the pursuit of schooling in South Korea*. Honolulu, HI: University of Hawaii Press.
Shin, H.-S. (2009). Gyowon yangseong cheje-ui gaepyon banghyang-gwa jeonryak-ui tamsaek [Policy directions and strategies to reform pre-service teacher education]. *Journal of Korean Education, 36*(3), 53–78.
Shin, H. Y., & Lee, J. W. (2004). Suhakgyosa-ui jisik-e kwanhan yeongu [A study on mathematics teacher knowledge]. *Journal of Korean Society of Mathematics Education Series E: Communications of Mathematical Education, 18*(1), 297–308.
Shulman, L. S. (1986). Those who understand: Knowledge growth in teaching. *Educational Researcher, 15*(2), 4–14.
Shulman, L. S. (1987). Knowledge and teaching: Foundations of the new reform. *Harvard Educational Review, 57*(1), 1–22.
Sorensen, C. W. (1994). Success and education in South Korea. *Comparative Education Review, 38*(1), 10–35.
Stigler, J. W., & Hiebert, J. (1999). *The teaching gap: Best ideas from the world's teachers for improving education in the classroom*. New York, NY: Free Press.
Tatto, M. T. (2008). Teacher policy: A framework for comparative analysis. *Prospects, 38*(4), 487–508.
Tatto, M. T., Schwille, J., Senk, S., Ingvarson, L., Peck, R., & Rowley, G. (2008). *Teacher education and development study in mathematics (TEDS-M): Conceptual framework*. East Lansing, MI: Teacher Education and Development International Study Center/College of Education/Michigan State University.
Wang, J., & Lin, E. (2005). Comparative studies on U.S. and Chinese mathematics learning and the implications for standards-based mathematics teaching reform. *Educational Researcher, 34*(5), 3–13.
Wayne, A. J., & Youngs, P. (2003). Teacher characteristics and student achievement gains: A review. *Review of Educational Research, 73*(1), 89–122.
Weick, K. E. (1976). Education organizations as loosely coupled systems. *Administrative Science Quarterly, 21*(1), 11–19.
Wilson, S. M., Floden, R. E., & Ferrini-Mundy, J. (2001). *Teacher preparation research: Current knowledge, gaps, and recommendations*. Seattle, WA: Center for the Study of Teaching and Policy/University of Washington.
Wise, A. E. (1999). Moving toward performance-based accreditation systems for the teaching profession. In R. A. Roth (Ed.), *The role of the university in the preparation of teachers* (pp. 151–165). Philadelphia, PA: Falmer Press.
Wiseman, A. W., & Baker, D. P. (2006). The symbiotic relationship between empirical comparative research on education and neo-institutional theory. In D. P. Baker & A. W. Wiseman (Eds.), *The impact of comparative education research on institutional theory* (pp. 1–26). Oxford: Elsevier.
Woo, Y.-J., & Ahn, H.-S. (2006). Geundae-jeok gyowon yangseong jedo-ui byeoncheon-gwa sabeomdaehak-ui seolrip [Teacher training system from the Hanseong Normal School to the college of education of Seoul National University]. *Asian Journal of Education, 7*(4), 187–215.
Zeichner, K. M., Grant, C., Gay, G., Gillette, M., Valli, L., & Villegas, A. M. (1998). A research informed vision of good practice in multicultural teacher education: Design principles. *Theory into Practice, 37*(2), 163–171.

Chapter 8
Pre-service Teacher Training for Secondary School Mathematics in Japan and Korea

Masataka Koyama and Hee-chan Lew

Abstract It is well recognized that there are many similarities between Japan and Korea in terms of school mathematics curriculum policy, students' high achievement in international assessments such as the Trends in International Mathematics and Science Study (TIMSS) and the Programme for International Student Assessment (PISA), and negative attitudes toward school mathematics (Wong, Koyama, & Lee, 2013). On the other hand, not much has been written about the guidelines and contents of undergraduate mathematics education curricula in the two education systems, including teaching practicums for pre-service secondary school mathematics teachers. In this chapter, we attempt to fill this gap by identifying similarities and differences in Japanese and Korean approaches to supporting pre-service teachers' awareness and knowledge development in the department of mathematics education. We will mainly focus on two aspects of pre-service teacher training for secondary school mathematics: (a) prescribed conditions for teaching certificates and (b) undergraduate curricula for training secondary school mathematics teachers in Japan and Korea. To illustrate these aspects we examine Hiroshima University in Japan and Korea National University of Education in Korea, because their pre-service programs are typical models of each country's mathematics teacher training programs. We will identify similarities and differences between the programs, and discuss some issues to improve the guidelines and contents of undergraduate curriculum in training pre-service teachers for secondary school mathematics. Further comparative study on pre-service mathematics teacher training can be stimulated through more in-depth international analyses of these elements.

M. Koyama (✉)
Hiroshima University, Higashi-Hiroshima, Japan
e-mail: mkoyama@hiroshima-u.ac.jp

H.-c. Lew
Korea National University of Education, Cheongju, Korea

J.-W. Son et al. (eds.), *What Matters? Research Trends in International Comparative Studies in Mathematics Education*, Research in Mathematics Education, DOI 10.1007/978-3-319-51187-0_8

Keywords Mathematics teacher training • Secondary school • Teaching certificate • Teacher employment examination • Undergraduate curriculum • Japan • Korea

Introduction

Education is a key factor in understanding Japanese and Korean societies. Both societies have substantial interest in children's education, so parents and governments make great investments into their children's education to provide them with a better life. The two societies likewise have a long tradition of treating teachers with great respect, perhaps influenced by Confucian heritage which emphasizes education and devotion to one's family. Both countries' universities also feature high-level teacher training programs, so teachers are believed to be well equipped with pedagogical and subject content knowledge. Such keen interest in education and the quality of teachers is considered one of the main reasons why Japanese and Korean students perform so well in such international comparative studies as the Trends in International Mathematics and Science Study (TIMSS), the Third International Mathematics and Science Study-Repeat (TIMSS-R), and the Programme for International Student Assessment (PISA) (Mullis, Martin, Foy, & Arora, 2012; National Institute for Educational Policy Research, 2001, 2004; Organization for Economic Co-operation and Development, 2010). But it has been reported that Japanese and Korean students have a low interest, value, and confidence in mathematics despite ranking in the highest grades on international mathematics tests (Mullis et al., 2012). Believing that this mismatch arises from traditional teaching methods such as memorizing and solving problems without understanding of mathematical concepts, the Ministries of Education (MOEs) of the two countries have carried out various programs and policies to improve the quality and capability of teachers, under the philosophy that the quality of education cannot exceed the quality of teachers.

This chapter characterizes these efforts by looking at curricula of the department of mathematics education of Hiroshima University in Japan and Korea National University of Education in Korea, both of which are among the top education programs in their respective countries. The two universities were selected as typical models of mathematics teacher training in Japan and Korea, based on their pre-service teacher training programs as well as their research on teacher education.

Prescribed Conditions for Teaching Certificates

In both Japan and Korea, the pre-service teacher training for teaching certificates is undertaken by universities which have 4-year programs approved by the MOE in partnership with university-attached schools, or with local schools if the university has no attached school. Teaching practicums are mainly undertaken at university-attached schools or local schools. Training for secondary school teachers is provided not only at national universities of teacher education/faculties of education but also at other noneducational national, public, and private universities/faculties with certificate coursework approved by the MOE. This section describes the prescribed conditions for teaching certificates in Japan and Korea, respectively.

Required Credits for Teaching Certificates

Table 8.1 shows the prescribed conditions for secondary school teaching certificates according to the regulations for teaching certificates in Japan, in terms of the minimum credits required in 4-year undergraduate program at MOE-authorized universities (Koyama, 2008). Additionally, to earn teaching certificates university students have to take courses in four areas: the Constitution of Japan (2 credits), Health and Sports (2 credits), Foreign Language Communication (2 credits), and Information Studies (2 credits), all in the category of Liberal Arts. In terms of the prescribed conditions, there is a little difference between lower and upper secondary school teaching certificates. University students pursuing the lower secondary school teaching certificate need more credits in Curriculum and Teaching Methodology courses, including a course in Teaching Methodology of Moral Education and a Teaching Practicum in the category of Teaching Profession, compared to those pursuing the upper secondary school teaching certificate.

Table 8.2 shows the undergraduate curriculum framework for the program in secondary mathematics education at Hiroshima University (HU) in Japan. The framework consists of two main education categories: Liberal Arts Education (40 credits) and Specialized Education (88 credits). The framework fulfils the requirements shown in Table 8.1 for both lower and upper secondary school teaching certificates of mathematics. Specialized Education is divided into four

Table 8.1 Prescribed conditions for secondary school teaching certificates in Japan

Category	Course	Lower secondary	Upper secondary
Teaching profession	Significance of Teaching Profession	2	2
	Basic Theory of Education	6	6
	Curriculum and Teaching Methodology	12	6
	Student Counselling	4	4
	Teaching Practicum	5	3
	Practical Seminar for Teaching Profession	2	2
	Subtotal	31	23
School subjects	Algebra, Geometry, Analysis, Probability and Statistics, Computer Science	20	20
Electives	Courses in Teaching Profession or School Subjects	8	16

Notes: In general, earning 1 credit requires attending fifteen 45-min classes

Table 8.2 Credits required to complete the program in secondary mathematics education at HU

<table>
<tr><th colspan="4">Category of education and courses</th><th colspan="3">Credits</th></tr>
<tr><td rowspan="9">Liberal arts education</td><td rowspan="3">Core Courses</td><td colspan="2">Introductory Seminar for First Year Students</td><td colspan="2">2</td><td rowspan="9">40</td></tr>
<tr><td colspan="2">Peace Science Courses</td><td colspan="2">2</td></tr>
<tr><td colspan="2">Integrated Courses</td><td colspan="2">6</td></tr>
<tr><td rowspan="5">Common Courses</td><td rowspan="2">Foreign Languages</td><td>English</td><td colspan="2">8</td></tr>
<tr><td>Others</td><td colspan="2">4</td></tr>
<tr><td colspan="2">Information Courses</td><td colspan="2">2</td></tr>
<tr><td colspan="2">Area Courses</td><td colspan="2">6</td></tr>
<tr><td colspan="2">Health and Sports Courses</td><td colspan="2">2</td></tr>
<tr><td colspan="3">Foundation Courses</td><td colspan="2">8</td></tr>
<tr><td rowspan="9">Specialized education</td><td rowspan="6">Basic Specialized Courses</td><td colspan="2">Mathematics Education</td><td>8</td><td rowspan="6">26</td><td rowspan="9">88</td></tr>
<tr><td colspan="2">Algebra</td><td>4</td></tr>
<tr><td colspan="2">Geometry</td><td>4</td></tr>
<tr><td colspan="2">Analysis</td><td>4</td></tr>
<tr><td colspan="2">Probability and Statistics</td><td>4</td></tr>
<tr><td colspan="2">Computer Science</td><td>2</td></tr>
<tr><td colspan="3">Specialized Courses</td><td colspan="2">20</td></tr>
<tr><td colspan="3">Elective Specialized Courses and Free Elective Courses</td><td colspan="2">36</td></tr>
<tr><td colspan="3">Graduation Research</td><td colspan="2">6</td></tr>
<tr><td colspan="4">Total credits for graduation</td><td colspan="3">128</td></tr>
</table>

categories: Basic Specialized courses (26 credits), Specialized courses (20 credits), Elective Specialized courses and Free Elective courses (36 credits), and Graduation Research (6 credits).

In Specialized Education, students take specialized (basic) courses from professors in the department of mathematics education in order to study both mathematics education and mathematics. Additionally, they take elective specialized courses to study the significance of teaching as a profession, basic theories of education, and student counselling, from professors in the department of education and the department of psychology. Teaching Practicum courses are included in this category. Depending on their interests, students can also take other specialized courses in either the faculty of education or the faculty of science. Finally, they have to do their own graduation research on mathematics education and submit a graduation thesis.

At HU, the department of mathematics education in the faculty of education accepts about 22 undergraduate students every year through highly competitive entrance examinations. Almost all students graduate with both lower and upper secondary mathematics teaching certificates, though in principle they can graduate the program without earning any teaching certificate. On the other hand, students of mathematics in the faculty of science and the faculty of integrated arts and sciences can get both or either certificate(s) when they take some Teaching Profession courses provided by the faculty of education for teaching certificates. Therefore,

Table 8.3 Prescribed conditions for secondary school teaching certificates in Korea

Category	Course	Minimum credits required
Teaching profession	Introduction to Education (2) Educational Philosophy and History of Education (2) Theory of Curriculum (2) Educational Evaluation (2) Educational Methodology and Technology (2) Educational Psychology (2) Educational Sociology (2) Teaching Practicum (4)	22
Basic mathematics subjects	Theory of Mathematics Education (3) Number Theory (3) Complex Analysis (3) Advanced Calculus (3) Linear Algebra (3) Modern Algebra (3) Differential Geometry (3) Introduction to Geometry (3) Topology (3) Probability and Statistics (3) Combinatorics and Graph Theory (3)	21
Elective Mathematics Subjects	Courses in other Mathematics Subjects besides Basic Mathematics Subjects	30

Notes: In general, earning 1 credit requires attending fifteen 50-min classes

about 60 students earn one or both certificates every year at HU. Although these certificates are issued by the Hiroshima Prefecture Board of Education, they are valid anywhere in Japan.

In Korea, secondary mathematics teachers are trained mainly in the department of mathematics education at colleges of education, and partially in the department of mathematics or the department of statistics at comprehensive universities and graduate schools of education. In general, the curriculum for all pre-service secondary teachers, including mathematics teachers, sets the graduation credit requirements between 130 and 150 credits of liberal arts courses, major courses, and free elective courses (Cho & Choe, 2012). Table 8.3 shows the major courses required by the Korean MOE for secondary school teaching certificates. Course categories include the Teaching Profession (22 credits) and Mathematics Subjects (51 credits), which includes seven of the eleven Basic Mathematics Subjects: Theory of Mathematics Education, Number Theory, Complex Analysis, Advanced Calculus, Liner Algebra, Modern Algebra, Differential Geometry, Introduction to Geometry, Topology, Probability and Statistics, and Combinatorics and Graph Theory.

The main characteristic of the Korean curriculum for secondary mathematics teachers is a strong emphasis on subject matter knowledge such that there is almost no curricular difference between the department of mathematics and the department of mathematics education. This means that most of the educational institutions attempt to maintain a high level of mathematics knowledge among future teachers, and that Korean secondary mathematics teachers are expected to possess strong mathematical knowledge. However, recently there have been curricular reforms in the pre-service teacher institutions, aimed at enhancing the correlation between teachers' subject matter knowledge of school mathematics and their actual teaching practice (Kim et al., 2010). A number of universities offer programs to develop the teaching competencies of pre-service teachers and to make connections between theories and teaching practice. Through these programs, pre-service teachers are expected to develop their expertise in implementing curriculum, by applying and integrating the core structure of their subject matter knowledge to school curriculum (Kim et al., 2010).

Table 8.4 shows the undergraduate curriculum framework for the program in secondary mathematics education at Korea National University of Education (KNUE) in Korea. The framework consists of five main education categories: Liberal Arts Education (21 credits), Teaching Profession courses (24 credits), Mathematics Major courses (57 credits), Free Elective courses (38 credits), and a Dormitory Education Program (Pass/Fail). Liberal Arts Education consists of three types of programs peculiar to KNUE as a comprehensive university to cover all school levels from kindergarten to senior high school. Teaching Profession courses also include three types of programs to bring together theory and practice in the teaching profession. A unique aspect of the undergraduate curriculum for teaching certificates in Korea is that pre-service teachers have to volunteer for a minimum of 60 hours in their educational communities, in addition to the regular teaching

Table 8.4 Credits required for graduation in secondary mathematics education at KNUE

Category of education and courses			Credits: Major	Credits: Double major	
Liberal arts education	KNUE vision		3		
	KNUE Special	Leadership	6		
		Character			
		Globalization Manner			
	KNUE general		12		
Teaching profession	Theory of Teaching Profession		12		
	Basic Knowledge of Teaching Profession		6		
	Teaching Practicum	School Teaching Practicum	4	2	
		Educational Service	2		
Mathematics major courses	Mathematics Education	Compulsory	3	3	57
		Elective	6	6	
	Mathematics	Compulsory	18	18	
		Elective	30	30	
Free elective			38		
Dormitory Education Program			P/F		
Total credits for graduation			140	59	

practicum performed at the attached schools or the partner schools. This regulation was introduced to enhance pre-service teachers' general knowledge of the educational community. In the mathematics major courses, students in the department of secondary mathematics education take at least 57 credit hours of specialized courses from professors in the department of mathematics education. Additionally, they have to take 38 more credits from any area in the curriculum; usually they take mostly mathematics major courses to increase their abilities in pedagogical content knowledge and subject matter knowledge.

Table 8.4 also shows the undergraduate curriculum framework for the program in secondary mathematics education for double majors who want to earn certification in both mathematics and another subject like physics or technical education. All students who want to take mathematics as a double major have to take Mathematics Major courses (57 credits) and Teaching Practicum (2 credits) within the program in secondary mathematics education, besides the courses for their major subjects. Students who want double majors in mathematics and another subject do not need to take the Liberal Arts education, Teaching Profession, Free Elective, and Dormitory Education Program twice, except for taking 2 credits of Teaching Practicum for their minor subject.

The Teacher Employment Examination

In Japan, public school teachers are local prefectural or municipal public officials, and are employed by the respective local prefectural or municipal boards of education in which the schools are located. Public school teachers are selected for employment through competitive examinations under the provision of the Local Public Service Personnel Law. This means that students with teaching certificates cannot always become teachers after graduating from a university. In the summer, usually July and August, fourth-year students who want be teachers have to take two-step employment examinations for lower secondary or upper secondary school teachers in a certain prefecture/city. The first step is usually done through paper and pencil tests in psychology, liberal arts, education laws and regulations, and their major school subjects (here, mathematics education and mathematics). After passing the first step, students must take the second step examination, which usually consists of micro teaching, group discussion, and interviews. If students pass the second test, they will be able to become school teachers starting in April of the next academic year. If students unfortunately fail the employment examinations, they have to wait and take them again the next year. In the case of private schools, the institution itself selects teachers using its own employment examination.

The test items and discussion/interview questions for teacher employment examinations are developed by the individual local board of education without any involvement of university professors. The number of newly employed teachers is not stable but changes every year, based on prefecture/city recruitment and target figures deduced from several relevant factors, such as the increase/decrease in the number of school students and the current age profile of teachers in its region.

Generally speaking, the employment rate of university graduates as school teachers is low. Examination pass rates are available, but the data does not specify multiple tries. For example, the AY 2014 data for the employment examinations for secondary school mathematics teachers in Hiroshima Prefecture and Hiroshima City shows that 28 out of 129 examinees passed the second test for lower secondary school mathematics teachers, while 19 out of 146 examinees passed the second test for upper secondary school mathematics teachers to be employed in Hiroshima Prefecture or Hiroshima City (Hiroshima Board of Education, 2014). Those who pass the examination are assigned to a school in the area by the Local Board of Education.

Like in Japan, completing the 4-year teacher education in Korea does not in itself guarantee a teaching position in public schools. Upon their graduation from the college of education, teacher candidates are conferred the "second-level" secondary school teacher qualification, which makes them eligible to apply for the national Teacher Employment Test (TET) for teaching in secondary schools. Each of 17 provincial superintendents selects teachers from those who hold teaching certificates through open competition, to hire teachers in their local public schools. If teacher candidates want to teach in a public school, they must pass the national examination with a high competition rate, ordinarily over 20 to 1. In the case of private schools, the institution itself selects its own teachers. The first open competition examination for the selection of secondary public school teachers was conducted in November 1991. The Provincial Offices of Education forms "the committee for teacher recruitment" and entrusts a research institute such as Korea Institute of Curriculum and Evaluation (KICE) to develop and score the TET. Table 8.5 shows the structure of the current TET for elementary and secondary school teachers in Korea (Lew, 2015).

Table 8.5 Structure of the current TET for elementary and secondary school teachers in Korea

Phase	Applicants	Test	Test Contents	Examination Type	Secondary School	Elementary School
Phase 1	All applicants	Pedagogy	Pedagogy	Essay Type	60 min	60 min
		Major A	Pedagogical Content Knowledge (25–35%) Subject Matter Knowledge (65–75%)	Short Answer Type	90 min	70 min
		Major B		Sentence Completion Type Descriptive Type	90 min	70 min
		History of Korea		Substitution by Korean History Ability Test		
Phase 2	Successful Candidates of Phase 1	In-depth Interview of PCK			10 min	60 min
		Making a Teaching Plan			60 min	20 min
		Teaching Demonstration			20 min	10 min
		English Interview				10 min
		English Teaching Demonstration				10 min

The TET for being a public school teacher has had a great influence on teacher quality as well as the curriculum of teacher education programs. Teacher training universities are sensitive to the TET in the management of their teacher training programs in order to increase the pass rates of their graduates. The TET consists of two phases: a written examination, and an interview and teaching demonstration. The first phase screens 150% of the number of final candidates through a general education theory examination, two types of mathematical examinations, and a Korean History Ability Test. The final successful candidates are chosen by the second phase, consisting of three kinds of interviews and demonstration processes: an In-depth Interview of Pedagogical Content Knowledge (PCK), Making a Teaching Plan, and a Teaching Demonstration. Unlike secondary teachers, for elementary teachers an English Interview and English Teaching Demonstration are also required, because English is introduced into the elementary school curriculum starting in the 3rd grade.

Undergraduate Curricula for Training Secondary School Mathematics Teachers

In Japan and Korea, mathematics teachers teach mathematics to their students based on the national standards, using mathematics textbooks approved/authorized by the MOEs according to the standards. The national standards for K-12 education in Japan and Korea are prescribed in the Course of Study or National Curriculum issued by the MOE, each of which provides the overall objectives of each subject and the objectives and contents of teaching for each grade level (Koyama, 2010; Wong, Koyama, & Lee, 2013). These Courses of Study and National Curricula are closely related with the curricula of the teacher training courses in both countries. This section begins with background information on secondary school mathematics to describe the curricula of the teacher training courses at HU and KNUE.

Background Information on Secondary School Mathematics

In Japan, the current Courses of Study for lower and upper secondary schools emphasize students' mathematical activities in the teaching and learning of mathematics so that through their self-directed mathematical activities students acquire fundamental mathematical knowledge and skills, cultivate their thinking-judging-representing ability, and foster a positive attitude toward learning mathematics (MOE Japan, 2008, 2009). Notably, for the first time the current Courses of Study have incorporated mathematical activities into the mathematics curriculum for grades 7–10 as "content" to be taught and learned, like in other content areas. For example, when learning each of four content areas and the connections among them, lower secondary school students should be provided with opportunities to do mathematical activities like the following: (a) activities for finding out and developing the properties of numbers and geometric figures based on previously learned mathematics, (b) activities for making use of mathematics in daily life and society, and (c) activities for explaining and communicating with each other in an evidenced, coherent, and logical manner by using mathematical representations (MOE Japan, 2008). The secondary school mathematics teachers are also expected to contribute to students' character building through teaching mathematics in schools.

In lower secondary schools, all students learn mathematics for 3 years: 4 unit-hours per week in Grade 7, 3 unit-hours per week in Grade 8, and 4 unit-hours per week in Grade 9 (1 unit-hour is 50 min in lower secondary school). Lower secondary school mathematics consists of four content areas: "A. Numbers and Algebraic Expressions," "B. Geometric Figures," "C. Functions," and "D. Making Use of Data." The upper secondary school mathematics is organized around the idea of a "core-option" such that Mathematics I (3 credits), Mathematics II (4 credits), and Mathematics III (5 credits) are core subjects, while Mathematics A (2 credits), Mathematics B (2 credits), and Application of Mathematics (2 credits) are electives (1 credit in upper secondary school means one 50-minute class per week for a year). In upper secondary schools, students do not necessarily learn all mathematics subjects. In fact, Mathematics I (3 credits) is the only compulsory subject for all students. At the other extreme, those students who intend to study science at a university take all the available mathematics classes during their 3 years of upper secondary school.

In Korea, the current school mathematics curriculum was revised in 2011 (MOE Korea, 2012). This creativity-focused curriculum rejects rote learning and emphasizes mathematical process including manipulation activities; the connection between mathematics and everyday life; independent problem solving; reasoning, communication, and justification by students' intuitive understanding; and established knowledge and thinking skills (Lew, Cho, Koh, Koh, & Paek, 2012). In order to actualize such new instructional directions, the content that had been taught mechanically was substantially eliminated, which led to a significant reduction in students' workload (Hwang et al., 2011).

In lower secondary school, all students learn mathematics for 3 years: 3 unit-hours per week in Grade 7, 4 unit-hours per week in Grade 8, and 3 unit-hours per week in Grade 9. The curriculum is comprised of five areas: Numbers and Operations, Letters and Expressions, Functions, Probability and Statistics, and Geometry. Throughout the five areas the curriculum recommends the use of computer technology, for example, in drawing geometric figures and graphs, calculating, or manipulating data.

In upper secondary school, the mathematics subjects consist of six 5-unit elective courses: Math I, Math II, Calculus I, Calculus II, Geometry and Vectors, and Probability and Statistics. Math I and Math II include basic content needed for covering calculus, and Calculus I and Calculus II consist of basic calculus, calculus of polynomial functions, and calculus of transcendental functions. Additionally, as knowledge of probability and statistics is widely required in modern society, Probability and Statistics is provided as one independent subject, and Geometry and Vectors has also been placed as an independent subject. In addition, the curriculum considers various levels of students by providing level-specific courses like Basic Math, Advanced Math I, and Advanced Math II. Basic Math is an introductory course designed for students who are not equipped with a sound foundation in lower secondary school mathematics. Some students can select Advanced Math I and II in order to reach a systematic understanding of high-level mathematical concepts needed for studying at the Science High Schools.

Undergraduate Curricula for Teaching Certificates

In the following sections, as a case study, we compare the undergraduate curricula for training secondary school mathematics teachers at HU and KNUE in order to identify similarities and differences between them. For that purpose, we decompose the two undergraduate curricula into four areas: mathematics education, mathematics, the teaching profession, and school-based work.

Mathematics Education

Table 8.6 shows the mathematics education curricula at HU and KNUE. At HU, for teaching certificates in secondary school mathematics, students are required to take two courses in Introduction to Mathematics Education in Year 2. However, for graduation, the students in the secondary mathematics education major are required to take all the mathematics education courses in Years 1 and 2, and required to take almost all of those in Years 3 and 4, in order to deepen their pedagogical content knowledge. This is designed to enhance their teaching skills and research abilities, which are needed not only for teaching mathematics in secondary schools but also for doing in-depth study/research on mathematics education in Master's and PhD education programs after completing the undergraduate program.

At KNUE, for teaching certificates in secondary school mathematics and for overall graduation, students are required to take Introduction to Mathematics Education, Researches in Teaching Materials, and Essay Writing in Mathematics. However, students are also required to take five courses in mathematics education in order to deepen their pedagogical content knowledge and enhance their teaching skills. Interestingly, Essay Writing in Mathematics is an integrated subject which is taken in the 7th semester (Year 4). It aims to foster students' critical writing abilities by integrating mathematics and pedagogy. In order to attain this end, students are taught to compose their thoughts logically based on their knowledge of algebra, mathematical analysis, geometry, topology, and statistics, and of curriculum, methodology, assessment, history of mathematics, and mathematical philosophy.

Table 8.6 Mathematics education courses in the HU and KNUE curricula (credits)

Year	Semester	HU	KNUE
1	1		
	2	Methodology of Mathematics Education (2)	
2	3	Introduction to Mathematics Education I (2)	Introduction to Mathematics Education (3)
	4	Introduction to Mathematics Education II (2) Design of Mathematics Education (2)	Research in Teaching Materials (3) Instructional Technology for Mathematics (3)
3	5	Curriculum in Mathematics Education (2) Evaluation in Mathematics Education (2)	Theory of Mathematics Curriculum (3) Teaching Methods of Secondary School Mathematics (3)
	6	Research on Mathematics Education (2) Research on Methodology in Mathematics Education (2)	
4	7	History of Mathematics Education (2) Practice of Mathematics Education (2) Practical Research on Mathematics Education (2) Practical Research on Methodology in Mathematics Education (2)	Essay Writing in Mathematics (3)
	8	Special Study for Graduation (6)	

Mathematics

Table 8.7 shows the curricula of the mathematics area at HU and KNUE. At HU, to earn a teaching certificate in secondary school mathematics, students are required to take at least 20 credits in mathematics. For graduation, the students in the secondary mathematics education major are required to take all the mathematics courses in Years 1 and 2, and almost all of those in Years 3 and 4, in order to deepen their content knowledge and enhance their teaching skills and research abilities. At KNUE, for both the secondary teaching certificate and for graduation, students must take 18 credits in 6 required courses (Analysis I, Complex Analysis I, Modern Algebra I, Differential Geometry I, Topology I, and Probability and Statistics I) and 30 credits in 10 elective courses. Algebra for Teachers and Probability and Statistics are the courses designed to connect university algebra and school algebra and university Probability and Statistics with their school counterparts. Applied Mathematics and Practice I and II are comprehensive courses for senior students who are preparing for the Teacher Employment Test.

Table 8.7 Mathematics courses in the HU and KNUE curricula (credits)

Year	Semester	HU	KNUE
1	1	Introduction to Mathematics (2) Calculus and Its Practice (2)	Calculus and Practice I (3) Set Theory (3) Number Theory (3)
	2	Matrix Theory and Its Practice (2)	Calculus and Practice II (3) Modern Geometry (3) Discrete Mathematics (3)
2	3	Introduction to Algebra (2) Introduction to Geometry (2) Introduction to Analysis (2) Introduction to Statistics (2) Computer Practice I (2)	Analysis I (3) Linear Algebra (3) Mathematical Modeling and Practice I (3) Differential Equation (3)
	4	Practice in Introduction to Algebra (2) Practice in Introduction to Geometry (2) Practice in Introduction to Analysis (2) Computer Practice II (2)	Analysis II (3) Modern Algebra I (3) Mathematical Modeling and Practice II (3)
3	5	Research Methods in Algebra (2) Research Methods in Geometry (2) Research Methods in Analysis (2) Research Methods in Probability and Statistics (2)	Complex Analysis I (3) Differential Geometry I (3) Numerical Analysis and Practice (3) Modern Algebra II (3) Mathematical Planning (3)
	6	Study of Instructional Materials in Algebra (2) Study of Instructional Materials in Geometry (2) Study of Instructional Materials in Analysis (2) Study of Instructional Materials in Mathematical Statistics (2)	Topology I (3) Probability and Statistics I (3) Complex Analysis I (3) Differential Geometry I (3)
4	7	Study of Instructional Materials in Algebra Education (2) Study of Instructional Materials in Geometry Education (2) Study of Instructional Materials in Analysis Education (2) Study of Instructional Materials in Mathematical Statistics Education (2)	Algebra for Teachers (3) Topology II (3) Probability and Statistics II (3) Applied Mathematics and Practice I (3)
	8	Special Study for Graduation (6)	Probability and Statistics for Teachers (3) Applied Mathematics and Practice II (3) Mathematical History and Philosophy (3)

The Teaching Profession

Table 8.8 shows the curricula of the teaching profession area at HU and KNUE. At HU, ten courses in Years 2 and 3 are provided for the students to learn about various topics in education and psychology, and are required for teaching certificates. These

Table 8.8 The teaching profession courses in the HU and KNUE curricula (credits)

Year	Semester	HU	KNUE	
			Theory	Practical Knowledge
1	1		Introduction to Educational Theory (2)	
	2		Educational Psychology (2)	
2	3	Guidance on the Teaching Profession (2) Principles of Education (2)	Curriculum Theory (2) Educational Evaluation (2) Educational Methodology and Technology (2)	Introduction to Special Education (2)
	4	Education, Society and Educational System (2) Methodology of Special Activities (2) Psychology of Education and Vocational Guidance (2)	Educational Philosophy and History (2)	
3	5	Developmental Psychology of Children and Adolescents (2) General Curriculum Theory (2) Methodology on Moral Education (2)	Guidance of School Life (2)	Business Experience in the Teaching Profession (2)
	6	Educational Methods and Teaching Skill (2) Educational Counseling (2)	Educational Sociology (2) Educational Administration and Management (2)	Introduction to and Practice of Protection from School Violence (2)
4	7			
	8	Practical Seminar for Teaching Profession (2)		

courses are provided by professors in the department of education and the department of psychology in the faculty of education. In their 8th semester (Year 4), all students must take the Practical Seminar for Teaching Profession course, to reflect on and enhance the knowledge, skills, and abilities they have acquired. In this course they use their own teaching certificate portfolios accumulated during the previous 3 years, in accordance with the eight standards originally established for HU teacher training programs. In the case of secondary school mathematics teaching certificates, these include mathematics education and mathematics as well as education, psychology, and the teaching profession. This course is taught by a team of professors in the faculty of education and local education practitioners such as principals, head teachers, and experienced teachers from secondary schools.

At KNUE, there are two kinds of teaching profession courses: Theory of Teaching Profession and Practical Knowledge of Teaching Profession. In the theory of teaching profession area, nine 2-credit courses are provided for the students to learn about various topics in education and psychology, and are required for teaching certificates. These courses are provided by professors in the department of education. Students must choose six courses among these nine courses. In the practical knowledge of teaching profession area, three 2-credit courses are provided, and students must take all of them. This area includes a course on "Introduction to and Practice of Protection from School Violence," to learn how to take precautions against any violence in school, which has become a social matter of grave concern.

School-Based Work

Table 8.9 shows the curricula in the school-based work area at HU and KNUE. HU offers Introduction to Teaching Practicum for Lower and Upper Secondary Schools in the 1st semester of Year 1, for the students in the faculty of education. The main aim of the course is for the students to change from a learner's perspective to a

Table 8.9 School-based work courses in the HU and KNUE curricula (credits)

			KNUE	
Year	Semester	HU	Teaching Practicum	Educational Voluntary Service
1	1	Introduction to Teaching Practicum for Lower and Upper Secondary Schools (2)		Over 60 hours of service in any level of school during the period of enrollment (2)
	2			
2	3	Introduction to Care Practice (1) Care Practice (Required)		
	4	Classroom Observation at Secondary Schools (2)		
3	5	Preparation for the Teaching Practicum at Secondary Schools (2)		
	6	Teaching Practicum at Secondary Schools (4 or 2)	Teaching Practicum I (2)	
4	7		Teaching Practicum II (2)	
	8			

teacher's perspective. During this course they have a chance to observe mathematics lessons at HU-attached secondary schools, followed by group discussions on the observed lessons. The Year 2 courses on Introduction to Care Practice and Care Practice are required for lower secondary school certificate students to understand the situations, feelings, and behaviors of people concerning care practice, and to increase their awareness of how to interact with people in special needs schools and social welfare facilities. The students in the secondary mathematics education major take courses related to Teaching Practicum, from the observation of mathematics lessons in Year 2 to the preparation of a teaching practicum in Year 3. In the sixth-semester course called Teaching Practicum at Secondary Schools, these students do teaching practicums at HU-attached secondary schools. The practicums take four weeks (and earn 4 credits) for students pursuing both lower and upper secondary school teaching certificates, or two weeks (2 credits) for just the upper secondary school teaching certificate.

The teaching practicum is conducted using lesson study, such that the students make lesson plans with their mentors' advice, teach mathematics in classrooms, and observe peers' lessons, followed by reflections on each lesson during peer group discussion with their mentors. The secondary mathematics education majors receive regular reviews of their progress by their mentors, and some professors also visit periodically to check the students' progress. The final evaluation of each student's teaching practicum in schools is authorized by the committee on teaching practicum in the faculty of education, based on a report from the schools.

At KNUE, teaching certificate students have to complete both Teaching Practicum and Educational Voluntary Services activities. Class Observation, conducted during the first Teaching Practicum period in the 6th semester (Year 3), focuses on understanding the educational field as a classroom and a school, and improving the pre-service teachers' temperaments through observing educational activities, including teaching performed by a guidance teacher. Its objectives include school life guidance for normal students, extracurricular activity guidance for special students, and class management conducted by teachers and school administrators. In the second Teaching Practicum, taken in the 7th semester (Year 4), students conduct practically real classroom teaching experiences under the guidance of school teachers. This teaching practicum includes making lesson plans, teaching preparation with various educational media, imposing daily assessments, and an open class by a student teacher representative. Voluntary educational services of at least 60 h help pre-service teachers learn and understand lots of activities related to their educational community by supporting school curriculum and events, providing academic advising to underachievers, and guiding students with special needs. To confirm their service, students must get official certification from the organizational institute, and send it to the relevant dean of colleges before the semester ends. The dean decides whether the student passes or fails the course. Seventy percent of students' total score for the teaching practicum comes from field points, and 30% from a university advisor's evaluation. Grades are assessed on an absolute scale.

Similarities and Differences

Japanese and Korean societies have had a long tradition of treating teachers with great respect, so students in both countries have relatively high motivation to become teachers. Based on the analytical results from our comparative case study, in this section we identify several similarities and differences between Japan and Korea, as represented by HU and KNUE, in terms of the prescribed conditions for teaching certificates and the undergraduate curricula for training secondary school mathematics teachers.

Main Similarities

Japan and Korea share significant similarities in three aspects.

First, in their prescribed conditions for teaching certificates as shown in Tables 8.1, 8.2, 8.3, and 8.4, the two countries have similar conditions for teaching certificates prescribed by the MOEs in terms of liberal arts, the teaching profession, and the school subject, and similar 4-year courses for pre-service teachers seeking teaching certificates.

Second, in both countries a 4-year teacher education does not in itself guarantee a teaching position in secondary schools. The teacher candidates are selected through highly competitive teacher employment examinations. However, Korea has a national employment test entrusted to a research institute, while in Japan the individual local boards of education develop their own employment examinations without any involvement from a research institute or university professors. In any case, in both countries the employment rate of university graduates as school teachers is very low. This implies that local boards of education and governments have a high probability of hiring the most highly qualified candidates, all of whom have worked very hard at teacher training universities.

Third, in terms of the curricula for pre-service teacher training, the secondary school mathematics teacher training courses at HU and KNUE both seek to educate students in fundamental concepts of pure mathematics and school mathematics, as well as in mathematics education, aiming to foster qualified secondary school mathematics teachers. In order to become a respectable mathematics teacher, one must be equipped with knowledge of mathematical content, mathematics education, and teaching skills that can support teaching mathematics based on a desirable inclination to teaching and educational philosophy. To attain this goal, these programs provide knowledge of both mathematical content (such as algebra, geometry, analysis, probability and statistics, and applied mathematics) and mathematics education (such as mathematics educational philosophy, mathematics educational psychology, mathematics curriculum, and mathematics methodology and assessment). Moreover, the two universities have similar teaching practicums, through which students gain teaching experience and clinical experience and discuss problems related to mathematics education in schools with experienced mathematics teachers.

Main Differences

On the other hand, we identify two differences between the curricula of HU and KNUE for training pre-service secondary school mathematics teachers. First, KNUE emphasizes mathematics content knowledge, as indicated by the fact that there is almost no curricular difference in mathematics content knowledge between the department of mathematics and the department of mathematics education. In contrast, at HU the department of mathematics education emphasizes mathematics pedagogical content knowledge, because it is separate from the department of mathematics in the faculty of science.

Second, there is a difference in the teaching practicum period. The secondary school teacher training courses at HU have a relatively weak teaching practicum; in the 6th semester, students seeking lower secondary school teaching certificates only need four weeks of teaching practicum in schools, while those seeking upper secondary school teaching certificates need just two weeks (Koyama, 2011). At KNUE, since 2009 students have taken an eight-week practicum instead. Concurrent with the extension of the teaching practicum period, the links between the university and practicum sites were reinforced to promote educational interactions between the attached school or partner school mentor teachers and professors, who take charge of the practicum to solve various difficulties related to the practicum. In contrast, HU provides more systematic preparation steps toward the sixth-semester teaching practicum. This leads to a difference in the variety of activities students experience during their teaching practicums. At HU, based on the systematized teaching practicum program shown in Table 8.9, pre-service teachers have the chance to observe mathematics classroom in schools, prepare for the teaching practicum by shadowing their peers in teaching mathematics, and teach mathematics using lesson study with the assistance of mentor teachers. At KNUE, based on the longer period of teaching practicum, pre-service teachers have the chance to experience various activities related to school management besides teaching mathematics.

Conclusion

In this chapter, we mainly focused on two aspects of pre-service teacher training for secondary school mathematics in Japan and Korea: (a) prescribed conditions for teaching certificates and (b) undergraduate curricula for training secondary school mathematics teachers. Our goal was to identify similarities and differences in educational approaches to supporting pre-service teachers' awareness and knowledge development in the respective departments of mathematics education. These two countries share the philosophy that the quality of education cannot exceed the quality of teachers, even though there is a curricular difference between the two universities examined here. The philosophy has a significant effect on efforts in both countries to improve pre-service teacher quality.

As a case study, we looked at curricula from the department of mathematics education of Hiroshima University in Japan and Korea National University of Education in Korea. The two universities were selected because they are leading universities in terms of pre-service teacher training as well as research on teacher education, and because their pre-service programs are typical models of mathematics teacher training programs in the two countries. This case study remarkably illustrates the importance of educating students in fundamental concepts of pure mathematics and school mathematics, as well as improving their knowledge of mathematics education. In order to become a respectable mathematics teacher, one must be equipped with knowledge of mathematical content, mathematics education, and teaching skills that can support teaching mathematics based on a desirable inclination to teaching and educational philosophy.

However, there are still two remaining issues to be solved for further comparative study of curricula and practices in pre-service teacher training for secondary school mathematics. One issue is to form an appropriate balance between mathematics content knowledge and mathematics pedagogical content knowledge in a 4-year undergraduate course for secondary school mathematics teachers. The other is to realize an effective teaching practicum for pre-service mathematics teachers in the process of improving their competence as mathematics teachers. These issues are not easy to solve. We need in-depth international analyses and considerations to improve the quality of pre-service teacher training for secondary school mathematics, particularly in terms of activities and teaching methods for pre-service teachers.

References

Cho, S., & Choe, S. H. (2012). Mathematics teacher education in Korea. *National presentation of Korea. Pre-proceedings of the 12th International Congress on Mathematical Education*, 8059–8082.

Hiroshima Board of Education. (2014). *Rate of teacher employment examinations in Hiroshima Prefecture and Hiroshima City (in Japanese)*. Retrieved September 24, 2015, from http://kyousai.info/bairitu/45.html

Hwang, S. W., Hwang, H. J., Paik, S. Y., Lew, H. C., & Park, H. S. (2011). *Draft of a creativity-focused mathematics curriculum for the future (in Korean)*. Seoul: Korea Foundation for the Advancement of Science & Creativity.

Kim, H., Kim, J., Park, Y., Lee, K., Lee, H., Jo, S., Kim, E., et al. (2010). *A study on teacher training programs to develop pre-service teacher's instruction competency (in Korean)*. Korea Institute for Curriculum and Evaluation Research Paper, RRI 2010–16.

Koyama, M. (2008). Mathematics teacher training in Japan. In D. Burghes (Ed.), *International comparative study in mathematics teacher training: Perspective* (pp. 26–28). UK: CfBT Education Trust.

Koyama, M. (2010). Mathematics curriculum in Japan. In F. K. S. Leung & Y. Li (Eds.), *Reforms and issues in school mathematics in East Asia: Sharing and understanding mathematics education policies and practices* (pp. 59–78). Rotterdam: Sense Publishers.

Koyama, M. (2011). Country reports: Japan. In D. Burghes (Ed.), *International comparative study in mathematics teacher training: Enhancing the training of teachers of mathematics* (pp. 35–37). UK: CfBT Education Trust.

Lew, H. C. (2015). Leadership and autonomy driving mathematics teacher education in South Korea. *International Journal of Education Sciences, 8*(1), 177–187.

Lew, H. C., Cho, W. Y., Koh, Y. M., Koh, H., & Paek, J. S. (2012). New challenges in the 2011 revised middle school curriculum of South Korea: Mathematical process and mathematical attitude. *ZDM-International Journal of Mathematics Education, 44*(2), 109–119.

Ministry of Education, Culture, Sports, Science and Technology, Japan. (2008). *Guidebook for lower secondary school mathematics in the Course of Study (2008) (in Japanese)*. Tokyo: Kyoiku Publisher.

Ministry of Education, Culture, Sports, Science and Technology, Japan. (2009). *Guidebook for upper secondary school mathematics in the Course of Study (2009) (in Japanese)*. Tokyo: Jikkyo Publisher.

Ministry of Education, Science and Technology, Korea. (2012, January 11). *Policy on advanced mathematics education*. Press release.

Mullis, I. V. S., Martin, M. O., Foy, P., & Arora, A. (2012). *TIMSS 2011 international results in mathematics*. Chestnut Hill, MA: TIMSS & PIRLS International Study Center, Boston College.

National Institute for Educational Policy Research, Japan. (2001). *International comparison in mathematics and science education: A report on the TIMSS-R (in Japanese)*. Tokyo: Gyosei Corporation.

National Institute for Educational Policy Research, Japan. (2004). *Report on the international result of the OECD 2003 PISA survey (in Japanese)*. Tokyo: Gyosei Corporation.

Organization for Economic Co-operation and Development. (2010). *PISA 2009 results: What students know and can do: student performance in reading, mathematics and science* (Vol. 1). Paris: Author.

Wong, K. Y., Koyama, M., & Lee, K. H. (2013). Mathematics curriculum policies: A framework with case studies from Japan, Korea, and Singapore. In Y. Li & G. Lappan (Eds.), *Mathematics curriculum in school education* (pp. 79–91). The Netherlands: Springer.

Chapter 9
Predictors of the Teaching Readiness of Future Secondary Mathematics Teachers: A Comparison of Singapore, Taiwan, and the United States

Ting-Ying Wang and Feng-Jui Hsieh

Abstract This study explores and compares the predictors for the readiness of future secondary teachers to teach mathematics in Singapore, Taiwan, and the United States, using a hierarchical linear model and multivariate linear regression model. On the basis of data from the Teacher Education and Development Study in Mathematics at the individual and institutional levels, it examines the relationships between the teaching readiness of future teachers and their demographics, entrance quality, motivation, knowledge performance, opportunities to learn (OTLs), and teacher education quality. The findings include that the intrinsic motivation of future teachers to become teachers and their OTL in general education positively predicts their teaching readiness across the three countries. Course consistency in a university and the continuity between university instruction and practicum instruction also positively predict teaching readiness in Taiwan and the United States, respectively.

Keywords Teaching readiness • Preservice mathematics teachers • Teacher education quality • Demographic characteristic • Cognitive characteristic • Affective characteristic

Introduction

Teaching quality is a crucial factor influencing student learning (Hill, Rowan, & Ball, 2005). The results of international comparison studies on student mathematics achievement, the Trends in International Mathematics and Science Study (TIMSS), and the Programme for International Student Assessment (PISA) have revealed significant differences among countries (National Center for Education Statistics

T.-Y. Wang • F.-J. Hsieh (✉)
Department of Mathematics, National Taiwan Normal University, Taipei, Taiwan
e-mail: ting@abel.math.ntnu.edu.tw; hsiehfj@math.ntnu.edu.tw

J.-W. Son et al. (eds.), *What Matters? Research Trends in International Comparative Studies in Mathematics Education*, Research in Mathematics Education, DOI 10.1007/978-3-319-51187-0_9

[NCES], 2009; Organisation for Economic Co-operation and Development [OECD], 2007). The findings of the TIMSS 1999 video study (NCES, 2003) further suggested teacher quality as a possible cause of these differences. Past literature also indicated teacher quality as the most critical school-related factor to influence student achievement (Kaplan & Owings, 2001; Rice, 2003). Thus, a group of Michigan State University scholars, with the support of the National Science Foundation in the United States, launched the first data-based international comparison study of mathematics teacher education: Mathematics Teaching in the twenty-first century. It was followed by the Teacher Education and Development Study in Mathematics (TEDS-M), sponsored by the International Association for the Evaluation of Educational Achievement (IEA).

The TEDS-M collected data on nationally representative samples of future secondary mathematics teachers from 15 countries. The study explored the components that reflected the quality of future teachers, such as their mathematics content knowledge (MCK), mathematics pedagogical content knowledge (MPCK), mathematics teaching readiness, and mathematics teaching-related beliefs. It also examined the backgrounds of future teachers, opportunities to learn (OTLs) provided by teacher preparation institutions, and future teachers' perceptions of the quality of their teacher education. Numerous studies have examined these future teachers' MCK, MPCK, and beliefs, along with their relationships with the future teachers' backgrounds and OTL (e.g., Blömeke, Hsieh, Kaiser, & Schmidt, 2014; Hsieh et al., 2011; Laschke, 2013).

In addition to MCK, MPCK, and beliefs, whether teachers are ready to execute tasks that are central to mathematics teaching is a critical indicator reflecting their preparation quality. The professional teaching standards for mathematics teachers not only include the knowledge and skills that teachers should possess but also delineate the actions teachers should undertake during teaching (e.g., Council of Chief State School Officers, 2010; National Board for Professional Teaching Standards, 2001; National Council of Teachers of Mathematics, 2000). For example, the performance-based standards of the Council of Chief State School Officers account for actions. Previous studies have also focused on the teaching readiness of mathematics teachers, and some of them have reported that some teachers did not feel well prepared (Casey, 2011; Mentzer & Becker, 2010; Swars, Smith, Smith, & Hart, 2009).

How well teachers are prepared to execute tasks that are central to mathematics teaching (hereafter referred to as "teaching readiness") influences their teaching and their students' learning. Additionally, teachers who perceived insufficient teaching readiness to execute certain tasks, such as facilitating student engagement, employing various instructional strategies, and managing classrooms, had negative emotions, low value, and low commitment toward the teaching profession; thus, they tended to drop out, which caused a waste of resources for teacher education (Hong, 2010). The TEDS-M also investigated the teaching readiness levels of future secondary teachers (Tatto et al., 2012). Various facets of teaching readiness were measured in the TEDS-M, including instructional planning (e.g., setting up mathematics learning activities to help pupils achieve learning goals), instructional strategies (e.g., using computers and ICT to aid in teaching mathematics), assessment (e.g., developing assessment tasks that promote learning in mathematics), learner development (e.g., challenging pupils to engage in critical thinking about mathematics), learning differences (e.g., having a positive influence on difficult or unmotivated pupils), learning environment (e.g., establishing a supportive environment for learning mathematics), and collaboration (e.g., working collaboratively with other teachers). Tang and Hsieh (2012) found that future secondary mathematics teachers from different countries reported different levels of teaching readiness. However, we still lack studies regarding which characteristics affect the teaching readiness of future teachers and whether these characteristics are identical in various countries, so these issues need to be explored further. Determining the characteristics that affect teaching readiness is crucial for teacher education institutions to develop their training programs, and to establish criteria for recruiting and screening future teachers according to reliable references.

The present study applies TEDS-M data to explore and compare several predictors of teaching readiness among future teachers in three countries: Singapore, Taiwan, and the United States. These countries have achieved MCK and MPCK scores that exceed the international mean of 500, and will hereafter be called "higher-achieving countries" (Hsieh et al., 2011). Specifically, the study considers possible individual-based predictors, such as future teacher demographics, motivations to be a teacher, and quality before entering teacher preparation programs, along with institution-based predictors, such as the quality of courses and the quality of educators in teacher preparation programs.

The United States is an advanced Western country with Greek/Latin/Christian traditions, which emphasize how to improve education quality (Leung, 2006). Hence, its students' mathematics achievements in TIMSS and PISA and the knowledge performance of its future mathematics teachers in the TEDS-M are usually near the international means. Researchers in the United States have explored the characteristics affecting mathematics teacher quality, in order to recruit and train well-prepared teachers who can teach students more effectively. Singapore and Taiwan are two East Asian countries sharing Chinese/Confucian culture, and both are also influenced by Western educational ideas and philosophies (Siu, 2009). In addition, like the United States, Singapore uses English as its formal language and the written language in textbooks for students and future teachers. Comparing to Taiwan, Singapore could be considered a more Westernized country. International comparison studies on mathematics performances of students and teachers have ranked Singapore and Taiwan at the top.

Regarding teachers' social status and context, each of the three countries has its own country-specific characteristics. These characteristics may influence the quality of teacher candidates recruited by teacher preparation programs, the various motivations of teacher candidates to enroll in such programs, how much effort the candidates are willing to expend, and how well they can be prepared. In Confucian cultures, a prevailing idea is that educational performance is crucial for an individual to obtain an outstanding career, fulfilling life, and high social position in the future; thus, education and academic excellence are particularly valued (Salili,

1995; Tan & Yates, 2011). Teachers in Singapore and Taiwan have a higher social status and enjoy higher respect than teachers in many Western countries (Dolton, Marcenaro-Gutierrez, Pota, Boxser, & Pajpani, 2013; Hsieh, Lin, Chao, & Wang, 2013). Teaching in these two countries is a career-based occupation. Teachers in public schools are civil servants and stably appointed. They enjoy generous salaries, health insurance, retirement pensions, and career development opportunities. Singaporean teachers receive salaries even when they are undergoing training during teacher preparation programs, so teaching is an attractive career in both countries. The traditional high status of teachers in Taiwan has gradually eroded as the society has become increasingly open in recent years; however, becoming a teacher in Taiwan remains a competitive process (Hsieh et al., 2013; Wong et al., 2013). In the United States, teaching is a position-based occupation, in which a market-based approach is used for recruiting highly qualified candidates. However, teaching is not an attractive career. Teaching salaries are lower than those of other equivalent academic professions. Moreover, American teachers are not satisfied with their wages, benefits, and opportunities for promotion (Youngs & Grogan, 2013).

Conceptual Framework

Similar to studies pertinent to student achievements, the sociodemographic, cognitive, and affective characteristics of future teachers, and the characteristics of the institutions that train future teachers, were used to predict the quality of future teachers. To identify the characteristics having an average effect on the MCK and MPCK of future teachers, Blömeke, Suhl, Kaiser, and Döhrmann (2012) used the TEDS-M data on the primary level among the 15 countries. The results indicated that the future teachers' gender, parental education, achievements in high school, and motivations were salient individual-level predictors, and OTL in mathematics was an institution-level predictor of MCK or MPCK. Schmidt, Cogan, and Houang (2011) focused on the situation in the United States at both the primary and the secondary levels and found that OTLs in mathematics and mathematics education affected the MCK or MPCK of future teachers. Laschke (2013) compared the situations in Germany and Taiwan at the secondary level and observed that the characteristics influencing future teachers' MCK and MPCK differed between the two countries. For example, gender, prior knowledge, motivations, study circumstances, and OTLs in mathematics predicted MCK in Germany, whereas prior knowledge and study circumstances were not predictors in Taiwan. The literature lacks studies pertinent to which characteristics affect the teaching readiness of mathematics future teachers. Considering that teaching readiness is as critical an indicator of teacher quality as MCK and MPCK, adopting the predictors of MCK and MPCK to predict teaching readiness is reasonable. The present study thus adopts ideas from the aforementioned studies to examine and compare possible relationships between individual-based and institutional-based characteristics and the teaching readiness of future teachers among the three countries.

Sociodemographic Characteristics

Gender differences in mathematics achievements constantly draw attention from researchers. Between 2003 and 2011, the TIMSS reported several gender differences in the mathematics achievements of fourth and eighth graders in Singapore (in favor of females) and the United States (in favor of males; Martin, Mullis, & Foy, 2008; Mullis, Martin, Gonzalez, & Chrostowski, 2004; Mullis, Martin, Foy, & Arora, 2012). However, no significant gender differences in mathematics achievements were observed in Taiwan. Blömeke, Suhl, and Kaiser (2011) provided evidence that gender differences in mathematics-related competence were also present in teacher education. They observed that male future primary teachers outperformed their female counterparts in MCK in Singapore, Taiwan, and the United States, whereas no significant difference was observed in MPCK in the three countries. Hsieh et al. (2010) revealed similar results for Taiwanese future secondary teachers.

Home language can represent immigrant status in the United States and serve as an indicator of the educational disadvantage faced by immigrants, for whom a lack of language skills is the greatest barrier to attaining achievement scores (Parsons & Smeeding, 2006). The majority of the US TEDS-M sample at the secondary level reported that they always used the official language at home. However, only 13.6 and 15.5% of the future secondary mathematics teachers in Singapore and Taiwan, respectively, did so. In Singapore, this practice reflects its constitution of multiple ethnic groups and the bilingual combination of the home language and English (Dixon, 2005). However, every future teacher is required to undergo training on "language enhancement and academic discourse skills" to develop the skills of using English for communication, especially for academic professional purposes (Wong et al., 2013, p. 201). In Taiwan, families typically use Mandarin and Taiwanese in parallel.

Socioeconomic status (SES) also reflects access to resources that are critical to learning, such as education and wealth (Stevenson & Baker, 1992). Blömeke et al. (2012) reported that the MCK of TEDS-M future primary teachers and the education levels of their parents (used as indicators of SES) were positively related. However, Laschke (2013) stated that the educational backgrounds of the parents of the TEDS-M future secondary mathematics teachers in Taiwan did not predict their MCK or MPCK when other individual-based characteristics were controlled for.

Entrance Quality

Blömeke et al. (2012) and Laschke (2013) used two indicators related to the cognitive characteristics of future teachers to predict MCK and MPCK, by considering the highest grade level of secondary mathematics that the future teachers had studied, and their secondary school marks/grades compared with their same year-level peers. These two indicators represent the academic quality of future teachers when entering the teacher preparation programs (Hsieh et al., 2010). It is reasonable to assume that the entrance quality of future teachers will influence their final performance, including their teaching readiness.

In Taiwan, students learn mathematics during all 3 years of senior high school. Almost all TEDS-M future secondary mathematics teachers have studied 12th-grade mathematics. A total of 90.8% of future teachers have studied 12th-grade Mathematics A, which is designed for students with a science orientation, and 8.6% have studied 12th-grade Mathematics B, which is designed for students with a literature and arts orientation. Another 0.6% of future teachers have studied vocational mathematics, which is equivalent to 11th-grade mathematics. Similarly, an overwhelming majority (86.9%) of Singaporean future secondary teachers have studied 12th-grade mathematics or higher. In the United States, approximately three-fourths and one-fifth of the TEDS-M future secondary teachers have studied 12th- and 11th-grade mathematics.

In Singapore and Taiwan, high-achieving graduates from secondary schools can generally be admitted to teacher preparation programs because the teaching profession is attractive (Hsieh et al., 2013; Wong et al., 2013). The situation in the United States is different. According to the analysis of Schmidt, Houang and Cogan (2012), the United States must recruit its future teachers from above the 75th percentile of the national distribution for the teachers to be comparable with those from Singapore and Taiwan, if these two countries were to draw their future teachers from among the average eighth graders. However, the teaching profession is not as attractive in the United States.

Opportunity to Learn

The real exposure to learning courses is a critical type of OTL for both students and teachers (McDonnell, 1995; Tatto et al., 2012). Following IEA studies at the school student level, the TEDS-M probed the OTLs of future teachers regarding their course study in the fields of university-level mathematics, school-level mathematics, mathematics education, and general education, on the basis of potential relationships between OTLs and achievements (Tatto et al., 2012). For teacher education, the opportunities for future teachers to be exposed to various courses and the extent of exposure can be regarded as the level of the knowledge equipment of an individual, and reflect the philosophical and educational viewpoints of an institution or country on how to shape future teachers into qualified professionals (Hsieh et al., 2010; Schmidt, Cogan, & Houang, 2011; Schmidt et al., 2008). Thus, this study assumed that OTLs will affect teaching readiness at both individual and institutional levels.

Knowledge Achievements

Researchers generally agree that teachers require adequate knowledge to teach effectively (e.g., Ernest, 1989; Grossman, 1995; Shulman, 1986). Literature on mathematics education often discusses teacher MCK and MPCK, both of which are typically related to mathematics teaching practice (Ball, Thames, & Phelps, 2008; Hsieh, 2009). Several researchers have concluded that some mathematics teachers do not possess adequate MCK or MPCK, which results in their inability to teach mathematics effectively (Ball, 1991; Borko et al., 1992; Leinhart & Smith, 1985). These results highlight the necessity for understanding the relationships among the MCK, MPCK, and teaching readiness levels of future teachers.

MCK and MPCK were also two knowledge achievements that the TEDS-M investigated in future teachers. The scores of US future teachers were approximately equal to the international mean for both MCK and MPCK, but they were outperformed by students in the two East Asian countries. Singaporean and Taiwanese eighth graders performed at approximately the same level in the series of the TIMSS; however, Taiwan outperformed Singapore by almost one standard deviation (100 points) in both MCK and MPCK (Hsieh, 2012; Hsieh & Wang, 2012).

Motivation

Motivation is a critical affective characteristic that affects not only students' mathematics achievements (Eklöf, 2010) but also teacher performance. The TEDS-M thus investigated the factors that motivated future teachers to pursue a teaching career. Studies have revealed that the intrinsic motivation and empathy from prior learning experiences positively predict future teachers' knowledge achievements, whereas extrinsic motivation negatively predicts knowledge achievements (Blömeke et al., 2012; Hsieh et al., 2010; Laschke, 2013). Both East Asian and Western societies highly value intrinsic motivation (Zhu & Leung, 2011); however, the approaches to it differ in various cultures. In East Asian cultures, intrinsic motivation involves attempting to master practices and gaining assurance from others, reflecting a "social orientation," whereas in Western cultures, intrinsic motivation relates closely to individual interest and fulfillment, reflecting an "individual orientation" (Hofstede, 1986; Markus & Kitayama, 1991). Western society does not encourage extrinsic motivation in terms of student learning; however, the value of academic excellence, which is typically related to outperforming others in East Asia, makes extrinsic motivation a force for pursuing knowledge (Laschke, 2013).

Teacher Education Quality

School education quality is typically assumed to influence student achievements (Akyüz & Berberoğlu, 2010). Thus, the present study hypothesized that teacher education quality also affects the performance of future teachers at the institutional level. Hsieh et al. (2011) proposed a framework for examining teacher education quality, which includes course and personal qualities. Two indicators were designed for course quality. One was course arrangement, which was a measure of the consistency of courses and content within a university, and the other was teaching coherence, which was a measure of the connection and continuity between the instruction in universities and that in schools. Four indicators were designed to measure personal quality. MR-instructor and SB-supervisor indicated the effectiveness of instructors, and respectively address the effectiveness of educators responsible for teaching mathematics-related courses and that of those responsible for supervising the school-based experiences of future teachers. The other two indicators, MCK and MPCK, measured the achievements of future teachers. According to this framework, the present study hypothesized that MCK and MPCK also influence teaching readiness at the institutional level. On the basis of the discussion above, this study proposes the framework shown in Fig. 9.1.

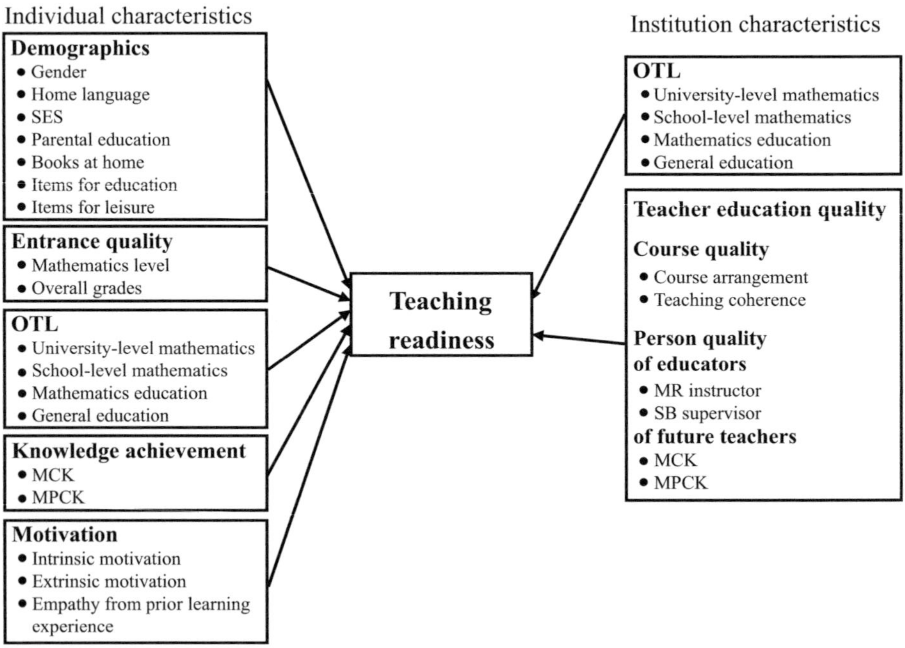

Fig. 9.1 Conceptual framework of the study

Research Method

Participants and Data Collection Process

The present study used the TEDS-M samples of future secondary mathematics teachers in Singapore, Taiwan, and the United States. The future teachers were in their final year of teacher education and would receive a license to teach mathematics in secondary schools. The TEDS-M used a stratified multistage probability sampling design. Future teachers were randomly selected from teacher preparation institutions, which were also randomly chosen. The samples reflected the distribution of future teachers at the end of their training in each country (Tatto et al., 2012).[1]

The samples for the present study included 393 Singaporean, 365 Taiwanese, and 607 American future teachers at 1, 19, and 46 institutions, respectively. Two training programs are offered for training secondary future teachers in Singapore. The teachers are trained to teach two subjects from Grades 7 to 8 or 7 to 12. Only one program is offered for training secondary future teachers in Taiwan, and the teachers are trained to teach a single subject from Grades 7 to 12. In the United States, teachers are trained as specialists for teaching one subject, either to Grades 4 or 5 through 8 or 9 or 6 or 7 through 12 (Tatto et al., 2012).

The surveyed future teachers had to fill out their questionnaires within 90 min. The questionnaires included items in five categories: general background, OTL, MCK, MPCK, and mathematics teaching-related beliefs. The answer choices for general background, OTL, and beliefs were multiple-choice. The tests for the MCK and MPCK employed a balanced incomplete block design and included both multiple-choice items and open-ended items. The survey process in each country followed the regulations set by the international TEDS-M study team (Tatto et al., 2008).

Measures

Teaching readiness was captured using a set of 11 items, prompted by this question: "Please indicate the extent to which you think your teacher education program has prepared you to do the following when you start your teaching career." All the items were rated on a 4-point Likert scale: 1 = *not at all*, 2 = *a minor extent*, 3 = *a moderate extent*, and 4 = *a major extent*. The TEDS-M applied a partial-credit model to estimate the logit scores of future teachers on the scale. A score of 10 was associated with the neutral position, and higher scores represented higher self-evaluated teaching readiness (Tatto et al., 2012).

[1]A census was conducted in Taiwan. The United States limited its participation to public institutions.

In the TEDS-M, the question probing the teaching readiness of future teachers involved self-reporting. Self-reported data may be skewed because participant self-impressions may deviate from reality; however, self-reporting is a simple and economical method that enables a high number of respondents from various cultures with various languages to be surveyed and compared. Moreover, self-evaluation of teaching readiness by future teachers who were in their final year of teacher preparation is an effective indicator, because it incorporates components of future teachers' individual self-reflections and their practical field experience in practicum into the survey process (Wang, Hsieh, & Tang, 2014).

Demographics. Gender was a dichotomous item with 0 = *male* and 1 = *female*. Home language was examined according to the frequency of speaking the official language used in teacher education at home. The item was rated on a 4-point Likert scale from 1 = *never* to 4 = *always*. Four variables were used to estimate respondents' socioeconomic status. The first variable was parental education, which was captured by the highest educational level held by the respondent's father or mother. The codes for this variable were from 1 = *primary* to 7 = *beyond* ISCED (The International Standard Classification of Education) 5A. The second variable was the number of books at home, which was coded from 1 = *none or few* (0–10 books) to 5 = *enough to fill three or more bookcases* (more than 200 books). The third variable was the quantity of educational items at home, which combined five dichotomous items asking whether future teachers have a calculator, computer, study desk, dictionary, or encyclopedia at home. The fourth variable was the quantity of items for leisure at home, which combined three dichotomous items asking whether future teachers have any game systems, DVD players, or more than three cars.

Entrance Quality. The secondary mathematics level was measured according to the highest mathematics grade level that the respondents studied in secondary school. A 5-point Likert scale was used: 1 = *below Year 10* to 5 = *advanced level of Year 12*. The overall grades were captured by measuring the secondary school achievements of future teachers compared with those of their age cohort. The item was also graded using a 5-point Likert scale from 1 = *generally below average* to 5 = *always at the top*.

Opportunity to Learn (OTL). The OTLs of future teachers were surveyed by asking them to report whether they had studied a list of topics in four fields during their teacher preparation programs. Tertiary-level mathematics included 19 topics such as calculus, linear algebra, set theory, and discrete mathematics. School-level mathematics was composed of seven topics such as numbers, geometry, and probability. Mathematics education contained eight topics such as context of mathematics education, development of mathematics ability and thinking, and mathematics instruction. General education also included eight topics such as philosophy of education, educational psychology, and theories of schooling. The sums of topics in each field were used as predictors in this study.

Knowledge Achievement. The TEDS-M measured the MCK and MPCK of future teachers by using 76 and 27 items, respectively, and applied a balanced incomplete block design with three booklets. The scaled scores for MCK and MPCK were obtained by applying item response theory. The scores were standardized to a mean of 500 and a standard deviation of 100 in the TEDS-M to facilitate understanding them (Tatto et al., 2012). In the present study, the MCK and MPCK scores used were further standardized to a mean of 0 and a standard deviation of 1.

Motivation. The TEDS-M measured the motivations of future teachers to pursue teaching by using nine items. The items were rated on a 4-point Likert scale from 1 = *not a reason* to 4 = *a major reason*. Hsieh et al. (2010) conducted a factor analysis and extracted three aspects: intrinsic motivation, salary and job security, and empathy from prior learning experiences. Intrinsic motivation included four items such as "I like working with young people." Salary and job security contained three items such as "I am attracted by teacher salaries." Empathy from prior learning experience included two items such as "I love mathematics." The average of the rating points within each aspect was employed in the present study.

Teacher Education Quality. Regarding course quality, course arrangement and teaching coherence contained six and five items, respectively. All the items were rated on a 4-point Likert scale from 1 = *disagree* to 4 = *agree*. Course arrangement contained items such as "The program was organized in a way that covered what I needed to learn to become an effective teacher" and "The courses seemed to follow a logical sequence of development in terms of content and topics," while teaching coherence contained items such as "I learned the same criteria or standards for good teaching in my courses and in my field experiences/practicum" and "In my field experience/practicum, I had to demonstrate to my supervising teacher that I could teach according to the same criteria/standards used in my university/college course." Regarding the person quality of educators, MR-instructor and SB-supervisor were determined on the basis of six and four items, respectively. MR-instructor involved using a 6-point Likert scale from 1 = *strongly disagree* to 6 = *strongly agree*. It contained items such as "Model good teaching practices in their teaching" and "Draw on and use research relevant to the content of their courses." SB-supervisor used a 4-point Likert scale from 1 = *disagree* to 4 = *agree*. It contained items such as "The feedback I received from my supervising teacher helped me to improve my understanding of pupils" and "The feedback I received from my supervising teacher helped me improve my teaching methods." The TEDS-M obtained logit scores for these four scales by applying a partial-credit model and set 10 as the neutral position (Tatto et al., 2012).

Data Analysis

The TEDS-M data are based on a nested sample structure. Thus, the present study employed two-level hierarchical linear models (HLM) to analyze the relationships of the teaching readiness of Taiwanese and US future teachers to their demographic, cognitive, and affective characteristics, and to the teacher education quality of their teacher training institutions. The analyses were conducted using HLM 6.08. Because only one teacher preparation institution exists in

Singapore—the National Institute of Education—a multivariate linear regression model was used to examine the relationships in this country. The analyses were conducted using SPSS.

When using the two-level HLM, the influence of individual characteristics (Level 1) on teaching readiness was first examined. To isolate the effects of individual characteristics from those of institutional characteristics (Level 2), Level 1 predictors were introduced using group centering (centered on the arithmetic mean of the institution). The effects of Level 2 predictors were then examined by controlling the Level 1 predictors. Regarding the data on Taiwan and the United States, institutions with fewer than six future teachers were excluded to ensure robust estimates. After the adjustment, the Taiwanese data set consisted of 361 future teachers from 18 institutions, and the US data set included 563 future teachers from 32 institutions. The weights of future teachers and institutions provided by the TEDS-M were incorporated to reflect selection probabilities and nonresponse rates.

Results

To examine which characteristics influenced the teaching readiness of future teachers, the predictors were introduced block by block. For Level 1, demographics were included in the model as the first three steps. Gender was introduced first, followed by home language, and then the four SES indicators were included. Entrance quality, motivation, knowledge achievements, and OTLs were introduced in that order. Controlled Level 1 predictors, the Level 2 predictors, future teacher personal quality, OTLs, educator personal quality, and course quality were then added to the model, in that order.

Individual Characteristics

The relationships between the demographic characteristics of future teachers and their teaching readiness differed among the three countries (Table 9.1). As expected, no demographic characteristics were predictive in Taiwan when other predictors were controlled for. Consistent with the results of the TIMSS for school students, gender did not affect the teaching readiness of Taiwanese future teachers. Although only 15.5% future teachers used the official language at home, the parallel use of Mandarin and Taiwanese was common. Thus, home language was not a barrier for future teachers to develop their teaching readiness. Furthermore, as in other East Asian countries, a prevailing belief in Taiwan is that an individual can succeed through hard work regardless of his or her socioeconomic background. Therefore, the indicators pertinent to the SES were not predictive. In Singapore as well, the home language and SES indicators did not predict the teaching readiness

Table 9.1 Teaching readiness of future teachers regressed on individual and institution characteristics

	Singapore		Taiwan		United States	
	Estimate	*SE*	Estimate	*SE*	Estimate	*SE*
Individual predictors						
Demographics						
Gender	0.34*	0.14				
Home language					−0.40†	0.23
Parental education						
Books at home						
Items for education						
Items for leisure					0.20†	0.10
Entrance quality						
Secondary mathematics level			0.41*	0.18		
Overall grades						
Motivation						
Intrinsic motivation	0.47**	0.12	0.53**	0.10	1.22**	0.16
Salary and job security						
Empathy from prior learning experience	0.18†	0.10				
Knowledge achievement						
MCK						
MPCK						
OTL						
University-level mathematics			0.06*	0.03		
School-level mathematics						
Mathematics education			0.22**	0.04	0.35**	0.07
General education	0.26**	0.04	0.09*	0.04	0.14*	0.07
Institutional predictors						
Person quality of future teachers						
MCK	–				−1.07*	0.41
MPCK	–				1.25*	0.53
OTL						
University-level mathematics	–		0.12†	0.06		
School-level mathematics	–					
Mathematics education	–				0.22†	0.13
General education	–		0.26†	0.12		
Person quality of educators						
MR-instructor	–					
SB-supervisor	–					
Course quality						
Course arrangement	–		0.49**	0.12		
Teaching coherence	–				0.50**	0.18
R^2 of Level 1			31.3%		28.7%	
R^2 of Level 2			42.8%		50.1%	

Note. *SE* = standard error. – = not applicable

†$p < .1$. *$p < .05$. **$p < .01$

of future teachers. However, gender differences existed. Consistent with the findings of the TIMSS, in which female students outperformed their male counterparts, Singaporean female future teachers outperformed their male counterparts. In the United States, the indicator items for leisure predicted the readiness of future teachers. Unexpectedly, home language exhibited negative effects in the United States as it represented immigrant status (Dixon, 2005). Further in-depth analysis is required.

Regarding entrance quality, secondary mathematics level positively affected the teaching readiness of Taiwanese future teachers. The three forms of the highest level of mathematics offered in Taiwan were 12th-grade Mathematics A, 12th-grade Mathematics B, and vocational mathematics. The concepts and skills introduced and the difficulty levels of these three forms differed substantially (Hsieh et al., 2010). In Taiwan, future teachers who were more equipped with secondary-level mathematics evaluated themselves as being more ready to execute tasks that are central to mathematics teaching. However, neither secondary-level mathematics nor overall grades affected teaching readiness in Singapore and the United States when other predictors were controlled for.

Regarding motivation, intrinsic motivation positively affected the teaching readiness of future teachers to a substantial degree in all three countries, especially in the United States. This finding is favorable because actions triggered by intrinsic motivation are highly valued in both East Asian and Western cultures (Zhu & Leung, 2011). Extrinsic motivation was not predictive in any of the three countries when controlling for other individual and institutional characteristics. The attractiveness of salary and job security did not contribute to the teaching readiness of future teachers. Empathy from prior learning experiences was not influential in the United States, and its influence on teaching readiness differed in the two Asian countries. In Taiwan, teacher training and the teaching profession are competitive. Generally, only students who were competent at the secondary level were admitted to the teacher preparation program. Thus, the cognitive and affective support resulting from the empathy from the prior learning experiences of future teachers was not particularly crucial, and did not affect their teaching readiness. Teacher training and teaching jobs are also attractive and competitive in Singapore. However, in this case empathy from prior learning experiences was a predictor. The reasons underlying this require further study.

OTLs in general education predicted teaching readiness in all three countries. This was the only influential OTL predictor at the individual level in Singapore. OTLs in mathematics education explained a more substantial proportion of variance in the teaching readiness of Taiwanese future teachers than did OTLs in university-level mathematics and general education, the coefficients of which did not reach the small effect size (0.1; Cohen, 1992). Similar to Taiwan, OTLs in mathematics education were also the most powerful OTL predictors in the United States, whereas the OTLs in university-level mathematics were not predictive. Through analyzing the items, teaching readiness explored by the TEDS-M covered the facets pertinent to learner and learning, instructional practice, and professional responsibility (CCSSO, 2010); MCK was not directly tested through these items.

Thus, it was not difficult to determine why OTLs in mathematics education and general education were more influential predictors than OTLs in university-level mathematics and school-level mathematics. Furthermore, this explains why MCK was not predictive in the three countries. However, unexpectedly, MPCK was also not influential. This could be because the MPCK items tested for the secondary level in the TEDS-M were insufficient (Hsieh, 2013).

Institution Characteristics

As described in the research method, only one teacher preparation institution exists in Singapore; therefore, institutional predictors were only applicable for Taiwan and the United States. Similar to the individual level, neither MCK nor MPCK predicted the teaching readiness of Taiwanese future teachers when other individual and institutional characteristics were controlled for. In the United States, MPCK positively predicted teaching readiness, whereas MCK negatively predicted teaching readiness. This indicates that differences in the knowledge achievement levels of future teachers among the institutions were related to differences in teaching readiness, after controlling for the relationships at the individual future teacher level. The reason underlying these opposite aggregate effects is not clear. Whether this phenomenon relates to the limitations of training time and personal energy, which make comprehensive development difficult, requires further exploration.

OTLs in university-level mathematics and general education were predictive in Taiwan, whereas those in mathematics education were predictive in the United States. The differences in OTLs among the institutions were related to differences in teaching readiness after controlling for the relationships at the individual level. The variations in courses taken at the institutional level may relate to the differences in course requirements, courses offered, and informal expectations from students to take a certain number of courses in various institutions (Schmidt, Blömeke et al., 2011). Combining the institutional effect size with the individual effect size revealed that OTLs in mathematics education were strongly related to teaching readiness in the United States.

Course quality influenced the teaching readiness of future teachers in Taiwan and the United States. However, the specific predictors varied: course arrangement was predictive in Taiwan, whereas teaching coherence was predictive in the United States. In Taiwan, when teaching coherence was introduced as a single course quality predictor, it significantly and positively affected teaching readiness (coefficient, 0.57; $p < .1$). However, when both course quality predictors were introduced in the model, the effect of teaching coherence became nonsignificant. This phenomenon indicated that in Taiwan, most of the variance explained by teaching coherence overlapped with that explained by course arrangement, and teaching coherence affected teaching readiness through course arrangement. In the United States, SB-supervisor predicted teaching readiness when it was singularly introduced in the model among the indicators pertinent to the personal quality of educators and the course quality (coefficient, 0.42; $p < .05$), after controlling for other individual- and institutional-level predictors. However, when teaching coherence was simultaneously introduced in the model, the effect of SB-supervisor became nonsignificant. In the United States, SB-supervisor influenced teaching readiness through teaching coherence between universities and schools.

Conclusion

The level of teaching readiness substantially affects the success of instruction that a new teacher may provide. Determining how to develop teaching readiness is thus a meaningful problem for teacher education programs. The present study used the first large-scale, nationally representative, international comparison dataset from the TEDS-M to examine the relationship between the teaching readiness of future teachers and their individual and institutional characteristics in two East Asian countries and one Western country—Singapore, Taiwan, and the United States. The three countries explored are of two historical roots. However, the results did not enable categorizing the three countries according to the East Asian or Western cultures. The degree of Westernization in the two East Asian countries varied and their results were also different, which was consistent with former findings (Hsieh et al., 2013).

Before discussing the conclusions, some limitations regarding the methodology should be pointed out. TEDS-M was a cross-national study which employed self-report methods on several items. Although self-reporting was an efficient method to collect large-scale one-point-in-time data, retrospective reports in the items regarding OTL and entrance quality, or self-evaluations of teaching readiness, could be biased from the reality. Further studies to develop better measures are required. To develop a full model revealing teacher quality in each country, future studies could include affective outcomes such as teaching-related beliefs, in addition to the cognitive teacher education outcomes used in this study. However, owning to correlations between various characteristics which may increase the risk of Type I errors, appropriate statistical methods to construct models should be considered.

The notion of SES has been considered a crucial problem in school education in Western societies (Coleman et al., 1966; Schmidt, Cogan, Houang, & McKnight, 2011); thus, it has been covered in most international comparison studies. The results of the present study reveal that the items for leisure predicted the teaching readiness of US future teachers, whereas the SES is not a concern in the two East Asian countries at the teacher learner level. These results reflect a prevailing belief that an individual can achieve success and become outstanding through hard work, regardless of his or her socioeconomic background. None of the demographic characteristics, including the SES factors, were predictive in Taiwan. However, the situation was different in Singapore. A gender difference, with females being favored, existed in Singapore, which was consistent with the research findings of the TIMSS for school students. TIMSS results also revealed that male school students in the United States outperformed their female counterparts in mathematics achievements; however, this did not apply to the readiness of future teachers to execute tasks that are central to mathematics teaching.

Intrinsic motivation was the strongest individual-level predictor of teaching readiness in all three countries, and especially crucial in the United States. This result is favorable because intrinsic motivation is highly valued for triggering the actions of individuals in both East Asian and Western cultures. However, the approaches for stimulating intrinsic motivation may vary, as mentioned in the conceptual framework section—intrinsic motivation was pertinent to attempts at mastering practices and gaining assurance from others in East Asia, whereas it was related to individual interest and fulfillment in the West.

Regarding cognitive characteristics, the entrance quality of future teachers—the highest level of secondary mathematics studied—predicted their teaching readiness in Taiwan, but not in the other two countries. In Taiwan, three forms of the highest level of mathematics were offered: 12th-grade Mathematics A, 12th-grade Mathematics B, and vocational mathematics. The difficulty levels of the concepts and skills introduced differed substantially. Particularly, vocational mathematics was only equivalent to 11th-grade Mathematics. The future teachers studying 12th-grade Mathematics A were science-oriented, whereas those studying 12th-grade Mathematics B were literature- and arts-oriented. Science-oriented senior high school students typically performed higher in mathematics than literature- and arts-oriented students did. Thus, various mathematics backgrounds may influence the evaluation by future teachers of their own teaching readiness from cognitive and affective facets.

OTLs after enrolling in teacher training were influential in all three countries, although the fields varied. General education was crucial in Singapore, whereas mathematics education was critical in Taiwan and the United States. Combining the institutional effect size with the individual effect size revealed the extraordinary importance of OTLs in mathematics education to teaching readiness in the United States. Why MPCK does not affect teaching readiness in any of the three countries remains unanswered. Whether the answers relate to the appropriateness of the MPCK items of TEDS-M, for example, the quantity of the items and their context, difficulty level, and degree of reflecting teaching competence, requires further exploration. The aggregate effects of knowledge achievements were only observed in the United States. The opposite effects of MCK and MPCK were unexpected. Whether this phenomenon is related to the limitations of training time and personal energy, which make comprehensive development difficult, remains unclear. Hence, additional studies are necessary. Regarding course quality at the institutional level, course arrangement and teaching coherence predicted the teaching readiness of future teachers in Taiwan and the United States, respectively. Teaching coherence was also influential in Taiwan, but its influence overlapped with that of course arrangement. Teaching coherence probably affected the teaching readiness of Taiwanese future teachers through course arrangement. SB-supervisor values in the United States yielded similar results.

The findings of the present study provide potential criteria to recruit students who have the potential to become high-quality teachers. It is suggested that teacher educators and teacher preparation programs should consider students' intrinsic motivation for being a teacher when screening and selecting teacher candidates into the programs. In Taiwan, the highest grade level of secondary mathematics students studied should also be considered. This study offers potentially useful information for countries in which teacher training and teacher profession are competitive, like Taiwan or Singapore, and for countries that require improving teacher quality, like the United States (Center for Research in Mathematics and Science Education, 2010). Moreover, the present study provides information on what to focus on when preparing future mathematics teachers for the three countries—the types of OTL to provide, the degree of the consistency of courses and content arranged within a university, or the degree of the connection and continuity between the instruction in universities and that in schools. One thing worth pointing out is that this study used composite indicators for OTLs. In these indicators, courses with different difficulty levels or various orientations were combined. Some OTL effects may not be detected. Further research is required to screen out influential topics or topic groups for far-reaching conclusions. The approaches to increase the degree of the consistency of courses and content arranged within a university, and the degree of the connection and continuity between the instruction in universities and that in schools should also be explored, to raise the preparation quality of teacher education programs.

Acknowledgments We gratefully acknowledge the following: the IEA, the ISC at MSU, the DPC, the ACER, the US NSF, and all the TEDS-M national research coordinators, for sponsoring the international study and providing information and data.

References

Akyüz, G., & Berberoğlu, G. (2010). Teacher and classroom characteristics and their relations to mathematics achievement of the students in the TIMSS. *New Horizons in Education, 58*(1), 77–95.

Ball, D. L. (1991). Research on teaching mathematics: Making subject-matter knowledge part of the equation. In J. Brophy (Ed.), *Advances in research on teaching, Teachers' knowledge of subject matter as it relates to their teaching practice* (Vol. 2, pp. 1–48). Greenwich: JAI Press.

Ball, D. L., Thames, M. H., & Phelps, G. (2008). Content knowledge for teaching: What makes it special? *Journal of Teacher Education, 59*(5), 389–407.

Blömeke, S., Hsieh, F.-J., Kaiser, G., & Schmidt, W. H. (Eds.). (2014). *International perspectives on teacher knowledge, beliefs and opportunities to learn.* New York, NY: Springer.

Blömeke, S., Suhl, U., & Kaiser, G. (2011). Teacher education effectiveness: Quality and equity of future primary teachers' mathematics and mathematics pedagogical content knowledge. *Journal of Teacher Education, 60*(2), 154–171.

Blömeke, S., Suhl, U., Kaiser, G., & Döhrmann, M. (2012). Family background, entry selectivity and opportunities to learn: What matters in primary teacher education? An international comparison of fifteen countries. *Teaching and Teacher Education, 28*, 44–55.

Borko, H., Eisenhart, M., Brown, C. A., Underhill, R. G., Jones, D., & Agard, P. C. (1992). Learning to teach hard mathematics: Do novice teachers and their instructors give up too easily? *Journal for Research in Mathematics Education, 23*(3), 194–222.

Casey, C. (2011). Teacher education admission criteria as measure of preparedness for teaching. *Canadian Journal of Education, 34*(2), 3–20.
Center for Research in Mathematics and Science Education. (2010). *Breaking the cycle: An international comparison of U.S. mathematics teacher preparation*. East Lansing: Michigan State University.
Coleman, J. S., Campbell, E. Q., Hobson, C. J., McPartland, J., Mood, A. M., Weinfeld, F. D., et al. (1966). *Equality of educational opportunity*. Washington: US Gov Printing Office.
Cohen, J. (1992). A power primer. *Psychological Bulletin, 112*(1), 155–159.
Council of Chief State School Officers. (2010). *Interstate teacher assessment and support consortium (InTASC) Model core teaching standards: A resource for state dialogue (Draft for public comment)*. Washington, DC: Council of Chief State School Officers.
Dixon, L. Q. (2005). The bilingual education policy in Singapore: Implications for second language acquisition. In J. Cohen, K. T. McAlister, K. Rolstad, & J. MacSwan (Eds.), *Proceedings of the 4th International Symposium on Bilingualism* (pp. 625–636). Somerville, MA: Cascadilla Press.
Dolton, P., Marcenaro-Gutierrez, O., Pota, V., Boxser, M., & Pajpani, A. (2013). 2013 Global teacher status index. Retrieved from https://goo.gl/4N6MJU.
Eklöf, H. (2010). Student motivation and effort in the Swedish TIMSS Advanced field study. Paper presented at the IEA International Research Conference. Retrieved from http://www.iea.nl/fileadmin/user_upload/IRC/IRC_2010/Papers/IRC2010_Eklof.pdf.
Ernest, P. (1989). The knowledge, beliefs and attitudes of the mathematics teacher: A model. *Journal of Education for Teaching: International Research and Pedagogy, 15*(1), 13–33.
Grossman, P. L. (1995). Teachers' knowledge. In L. W. Anderson (Ed.), *International encyclopedia of teaching and teacher education* (Vol. 2, pp. 20–24). Oxford, UK: Pergamon Press.
Hill, H. C., Rowan, B., & Ball, D. L. (2005). Effects of teachers' mathematical knowledge for teaching on student achievement. *American Educational Research Journal, 42*(2), 371–406.
Hofstede, G. (1986). Cultural differences in teaching and learning. *International Journal of Intercultural Relations, 10*, 301–320.
Hong, J. Y. (2010). Pre-service and beginning teachers' professional identity and its relation to dropping out of the profession. *Teaching and Teacher Education, 26*(8), 1530–1543.
Hsieh, F.-J. (2009). *Research on the development of the professional ability for teaching mathematics in the secondary school level (3/3)* (Rep. No. NSC 96-2522-S-003-008). Taiwan: National Science Council (in Chinese).
Hsieh, F.-J. (2012). Mathematics teaching competence of future secondary mathematics teachers. In F.-J. Hsieh (Ed.), *Taiwan TEDS-M 2008* (pp. 119–142). Taipei: Department of Mathematics, National Taiwan Normal University (in Chinese).
Hsieh, F.-J. (2013). Assessing strengths and weaknesses of teacher knowledge in Asia, Eastern Europe, and Western countries-differential item functioning in TEDS-M. *International Journal of Science and Mathematics Education, 11*(4), 923–947.
Hsieh, F.-J., Law, C. K., Shy, H. Y., Wang, T. Y., Hsieh, C. J., & Tang, S.-J. (2011). Mathematics teacher education quality in TEDS-M—Globalizing the views of future teachers and teacher educators. *Journal of Teacher Education, 60*(2), 172–187.
Hsieh, F.-J., Lin, P.-J., Chao, G., & Wang, T.-Y. (2013). Chinese Taipei. In J. Schwille, L. Ingvarson, & R. Holdgreve-Resendez (Eds.), *TEDS-M encyclopedia, a guide to teacher education context, structure, and quality assurance in 17 countries: Findings from the IEA Teacher Education and Development Study in Mathematics (TEDS-M)* (pp. 69–85). Amsterdam: IEA.
Hsieh, F.-J., & Wang, T.-Y. (2012). Mathematics competence of future secondary mathematics teachers. In F.-J. Hsieh (Ed.), *Taiwan TEDS-M 2008* (pp. 93–117). Taipei: Department of Mathematics, National Taiwan Normal University (in Chinese).
Hsieh, F.-J., Wang, T.-Y., Hsieh, C.-J., Tang, S.-J., Chao, G., Law, C.-K. et al. (2010). A milestone of an international study in Taiwan teacher education–An international comparison of Taiwan mathematics teacher preparation (Taiwan TEDS-M 2008). Retrived from http://tedsm.math.ntnu.edu.tw/result.htm.

Kaplan, L. S., & Owings, W. A. (2001). Teacher quality and student achievement: Recommendations for principals. *NASSP Bulletin, 85*(628), 64–73.
Laschke, C. (2013). Effects of future mathematics teachers' affective, cognitive and socio-demographic characteristics on their knowledge at the end of the teacher education in Germany and Taiwan. *International Journal of Science and Mathematics Education, 11*(4), 895–921.
Leinhart, G., & Smith, D. A. (1985). Expertise in mathematics instruction: Subject matter knowledge. *Journal of Education Psychology, 77*(3), 247–217.
Leung, F. K. S. (2006). Mathematics education in East Asia and the West: Does culture matter? In F. K. S. Leung, K.-D. Graf, & F. J. Lopez-Real (Eds.), *Mathematics education in different cultural traditions—A comparative study of East Asia and the West* (pp. 21–46). New York: Springer.
Markus, H. R., & Kitayama, S. (1991). Culture and the self: Implications for cognition, emotion, and motivation. *Psychological Review, 98*(2), 224–253.
Martin, M. O., Mullis, I. V. S., & Foy, P. (2008). *TIMSS 2007 international mathematics report—Findings from IEA's Trends in International Mathematics and Science Study at the fourth and eighth grades.* Chestnut Hill, MA: TIMSS & PIRLS International Study Center, Lynch School of Education, Boston College.
McDonnell, L. M. (1995). Opportunity to learn as a research concept and a policy instrument. *Educational Evaluation and Policy Analysis, 17*(3), 305–322.
Mentzer, N., & Becker, K. (2010). Academic preparedness as a predictor of achievement in an engineering design challenge. *Journal of Technology Education, 22*(1), 22–42.
Mullis, I. V. S., Martin, M. O., Gonzalez, E. J., & Chrostowski, S. J. (2004). *TIMSS 2003 international mathematics report—Findings from IEA's Trends in International Mathematics and Science Study at the fourth and eighth grades.* Chestnut Hill, MA: TIMSS & PIRLS International Study Center, Lynch School of Education, Boston College.
Mullis, I. V. S., Martin, M. O., Foy, P., & Arora, A. (2012). *TIMSS 2011 international mathematics report—Findings from IEA's Trends in International Mathematics and Science Study at the fourth and eighth grades.* Chestnut Hill, MA: TIMSS & PIRLS International Study Center, Lynch School of Education, Boston College.
National Board for Professional Teaching Standards. (2001). *NBPTS early childhood generalist standards* (2nd ed.). Arlington, VA: NBPTS.
National Center for Education Statistics. (2003). *Teaching mathematics in seven countries results from the TIMSS 1999 video study (NCES 2003-013).* Washington, DC: U.S. Department of Education Retrieved from http://nces.ed.gov/pubs2003/2003013.pdf.
National Center for Education Statistics. (2009). Trends in International Mathematics and Science Study (TIMSS). Retrieved from http://nces.ed.gov/timss/video.asp.
National Council of Teachers of Mathematics. (2000). *Principles and standards for school mathematics.* Reston, VA: NCTM.
Organisation for Economic Co-operation and Development. (2007). OECD Programme for International Student Assessment (PISA). Retrieved from http://www.pisa.oecd.org/pages/0,2987 , en_32252351_32235731_1_1_1_1_1,00.html.
Parsons, C. A., & Smeeding, T. M. (2006). *Immigration and the transformation of Europe*. Leiden: Cambridge University Press.
Rice, J. K. (2003). *Teacher quality understanding the effectiveness of teacher attributes.* Washington, DC: Economic Policy Institute.
Salili, F. (1995). Explaining Chinese students' motivation and achievement: A socioculutral analysis. In M. L. Maehr & P. R. Pintrich (Eds.), *Advances in motivation and achievement: A research annual* (Vol. 9, pp. 73–118). Greenwich, CT: JAI Press.
Schmidt, W. H., Blömeke, S., Tatto, M. T., Hsieh, F.-J., Cogan, L., Houang, R. T., et al. (2011a). *Teacher education matters: A study of middle school mathematics teacher preparation in six countries.* NY: Teacher College Press.
Schmidt, W. H., Cogan, L., & Houang, R. (2011b). The role of opportunity to learn in teacher preparation: An international context. *Journal of Teacher Education, 60*(2), 138–153.

Schmidt, W. H., Cogan, L. S., Houang, R. T., & McKnight, C. C. (2011c). Content coverage differences across districts/states: A persisting challenge for U.S. Education policy. *American Journal of Education, 117*(3), 399–427.

Schmidt, W. H., Houang, R., & Cogan, L. (2012). Preparing future math teachers. *Science, 332*, 1266–1267.

Schmidt, W. H., Houang, R. T., Cogan, L., Blömeke, S., Tatto, M. T., Hsieh, F.-J., & Paine, L. (2008). Opportunity to learn in the preparation of mathematics teachers: Its structure and how it varies across six countries. *ZDM - International Journal on Mathematics Education, 40* (5), 735–747.

Shulman, L. S. (1986). Those who understand: Knowledge growth in teaching. *Educational Researcher, 15*(2), 4–14.

Siu, M. K. (2009). Mathematics education in East Asia from antiquity to modern times. In K. Bjarnadottir, F. Furinghetti, & G. Schubring (Eds.), *Dig where you stand: Proceedings of the conference on on-going research in the history of mathematics education* (pp. 197–208). Reykjavik: School of Education of University of Iceland.

Stevenson, D. L., & Baker, D. (1992). Shadow education and allocation in formal schooling: Transition to university in Japan. *American Journal of Sociology, 97*(6), 1639–1657.

Swars, S. L., Smith, S. Z., Smith, M. E., & Hart, L. C. (2009). A longitudinal study of effects of a developmental teacher preparation program on elementary prospective teachers' mathematics beliefs. *Journal of Mathematics Teacher Education, 12*(1), 47–66.

Tan, J. B., & Yates, S. (2011). Academic expectations as sources of stress in Asian students. *Social Psychology of Education: An International Journal, 14*(3), 389–407.

Tang, S.-J., & Hsieh, F.-J. (2012). The beliefs of future secondary mathematics teacher. In F.-J. Hsieh (Ed.), *Taiwan TEDS-M 2008* (pp. 221–252). Taipei: Department of Mathematics, National Taiwan Normal University (in Chinese).

Tatto, M. T., Schwille, J., Senk, S. L., Ingvarson, L., Rowley, G., Peck, R., et al. (2012). *Policy, practice, and readiness to teach primary and secondary mathematics in 17 countries: Findings from the IEA Teacher Education and Development Study in Mathematics (TEDS-M)*. Amsterdam: IEA.

Tatto, M. T., Schwille, J., Senk, S., Ingvarson, L., Peck, R., & Rowley, G. (2008). *Teacher Education and Development Study in Mathematics (TEDS-M): Policy, practice, and readiness to teach primary and secondary mathematics conceptual framework*. East Lansing: Teacher Education and Development International Study Center, College of Education, Michigan State University.

Wang, T.-Y., Hsieh, F.-J., & Tang, S.-J. (2014). Predictors of future mathematics teachers' readiness to teach: A comparison of Taiwan, Germany, and the United States. In *Proceedings of the 38th Conference of the International Group for the Psychology of Mathematics Education and the 36th Conference of the North American Chapter of the Psychology of Mathematics Education, Canada* (Vol. 5, pp. 345–352).

Wong, K. Y., Lim-Teo, S. K., Lee, N. H., Boey, K. L., Koh, C., Dindyal, J., & Cheng, L. P. (2013). Singapore. In J. Schwille, L. Ingvarson, & R. Holdgreve-Resendez (Eds.), *TEDS-M encyclopedia, a guide to teacher education context, structure, and quality assurance in 17 countries: Findings from the IEA Teacher Education and Development Study in Mathematics (TEDS-M)* (pp. 193–208). Amsterdam: IEA.

Youngs, P., & Grogan, E. (2013). The United States of America. In J. Schwille, L. Ingvarson, & R. Holdgreve-Resendez (Eds.), *TEDS-M encyclopedia, a guide to teacher education context, structure, and quality assurance in 17 countries: Findings from the IEA Teacher Education and Development Study in Mathematics (TEDS-M)* (pp. 255–272). Amsterdam: IEA.

Zhu, Y., & Leung, F. K. S. (2011). Motivation and achievement: Is there an East Asian model? *International Journal of Science and Mathematics Education, 9*, 1189–1212.

Chapter 10
Similarities and Differences in Programs for Prospective Secondary Mathematics Teachers

Jeremy Kilpatrick

Keywords Knowledge expectations • Preparation programs • Readiness • Requirements • Certification

Each of the three chapters in Sect. "Program Requirements" addresses issues arising from similarities and differences in programs that prepare prospective teachers of secondary school mathematics, analyzed within and across two or three countries. Chapter 7, by Kim and Ham, addresses knowledge expectations, as expressed in official aims and curricula, in 49 programs in Korea and the United States. Chapter 8, by Koyama and Lew, compares two programs, one in Japan and one in Korea, with respect to their official curricula and certification requirements. Chapter 9, by Wang and Hsieh, examines self-assessed readiness to teach secondary mathematics in light of future teachers' individual and institutional characteristics, at 1, 18, and 32 institutions in Singapore, Taiwan, and the United States, respectively. The authors of all three chapters stress the exploratory nature of their work and the difficulty of drawing strong implications for policy and practice from their findings.

Knowledge Expectations

Arguing that expectations for the knowledge possessed by graduates of teacher education programs can be seen simultaneously as contextualized by the national culture and decontextualized by an evolving transnational discourse on education, Kim and Ham seek to explore teacher knowledge expectations from a comparative perspective. The authors sampled 28 US teacher-preparation institutions in 22 states

J. Kilpatrick (✉)
Department of Mathematics and Science Education, University of Georgia, Athens, GA, USA
e-mail: Jkilpat@uga.edu

J.-W. Son et al. (eds.), *What Matters? Research Trends in International Comparative Studies in Mathematics Education*, Research in Mathematics Education, DOI 10.1007/978-3-319-51187-0_10

that have programs approved by the National Council for the Accreditation of Teacher Education (NCATE), and 21 Korean teacher-preparation institutions that have programs approved by the Ministry of Education and that were distributed proportionately between Seoul and the rest of the country. In both countries, the authors focus on undergraduate-level teacher preparation programs with a single major in secondary mathematics education.

Kim and Ham analyze official documents from the institutions in both countries, especially the introductory sections of handbooks for the teacher education programs or equivalent information on official websites. Embedded in the documents' expressions of official aims for the programs, they find five key ideas: (a) a wide range of disciplinary knowledge, (b) deep content knowledge, (c) knowledge about instructional methods, (d) knowledge about how to deal with issues of student equity and diversity, and (e) knowledge gained from situated experience. They likewise classify the courses offered in each program according to the type of teacher knowledge they saw as primarily being developed in the course: (a) content knowledge, (b) pedagogical content knowledge, (c) pedagogical knowledge, (d) general knowledge, and (e) field experience (practice-based courses in a classroom setting).

In both Korea and the United States, all five key ideas were well represented in the statements of program aims with the exception of knowledge of issues of student equity and diversity, which was seldom mentioned in Korea. The other key ideas were mentioned by at least one-fifth of the institutions in each country, and roughly three-fifths of the institutions in each country mentioned teacher knowledge of instructional methods. Deep content knowledge was an aim that predominated in Korea, whereas along with knowledge of equity and diversity issues, the aims of a wide range of disciplinary knowledge and knowledge gained from situated experience predominated in the United States.

Regarding curricular structures, similarities in required courses between the two countries were striking. All five types of knowledge appeared in required courses in nine out of ten programs in the United States and virtually all the Korean programs. There was also agreement between countries on the ratio (about 2 to 1) of general and non-pedagogy-related knowledge to pedagogy-related knowledge, around which curricular requirements were organized. Differences between the two countries' curricular structures followed the differences in aims: Required courses in Korea emphasized content knowledge much more than in the United States, whereas US courses tended to emphasize general knowledge, pedagogical content knowledge, and field experience much more than Korean courses did.

On the one hand, Kim and Ham argue that similarities between the two countries in aims and required courses suggest the presence of "some common international models of teacher quality and teacher qualifications" (p. 24). On the other hand, they claim that several striking differences in aims and requirements between the countries reflect their different socio-historical contexts, with the United States emphasizing strong preparation to deal with diversity in educational settings and Korea emphasizing great competence in academic mathematics. They urge scholars and policymakers in the United States to address the subject-matter preparation of mathematics teachers and scholars, and policymakers in Korea to explore more opportunities for prospective teachers to learn from practice.

Program Requirements

Koyama and Lew studied the undergraduate teacher education curricula and certification requirements of two institutions selected as providers of typical teacher education programs in mathematics in their respective countries: Hiroshima University (HU) in Japan and Korea National University of Education (KNUE) in Korea. Japan and Korea are similar in school mathematics curriculum policy, students' high achievement in international assessments, and students' negative attitudes toward school mathematics, but little is known about how their teacher education programs are organized. A detailed look at the characteristics of the 4-year undergraduate programs at these two universities has the potential to reveal strengths and weaknesses of each approach.

Both HU and KNUE put great emphasis on mathematics and mathematics education as major parts of the prospective mathematics teachers' undergraduate curriculum. At HU, apparently because the mathematics education department is separate from the mathematics department, there is a greater emphasis on pedagogical content knowledge than at KNUE. HU admits a small number of candidates for its program each year; it is not clear, however, how selective the program is, or whether the KNUE program is equally selective. Both programs have relatively short practicums: the practicum at KNUE has recently been extended to 8 weeks, whereas the practicum at HU is 4 weeks for prospective lower secondary teachers and 2 weeks for prospective upper secondary teachers. Further, prospective teachers in Korea are required to participate in at least 60 h of volunteer work in their local education communities in addition to their practicum.

In both countries, students must pass a highly selective set of examinations before they receive a teaching certificate, with examinations set nationally in Korea and locally in Japan. Koyama and Lew point out that local boards of education and governments in the two countries are able to choose which teachers to employ from a huge pool of applicants. The authors do not, however, comment on the social costs of having so many people in each country prepared for a profession they cannot pursue.

Koyama and Lew identify two issues to be dealt with in subsequent research. The first is that of finding a good balance between content knowledge and pedagogical content knowledge for prospective mathematics teachers. It is not obvious from this study which of the two universities has struck the better balance. The second issue concerns ways to make the practicum experience more effective. The two universities have organized the practicum in different ways, but the relative effect of those differences is not known.

Teaching Readiness

Wang and Hsieh use data from the Teacher Education and Development Study in Mathematics (TEDS-M) to explore and compare predictors of the readiness of future secondary teachers to teach mathematics in Singapore, Taiwan, and the United States. Readiness was assessed with an 11-item self-report questionnaire, filled out by prospective teachers of secondary mathematics who were in the final year of a teacher preparation program. They were asked to rate on a Likert scale the extent to which they thought the program had prepared them to execute various tasks central to mathematics teaching. The predictors, identified after a review of the relevant literature, consist of sociodemographic, cognitive, and affective characteristics of future teachers, as well as characteristics of the institutions that train them that had been assessed as part of the TEDS-M study. They include demographic variables (gender, home language, and socioeconomic status), secondary mathematics preparation (course level and grades received), opportunity to learn (topics in tertiary-level mathematics, school-level mathematics, mathematics education, and general education), mathematics content knowledge and mathematics pedagogical content knowledge (measured by TEDS-M test items), motivation to teach (reported on a Likert scale to rate intrinsic motivation, salary and job security, and empathy from prior learning experiences), and teacher education program quality (reported on a Likert scale to rate course quality, course arrangement, program coherence, and quality of instructors in the mathematics-related courses and supervisors in the school-based experiences).

The prospective teachers in the sample were randomly selected from teacher preparation institutions, which were also randomly chosen. Singapore was a special case because it has only one such institution: the National Institute of Education. There were 393 prospective teachers in the Singapore sample. In Taiwan and the United States, an institution had to have data from at least six participants in the TEDS-M sample in order to be included in the present study. That requirement reduced the Taiwan sample by one institution (from 365 participants at 19 institutions to 361 at 18) and the US sample by 14 institutions (from 607 participants at 46 institutions to 563 at 32).

The data from Taiwan and the United States were analyzed using two-level hierarchical linear modeling, with individual characteristics taken as the first level and institutional characteristics as the second level. Because there was only one teacher education institution in Singapore, those data were analyzed using multivariate linear modeling of individual characteristics. The only individual characteristics to predict teaching readiness in all three countries were the prospective teacher's self-reported intrinsic motivation to teach and his or her self-reported opportunity to learn topics in general education. Otherwise, the predictors of readiness were not consistent across countries. For example, the only gender difference was in Singapore, where female prospective teachers rated themselves as more ready to teach than male prospective teachers did. The more likely US prospective teachers were to speak a language other than English at home, the less highly they rated themselves as ready to teach. The level of mathematics studied in secondary school predicted teaching readiness only for Taiwanese prospective teachers. In both Taiwan and the United States, self-reported opportunity to learn topics in mathematics education predicted readiness to teach mathematics.

At the institutional level, there were fewer significant predictors of teaching readiness in Taiwan and the United States than there were at the individual level. In Taiwan, the organization of courses in the program predicted teaching readiness, whereas in the United States it was the coherence between university instruction and practicum instruction. The most inexplicable result was that in the United States across institutions, prospective teachers displaying less mathematics content knowledge rated themselves as more ready to teach mathematics than their more knowledgeable colleagues, whereas those displaying more mathematics pedagogical content knowledge rated themselves as more ready to teach mathematics than their less knowledgeable colleagues. Wang and Hsieh observe that "the reason underlying these opposite aggregate effects is not clear" (p. 28), suggesting that it might be attributable to limitations of time and energy.

Research Issues

Several issues are raised by the three chapters in this section, particularly the limitations of cross-sectional data, the use of limited measures, and insufficient phases of analysis. These issues are far from fatal to the value of the studies, but they do raise points that readers seeking to use the studies' results should consider.

Cross-sectional data. All three chapters in this section report on secondary mathematics teacher education programs as they currently exist, contrasting program differences within and across institutions and countries. Although the authors attempt at times to suggest how these programs might change, they are limited by the cross-sectional nature of the program characteristics they studied. None of the authors are in a position to make causal inferences from their correlational data.

Kim and Ham look at social expectations for teacher knowledge as products of national context and international discourse—an insightful explanation for what might appear to be contradictory forces at the national and international levels—but by taking knowledge expectations as products, they miss the opportunity to assess the possible effects of such expectations on program characteristics. It is impossible to know how much or how well the expectations for curriculum and instruction embodied in the stated aims and course descriptions of each program were realized in the program as experienced by prospective teachers of secondary school mathematics.

Koyama and Lew argue that the location of the Hiroshima University department of mathematics education in the faculty of education accounts, at least in part, for the program's emphasis on mathematics pedagogical content knowledge, compared with the Korea National University of Education program's emphasis on mathematics content knowledge. Certainly a structural arrangement of faculty can account for program emphasis, but the opposite can occur, too. Without sufficient historical data on how each university has created and staffed its programs in mathematics and mathematics education, researchers cannot establish the role that university organizational structure might have played in program structure and vice versa.

Wang and Hsieh use a set of assorted variables to predict teaching readiness, but inasmuch as measures of the variables in TEDS-M were collected at roughly the same time, readiness could just as easily have been used as a predictor instead, with—for example—the measures of mathematics content knowledge and mathematics pedagogical content knowledge taken as the response variables. In other words, without more and different data, we cannot know how readiness influences or is influenced by other program characteristics. We cannot even know that the teacher education programs were in a steady state or that sufficient relevant variables were included in the predictive models.

Limited measures. International studies such as those presented in Sect. "Program Requirements" ordinarily involve large data sets even when, as in the last two chapters, the number of institutions is small: Koyama and Lew looked at only two institutions, and in Singapore, Wang and Hsieh studied only one. Although the data sets for studies of teacher education programs may be large, they typically address only a few aspects of the program. They often make use of official documents, public records, and questionnaire surveys of program faculty or students, and seldom do they include observations of a program in action or interviews with program participants.

All of the data obtained by Kim and Ham came from the text of official documents or websites describing program aims and curricula. Similarly, Koyama and Lew used data from published regulations for teacher certificates, program descriptions for university courses, and success rates in employment examinations. These data give an official's-eye picture of each program, but they do not provide measures of how the program functions, how participants respond to it, or what they learn from it.

The data analyzed in the third study, in contrast, came from other sources. The TEDS-M study included surveys of teacher educators and prospective teachers in the participating countries as well as questionnaires and knowledge assessments of prospective teachers. Wang and Hsieh confined their attention to the surveys, questionnaires, and assessments of the prospective secondary mathematics teachers in the three countries whose programs were studied. Those data were limited in the sense that they dealt only with characteristics and responses of program participants. They did not include teacher educators' views of the programs, official descriptions of the programs, or observations of program activities.

Insufficient analysis phases. Kim and Ham appear to have made a strong effort to pursue different phases of analysis by looking at similarities and then differences in program features between countries. They did not, however, address similarities or differences across institutions within countries very deeply, essentially assuming that accreditation procedures yield a certain uniformity across programs. They report the percentages of programs whose official aim statements mentioned certain key ideas and those whose credit requirements addressed different types of teacher knowledge, but they apparently did not attempt to partition the programs in a country into types sharing similar features. Such an analysis might have helped Kim and Ham uncover ways in which socio-historical contexts are less homogeneous than they assumed.

Because Koyama and Lew did two case studies of programs, they were not in a position to examine variation within and between countries, but they might have gone further than they did to analyze the courses and credits they list as required for certification in each country or graduation from each institution. For example, they could have looked at key ideas and knowledge types using categories like those developed by Kim and Ham. Such an analysis would likely have been more easily interpreted than their current comparison of mathematics education studies and mathematics studies at the two institutions, which consists only of side-by-side listings of course titles (Tables 6 and 7).

Wang and Hsieh partition their analyses into an analysis of individual characteristics followed by an analysis of institutional characteristics. Their inclusion of Singapore—which has only one institution—made the partitioning somewhat awkward to report. Moreover, the analysis of individual characteristics conflates sociodemographic, knowledge achievement, and self-report (opportunity to learn, motivation, program quality, etc.) variables as predictors, with self-reported preparation as the only outcome variable. That analysis might have instead been conducted in separate phases, so as to take into account what are likely to have been the strong intercorrelations among the self-report variables. Prospective teachers who rate themselves as strongly motivated to become teachers and as having had the opportunity to learn topics in mathematics and mathematics education are quite likely to rate themselves also as having been well prepared by their teacher education program, and disentangling cause and effect among those variables is a nontrivial task. It might have helped to begin the analysis with a phase in which the self-report variables were analyzed before throwing the predictors into the same pot.

Research Directions

Despite various limitations, which are well-recognized by the authors, the three chapters in Sect. “Program Requirements” make a fine start on investigating the institutionalization of teacher education programs in secondary school mathematics. Because the field is so undeveloped, it can grow in many directions. Three directions that seem especially promising are to (a) expand the number of countries represented in a study, (b) develop stronger theoretical frameworks, and (c) conduct long-term studies of change. Each chapter suggests why these might be promising directions.

More countries. One of the unfortunate limitations of the studies reported in Sect. “Program Requirements” is that no country was represented in all three chapters and that only Korea and the United States appear more than once. Korea participated in the first data-based international comparison study of mathematics teacher education—Mathematics Teaching in the twenty-first century, which was the precursor to TEDS-M—but not in TEDS-M itself. If it had, then Kim and Ham might have used TEDS-M data along with the data they had on aims and curricula in Korea and the United States to supplement the picture they obtained from official documents. Similarly, if Japan had been included in the study by Kim and Ham, then Koyama and Lew might have been able to augment their case studies of the Hiroshima University and the Korea National University of Education programs with data from documents describing aims and curricula in the two countries. In any event, by adding several more countries to the ones they chose to study, each group of authors could test the generalizability of the conclusions they have drawn from the data they have. Depending on the countries from which additional data might be obtained, researchers could test hypotheses about the typicality of models of teacher education in different countries.

Stronger frameworks. The chapter by Kim and Ham is the only one in the section with explicit research questions and perhaps not surprisingly, the only one with an explicit theoretical framework—one dealing with contextualized and decontextualized dimensions of teacher knowledge. The other two chapters report studies that are essentially explorations of the characteristics of programs to educate secondary school mathematics teachers, without invoking explicit frameworks to guide the nature and selection of measures of those characteristics. A clear direction for future research is the development and employment of stronger frameworks for framing international studies of teacher education programs in school mathematics. Even research that follows the Kim and Ham work would undoubtedly benefit from some elaboration of the culturally contextualized and semantically decontextualized dimensions of teacher knowledge, given that it is not clear why between-country differences are necessarily culturally contextualized and between-country similarities are necessarily semantically decontextualized.

Long-term studies. Eventually, research into the institutionalization of programs to educate mathematics teachers in different countries will need to address how change occurs in such programs and what might be responsible for that change. Studies of change require research conducted over enough time for that change to be not only apparent but also measurable. Until such long-term studies are conducted, however, causal inferences about the effects of teacher education programs will not be warranted. Such studies are likely to be expensive and difficult to conduct, but they are necessary if international comparative studies are to make a difference in the practice of teacher education.

Conclusion

The studies reported in the three chapters of the present section constitute an important first step in establishing a research basis for studying the effectiveness of mathematics teacher education programs around the world. Unfortunately, research into teacher education programs in mathematics—whether comparative or not—lacks agreed-upon measures of program outcomes, a situation that needs to change if progress is to be made in improving the effectiveness of such programs. Researchers should develop and assess measures of program outcomes concerning graduates' knowledge, skill, beliefs, and performance—in other words, their teaching proficiency—that will predict the quality of their subsequent teaching of mathematics. When such measures are available, studies can be done of the ability of program characteristics to predict graduates' teaching proficiency in mathematics. At that point, program differences within or between countries can be used to explore those characteristics that lead to greater or less teaching proficiency upon completion of the program, as well as effective teaching of mathematics thereafter.

Part III
Research on Improving Teacher Knowledge and Pedagogical Approaches

Chapter 11
Cross-Cultural Lesson Planning Between the United States and South Korea

Woong Lim and Ji-Won Son

Abstract Our chapter examines a cross-cultural learning project in which teachers from the United States and South Korea collaborated and shared feedback on writing lesson plans using asynchronous communication tools. In our documentation and discussion of participants' perceived strengths and weaknesses in lesson planning, we found that their cross-cultural experiences facilitate collaborative lesson planning by fostering cross-cultural perspectives on teaching and provide insight into curricular ideas that have the potential to narrow the teaching gaps between countries.

Keywords Lesson planning • Comparative study • Preservice teacher education

Introduction

Through the process of examining their own culture and others' cultures, teachers can develop an intercultural understanding of teaching students—appreciating the different ways in which individual, group, and national identities are reflected in classrooms, as well as the diverse and changing nature of pedagogy (Cai, Mok, Reddy, & Stacey, 2016; Lim, Kim, Stallings, & Son, 2015). Cross-cultural learning opportunities serve teachers to this effect, engaging them in learning and approaching diverse pedagogical cultures by recognizing commonalities and

---

This chapter uses the same data in a study (Lim & Son, 2013) first reported in *New Waves—Educational Research and Development.*

W. Lim (✉)
Department of Teacher Education, Educational Leadership & Policy, University of New Mexico, Albuquerque, NM, USA
e-mail: woonglim@unm.edu

J.-W. Son
Department of Learning and Instruction, University at Buffalo–The State University of New York, Buffalo, NY, USA
e-mail: jiwonson@buffalo.edu

J.-W. Son et al. (eds.), *What Matters? Research Trends in International Comparative Studies in Mathematics Education*, Research in Mathematics Education, DOI 10.1007/978-3-319-51187-0_11

differences, creating connections with others, and cultivating mutual respect in teaching professions (Miller, Smith, Zhu, & Zhang, 1995; Stigler & Hiebert, 1999).

The integration of multicultural approaches in mathematics instruction has long helped us identify appropriate attitudes about mathematics classrooms and their cultural relevancy in education (Stigler & Hiebert, 1999). While modern mathematics uses universal symbols, mathematics instruction can find its origins in many cultures. What's more, new technologies and digital learning environments can provide interactive contexts for the exploration of multiple perspectives in mathematics instruction, providing experiences through which teachers can identify and resolve teaching and learning issues in diverse classrooms.

In this chapter we introduce an international collaborative project wherein participants from the United States and South Korea created lesson plans together, experienced various approaches to curriculum planning, and partook in opportunities to broaden their perspectives on teaching mathematics. Our study was guided by the following research question: How does cross-cultural lesson planning help prospective teachers increase their awareness of multiple perspectives of pedagogy and the reflection of their teaching practices?

Relevant Literature Review

The premise of this study is that different cultural traditions can remarkably impact mathematics teaching (An, Kulm, & Wu, 2004; Cai, Ding, & Wang, 2013; Cai & Wang, 2009; Ma, 1999; Stigler & Hiebert, 1999). In fact, a wide range of international comparative studies have revealed the impact of cultural influence on mathematics education in different traditions (Cai et al., 2016), particularly comparisons between (but not limited to) China and the United States (Cai, 2005; Cai & Wang, 2009), Japan and the United States (Jacobs, Yoshida, Stigler, & Fernandez, 1997; Whitman & Lai, 1990), Hungary and England (Andrews, 1999; Hatch, 1999; Harries, 1997), France and Britain (Jennings & Dunne, 1996), and China, Hong Kong, and Britain (Leung, 1995). Cross-cultural studies comparing the mathematical achievements of the United States and East Asian countries (i.e., China, Japan, Korea, and Singapore) affirm that East Asian students have consistently outperformed American students in almost every area of mathematical knowledge (Geary, Fan, & Bow-Thomas, 1992; Lemke et al., 2004; Stevenson & Stigler, 1992).

To explain such learning gaps, researchers have explored and studied different systems of numerals (Fuson & Kwon, 1991; Miller et al., 1995; Miller & Stigler, 1987), cultural differences (e.g., parental expectations, student motivation, beliefs, and effort), school organization (e.g., time spent on learning mathematics in school), classroom practice (Cai & Wang, 2009; Yang & Cobb, 1995), and the content and organization of mathematics curricula (Geary, Bow-Thomas, Fan, & Siegler, 1993; Son & Senk, 2010; Stevenson & Stigler, 1992; Sutter, 2000). The outcomes of this body of research can have a potentially large bearing on current mathematics education in the United States. Yet without a careful exploration of cultural influence from the first-hand experiences of other cultural systems, it is difficult to achieve an understanding clear enough to promote learning from different educational systems (An, 2004; Wang & Lin, 2005).

Listening to the voices of teachers (practicing classroom teachers as well as prospective teachers) helps researchers to examine important education issues from diverse perspectives (Cai et al., 2013; Cai & Wang, 2009). Over the past two decades teacher conceptions of mathematics and views toward mathematics teaching have continued to interest many research communities, as they "play a significant role in shaping teachers' characteristic patterns of instructional behavior" (Thompson, 1992, p. 130). According to Stigler and Hiebert (1999), the integration of multicultural approaches in mathematics instruction helps identify appropriate attitudes toward teaching and learning mathematics. Numerous studies also confirm that cross-cultural studies in mathematics education provide mathematics educators with opportunities to increase their awareness of alternative teaching and learning, as well as to promote reflection on their teaching practices (An, 2004; Lemke et al., 2004; Stigler & Hiebert, 1999; Stigler & Perry, 1988). Unfortunately, there exists little research exploring whether prospective mathematics teachers who interact with peers on tasks in different cultural settings develop new teaching perspectives.

Theoretical Perspectives

Fundamental to this study's analysis is the concept of experiential learning by Kolb and Fry (1975). According to Kolb and Fry, experiential learning occurs through direct participation in which an individual experiences, reflects on, abstracts, and tests his or her learning in a new situation from primary experience (Jarvis, 1995). Later, An (2004) proposed three stages of observation, experience, and reflection, all experienced first-hand by teachers learning in a given cultural setting. Similarly, a body of prior research indicates that the integration of experiential learning for teachers in multicultural education benefits prospective teacher learning through cultural immersion in cross-cultural settings (e.g., Spalding, Wang, Lin, & Butcher, 2005; Stachowski & Mahan, 1998; Wiest, 1998; Willison, 1994). With that said, more studies are needed in the area of prospective teachers' learning to accomplish subject-specific and significant teaching tasks, such as lesson planning, delivery, or assessment, through cross-cultural experiences.

This study addresses the learning of US prospective mathematics and science teachers in mathematical lesson planning with international peers. We attempt to illuminate the role of teacher learning in a different cultural setting by focusing on the accounts of prospective teachers; this includes teachers' first-hand experiences of working with Korean teachers in preparing, refining, and reflecting on lesson planning, as well as developing reflective attitudes through experiential learning.

Methods

Project Description

This project originated from a lunch conversation in 2008 between two mathematics teacher educators—both had attended the 32nd Conference of the International Group for the Psychology of Mathematics Education, after which they agreed to implement the same course assignment at the course level as part of an international partnership. As the educators worked in different teacher preparation programs, they noticed the positive interactions among their graduate advisees between Korea and the United States. Prior to the study they revised their education courses, adding an international project as a major assignment. Thus, the project launched an international partnership in teacher education at the course level from spring 2009 to summer 2010, and teachers from the United States and South Korea participated in the project to build a community of science and mathematics teachers.

Participants

The participants consisted of two course instructors and their students, all of whom were teacher candidates and practicing teachers enrolled in science and mathematics methods courses. Participants collaborated in lesson planning and exchanged peer feedback through asynchronous communication tools, in the context of graduate-level education courses in the United States and South Korea. In the United States, the course "Middle School Curriculum Design" was offered at a state university as part of a Master of Arts in Teaching (MAT) program—participants were teacher candidates ($n = 27$) in their junior or senior year of the program in middle school mathematics and science, with no full-time teaching experience. The counterpart course in South Korea was titled "Curriculum and Instruction" and was offered in the university's MAT program. The Korean participants ($n = 40$) were math/science middle school teachers in the first year of their Master's degree program, and had 4–8 years of teaching experience. They were also taking an English conversation course as a co-requisite in the degree plan.

Descriptions of Lesson Planning Assignment

Due to differences in curriculum methods and perspectives between the two countries, researchers in the project examined course syllabi in order to identify learning goals and related tasks/assignments common to both courses. Students did not participate in the planning process of this project. Table 11.1 below shows an

Table 11.1 Outline of the project assignment

Assignment (duration)	Product (responsible party)	Communication tools
1. Create small groups (each group has 2–4 members) (<30 min)	• Group title and brief bios of members (Students)	N/A
2. Select a mathematics topic for a lesson plan (<30 min)	• Math topic and a description of big ideas (Students; Instructors provided a list of topics)	N/A
3. Partner with an international collaborator group (<3 days)	• List of collaborator groups with the common math topics (Instructors)	Email (or Dropbox)/ Video Conferencing
4. Produce welcome greetings to collaborator group (<2 days)	• Video file or written messages (Students)	Camcorder; Email (or Dropbox) video files/ document attachment
5. Identify a target grade/level of middle grade students and standards (<2 days)	• Description of student population and aligned standards (Students)	Email (or Dropbox)
6. Produce a lesson plan/ report progress/videotape meetings (optional) (<1.5 weeks)	• Lesson plan + Discussion forum entries (Students) • Discussions focused on synthesizing different content standards and infusing various pedagogical strategies	Email (or Dropbox) video files and documents; Online Discussion forum
7. Provide peer feedback on lesson plan (<1 week)	• Feedback form/video commentary (optional)(Students; Instructors provide feedback form)	Camcorder; Email (or Dropbox) video files and documents
8. Teaching demo of lesson plan (30 min) (<1 week)	• Videotaped teaching demo (Students)	Camcorder; Email (or Dropbox) video files
9. Provide peer feedback on teaching (<1 week)	• Feedback form/video commentary (optional)(Students; Instructors provide feedback form)	Camcorder; Email (or Dropbox) video files and documents
Repeat the entire process with the same collaborator group (<3 weeks)		
10. Reflection and Recommendations (2 h)	• Videotaped classroom discussion; online discussion forum (Students and Instructors)	Camcorder; Online Discussion forum

outline of the collaborative international learning task that instructors incorporated into their course syllabi. In order to build international collaboration at the course level, participants were asked to construct a lesson plan (on the same mathematical topic) addressing learning objectives, anticipatory sets, lecture/activities, application/problem-solving, closure, and assessment; the target grade levels were 6th, 7th, and 8th grades. A group of Korean mathematics and science teachers was paired with a group of American mathematics and science teachers. The project required the collaborator group to have at least one Korean mathematics teacher, one Korean science teacher, one American math teacher candidate, and one

American science teacher candidate, respectively. Two US participants opted out of the second lesson planning assignment, and instead submitted individual work by completing the same assignments without the international collaborative project.

Data Collection and Analysis

For the project investigation, student responses to surveys, reflective writings, comments in focus group discussions, video clips, and archived emails were used for analysis. Three rounds of surveys were administered: the first surveys were administered at the beginning of the course, the second surveys at the midpoint before the second lesson plan assignment, and the third during the last meeting of the course. These surveys solicited a critical perspective on program outcomes by addressing the areas of strengths, weaknesses, improvement, and individual reflections on change and growth. To evaluate student performances, instructors initially asked individual students to provide a narrative of their contributions to lesson planning and peer feedback, linking the evidence with examples and actual products. To elicit responses related to these themes, the following items were included in the survey and writing assignments:

- Describe in detail how the Korean/American partners were similar/dissimilar in constructing a lesson plan. In your description, please address each part of the lesson plan (Learning objective, Anticipatory sets, Lecture/activities, Application/problem-solving, Closure, and Assessment).
- How would you describe the following topics based on your experience of working with international partners?
  - Teaching mathematics in Korea vs. America
  - Mathematical knowledge of Korean vs. American teachers
- How was your experience working with science (or mathematics) teachers? How did it affect your teaching knowledge and skills? How did group work help shape your view toward STEM education?

Focus group discussions were purposeful in that researchers were able to use the opportunity to confirm proposed findings and allow the participants (students and instructors) to confirm, clarify, or further elaborate.

In regard to data analysis, it first began in an explorative manner—with survey responses and reflective writing assignments in order to capture primary patterns—and progressed in a confirmatory manner, with two focus group discussions that helped establish findings. The data was independently coded by research assistants who completed two graduate courses in qualitative research methodology. Primary patterns were then triangulated and determined valid and meaningful once the finding was supported by at least two data sources. As this study did not have interventions or experiments designed to produce measurable outcomes, its findings may not be generalized.

Outcomes of the Project

From our analysis of student responses to surveys and comments from reflection assignments and focus group discussions, the following three themes emerged: (1) Growth in the perspectives of teaching mathematics, (2) Community of mathematics educators, and (3) Embracing cross-cultural engagement. By participating in cross-cultural lesson planning, participants in both countries reported that their awareness of alternative teaching and learning had increased, as well as their reflection on teaching practices.

Perceived Growth in the Perspectives of Teaching Mathematics

The majority of participants (82% of US participants and 74% of Korean participants) stated that the project helped them develop new perspectives on teaching and learning mathematics. The three most popular words used to describe their cross-cultural lesson planning were *exciting*, *productive*, and *rigorous*. One popular observation ($n = 48$) was that the international project provided the participants with opportunities to experience *different* teaching. Another was that through the project they began to realize how significant a role culture plays (i.e., in the United States or South Korea) in shaping curriculum design and content delivery. For example, one Korean participant said:

> "I never heard of equity in a classroom. It [international learning] has made me think about how I treat my students fairly whether they are from a rich family or not. I guess American teachers think about ensuring equal access to education regardless of students' race or nationality. But Koreans won't worry too much about cultural diversity because Korea is a very homogenous society. I began to think about how our culture and society influence [what we do] as teachers."

Kolb and Fry's learning cycle was confirmed as an appropriate model to illustrate transformation in attitude and thinking such as new perspectives through the participants' cross-cultural experience. Through the concrete experience of constructing mathematics lesson plans in collaboration with international peers, the participants were challenged to think abstractly, explore methods unique to certain cultures of mathematics teaching, and conceptualize ways to contribute to the global community of mathematics educators. As to the fourth part of the cycle of testing in new situations, the study did not have data for actual testing, so the participants' statements about their willingness to try new ideas were referenced instead. The following table describes each state of the cycle with corresponding common words (coded to identify the characteristics of the stage) and representative participant comments (Table 11.2).

Additionally, the project examined how participants perceived the different approaches of the two countries. The lesson plan assignment required the

Table 11.2 Kolb and Fry's cycle and participants' supporting comments

Kolb and Fry's cycle	Descriptive words/phrases identified in coding process	Representative comment
1. Concrete experience	Real experience; face to face meetings; learning by doing projects	"I read about how mathematics is taught in other countries but never imagined I'd be talking with them on Skype, sending emails, and hearing how much they liked my ideas. I think I had a real valuable experience."/"It was interesting that I actually wrote a lesson and my American partner reviewed it, and I felt quite [real about it]."
2. Observation and reflection	Peer review; watch how they teach; analyze teaching; think deeply about; honestly, I think	"I am not certain how great teachers they (American prospective teachers) will become. But the way they are willing to listen to children's comments and rephrase their thinking made me reflect on how I had been with my Korean students. I think I was too authoritarian and controlling instead of focusing on how much my children learn from working with me."/"Korean [practicing teachers] knew lots of mathematics. That made me think about endless possibilities I could bring to my class if I had as much knowledge in math as the Korean partners."
3. Forming abstract concepts	Feel strongly; believe; argue; makes more sense if; idea and concept	"I thought teaching mathematics should be different one country from another. So learning mathematics could be a different process depending on cultures and personalities or needs."/"The idea about teaching STEM can't materialize until people understand how math or science is taught and how much they need the other content to develop complete thinking."
4. Testing in new situations	My own teaching; in the future; the way it applies to my teaching	"My learning experience with Korean teachers was definitely empowering. We teach different kids, though I think all students are the same and different at the same time. When my way of doing math is not helping my students, I think I will try different methods I heard from the Koreans."/"One thing I realized in this course is your teaching never goes as you plan and your students are never the same type of students. The more you know, the more you remain flexible, the better teaching you can make."

participants to address learning objectives appropriate for middle grades (i.e., 6th–8th grades), anticipatory sets, lecture/activities, application/problem-solving, closure, and assessment in their lesson plans. Participants selected a wide range of mathematical topics, including arithmetic sequences and the fundamental counting principle to the distance formula, exponential functions, fitting a line to data, function notation, graphing a linear function, irrational numbers and radicals, the measures of central tendency, linear inequalities, quadratic equations, rational expressions, solving a system of equations, surface area and volume, transformations, x-y coordinates, and, finally, ordered pairs and slopes. Since the two countries have different national and state content standards, not to mention various pedagogical strategies, the differences were identified and presented to the participants as part of the course curriculum on international curricular issues in mathematics. To extend the learning of international curriculum, participants were also encouraged to synthesize, negotiate, and infuse differences in content standards and teaching ideas to address common mathematical topics. Table 11.3 indicates some of the strengths identified by collaborator groups relating to content and pedagogy that were demonstrated through various aspects of the lesson plans. The numbers in parentheses indicate how many times the same description was mentioned in responses to surveys and reflective writings by different participants—for American prospective teachers, comments mentioned more than four times were included; for Korean teachers, those mentioned more than seven times were included. Note that the findings in the table should not be taken as the general characteristics of mathematics teaching in each country, primarily because other factors contributing to the perceived differences, such as teaching experience or content knowledge, are liable to exist. Instead, these findings demonstrate the multiple layers of perceived differences and rising learning opportunities in cross-cultural interactions.

One of the unique ways the project gave opportunities for teacher exploration was by enabling science teachers to contribute to lesson planning for sections concerning application and problem-solving, which allowed mathematics teachers in both countries to have greater consideration for teaching mathematics in conjunction with teaching science. On one hand, American participants recognized that Korean teachers strongly challenged their students to use basic math skills and mathematical reasoning in application and problem-solving in physics, chemistry, and biology; on the other hand, Korean participants recognized that American teachers showed greater commitment to making mathematics learning relevant and engaging. One Korean participant wrote the following:

> "American math teachers seem to really try hard to show how math is useful outside the class when all the Korean teachers were trying to show that math exists in classrooms and science exists in the lab. When I saw most American teachers were excited for using computer games or magazine articles in the application section, I knew that their attitude and intention was to make students think and learn rather than to make students work, work, and work. I thought I would start reading some science magazines just to get some ideas on the use of math in science so that I can share with my students."

Table 11.3 Perceived differences in lesson plans

Stages in a lesson plan	Korean participants said American teachers were *different* concerning:	American participants said Korean teachers were *different* concerning:
• Learning objectives	• Writing explicit objectives w/ measurable verbs ($n = 12$)	• Using objectives to establish high expectations ($n = 8$)
	• Aligning well with state standards ($n = 14$)	• Presenting mathematically powerful ideas ($n = 7$)
	• Including the "doing" of mathematics instead of "being able to solve" math problems ($n = 8$)	• Not afraid to ask for abstract thinking ($n = 12$)
• Anticipatory sets	• Allowing students to do mathematics ($n = 21$)	• Clearly addressing the prerequisite skills necessary for the pertaining lesson ($n = 8$)
	• Brief and easy to increase motivation ($n = 18$)	• Active use of textbook materials ($n = 5$)
	• Involving student-teacher conversations ($n = 29$)	
• Lecture/ activities	• Providing details about procedures ($n = 10$)	• Not afraid to present difficult problems ($n = 18$)
	• Actively using PowerPoint slides ($n = 33$)	• Pervasive use of decimals and fractions ($n = 11$)
	• Using visualizations to represent mathematics ($n = 13$)	• Emphasizing writing math solutions as a cohesive body of procedures and concepts ($n = 9$)
	• Tasks encourage collaboration and student–teacher interactions ($n = 9$)	
• Application/problem-solving	• Creative presentations, such as use of interviews with scientists, computer games, science fiction, and movies ($n = 28$)	• Using textbook examples effectively ($n = 11$)
	• Connecting to real life situations ($n = 31$)	• Asking students to research; not afraid to challenge students ($n = 18$)
	• Willing to take risks by trying teachers' own ideas; not relying on prescribed curricular materials ($n = 20$)	• Science teachers' high content knowledge ($n = 20$)
	• Emphasizing modeling situations ($n = 11$)	• Posing high cognitive demand tasks ($n = 10$)
	• Using manipulatives, objectives, and visuals ($n = 34$)	• Emphasizing individual responsibility ($n = 9$)
	• Flexible and selective use of existing curricular materials ($n = 9$)	
• Closure	• Asking for student feedback (e.g., the muddiest point) ($n = 18$)	• Willing to omit closure if the lecture/activities extended ($n = 14$)
	• Interesting strategies such as exit slip, 3-2-1, 3 W's ($n = 19$)	• Effective teacher questioning ($n = 8$)
	• Encouraging student–student interactions (i.e., think/write/pair-share) ($n = 8$)	

(continued)

Table 11.3 (continued)

Stages in a lesson plan	Korean participants said American teachers were *different* concerning:	American participants said Korean teachers were *different* concerning:
• Assessment	• Using performance-based assessment ($n = 11$)	• Using standardized testing ($n = 7$)
	• Applying multiple ways to assess ($n = 10$)	• Sharing the same assessment in the department ($n = 8$)
	• Enforcing test accommodations ($n = 10$)	• Consistent grading using scoring rubrics ($n = 9$)

Of the participating math/science teachers, 75% of American prospective teachers ($n = 20$) and 82% of Korean teachers ($n = 32$) supplied evidence to demonstrate changes in their attitude and perspective on content and pedagogy. The evidence included comments on attitude and descriptions of perspectives gained in content and pedagogy. Additionally, the focus-group discussion informed the study that mathematics teachers were often willing to revise content by emphasizing particular concepts or providing more opportunities for math skills practice after hearing the science teachers share their experiences of using mathematics in their teaching. Simultaneously, the science teachers consulted with mathematics teachers about designing application and problem sections in their lesson plans so that their science content could include rich mathematical thinking and reasoning. In particular, 38% of the participants ($n = 25$) mentioned that they recognized problem-solving as an important mathematical activity and renewed their commitment to implement more problem-solving in their future lessons.

Community of Mathematics Educators

Though the project did initially offer collaborative opportunities in curriculum planning, it included few social activities to promote a global community of learners and educators directly. This changed over time, as the participants began to foster relationships through working online (video-conferencing) and off-line (email) together, in addition to the project requirements. They interacted by exchanging solutions to problem-solving, sharing teaching resources, and even learning to speak each other's languages. Though these activities were not counted for a grade, the participants continued to seek networking opportunities and deepen their interpersonal relationships. This outcome supports the view that students will respond positively to collaborative educational settings. In fact, the majority of participants in the survey (93% of the US participants and 91% of the Korean participants) expressed comfort in having a successful global community of educators, saying that they would seek similar international learning opportunities in the future. One American participant had this to say:

> "The Korean teachers knew so much math and science. I felt I was contributing a lot in a very serious professional organization of international teachers. I thought a lot about my own skills as science teacher. Yes, I do speak English and helped the Korean teachers with their writing. But I wish I had more experience and content knowledge to discuss real teaching. I've got a lot work to do!"

Embracing Cross-Cultural Engagement

Both instructors reported a substantial increase in student participation and performance throughout the study. In the US program, the average grade for a lesson plan project over the course of three school years prior to this project was 78%, or a letter grade of C. This international project produced an average grade of 85%, with significantly fewer students who dropped out of the course. In the Korean program, the course traditionally produced 10% A's and 25% B's with about 15% dropped students. The international project allowed for almost 15% A's and 45% B's with only one dropped student, about which a Korean instructor commented:

> "Students attended to the tasks, were committed to the tasks when there were no extrinsic rewards, [and] persisted in completing the task even when the work became difficult; most students said that they didn't want to quit and in fact, worked harder to impress the American teachers."

Indeed, a meaningful opportunity to engage with a community of collaborative learners can largely help in addressing the critical aspect of becoming a reflective educator. Other evidence, especially regarding the development of reflective thinking by participating in a community of learners, come from an analysis of commentaries on the online discussion forum. About 67% of the participants ($n = 45$) wrote that learning with peers had a positive impact on broadening their perspectives and helped them critically examine their current teaching knowledge and practice. In fact, more than two-thirds of participants from both countries mentioned at least twice in their commentaries that their learning engaged them in a way that intrinsically motivated them to participate and perform at a higher level. When participants were asked to list motivational factors for actively participating in the project, the most popular was the international learning opportunity (mentioned by 52 out of 67 participants), followed by grades ($n = 48$), relationships ($n = 34$), and the usefulness of assignments ($n = 27$), among others. Moreover, about 50% of the participants ($n = 35$) described how their international partners were instrumental in understanding diversity and enriching their views toward content and pedagogy. One American participant wrote the following:

> "It was clear to me the Koreans have superior content knowledge. However, do they use the knowledge and work hard, finding great teaching strategies to make the materials easy for the students? Some Korean teachers mentioned they had not ever thought about increasing student motivation in the ways I explained to them. It felt great to know that teachers need to know both content and pedagogy regardless of nationality, and different cultures help us understand we need to collaborate in unison in educating our children."

Implications

Through the project we were able to observe how the integration of cross-cultural experience in mathematics planning and instruction can help participants identify different methods and approaches to teaching and learning mathematics—and help them see the value of culture in education. The perceived growth in the participants' reflective nature, especially among the US participants without full-time teaching experience, demonstrates that appropriate learning experiences within a community of teachers with diverse cultural perspectives have the potential to enable prospective teachers to increase the awareness of various pedagogical ideas in their future teaching—in the daily demands of educating students from various cultural backgrounds. For example, one US participant made this observation:

> "My experience with Korean teachers allowed me to think a lot about my own teaching in the classroom where I might end up having students and parents from different countries, I mean any country, with different expectations of learning math. I want to be the teacher who appreciate the differences and embrace [students'] math knowledge and cultural backgrounds as part of what makes [students] excited for learning algebra with me... [The] more mathematics you get to learn with people from different countries, the more open-minded you become."

Although this reflection does not indicate how the international experience transformed the participant's teaching, it does reveal how international learning impacted his/her views on working with people of different cultural backgrounds. We hope that international projects such as this study, facilitated by teacher educators in an international partnership and enabled by low-budget technology, can transform the attitudes of those who have never traveled outside of their home countries.

The study does have its limitations, such as the discrepancies between the two groups of participants. For example, the US participants were preservice teachers while the Korean participants were practicing teachers. We did not dwell much on this difference in the analysis because the study focused on the cross-cultural phenomenon of lesson planning. Our analysis is not so much about delineating cultural differences in teacher knowledge as it is an opportunity to interact with international peers in teacher education and reflect on the exchange.

Nonetheless, future studies can certainly make a more layered analysis of the phenomenon, by comparing preservice teachers in countries or practicing teachers in countries, respectively. Likewise, as the data sources for this study were largely self-reported, future studies should consider data that portray actual teaching practice or change in teaching practice. In addition, future studies can use previous lesson plans as a comparison group to produce better analyses of teacher skills (besides perspective) that can be gained through collaborative learning opportunities. Ultimately, what we need is a line of research reporting pervasive lesson plan patterns and making sense of them through the analyses of different teacher cultures—especially when the international mathematics teacher education community seeks to learn from each other.

In response to the need of prospective teachers—with the potential to grow into classroom teachers with rich international perspectives—providing student teachers with meaningful learning opportunities where they can interact with international peers and reflect on their practice can be empowering. We hope that similar international projects or transformative collaborative cross-cultural learning projects continue to inform teachers and become a regular part of teacher preparation programs around the world.

References

An, S. (2004). *The middle path in math instruction: Solutions for improving math education.* Lanham, MD: Scarecrow Education.

An, S., Kulm, G., & Wu, Z. (2004). The pedagogical content knowledge of middle school mathematics teachers in China and the U.S. *Journal of Mathematics Teacher Education, 7*, 145–172.

Andrews, P. (1999). Looking behind the rhetoric: Some new insights from Hungary. *Mathematics Teaching, 167*, 6–10.

Cai, J. (2005). U.S. and Chinese teachers' constructing, knowing, and evaluating representations to teach mathematics. *Mathematical Thinking and Learning, 7*(2), 135–169.

Cai, J., Ding, M., & Wang, T. (2013). How do exemplary Chinese and U.S. mathematics teachers view instructional coherence? *Educational Studies in Mathematics, 85*(2), 265–280.

Cai, J., Mok, I. A. C., Reddy, V., & Stacey, K. (2016). *International comparative studies in mathematics: Lessons for improving students' learning*. Switzerland: Springer International Publishing AG.

Cai, J., & Wang, T. (2009). Conceptions of effective mathematics teaching within a cultural context: Perspectives of teachers from China and the United States. *Journal of Mathematics Teacher Education, 13*(3), 265–287.

Fuson, K. C., & Kwon, Y. (1991). Chinese-based regular and European irregular systems of number words: The disadvantage for English-speaking children. In K. Durkin & B. Shire (Eds.), *Language in mathematics education: Research and practice* (pp. 211–226). Buckingham, England: Open University Press.

Geary, D. C., Bow-Thomas, C. C., Fan, L., & Siegler, R. S. (1993). Even before formal instruction, Chinese children outperform American children in mental addition. *Cognitive Development, 8*(4), 517–529.

Geary, D. C., Fan, L., & Bow-Thomas, C. C. (1992). Numerical cognition: Loci of ability differences comparing children from China and the United States. *Psychological Science, 3*, 180–185.

Harries, T. (1997). Reflections on a lesson in Kaposvar. *Mathematics Teaching, 161*, 11–13.

Hatch, G. (1999). It wouldn't be like that here. *Mathematics Teaching, 168*, 26–31.

Jacobs, J. K., Yoshida, M., Stigler, J. W., & Fernandez, C. (1997). Japanese and American teachers' evaluations of mathematics lessons: A new technique for exploring beliefs. *Journal of Mathematical Behavior, 16*(1), 7–24.

Jarvis, P. (1995). *Adult and continuing education: Theory and practice*. London: Routledge.

Jennings, S., & Dunne, R. (1996). A critical appraisal of the national curriculum by comparison with the French experience. *Teaching Mathematics and its Applications, 15*(2), 49–55.

Kolb, D. A., & Fry, R. (1975). Toward an applied theory of experiential learning. In C. Cooper (Ed.), *Theories of group process* (pp. 33–57). London: Wiley.

Lim, W., Kim, H., Stallings, L., & Son, J. (2015). Celebrating diversity by sharing multiple methods. *Mathematics Teacher, 109*(4), 290–297.

Lim, W., & Son, J. (2013). A case of cultivating a global community of STEM teachers. *New Waves—Educational Research and Development, 16*(2), 40–54.
Lemke, M., Sen, A., Pahlke, E., Partelow, L., Miller, D., Williams, T., et al. (2004). *International outcomes of learning in mathematics literacy and problem solving: PISA 2003 results from the US perspective. (NCES 2005–003)*. Washington, DC: US Department of Education, National Center for Education Statistics.
Leung, F. K. (1995). The mathematics classroom in Beijing, Honk Kong and London. *Educational Studies in Mathematics, 29*, 297–325.
Ma, L. (1999). *Knowing and teaching elementary mathematics*. Mahwah, NJ: Erlbaum.
Miller, K. F., & Stigler, J. (1987). Counting in Chinese: Cultural variation in a basic cognitive skill. *Cognitive Development, 2*, 279–305.
Miller, K. F., Smith, C. M., Zhu, J., & Zhang, H. (1995). Preschool origins of cross-national differences in mathematical competence: The role of number-naming systems. *Psychological Science, 6*, 56–60.
Son, J., & Senk, S. (2010). How reform curricula in the USA and Korea present multiplication and division of fractions. *Educational Studies in Mathematics, 74*(2), 117–142.
Spalding, E., Wang, J., Lin, E., & Butcher, J. (2005). *Working globally: Preparing American teachers and teacher educators in Guangzhou, China*. Paper presented at the annual meeting of the American Association of Colleges for Teacher Education. Retrieved December 12, 2008, from http://www.allacademic.com/meta/p36182_index.html.
Stachowski, L. L., & Mahan, J. M. (1998). Cross-cultural field placements: Student teachers learning from schools and communities. *Theory into Practice, 37*, 155–162.
Stevenson, H. W., & Stigler, J. W. (1992). *The learning gap*. New York, NY: Summit Books.
Stigler, J. W., & Hiebert, J. (1999). *The teaching gap: Best ideas from the world's teachers for improving education in the classroom*. New York, NY: The Free Press.
Stigler, J. W., & Perry, M. (1988). Cross-cultural studies of mathematics teaching and learning: Recent finding and new directions. In D. Grouws & T. Cooney (Eds.), *Effective mathematics teaching directions* (pp. 194–223). Reston, VA: National Council of Teachers of Mathematics.
Sutter, L. E. (2000). Is student achievement immutable? Evidence from international studies on schooling and student achievement. *Review of Educational Research, 70*, 529–545.
Thompson, A. G. (1992). Teachers' beliefs and conceptions: A synthesis of the research. In D. Grouws (Ed.), *Handbook of research on mathematics teaching and learning*. New York, NY: Macmillan.
Wang, J., & Lin, E. (2005). Comparative studies on U.S. and Chinese Mathematics learning and the implications for standards-based mathematics teaching reform. *Educational Researcher, 34*(5), 3–13.
Whitman, N. C., & Lai, M. K. (1990). Similarities and differences in teachers' beliefs about effective teaching of mathematics: Japan and Hawaii. *Educational Studies in Mathematics, 21* (1), 71–81.
Wiest, L. R. (1998). Using immersion experiences to shake up preservice teachers' views about cultural differences. *Journal of Teacher Education, 49*, 358–365.
Willison, S. (1994). Community field experiences in the American Indian Project. In K. Zeichner & S. Melnick (Eds.), *The role of community field experiences in preparing teachers for cultural diversity*. East Lansing, MI: National Center for Research on Teacher Learning.
Yang, M. T. L., & Cobb, P. (1995). A cross-cultural investigation into the development of place-value concepts of children in Taiwan and the United States. *Educational Studies in Mathematics, 28*(1), 1–33.

Chapter 12
The Instructional Quality of Mathematics Student Teachers in the United States and Japan: The Possible Impact of the Structure of Student Teaching

Douglas Lyman Corey, Keith R. Leatham, and Blake E. Peterson

Abstract In this chapter we explore the instructional quality of four US student teachers in a novel student teaching structure. To overcome some of the common problems associated with student teaching documented in the research literature, we adapted a student teaching structure commonly used in Japan. We evaluate the instructional quality of the lessons by using the Mathematical Quality of Instruction (MQI) video coding protocol. We compare the instructional quality to a sample of Japanese student teachers and to a large sample of lessons from six large US school districts, utilizing the Measures of Effective Teaching (MET) study. We also illustrate the quantitative findings with vignettes from US and Japanese student teaching. The results show that given the right support and structure, student teachers in the USA can implement lessons that are similar in quality to Japanese student teachers and much richer than typical US mathematics instruction.

Keywords Student teaching • International comparison • Japanese mathematics teaching • Instructional quality

For years teacher educators have expressed disappointment in the traditional US student teaching experience (Cochran-Smith, 1991; Wilson, Floden, & Ferrini-Mundy, 2002; Zeichner, 1981), the structure of which is plagued by a focus on survival rather than on teaching techniques and has no clearly articulated purpose or curriculum (Leatham & Peterson, 2010a). In addition, many US cooperating teachers (CTs), sometimes called mentor teachers or supervising teachers, see the primary purpose of student teaching, sometimes referred to as practicum or professional experience, as learning how to manage student behavior in real classrooms (Leatham & Peterson, 2010b). Except in rare cases, little work is done to develop

D.L. Corey (✉) • K.R. Leatham • B.E. Peterson
Department of Mathematics Education, Brigham Young University, Provo, UT, USA
e-mail: corey@mathed.byu.edu; kleatham@mathed.byu.edu; peterson@mathed.byu.edu

J.-W. Son et al. (eds.), *What Matters? Research Trends in International Comparative Studies in Mathematics Education*, Research in Mathematics Education, DOI 10.1007/978-3-319-51187-0_12

the core skills of ambitious teaching, such as how to craft and carry out a lesson that engages students in meaningful mathematical activity and how to anticipate, elicit, use, and extend students' mathematical thinking. Although a very good CT might mediate some of these problems, the foundational problems will still exist because these problems seem to be associated with the student teaching structure itself. Efforts to improve student teaching tend to take the overall purpose and structure for granted and result in modest success. Promising strategies include efforts to engage CTs in special training to help facilitate the student teachers' (STs') learning (e.g., Cunningham, 2007; Wilson, Anderson, Leatham, Lovin, & Sanchez, 1999) and efforts to improve coordination and discussion between the student teaching triumvirate of ST, CT, and university supervisor (e.g., Baker & Milner, 2006; Bullough et al., 2003). The implied underlying structure and purpose of student teaching, however, tend to remain unchanged and thus often mitigate the success of such attempts.

As part of a study of mathematics student teaching in Japan the authors were exposed to a very different structure, one that seemed to focus the learning experience of STs on the core knowledge and skills of crafting and implementing high-quality lessons (Peterson, 2005). This exposure, along with our own dissatisfaction, served as an impetus to reflect on our own student teaching structure and how it might be changed to overcome the problems in the traditional structure and provide a better learning opportunity for our students. Based on a review of the features of the Japanese model as well as results of research on student teaching (Leatham & Peterson, 2010b), we reconceptualized the purpose and structure of our student teaching model (briefly outlined in the methods section; see Leatham & Peterson, 2010a for further details).

Analysis of data collected during the first two semesters of the newly structured student teaching experience demonstrated that our experiment was successful. Satisfaction of STs, CTs, and university supervisors was improved, and all involved were able to focus significantly more on students' mathematics and less on their behavior (Leatham & Peterson, 2013). In this paper we begin to document student teachers' performance within this structure by exploring the quality of their lessons. Although an ideal research design would have been to place randomly selected STs from our university in both the traditional and new model, given the success of the initial implementation we were unwilling to enact such a design. We refused to relegate some STs to a difficult learning environment for research purposes. Thus, because we did not have an ideal comparison group, we compared the instructional quality of our STs to two other groups of teachers: a sample of Japanese STs and a large sample of US middle and upper elementary school teachers in six large districts. The former comparison group was a natural comparison because our novel structure shares many similarities to a common student teaching structure in Japan. The US comparison group serves as a proxy for traditional US teaching. It is a natural choice for comparison because we use the results of the Measures of Effective Teaching (MET) study, which used some of the same instructional measures as we did (Kane & Cantrell, 2010).

Literature Review

In this section we review two core areas of the research literature that are pertinent to our study: research on the structure of student teaching, and research on instructional quality in both the USA and Japan.

The Structure of US Student Teaching

Efforts to improve US student teaching vary considerably, but the efforts largely focus on improving aspects within a given structure rather than the structure itself. The typical structure places a single ST with an experienced, full-time, public school teacher for 10–16 weeks, where the ST quickly builds up to teaching most (and often all) of the CTs' classes. A university supervisor visits periodically to evaluate the ST. Because most research takes its underlying purpose for granted, there is little research on the impact of the *structure* of student teaching. Some examples of improvement efforts include lengthening student teaching (Ronfeldt & Reininger, 2012), carefully training CTs (e.g. Cunningham, 2007; Wilson et al., 1999), implementing supplemental teaching seminars (Hodge, 2011), and changing the frequency or focus of university supervision (Borko & Mayfield, 1995).

The study by Ronfeldt and Reininger (2012) of over 1000 STs found that the length of student teaching had little or no impact on outcomes of feeling prepared to teach, teaching efficacy, and commitment to teaching for more years, although the perceived quality of student teaching, as reported by the ST, did. In traditional student teaching, that quality seems to be determined almost exclusively by the CT. Wilson et al. (1999) showed wide variability among CTs in their approaches to and goals for student teaching. That CTs did not agree on the purpose(s) of student teaching is not an isolated phenomenon (e.g., Leatham & Peterson, 2010b) and is likely related to a lack of consensus among teacher educators on these purposes. For example, there seems to be no consensus about the key outcomes of the student teaching experience. The outcome variables of studies which have evaluated student teaching include teaching efficacy, perceptions of instructional preparedness, commitment to a teaching career, teaching beliefs, satisfaction with the student teaching process, mathematical knowledge for teaching, teacher value-added measures, and pedagogy (Grossman, Ronfeldt, & Cohen, 2011; McIntyre, Byrd, & Foxx, 1996; Ronfeldt & Reininger, 2012). If the core purpose of student teaching, however, is to learn how to anticipate, elicit, and use students' mathematical thinking to develop and deepen students' mathematical understanding, which is how we view it, then few of the previous outcome variables actually measure these core practices. We contend that measuring the mathematical quality of instruction provides a means for measuring the desirable outcomes of our novel student teaching structure.

The Structure of Japanese Student Teaching

The structure of student teaching in Japan differs from the typical structure for student teaching in the USA (Borko & Mayfield, 1995; Cuenca et al., 2011; Peterson, 2005). Preservice teachers in Japan typically student teach for 7 weeks, broken up across the last 2 years of their program. Thus, in Japan student teaching is not necessarily a capstone experience like in the USA, where students typically take all other mathematics, education, and mathematics education courses before student teaching (with brief practicum experiences along the way). Rather, in Japan the student teaching experience is a context for coursework in the final semesters of the program. Japanese teachers have a carefully mentored first year of teaching, which is atypical in the USA. This structure is likely one reason why the duration of Japanese student teaching is shorter than what is typically seen in the USA.

STs in Japan teach very few lessons compared to their US counterparts (Peterson, 2005). It is common for Japanese STs to teach about one lesson each week rather than the multiple lessons a day typical of US STs. Furthermore, Japanese STs spend much more time preparing those few lessons than do their US counterparts. They meet with their CTs multiple times to discuss their plans for each upcoming lesson, revising their plans between meetings to address the CTs' feedback. Whereas US STs typically work with one CT, with little structured interaction with other CTs or STs, Japanese STs often work with multiple CTs and CTs often work with a group of STs (Peterson, 2005). The Japanese student teaching experience is centered on carefully planning, teaching, and reflecting upon lessons, in addition to observing and reflecting upon the lessons taught by their peers. At the end of the day, a ST who taught the lesson, the peers who observed the lesson, and the CT whose class was taught participate in a debriefing about the lesson called a *hanseikai* or reflection meeting (similar to those used in lesson study).

Comparing the Instructional Quality of Mathematics in Japan and the USA

Differences in instructional quality between Japan and the USA have been most carefully documented as part of the 1995 and 1999 TIMSS video studies (Hiebert et al., 2003; Stigler & Hiebert, 1999). The results of these studies revealed different models of mathematics teaching in each country. Although different models do not necessarily imply a difference in quality, the results showed large quality differences between the USA and Japan (and in general between the USA and high-achieving countries in the studies) when the lessons were analyzed with a focus on the quality of the mathematical work in each country. For example, the overall measure took into account the challenge, development, and coherence of the lesson, and lessons were measured on a three-level scale of overall mathematical quality: low, medium, and high. Almost all of the US lessons were rated at the lowest level, with none being rated at the highest level. By contrast, almost all of the Japanese lessons were rated as either medium or high.

The findings from these studies paint a stark difference between the quality and nature of eighth-grade mathematics instruction in the USA as compared to Japan. Stigler and Hiebert (1999) characterized typical Japanese mathematics instruction as "structured problem solving" (p. 27). By contrast, they summed up typical US instruction as "learning terms and practicing procedures" (p. 27). Similarly, and based on the results of the later TIMSS video study, Hiebert et al. (2005) characterized US instruction as "frequent reviews of relatively unchallenging, procedural oriented mathematics during lessons that are unnecessarily fragmented" (p. 116). Later studies confirmed that these stark differences persist (e.g., Clarke, Emanuelsson, Jablonka, & Mok, 2006).

We do not know how well these differences or characterizations fit the instruction of Japanese and US *student* teachers, but we assume that typical STs in a given country teach in a similar way to, but at a lower quality than, typical practicing teachers in that country. As such it would be expected that Japanese STs would exhibit a higher quality of mathematics instruction than their US counterparts. It is the purpose of this paper to examine the impact of changing the structure of student teaching in the USA by comparing the ST instructional quality of a special sample of STs in the USA with Japanese STs, as well as with experienced US teachers.

Theoretical Framework

We view instruction as a system of interacting features, with the core interactions happening between teachers and students around content (Cohen, Raudenbush, & Ball, 2003). As such we will measure instructional quality as determined by the nature of the interactions among students, teachers, and content. For this particular study we have chosen to focus largely on students' interactions with the mathematics, and teachers' mediation of this interaction, as the key measure of instructional quality. Focusing on students' interaction with the mathematical content, such as the mathematical activities in which they engage and the quality and depth of the mathematical work they produce, reveals key characteristics of the educational environment. These key characteristics capture the fundamental mathematical ideas about which students are most likely to be thinking. We have chosen a set of characteristics that have been used in analyses of international comparisons of mathematics lessons (Hiebert et al., 2005; Stigler & Hiebert, 1998) and aligned them with a robust measure of the instructional quality of mathematics lessons, the Mathematical Quality of Instruction (MQI) video coding protocol (Hill et al., 2008). The characteristics we use to judge the quality of the mathematical work include the richness of the mathematical work and discourse; attention to the correctness and clarity of mathematical ideas, language, and notation; and engaging students in a cognitively demanding task with opportunities to provide reasoning and explanations about their mathematical work. Mathematical work is "rich" if it develops connections between already known mathematical ideas and representations, provides students multiple methods or ways of reasoning through problems, and builds toward generalizations of ideas or procedures. Although there are other aspects of mathematics instruction that are important indicators of instructional quality, we posit that focusing on the quality of mathematical work is a fundamental indictor of instructional quality and students' opportunity to learn mathematics.

We view quality student teaching as an apprenticeship in the core practices of teaching mathematics: how to anticipate, elicit, and use students' mathematical thinking to develop and deepen students' mathematical understanding. Other purposes—such as learning how to manage student behavior or accumulating ideas about how to manage homework collection and grading—are secondary. This focus on the quality of the mathematical interactions between teachers and students fits our view that the primary purpose of student teaching is developing students' mathematical understanding by building on their mathematical thinking (Peterson & Leatham, 2009).

Methods

As mentioned previously, we compared the instructional quality of our STs in a new student teaching structure with the instruction quality of both a sample of Japanese STs and a sample of US upper elementary and middle school teachers. We make two basic assumptions in order to argue the value of comparing our US STs with these two groups. First, we assume that typical Japanese mathematics instruction is of much higher quality than typical US mathematics instruction (Hiebert et al., 2003; Stigler & Hiebert, 1999). Second, we assume that typical STs in a given country teach in a similar way to, but at a lower quality than, typical practicing teachers in that country, as suggested by some research (e.g., Borko & Livingston, 1989). Based on these assumptions, we conjecture that the teaching quality of our sample of Japanese STs is relatively high, although likely not of the quality of more experienced Japanese teachers. It would thus be a significant accomplishment were our STs to provide instruction at a level comparable with our group of Japanese STs, and likewise with our group of experienced US teachers.

Participants and Context

We analyzed the instruction of three Japanese STs and four US STs. Here we describe the STs and the student teaching environment for both our Japanese and US samples. We briefly compare and contrast the two samples, and then provide some background for the other comparison sample of public school teachers in the USA.

Japanese student teaching context. The Japanese sample for our study came from student teaching that took place at a junior high school (grades seven through nine) affiliated with a university in Southern Japan. In Japan, an affiliated school (*fuzoku chugakko*), similar to a lab school in the USA, is considered to be of excellent quality, is heavily relied upon for preservice teacher observations, and has teachers who are considered part of the university's college of education faculty. The school had 480 students, 160 students in each grade. There were seven mathematics STs, three of whom (two males and one female) were chosen as participants (for details of this selection process see Corey, Peterson, Lewis, & Bukarau, 2010). They worked with three mathematics CTs for 4 weeks in the middle of the school year. Each ST taught three lessons, one lesson at each grade level (seventh, eighth, and ninth), and each lesson involved a different CT. The two male CTs and the one female CT had been teaching junior high mathematics for 9–18 years. The female teacher had a master's degree and the male teachers had bachelor's degrees. Each CT taught four classes daily, two at one grade level and two at another.

The three STs were in the middle of their junior year (second to the last year) of college and were about 20 years old. The STs had completed at least 25 credits of their required coursework. These requirements consisted of 16 credits of basic pedagogy and psychology, 18 credits of methods for elementary subjects (this includes subjects beyond mathematics), 8 credits of content for elementary subjects (this also includes subjects beyond mathematics), 6 credits of methods for middle school mathematics, and 30 credits of mathematics content courses designed for middle school teachers.

No university supervisor observed the STs teach their three lessons. The only observed interaction with university professors came at the end of the student teaching experience, when one of the STs in our study taught a lesson that had been prepared by all 7 STs. The ST who taught this lesson had practiced teaching it to the other 6 STs and all three CTs several times, to receive feedback prior to teaching it in a real classroom. Because of the extensive preparation of and feedback on this special lesson prior to its use in the classroom, it was considered an outlier example and thus was not included in the study.

US student teaching context. The four US STs were all female and each had been placed in local middle schools for her student teaching experience. Although several aspects of the experience are typical of US student teaching (e.g., the student teaching took place during the last semester before graduation and lasted 15 weeks), the program itself is atypical of US programs in general. The US STs were majoring in mathematics education in the department of mathematics education at a large private university in the western United States. Its undergraduate requirements are similar to those in other states that do not require a graduate degree to become a certified secondary teacher. In addition to the typical core undergraduate mathematics classes (38 credit hours), and along with four general education courses (7 credit hours total), the program required five mathematics education courses (12 credit hours), leading up to the 15-week student teaching experience. The 12 credit hours of mathematics education courses is more than the

3–6 credit hours that seems to be typical of US programs (Conference Board of the Mathematical Sciences, 2001, 2012).

The program, meaning the 12 credit hours of mathematics education courses and the culminating student teaching experience, espoused a task-based approach to teaching mathematics with an emphasis on eliciting and using student mathematical thinking to develop students' understanding of the fundamental mathematical ideas of a lesson. In addition, the student teaching experience had been designed purposefully to build on this focus (see Leatham & Peterson, 2010a for student teaching structure details). Here we briefly describe several features of the student teaching experience that made it different from what is typical in the USA. STs were placed in pairs with each CT, and then clustered in neighboring schools. During weeks two through five of student teaching the STs prepared and taught one lesson a week. They spent the majority of their time during these 5 weeks in learning-to-teach activities such as collaborative lesson planning, focused observations of experienced teachers and fellow STs, student interviews, and reflections both in group discussions and in writing. During the last 10 weeks of student teaching each ST took on half (two or three classes) of the CTs' class loads.

The four STs were placed in classrooms where task-based teaching and rich mathematical discussions were common. Both of the US CTs were award-winning teachers and were experienced CTs, having supervised STs previously. One had been teaching for 7 years and the other had been teaching for 17 years. In our view the quality of these two US CTs was comparable to those of the Japanese CTs. Because the new student teaching structure was quite novel for the cooperating teachers (particularly placing the STs in pairs), an orientation was held to explain the structure and the desired focus on crafting lessons that elicited and used student mathematical thinking to develop mathematical understanding.

Comparing the two contexts. Although not deliberately set up as an experiment, as mentioned previously this study is best thought of as a small-scale trial to see how US STs perform when they are placed in a novel student teaching environment comparable to the typical Japanese student teaching environment. Table 12.1 compares key features of the Japanese student teaching experience with the first 5 weeks of the US student teaching experience. Although the situations are not identical (Japanese STs receive more feedback from the CT before the lesson, while US STs planned lessons with their ST partners), the supports and resources offered to both sets of STs are similar enough to warrant exploring differences and similarities in instructional quality.

Data Collection

The data for this study were videotapes of the STs' lessons. All three lessons for each of the three Japanese STs were videotaped, with the exception of the special lesson planned collaboratively by all seven of the mathematics STs. Thus there were eight videos in total for the Japanese STs. We videotaped the four lessons

Table 12.1 Comparison of key features in the two student teaching environments

Feature	Japan	USA
ST teaching frequency	About one lesson a week	About one lesson a week
ST school assignment	A single school with six other STs	A single class with one other ST and clustered with two other ST pairs in neighboring schools
CT/ST configuration	Each CT mentored all (six) STs and each ST worked with all (three) CTs	Each CT mentored two STs and each ST worked with one CT
CT experience	Experienced and highly regarded as teachers	Experienced and highly regarded as teachers
CT feedback on lesson plan prior to teaching	Extensive feedback	Little feedback
Lesson planning	STs may collaborate	STs prepare lessons as a pair
Feedback after teaching	All observing STs (2–6) and the CT participate in a post-lesson reflection meeting with the ST who taught the lesson	All observing STs (5), the CT, and the university supervisor participate in a post-lesson reflection meeting with the ST who taught the lesson

taught by each of the US STs during weeks two through five of the student teaching experience. Thus there were 16 videos in total for the US STs. The Japanese videos had been transcribed and translated into English, and had English subtitles. Because of the quality of the audio in the Japanese videos, it was difficult to hear what students said when the teacher was working with them individually or in small groups. We therefore restricted our analysis to whole class discussions in both the Japanese and US lessons.

Analysis

The main analysis tool was the Mathematical Quality of Instruction (MQI) video coding protocol, which is designed to capture important elements of the mathematical work in which teachers and students engage during a lesson (Hill et al., 2008; Learning Mathematics for Teaching Project, 2010). Table 12.2 shows the five MQI categories and the subcodes for each category.

Each scale is coded on a three-level scale (Low, Mid, or High). A rating of "Low" generally represents inappropriate use of the attribute being examined by the subcode, while a rating of "High" represents appropriate use of the characteristics in the subcode (for the error and imprecision codes, however, a "High" code means there were more errors or more imprecision).

The videos were separated into segments no longer than 5 min long. As mentioned previously, we only coded portions of the Japanese and US videos consisting of whole class mathematics instruction. Coding captured on average 49% of each Japanese lesson and 48% of each US lesson. The lessons in Japan were taught in

Table 12.2 MQI coding protocol categories

Category	Subcodes
Richness of the mathematics	Linking and connections
	Multiple procedures or solution methods
	Developing mathematical generalizations
	Mathematical language
	Overall richness of the mathematics
Working with students and mathematics	Remediation of student errors and difficulties
	Responding to student mathematical productions in instruction
	Overall working with students and mathematics
Errors and imprecision	Major mathematical errors
	Imprecision in language or notation (mathematical symbols)
	Lack of clarity in presentation of mathematical content
	Overall errors and imprecision
Student participation in meaning-making and reasoning	Students provide explanations
	Student mathematical questioning and reasoning
	Enacted task cognitive activation
	Overall student participation in meaning-making and reasoning

50-min periods while all lessons in the USA were taught in 80-min periods on a block schedule. The Japanese lessons varied from 30 to 73% of the lesson being spent in whole class instruction, while in the USA the whole class portion ranged from 36 to 63%.

Coders assigned an overall lesson rating after all lesson segments had been coded. The overall code (also on a three-point scale) is a measure of the whole lesson Mathematical Quality of Instruction (MQI). The overall lesson codes and the overall segment codes (e.g., Overall Richness of the Mathematics) were not formally calculated from the segment codes, but the coders took into account segment codes to help them assign the overall lesson and the overall segment scores.

The lessons were coded by four research assistants that had been trained on the MQI through the certification training run through Harvard University. To get reliable measures of instructional quality each lesson was coded by two coders. The inter-rater reliability of coders was calculated by the percent of exact matches from the total number of possible codes. The inter-rater reliabilities varied with scores of 65%, 75%, 77%, and 82%. Not all possible combinations of coder pairs occurred in the coding process. In the analysis of the MQI lessons, both sets of codes were used for each lesson.

To ensure the Japanese lessons were coded fairly (given that they had been translated) we hired a researcher fluent in both Japanese and English and trained in mathematics education to code a sample of half of the Japanese videos directly from Japanese. The researcher assigned scores for the five codes most closely tied to the mathematical language of the classroom: the mathematical language subcode and all four subcodes in the Errors and Imprecision category. There were no significant differences between the codes based on videos with English subtitles and codes based on videos with audio in Japanese, so all analyses were completed with the codes based on the subtitled videos.

We used a Proportional Odds Model (POM) in a Mixed Modeling (MM) framework to analyze differences between countries in the quantitative codes from the MQI. POMs test whether there is significant evidence that a group (e.g., teachers in country A) receive higher categorical scores than the categorical scores of another group (teachers in country B). These two aspects of our model account for the basic reality of our data: our outcomes are categorical and our codes are nested within teachers and cross-nested within coders. POMs were used to analyze the categorical outcomes and MM was used to account for the nested and cross-nested data structure.

MET comparison study. As mentioned previously, we assumed that ST instructional quality was lower than that of experienced teacher instructional quality in the same country. As such, we have chosen to compare the instructional quality of the US STs in our sample with a large group of experienced US teachers. A suitably large study of instructional quality was undertaken in the Measures of Effective Teaching (MET) study. The sample in that study included six large school districts from six different states in the USA (Kane & Cantrell, 2010). The MET study analyzed the teaching of a sample of over 3000 teachers (teaching from fourth to eighth grade) from these districts in language arts and mathematics, by videotaping and then coding the lessons. The mathematics lessons were analyzed using five different coding protocols, one of which was an abbreviated version of the MQI. We use the four common subscales from the MQI measure in the MET study and our study to make the comparison. We do not use any of the other scales from the MET study. Besides the grade level of the teachers, there are two other methodological differences between our study and the MET study: the MET researchers only coded 30 continuous minutes of the lesson, and just one researcher, not two, coded each lesson.

Results

In this section we present the quantitative results first and then share two qualitative vignettes, one from each country. The vignettes provide examples of typical mathematical work, discussion patterns, and content in the two samples. We use the vignettes as a way to give deeper meaning to the quantitative results.

Table 12.3 The percentages of high, medium, and low MQI content codes for the sample of US and Japanese lessons

	Japanese lessons				US lessons			
	Low (%)	Mid (%)	High (%)		Low (%)	Mid (%)	High (%)	
Whole lesson score								
Whole lesson MQI score	6.3	68.8	25		3.1	93.8	3.1	
Richness								
Linking and connections	60.4	21.9	17.7		66.1	29.4	4.4	
Teacher explanations	29.2	59.4	11.5		25.6	70.7	3.7	
Multiple solution methods	70.8	15.6	13.5		74.6	22.0	3.4	
Developing math generalizations	82.3	8.3	9.4	*	89.8	9.1	1.1	
Mathematical language	7.4	61.1	31.6	**	13.8	77.7	8.5	
Overall richness	36.8	44.2	18.9		37.9	57.6	4.5	
Working with students and math								
Remediation of student errors	75.0	10.4	14.6		72.3	22.5	5.2	
Responding to math productions	44.8	25.0	30.2		44.5	46.0	9.5	
Overall working with students' math	44.2	31.6	24.2		41.2	51.9	7.0	
Errors and imprecision								
Major math errors	93.8	4.2	2.1		91.3	8.4	0.4	
Imprecision in language or notation	90.6	9.4	0.0	*	80.1	19.9	0.0	
Lack of clarity	89.6	10.4	0.0	**	72.5	27.5	0.0	
Overall errors and imprecision	82.3	16.7	1.0	**	60.9	39.1	0.0	
Participation in meaning making								
Students provide explanations	51.0	41.7	7.3		33.8	64.6	1.7	*
Students' questioning and reasoning	72.9	24.0	3.1		55.2	43.9	0.8	**
Enacted task cognitive activation	47.9	40.6	11.5		34.8	62.2	3.0	
Overall participation in meaning making	53.1	38.5	8.3		30.6	66.9	2.4	**

Note: A single asterisk (*) indicates a p-value of 0.05 or below. A double asterisk (**) indicates a p-value of 0.01 or below.

Quantitative Results

MQI scores for the STs in the Two Countries. The bulk of our quantitative results are found in Table 12.3, which displays the distribution of codes in the US and Japanese lessons. The STs in our two samples have similar distributions of scores (no significant differences in distribution) on 10 of the 18 dimensions. We highlight that there was no significant difference in the Whole Lesson MQI Score, which is the coder's rating, after coding the lesson, of the overall level of the mathematical quality of instruction.

Japanese STs scored significantly higher than the US STs on five dimensions, four of which are closely related: Mathematical Language, Imprecision in Language or Notation, Lack of Clarity, and Overall Errors and Imprecision. The mathematical language use of the Japanese STs appears to be more clear and accurate than the mathematical language use of the US STs. One other dimension significantly favored the lessons taught by Japanese STs: Developing Mathematical Generalizations. This dimension captured the extent to which teachers developed, not just stated, mathematical generalizations using mathematical reasoning and at least two examples.

The US STs scored significantly higher on three subcodes, all within the category of Students' Participation in Meaning Making. The three subcodes are Students Provide Explanations, Students' Questioning and Reasoning, and Overall Participation in Meaning Making. US students were more likely to share an explanation about a solution to a problem, share their reasoning for a solution or a mathematical idea, and question other students about their mathematical work, reasoning, or process.

MQI scores of STs and a sample of US mathematics teachers. We report here the four common measures from the versions of the MQI used in this study and the MET study (see Table 12.4). These results suggest that our sample of US STs may have a higher prevalence of errors and imprecision than more typical US middle school/upper elementary school teachers. This may also be less of an indication of weaker mathematical knowledge than of a teaching practice by the STs that is more likely to reveal weakness in their mathematical knowledge. The bulk of lessons in the MET study fell in to the category of the US *school mathematics* tradition (Cobb, Wood, Yackel, & McNeal, 1992). The primary characteristics of such instruction are a teacher-led lecture where students are shown steps to memorize in order to solve problems with the same mathematical form. Instruction by the US STs in our sample falls in the *inquiry mathematics* category, as our results will continue to show. Teaching in the inquiry mathematics tradition relies on broader and deeper mathematical knowledge (Ball & Bass, 2000) while teaching in the *school mathematics* tradition may make it easier to hide mathematical weaknesses.

The other three categories, however, show some dramatic differences between the instruction of the MET study sample and the STs in our sample. A large majority of instructional segments, about 80%, in the MET study received a low score in the remaining three overall codes. However, less than 25% of the US and Japanese STs' instructional segments, in some cases much less, received a low score on these three overall codes. Furthermore, the table shows that the mathematical quality of instruction of US STs in our sample is much higher than the MET sample, at least along the last three dimensions. The figures also show how the instructional quality of our US STs is more similar to that of the Japanese STs than to the US teachers in the MET study. The quantitative results show that our sample of US STs has a much higher mathematical quality of instruction than the experienced teachers in the MET study, as measured on the overall indicators of the MQI, except perhaps on the Overall Errors and Imprecision dimension.

Table 12.4 Comparison of four overall measures

Overall student participation in meaning making			
	Low (%)	*Mid (%)*	*High (%)*
US sample	9	63	29
Japanese sample	8	39	53
MET	84	15	1
Overall Richness			
	Low	*Mid*	*High*
US sample	3	58	38
Japanese sample	19	44	37
MET	80	19	2
Overall working with students and mathematics			
	Low	*Mid*	*High*
US sample	15	48	38
Japanese sample	24	32	44
MET	78	21	2
Overall errors and imprecision			
	Low	*Mid*	*High*
US sample	60	38	2
Japanese sample	82	17	1
MET	77	18	3

Example Lessons

Although the quantitative tables reveal several similarities and some significant differences, they give limited insights into the nature of the lessons themselves. We thus use two episodes to highlight the overall quality of the lessons and to illustrate how some of the differences found in the statistical models played out in the classroom. We chose these examples because they seemed typical of the lessons we saw in each country. We begin with an eighth-grade lesson about solving simultaneous equations (taught by a male Japanese ST), then share a pre-algebra lesson about similar figures (taught by a female US ST).

Japanese lesson: Simultaneous equations. The teacher started class by writing the following problem on the board:

> There are two red boxes and one yellow box. You put a card with a number from 1 to 9 written on it in each box. However, both red-box numbers have to be the same. The total of the numbers in all 3 boxes combined was 13. What numbers could be in each box?

The class was reminded that the purpose of the question was to find out the card number in each box. The teacher gave the students about 20 s to think about this problem, then asked students to call out whatever numbers they thought could be in the boxes. Student responses included four possibilities—(3,3,7), (4,4,5), (5,5,3), and (6,6,1).

The teacher then explained that there are many correct answers to this question so you cannot know which exact cards are in the boxes. He then wrote the following on the board: "Condition 1: We leave the cards in the boxes the way they are, and take away 1 red box. The total of the numbers from 1 red box and 1 yellow box is 9." Almost immediately, the class yelled out (4,4,5) as the correct answer. The teacher agreed that (4,4,5) was the correct answer and then handed out a worksheet and instructed the class to either make an equation or use a table that would come up with the answer (4,4,5). He told the students to write their equations on the worksheet and to try to find many ways to solve the problem. The students then worked individually for about 16 min while the teacher walked around the room and helped various students.

The teacher asked three students to share their solutions (corresponding to substitution, elimination, and using a table) by writing them on the board and explaining their solution methods (see Fig. 12.1). The first two students talked through the steps of their solutions (labeled Blackboard 1 and Blackboard 2, respectively) for a minute and a half total. Before the third student presented, the teacher reminded the students that this next solution used a table, an approach they should remember from the previous year when they were learning about linear equations. After these students had explained their solutions, the teacher led a discussion on the advantages and disadvantages of different methods. The table, for example, is difficult to use if the numbers are large or decimal values are needed. The teacher also explained why the equations are called "simultaneous" equations.

The teacher reminded the students that they could only solve this problem because of the added condition. He then handed out a worksheet with new conditions. He left the original problem on the board, but changed Condition 2 to the following: "The cards in each box are left as they are, and there are 4 red boxes, and 2 yellow boxes. The total of the numbers on the cards in the boxes is 26." The teacher asked the students to think about how this condition would change the solution. He also asked them to consider what would happen if they replaced this condition with Condition 3: "The cards in each box are left as they are, and there are 2 red boxes and 1 yellow box. The total of the numbers on the cards in the boxes is 20." The teacher gave the students about 2 min to solve these problems, either by using a previous solution method or by finding a new strategy.

When the class came back together to discuss, one student said he could not solve the problem with Condition 2 because the answer became zero. A second student stated that there is a difference between getting 0 and having no solution, which the student later clarified as meaning that both sides of the equation equal zero when doing a technique like the one labeled "Blackboard 2" in Fig. 12.1. The teacher agreed that Condition 2 gave no more of a solution than the initial condition alone. He explained that even though they had added a condition, it did not mean that they could necessarily solve the problem: "In this case, the answer cannot be determined. This kind of simultaneous equation, you might already know, is called *indeterminate*. This means that the answer cannot be determined." The teacher then asked what the circumstance was in Condition 3, immediately stating that this condition created a contradiction. He explained to the class that we call this sort of

Fig. 12.1 Recreation of students' solutions displayed on the blackboard in the Japanese lesson

Blackboard 1
2X+(9-X)=13
X=4
9-4=5
The red, 4
The yellow, 5

Blackboard 2
X represents a card from the red box
Y represents a card from the yellow box

2X+Y=13—①
X+Y=9—②
①-②

2X+Y=13
X+Y=9
X=4

Put X=4 into ②
4+Y=9
Y=9-4
Y=5
The yellow card, 4
The red card, 5

Blackboard 3
This was solved by using the table.

The yellow box	1	2	3	4	5	6	7	8	9
The red box	X	9	7	5	3	1	X	X	X

contradiction an *inconsistency*. He explained that the purpose in giving them Conditions 2 and 3 had been to show that even though another condition is added, it does not guarantee that you will find a solution. The teacher summed up the lesson by reviewing the terms *ichigen ichiji hoteishiki* (one unknown, first degree equation), *nigen ichiji hoteishiki* (two unknowns, first degree equation), and *renritsu hoteishiki* (system of equations).

US lesson: Similar figures. The class started with students working individually to determine whether each of the three statements on the board (see Fig. 12.2) were true or false and to justify their answers. After about 8 min the teacher asked the class, "Who thinks they know the answer to number one?" She called on a student to come to the front of the class to explain his answer. The student believed the first statement was true because "to be a square, all four sides must have equal length and all four angles must be the same, where both sets are parallel lines." When the teacher asked about the two criteria for figures to be similar the student responded, "It has to have the same angles and have the corresponding lines be equal." The teacher questioned, "Be equal?" then the student clarified by saying, "Well, like not the same length, just the same scale factor needs to be used."

After the discussion with that student, the teacher asked if anyone thought the second statement was false. A student replied by saying, "I think it's true just because, no matter what, they both have parallel lines. And if you just draw sets of parallel lines you get a square." The teacher asked about what parallel lines had to do with being similar and the student responded, "You have to have two sets of parallel lines to be a square. So obviously, if it's a square, they both have parallel lines. So that's what similar means." The teacher asked a few clarifying questions, then drew a non-square rectangle and a square on the board and asked whether they were similar. The same student responded in the affirmative and added, "Similar

Fig. 12.2 List of statements about similarity written on the board

1. All squares are similar.
2. All rectangles are similar.
3. If the factor between two similar shapes is 1, then the two shapes are the same size.

Fig. 12.3 Criteria for similar figures written on the board

Similar
1. Corresponding lengths have same scale factor
2. Corresponding angles are the same

means they have to have something in common. The thing they have in common is parallel sides." The teacher responded, "It is true that when we're talking about something being similar they have to have something in common, but when we're talking about something being mathematically similar—." She was cut off by students calling out terms like *same angles* and *scale factor*. After leading a brief discussion about what it means to be mathematically similar the teacher wrote the criteria for similarity on the board (see Fig. 12.3).

The teacher asked the class what *corresponding lengths* meant and they did not seem to know. She turned to the drawing of the square and non-square (1×8) rectangle on the board, drew another rectangle, and labeled its width and length ½ and 4, respectively. The teacher used the two non-square rectangles to discuss which sides of the rectangles would be corresponding lengths. She then led a discussion with the class to clarify that having parallel sides is insufficient information to determine whether two figures are mathematically similar. The teacher then presented an example of non-similar rectangles. Each had widths of 1 but different lengths (2 and 4). The class agreed that those two rectangles were not similar because the scale factors of the corresponding sides were not equal. The teacher asked, "What's our scale factor between these two corresponding sides [pointing to the sides of length one on both rectangles]?" The students answered, "One." "What's our scale factor between these two corresponding sides [pointing to the sides of length 2 and 4]?" Students answered, "Two." The teacher then clarified (while pointing to the criteria), "So, the corresponding sides don't have the same scale factor. So you're right. This is false."

The teacher did not discuss the third statement, instead moving on in the lesson with an activity that allowed students to make conjectures about the relationships between the scale factor of similar figures and the areas and perimeters of those figures. At one point near the end of the lesson the students disagreed about whether the perimeter of a scaled figure (with a scale factor of 2) was two times or four times larger than the original figure. The teacher tried to resolve the disagreement by highlighting the original rectangle in a similar rectangle made of four rectangles the size of the original rectangle (2 wide and 2 long): "Here I have my original figure. How many sides do I have?" Students respond, "Four." The teacher continued, "Four. We are just going to ignore the fact that they are different lengths for now." She then showed that there are eight "sides" in the larger figure and asked if "it" was two times or four times bigger. The students respond by saying, "Two." The class then continued trying to make connections between the scale factor of similar figures and their perimeters and areas. They concluded that the ratio of areas in similar figures is the square of the scale factor between the figures, while the ratio of perimeters was the scale factor itself.

Comparison of lessons. These lesson vignettes illustrate the key findings and differences revealed in the quantitative analyses. The two lessons show students that are engaged in higher-level mathematical tasks with teachers who are trying to make strong connections to what the students already know, and to build off student work to achieve the mathematical goals of the lessons. These aspects of the lesson alone set the instruction of both lessons apart from the kind of mathematics instruction that is more typical in US middle schools (Stigler & Hiebert, 1999).

The development of the mathematics in the US lesson involved more contributions from students than the Japanese lesson, with the students often commenting or questioning about the mathematical work or statements shared by other students. The US teacher often turned mathematical issues to be resolved back to the students, although she did step in with her own explanations as well. Although the Japanese students' solutions were shared on the blackboard, they were not asked to explain their reasoning further, nor were other students asked to respond to their solution strategies. These aspects of the lesson illustrate how the US STs scored higher than the Japanese STs with respect to involving students in meaning making through student explanations, questioning, and reasoning.

A portion of the US lesson also shows the particular weakness of the US STs in our sample compared to our sample of Japanese STs: errors, imprecision, and lack of clarity. When trying to understand the relationship between scale factor and the perimeters of similar figures, the ST used an explanation that ignored the length of the sides and merely counted the number of sides, a strategy that would not work on other examples, even some that the class had developed. The teacher also lacked precision in some of her language, asking how much bigger "it" was at a point in the conversation when the referent of "it" was unclear.

Discussion and Conclusion

Our purpose was to explore the instructional quality of student teachers in a novel structure for student teaching. We had previously documented how the conversations in the new structure focused much more on students' mathematical thinking and much less on classroom management than is typical (Leatham & Peterson, 2013). Was this change in focus associated with a change in instructional quality? Our results show that our sample of STs in the novel student teaching structure did provide high-quality mathematical lessons, as measured by the MQI. The instruction of our STs looked much more like the instruction of a small sample of Japanese STs than that of typical US mathematics instruction (from the MET study). These results suggest that the student teaching structure may have been a key component in supporting these STs in developing and implementing mathematics instruction focused on facilitating mathematics learning, through using rich tasks and eliciting and building on student mathematical thinking.

There was no significant difference between the whole lesson MQI scores for the sample of US and Japanese STs. This result is a stark contrast to the difference in the quality of mathematics instruction of practicing teachers found in the TIMSS video studies, and is contrary to our initial assumption that the instructional quality of the sample of Japanese STs would be higher than that of the sample of US STs. We failed to see the wide gap exhibited in nearly every dimension of instruction described in the *Teaching Gap* study (Stigler & Hiebert, 1999). One poignant example comes from the three-level scale of overall quality that was given in our study as well as in the *Teaching Gap* study. In the *Teaching Gap* only 11% of US lessons were scored as medium (and none were scored at high), while 90% of the Japanese lessons were scored at medium (51%) or high (39%). We also had a three-level measure of overall mathematical quality of instruction. Both the Japanese and the US STs had over 90% of their ratings in the medium or high category, although the Japanese STs did have 25% of their lesson ratings at the highest level compared with only 3% for the USA.

We cannot claim, on the basis of our results, a causal relationship between the structure of student teaching and quality of instruction. We have no control group. It is also possible that the key contributor was not the structure itself, but the previously documented change in the content of the conversations that STs had with each other, and that the cooperating teachers and university supervisors had with the STs (Leatham & Peterson, 2013). However, we do believe that the structure was at least a key piece in contributing to the results we have documented here. Before and after the change in structure, our work as university supervisors with student teachers and our program emphasis had the same focus: rich mathematical tasks, developing mathematical understanding based on student thinking, and reasoning and sense making. But our experience with student teaching before the structure change was the same struggle we had been experiencing for years, and the same that has been documented in the literature: the ideas we focused on were pushed aside as student teachers worked on survival and produced mathematically weak lessons with little intellectual engagement.

It is also important to note that the structure of student teaching was not changed with specific features to focus the experience on core knowledge and skills for crafting and implementing high-quality lessons. Rather, the structure was changed to try to remove the features that drew the focus of STs to survival tactics and classroom management. Thus, the revised student teaching structure seemed to afford STs the time, resources, and support to implement lessons that aligned with the focus of their teacher education program. The overall purpose, from our perspective, was the same. But now the student teachers had time to craft well-thought-out tasks and plan on how to build mathematical ideas on student mathematical thinking. They also had strong support from another like-minded student teacher. There was significant space provided for discussion of their lessons, and the structure of those discussions helped to maintain a focus on the craft of mathematics teaching rather than issues of classroom management.

Although we cannot offer a causal claim (the small sample size alone might prevent generalization), this study can be viewed as an existence proof. It is possible to develop STs in US mathematics teacher education programs who are capable of implementing high-quality mathematics instruction even though the mathematics teaching culture in the USA as a whole remains unchanged (Stigler & Hiebert, 2009). These STs were also able to perform better than their more experienced counterparts in the USA. The ability of US mathematics education programs to produce such well-prepared graduates, on par with a country documented to have some of the best instruction in the world, is a significant accomplishment. A key component in this success may be a student teaching structure that gives student teachers the space, support, and resources to implement such instruction. If so, this is a large step forward in understanding how we might eliminate the teaching gap that exists among typical practicing teachers in the two countries.

References

Baker, R. S., & Milner, J. O. (2006). Complexities of collaboration: Intensity of mentors' responses to paired and single student teachers. *Action in Teacher Education, 28*(3), 61–72.

Ball, D. L., & Bass, H. (2000). Interweaving content and pedagogy in teaching and learning to teach: Knowing and using mathematics. In J. Boaler (Ed.), *Multiple perspectives on mathematics teaching and learning* (pp. 83–104). Westport, CT: Ablex.

Borko, H., & Livingston, C. (1989). Cognition and improvisation: Differences in mathematics instruction by expert and novice teachers. *American Educational Research Journal, 26*, 473–498.

Borko, H., & Mayfield, V. (1995). The roles of the cooperating teacher and university supervisor in learning to teach. *Teaching and Teacher Education, 11*, 501–518.

Bullough Jr., R. V., Young, J., Birrell, J. R., Clark, D. C., Egan, M. W., Erickson, L., et al. (2003). Teaching with a peer: A comparison of two models of student teaching. *Teaching and Teacher Education, 19*, 57–73.

Clarke, D. J., Emanuelsson, J., Jablonka, E., & Mok, I. A. C. (Eds.). (2006). *Making connections: Comparing mathematics classrooms around the world.* Rotterdam, The Netherlands: Sense Publishers.

Cobb, P., Wood, T., Yackel, E., & McNeal, B. (1992). Characteristics of classroom mathematics traditions: An interactional analysis. *American Educational Research Journal, 29*, 573–604.

Cohen, D., Raudenbush, S., & Ball, D. (2003). Resources, instruction, and research. *Educational Evaluation and Policy Analysis, 25*(2), 1–24.

Cochran-Smith, M. (1991). Reinventing student teaching. *Journal of Teacher Education, 42*, 104–118.

Conference Board of the Mathematical Sciences. (2001). *The mathematical education of teachers.* Washington, DC: American Mathematical Society and Mathematical Association of America.

Conference Board of the Mathematical Sciences. (2012). *The mathematical education of teachers II.* Washington, DC: American Mathematical Society and Mathematical Association of America.

Corey, D., Peterson, B. E., Lewis, B. M., & Bukarau, J. (2010). Are there any places that students use their head? Principles of high-quality Japanese mathematics instruction. *Journal for Research in Mathematics Education, 41*, 434–478.Cuenca, A., Schmeichel, M., Butler, B., Dinkelman, T., & Nichols, J. Jr. (2011). Creating a "third space" in student teaching:

Implications for the university supervisor's status as outsider. *Teaching and Teacher Education*, 27, 1068–1077.

Cuenca, A., Schmeichel, M., Butler, B., Dinkelman, T., & Nichols, J. Jr. (2011). Creating a "third space" in student teaching: Implications for the university supervisor's status as outsider. *Teaching and Teacher Education, 27*, 1068–1077.

Cunningham, B. (2007). All the right features: Towards an "architecture" for mentoring trainee teachers in UK further education colleges. *Journal of Education for Teaching, 33*, 83–97.

Grossman, P., Ronfeldt, M., & Cohen, J. (2011). The power of setting: The role of field experience in learning to teach. In K. Harris, S. Graham, T. Urdan, A. Bus, S. Major, & H. L. Swanson (Eds.), *American Psychological Association [APA] Educational Psychology Handbook, Applications to learning and teaching* (Vol. 3, pp. 311–334). Washington, DC: APA.

Hiebert, J., Gallimore, R., Garnier, H., Givvin, K. B., Hollingsworth, H., Jacobs, J., et al. (2003). *Teaching mathematics in seven countries: Results from the TIMSS 1999 Video Study (NCES Publication No. 2003–013)*. Washington, DC: U.S. Department of Education, National Center for Educational Statistics.

Hiebert, J., Stigler, J. W., Jacobs, J. K., Givvin, K. B., Garnier, H., Smith, M., et al. (2005). Mathematics teaching in the United States today (and tomorrow): Results from the TIMSS 1999 Video Study. *Educational Evaluation and Policy Analysis, 27*, 111–132.

Hill, H. C., Blunk, M., Charalambous, C., Lewis, J., Phelps, G., Sleep, L., & Ball, D. L. (2008). Mathematical knowledge for teaching and the mathematical quality of instruction: An exploratory study. *Cognition and Instruction, 26*, 430–511.

Hodge, A. (2011). The design and benefits of a teaching seminar for pre-service secondary mathematics teachers. *PRIMUS: Problems, Resources, and Issues in Mathematics Undergraduate Studies, 21*, 263–273.

Kane, T., & Cantrell, S. (2010). *Learning about teaching: Initial findings from the Measures of Effective Teaching project*. Seattle, WA: Bill and Melinda Gates Foundation.

Learning Mathematics for Teaching Project. (2010). Mathematical quality of instruction (MQI). *Learning Mathematics for Teaching (LMT)*. Retrieved from http://lmt.mspnet.org/index.cfm/20926

Leatham, K. R., & Peterson, B. E. (2013). Talking about pedagogy, students and mathematics. In M. V. Martinez & A. Castro Superfine (Eds.), *Proceedings of the 35th Annual Meeting of the North American Chapter of the International Group for the Psychology of Mathematics Education* (pp. 629–636). Chicago: University of Illinois at Chicago.

Leatham, K. R., & Peterson, B. E. (2010a). Purposefully designing student teaching to focus on students' mathematical thinking. In J. Luebeck & J. W. Lott (Eds.), *Mathematics teaching: Putting research into practice at all levels* (pp. 225–239). San Diego, CA: Association of Mathematics Teacher Educators.

Leatham, K. R., & Peterson, B. E. (2010b). Secondary mathematics cooperating teachers' perceptions of the purpose of student thinking. *Journal of Mathematics Teacher Education, 13*, 99–119.

McIntyre, D. J., Byrd, D. M., & Foxx, S. M. (1996). Field and laboratory experiences. In J. Sikula (Ed.), *Handbook of research on teacher education* (2nd ed., pp. 171–193). New York, NY: Macmillan.

Peterson, B. E. (2005). Student teaching in Japan: The lesson. *Journal of Mathematics Teacher Education, 8*, 61–74.

Peterson, B. E., & Leatham, K. R. (2009). Learning to use students' mathematical thinking to orchestrate a class discussion. In L. Knott (Ed.), *The role of mathematics discourse in producing leaders of discourse* (pp. 99–128). Charlotte, NC: Information Age Publishing.

Ronfeldt, M., & Reininger, M. (2012). More or better student teaching? *Teaching and Teacher Education, 28*, 1091–1106.

Stigler, J., & Hiebert, J. (1998). Teaching is a cultural activity. *American Educator, 22*(4), 1–10.

Stigler, J., & Hiebert, J. (1999). *The teaching gap: Best ideas from the world's teachers for improving education in the classroom*. New York, NY: Free Press.

Stigler, J., & Hiebert, J. (2009). Closing the teaching gap. *Phi Delta Kappan, 91*(3), 32–37.

Wilson, P. S., Anderson, D. L., Leatham, K. R., Lovin, L. H., & Sanchez, W. B. (1999). Giving voice to mentor teachers. In F. Hitt & M. Santos (Eds.), *Proceedings of the Twenty-First Annual Meeting of the North American Chapter of the International Group for the Psychology of Mathematics Education* (Vol. 2, pp. 811–817). Columbus, OH: ERIC Clearinghouse for Science, Mathematics, and Environmental Education.

Wilson, S. M., Floden, R. E., & Ferrini-Mundy, J. (2002). Teacher preparation research: An insider's view from the outside. *Journal of Teacher Education, 53*, 190–204.

Zeichner, K. M. (1981). Reflective teaching and field-based experience in teacher education. *Interchange, 12*(4), 1–22.

Chapter 13
Reflective Capabilities of Mathematics Education Systems in China, Japan, and the United States

Thomas E. Ricks

Abstract This chapter compares the reflective capabilities of three nation-states' mathematics education systems. Situated in the theoretical framework of complexity and using methods suited for theory generation, I describe the results of my research study comparing the US mathematics education system with those in China and Japan. In particular, I detail how nation-state educational systems exhibit intelligent self-perpetuation, and I delineate a rubric for measuring the reflective thinking manifest by each systemic totality. I propose reasons for why the US mathematics education system seems more resistant to reform than those in China and Japan, and I make concrete recommendations for US educational stakeholders in their reform efforts.

Keywords Collective reflection • Complexity • Internationalization • Mathematics education • Reform • Teacher training

Introduction

The internationalization of pedagogy over the last several decades has greatly facilitated cross-cultural mathematics education research (Hiebert et al., 2003; Huang & Li, 2009). A major international video study conducted in tandem with the Third International Mathematics and Science Study was the first of its kind to collect large amounts of video, interview, and other data about the teaching practices of diverse nation-states. Stigler and Hiebert (1999) summarized their research conclusions from this video study in the landmark book *The Teaching Gap*, which popularized the "teaching gap" phrase among US educators, educational researchers, and educational policy-makers.

T.E. Ricks (✉)
School of Education, Louisiana State University, Baton Rouge, LA, USA
e-mail: tomricks@lsu.edu

J.-W. Son et al. (eds.), *What Matters? Research Trends in International Comparative Studies in Mathematics Education*, Research in Mathematics Education, DOI 10.1007/978-3-319-51187-0_13

Stigler and Hiebert (1999) argued that at least three distinct "teaching gaps" existed between the US educational enterprise and those of other high-achieving nation-states. The first "teaching gap" referred to the gap in teaching methods between countries. For example, US mathematics instruction was characterized as mostly "learning terms and practicing procedures," German mathematics instruction as "developing advanced procedures," and Japanese mathematics instruction as "structured problem solving" (Stigler & Hiebert, 1999, p. 27; see also Stevenson & Nerison-Low, 1998). Others have characterized Chinese mathematics instruction as "one topic with variation" (Wang & Murphy, 2004). A second use by Stigler and Hiebert (1999) of the term "teaching gap" referred to the gap between the type of teaching needed by a nation-state to achieve its educational aspirations and the actual type of teaching occurring in its classrooms. Particularly in the USA, politicians and the media highlight this teaching gap, because despite US aspirations for having the best schools in the world, the type of teaching occurring in US classrooms (learning terms and practicing procedures) is not doing substantial mathematics (Stein, Smith, Henningsen, & Silver, 2000) and thus cannot achieve such excellence, as illustrated by decades of international comparisons.

The third use of the term "teaching gap" referred to the gap that existed between the meaningful real-time educational reform mechanisms of high-achieving nation-states like China and Japan versus the apparent lack of such reform capabilities in US education (Stigler & Hiebert, 1999). Other educational researchers have also noticed the resilient permanence of US teaching culture despite decades of diligent reform attempts (Kessel, 1990; Kilpatrick, Swafford, & Findell, 2001; Weeks, 1931). Dewey's (1904) concerns about US education are still with us a century on: We still prepare teachers in settings where theory is mostly divorced from praxis, where the apprenticeship forms the backbone of practical training, and where non-classroom-based experts develop models for teachers to mimic pedagogically. Additionally, the general US cultural method of instruction that Dewey deplored continues to dominate, and the most recent reform attempt—the Common Core Standards, perhaps the nation's most aggressive and unified effort at a major educational overhaul ever attempted—is already floundering.

This paper examines this third type of "teaching gap" by comparing the US mathematics education system's ability to reform with systems in China and Japan. In particular, the research I report was guided by my attempt to understand why the US educational system appears to resist reform more than other nations' educational systems. The specific research questions I investigated were: (1) Why does the US educational system appear so resistant to reform attempts, and (2) what can be done to help the US educational system be more open to improvement?

Theoretical Framework

I investigated these two questions using the theoretical framework of *complexity*. Complexity studies *complex systems*—emergent, self-organizing, adaptive, and holistic entities (Davis & Sumara, 2001; Ricks, 2007)—that develop when multiple agents produce various supra-agent phenomena through their collective interrelationships (Johnson, 2001). Colloquially, complexity investigates wholes that become greater than the sums of their parts (Ricks, 2009a, 2009b). Complex systems often nest within larger complex systems, or overlap and intertwine with them; as such, complex systems have blurred and nonlinearizable internal structures (Davis, Sumara, & Simmt, 2003). Often, complex systems manifest fractal self-similarity. They lack clear boundaries and descriptions of various interrelationships between components are by their very nature incomplete; at each level of complexity entirely new rules appear that govern emergent phenomena (Anderson, 1972).

The human body is a classic example of a complex system: individual cells in the human body join to make tissues, tissues merge to form organs, organs collect as systems (e.g., circulatory, nervous, digestive, and immune), and systems unify to constitute the whole physical human being. Finding the edges or limits of the nervous system, for example, and separating it from the cardiovascular system is nigh impossible, as the new field of neurocardiology shows. Transcendent notions of *mind* and *soul* complicate the boundary of the human being even more. Complexity also describes how individual people can interact in ways that make even higher-order complex systems—like families, neighborhoods, cities, and nation-states.

This study considers the nation-state educational system as a complex system capable of emergent, self-organizing, and adaptive behaviors (Ricks, 2010a; Stigler & Hiebert, 1999). The various subsystems of that system (like mathematics or science education) nest and overlap in integrated, nonlinear ways. Each nation-state educational system is further composed of smaller, intertwined complex systems like state, regional, or provincial educational systems—down to districts, schools, and classrooms. In the USA, for example, the Common Core State Standards movement can be thought of as a subsystem within the larger US educational system; so, too, can groups of individuals who write and approve textbooks, set curriculum standards, or implement assessment systems like standardized tests. And with the rise of the internationalization of pedagogy, nation-state educational systems are beginning to interact on a supra-national level with exciting possibilities.

Methods

This study used Glaser's (1965) four-part constant comparative method (comparing, integrating, delimiting, and writing) to develop a theory for wider application. These stages are not chronological and can reoccur cyclically, with a gradual progression toward integrating, delimiting, and writing theory. The constant comparative method is good at developing rather than testing theory because it "is concerned with generating and plausibly suggesting (not provisionally testing) many properties and hypotheses about a *general* phenomenon" (Glaser, 1965,

p. 438, emphasis in original). I have also found the constant comparative methodology to be particularly helpful in juxtaposing international pedagogical approaches because of its broad holistic approach (Ricks, 2012; Yuan & Ricks, 2011a, 2011b), something called for by a variety of educational researchers (Cai, 2008; Ricks, 2007).

Data for this study were drawn from a variety of sources, consistent with constant comparative methodology: (a) the literature, including publicly available mathematics and science lessons and videos of professional development from a variety of countries (e.g., www.TIMSSvideo.com); (b) various videotaped student/teacher interviews, videotaped classroom lessons, and videotaped professional development activities (including Chinese teaching research groups and model lessons) collected by my research group in both Chinese (Beijing, Tianjin, Hangzhou, Shanghai, and Chengdu) and US cities; and (c) my own life experience, an especially pertinent piece of constant comparative data (Lu & Ricks, 2012, 2013; Ricks, 2010a, 2010b, 2011a, 2011b; Ricks, Lu, & Fleener, 2009; Yang & Ricks, 2011, 2012). In the constant comparative method, such a wide data net contributes to robust theory creation and sharpens the clarity of the constant comparisons during data analysis. I chose the constant comparative method because it fit well with my desire to investigate a poorly understood phenomenon for which little theory existed (see also Ricks, 2014). In particular, I used the constant comparative methods' inherent "flexibility [to] aid the creative generation of theory" (p. 438). Theory development—when properly conducted—is the method of the mind itself, which Dewey (1904) argued is pure scientific method. This study thus develops theory in education, something called for by various researchers (Lather, 2004; Tierney, 2001).

Results

This international cross-cultural comparative study, juxtaposing the US mathematics education system with those in China and Japan, has resulted in the creation of a complex-oriented theory that considers nation-state educational systems as legitimate living entities that stretch over and encompass the educational subsystems of each particular nation-state. This perspective posits that all nation-state educational systems manifest the fundamental properties of all life forms, particularly (1) self-perpetuation and (2) intelligence.

Before I continue, I must acknowledge that claiming that a nation-state's education system is alive—as alive as you and I—might sound so radical as to border on the ridiculous. Complexity, however, has revolutionized a variety of mainstream scientific domains by enabling the recognition of phenomena that were previously invisible. The domain of bacteriology is one clear example: bacteriologists now consider the bacterial colony *in its entirety* to be the fundamental biological entity, not the individual bacterial cells that constitute it (Shapiro, 1998). So by way of metaphor, just as bacteriology now recognizes the total collection of individual bacterial cells (the colony) to be a legitimate life form in its own right (Ben-Jacob, 1998), the theoretical results of this study suggest similar paradigmatic progress in the area of mathematics education by considering "the total collection of educational participants in a given nation-state" as a legitimate life form. I detail below how each nation-state's educational system manifests its vitality through the two fundamental biological behaviors (self-perpetuation and intelligence) at the supra-individual level.

Self-Perpetuation

All life self-perpetuates. Similarly, the nation-state educational-system entity manifests the fundamental biological characteristic of self-perpetuation through reproducing its unique educational culture in the next generation of citizens. Teaching is a systemic cultural activity, which enculturates students into certain modes of operating and habits of thinking (Hiebert & Stigler, 2000; National Research Council and Institute of Medicine, 2000). Virtually all educational stakeholders—students, parents, teachers, educational administrators, special interest groups, policy-makers, educational researchers, politicians, and even the media—have been enculturated into their particular nation-state's educational culture while sitting in classrooms—the womb for reproducing educational culture in the next generation of citizens. Each nation's population thus possesses a unique and collective cultural expectation of what education looks like for that nation-state (Ma, 1999; Stigler & Hiebert, 1999).

Students that succeed in any particular nation-state's educational system have the opportunity to choose career paths as educators; almost without exception, those students who choose to spend their entire careers as teachers thrived during their educational enculturation years, during what Lortie (1975) called their "apprenticeship of observation" (p. 61). These new teachers are perfectly situated to help perpetuate the educational culture in their own classrooms for the next generation of students. In most developed nations, post-secondary teacher training programs act as further selective filters for candidates, so only the most culturally acceptable candidates become legitimized (and usually certified) teachers. Over multiple generations, a natural-selection-like process occurs where each new generation of teachers becomes more and more selected to reproduce the overall nation-state's educational culture.

Intelligence

In addition to self-perpetuation through the reproduction of educational culture, each nation-state educational-system entity demonstrates intelligence on the supra-human plane. All life forms demonstrate intelligence, the capacity to learn by adapting their internal structures and external behaviors to enhance their chances for survival (Brown & Duguid, 2000). This problem-solving ability is manifest even by the lowliest of species—the humble bacteria—when they hypermutate, quorum sense, form genomic webs, or task-differentiate in multi-specie biofilms (Gumbo, Ross, & Cloete, 2010; Shapiro, 2006). Intelligence manifests itself in different ways (Gardner & Hatch, 1989). Many scholars, like Bloom (1956) in his well-known taxonomy, posit a gradation of intelligence levels.

Dewey (1933) argued that the most powerful form of intelligence is *reflective thinking*: a multistage, often-cyclical process akin to the scientific method (Dewey, 1904) that involves interpreting and developing theory about an experience, then actually testing the theory "by overt or imaginative action" (Dewey, 1933, p. 107). In essence, Dewey considered reflective thinking to be a form of hypothesis formation, testing, and refinement that leads to robust theory generation grounded in practice (Ricks, 2011a). A critical component of this process is being exposed to different ways of interpreting experience, instead of jumping to spontaneous conclusions as the mind tends to do (Rodgers, 2002). Successful educational reform requires robust reflection (Shulman, 1986); for example, Cooney et al. (1998) noted that "the act of reflection is central to any reform process" (p. 312), and Morine-Dershimer and Kent (1999) agreed that "reflection ultimately determines the extent of professional development that occurs" (p. 41). I extend this principle of reflective thinking to the intelligence manifested by nation-states' educational systems: just as the human mind can manifest thinking at different levels, so can nation-state educational systems. I consider reflective manifestations of intelligence by educational systems to be the most powerful—and most helpful— prompts for successful educational reform.

Comparing Nation-State Educational Systems' Reflective Capacities

Thinking about nation-state educational systems as intelligent, self-perpetuating entities illuminate several interrelated reasons for why the US educational system manifests fewer reform capabilities compared to other nation-states' educational systems. Primarily, the unique US educational culture has resulted in a distinct US pedagogical approach and distinct educational structures that jointly limit the ability of the system as a whole to manifest reflective intelligence. Additionally, when reform attempts are made, the lack of reflective ability by the US's educational subsystems initiates a resistive, immune-like response that dampens the effect of any reform.

US Pedagogy Limits Reflection

In my comparison of different nation-states' educational systems, I have noticed a surprising and intriguing regularity: US pedagogical practice appears similar at all subsystem levels of the US educational system. Upon introspection, such paralleled pedagogical practice makes logical sense: as a complex system, an educational system naturally manifests fractal, self-similar structures. US teachers teach students in the culturally distinct US approach; having been students themselves of such approaches, teachers also expect to be taught in similar ways. Thus, US educational culture manifests itself in a unique pedagogical approach that happens at various subsystem levels: (1) in the classroom, (2) in professional development, and (3) in overall reform attempts.

US classroom pedagogy. The pedagogy in most US classrooms is culturally distinct from the pedagogies manifest in other countries' classrooms. Stigler and Hiebert (1999) characterized US classroom pedagogy as "learning terms and practicing procedures" (p. 27); other researchers have used similar terminology like absorption, memorization, regurgitation, repetition, recitation (Meyer & Wilkerson, 2011; Romberg & Carpenter, 1986), or consuming and reproducing content (Ricks, 2009b). This is in stark contrast to the more dynamic "structured problem solving" or "one topic with variation" pedagogies exhibited in Japanese and Chinese instruction, respectively (Wang & Murphy, 2004). The US system tends toward an expert-knower attempting to transmit knowledge to a novice (pupil); the role of the pupil is to absorb and faithfully replicate the expert's knowledge when asked (Kilpatrick et al., 2001). US students demonstrate a palpable aversion to struggle, an avoidance of challenge, and a desire for mimicry of expert practice. I characterize this pedagogy as *complacent procedural mimicry*. I use the term "complacent" because students often resist pedagogies that require them to use more challenging forms of thinking.

Similar US professional development pedagogy. This complacent procedural mimicry births certain cultural modes of thought manifested by all US educational stakeholders. Such a pedagogy of transmission from expert-to-novice (teacher-to-student) is well documented in US classrooms, and the US educational system attempts to train teachers in the same way. At the level of teacher preparation, expert-knowers (university professors, student teaching supervisors) transmit knowledge of "best practices" or "reform recommendations" to the neophytes (teacher candidates) (Hart, Alston, & Murata, 2011). There is a strong undercurrent belief that a new teacher needs to apprentice with an expert teacher, and that the teacher candidate should be given as many diverse training opportunities as possible to best prepare him or her for the real world. Teacher candidates thus prepare a wide range of lessons. In-service professional development is similar: the outside expert (professional developer) is brought in by district or school administrators to attempt to instill in teachers the latest and greatest in reform, often mere educational fads. US teachers manifest the same cultural modes of thought as their students:

> The "model lesson" of the teachers' institute and of the educational journal is a monument... of the willingness of our teaching corps to accept without inquiry or criticism any method or device which seems to promise good results. Teachers, actual and intending, flock to those persons who give them clear-cut and definite instructions as to just how to teach this or that. (Dewey, 1904, p. 16)

Teachers look to professional developers as the knowledge authorities on reform, and teachers want these reformers to simplify the complicated task of reform procedurally by breaking it into easy-to-solve substeps. The pedagogy of professional development thereby mirrors the pedagogy of the classroom (Ricks, 2010b).

Similar US reform pedagogy. The pedagogical approach used in the USA to reform the educational system parallels the one used to develop teachers professionally and to teach students in classrooms: complacent procedural mimicry (Hart, Alston, & Murata, 2011). The entire American nation has a cultural expectation of what education looks like, and this includes the "fixing" of the educational system by expert knowers like policy-makers, special interest groups, or government. This shared US pedagogy of complacent procedural mimicry does not encourage reflective thinking. Actively avoiding struggle and wanting an expert to simplify challenging tasks is not compatible with the "effort of thought" (Dewey, 1902, p. 4) required to complete the multistage reflective process advocated by Dewey.

By contrast, both China's and Japan's educational cultures engender pedagogies that promote rather than inhibit reflective thinking by their students, teachers, and wider educational systems (Schümer, 1999). For example, China's "one topic with variation" and Japan's "structured problem solving" pedagogical approaches require students to struggle in generating answers, and to consider various alternatives before acting in solving pedagogical tasks (Wang & Murphy, 2004; Yuan & Ricks, 2011b). These pedagogical methods also value the exposure to difference so necessary to reflective thinking. As I detail later in this chapter, both China and Japan have professional development pedagogies that parallel the pedagogies of their classrooms; Chinese and Japanese teachers are expected to struggle like students in generating and considering various alternatives before acting to solve pedagogical tasks, e.g., in lesson study. Their system as a whole does the same thing.

US Educational Structures Limit Reflection

Although the USA has the highest per-pupil educational expenditure (Leung, 1995) and "one of the most developed educational systems in the world" (Chen & Mu, 2010, p. 129), the US educational system has certain structural features that limit its collective reflective abilities, including (1) pervasive isolation, (2) lack of time for teacher reflection, (3) incoherent curricula, textbooks, and lessons, and (4) a lack of clustered office space and shared planning time (Kilpatrick et al., 2001).

Pervasive isolation at all levels. The US system does not have an integrated, networked educational system, so isolation among educational subsystems is severe, across all levels (Paine, 1997). This isolation severely inhibits reflection (Rodgers, 2002; Schön, 1983). US educational isolation manifests itself in the following ways: (a) Students are viewed by the educational system as unconnected individuals, (b) teachers work mostly alone, and (c) schools function as mostly autonomous entities.

Although US students might engage in some occasional partner or group work on scattered classroom tasks, they are viewed by the US educational system as being unconnected individuals that do not form a cohesive, sustained part of any collective. US pedagogical strategy does not harness collective power for individual or group learning (Ricks, 2007). For example, US classes mostly operate as ability-grouped *collections* of individuals rather than as a diverse *collective* of students (Ricks, 2014)—ever heard of "students-centered" (plural) instead of "student-centered" pedagogy? Classes in China and Japan, by contrast, have more collective characteristics, including operating as multi-grade-level cohorts with teacher looping, little to no ability-grouping, joint exercise during recesses, and student ownership of the classroom space (Liang, 2001; Yang & Ricks, 2012).

When compared against the working conditions of their peers, US teachers' isolation in their classrooms is severe (Ball, Lubienski, & Mewborn, 2001; Stepanek, 2000). US teachers rarely share subject-specific office space, collaboratively plan lessons, or observe each other teach, and typically lack administrative support for such activities. Public teaching exhibitions with classrooms of real students at teaching conferences or teaching competitions are also unheard of (An & Kulm, 2010; Chen & Mu, 2010; Li & Li, 2009). Schön (1983) said about the impact such isolation has on reflection: "The teacher's isolation in her classroom works against reflection-in-action. She needs to communicate her private puzzles and insights, to test them against the views of her peers" (p. 333).

Schools in the US are also more isolated than in other countries. Compare, for example, the growing US governmental strategy to shut down or take over "failing" schools instead of pairing a struggling school with a nearby successful school, as in Shanghai-China's Empowered Management Program (Jensen & Farmer, 2013). But whether it be student, teacher, or school isolation, the isolation manifest in the US educational system is problematic because it prevents the formation of unified reflective activity at each subsystem level and across such levels, which in turn limits the reflective thinking manifest throughout and by the entire US educational system. Rodgers (2002) explained:

> No [person] outgrows the need for others' perspectives, experience, and support... The community... serves as testing ground for an individual's understanding as it moves from the realm of the personal to the public. A reflective community also provides a forum wherein the individual can put form to what it is he or she was thinking—or feeling—in the first place. (p. 857)

Lack of time for US teachers to reflect. US teachers lack the time granted to other nations' educators for reflection (Darling-Hammond, Ancess, & Ort, 2002). The large number of student contact hours that US teachers have per day and the higher-than-global-norm student-teacher ratio (because US teachers, despite having smaller classes on average than other nations, teach more classes per day) also limit the time teachers have to reflect. Such a lack of time limits teachers' reflective abilities (Stanley, 1998). Additionally, when given extra time by administrators, teachers often lack the skills to use that time effectively for reflection (Ball, Lubienski, & Mewborn, 2001; Schleppenbach, Flevares, Sims, & Perry, 2007).

Incoherent curricula, textbooks, and lessons. US curricula are less coherent than curricula in other nations. Although US curricula have more content topics than other nations' curricula, they are fractured, diffuse, shallow, unconnected to other curricular domains, and are not focused on central, guiding goals (Kilpatrick, Mesa, & Sloane, 2006; Ministry of Education, P. R. China 2003; Patnam, 2013; Polikoff, 2012; Schmidt, Cogan, Houang, & McKnight, 2011; Schmidt & Houang, 2012; Thompson & Preston, 2004; Tienken, 2008). US curricula are also unnecessarily repetitive across grade levels—sixth grade being the worst example, with less than a third being new material (Hiebert et al., 2005; Porter, Polikoff, & Smithson, 2009; Schmidt et al., 2005; Wang & Lin, 2005). The USA also has less coherent textbooks, and less coherence between a textbook and its course outline (Cogan, Schmidt, & Wiley, 2001), perhaps explaining why US teachers are less willing to study and integrate textbook material into lessons. US teachers see the textbook as a problem-set booklet and choose problems more randomly for their lessons (Cai & Wang, 2010).

Less coherent curricula and textbooks lead to less coherent US lessons (Cai & Wang, 2010; Liao & Cao, 2010; Perry, 2000; Stevenson & Stigler, 1992), diminishing students' opportunities to learn and reflect on the subject matter (Carroll, 1963; Hiebert & Grouws, 2007; Schmidt & Maier, 2009). US lessons tend to be single-day affairs, rather than stretching across multiple days as in China or Japan (Liao & Cao, 2010; Stigler & Hiebert, 1999). Chinese and Japanese lessons are also more interconnected (content-wise) (Cai & Wang, 2010). The mathematics is presented "more as an integrated whole than as subject matter that can be parsed by topic area" (Kilpatrick et al., 2006, p. 3). Chinese and Japanese teachers are much better able to describe how underlying mathematical concepts of the lesson relate to its structure (Wang & Murphy, 2004). Their lessons are centralized on fewer tasks, providing more attention to the mathematical details of student solutions (Liao & Cao, 2010; Stigler & Hiebert, 1999).

In summary, the US educational culture has resulted in educational structures that "collectively do not form a cohesive and cumulative program" (Kilpatrick et al., 2001, p. 431). The US educational culture has resulted in a pedagogy of complacent procedural mimicry, and in various fractured educational structures, both of which combine to limit the reflective capabilities within and across the overall US educational system.

US Educational Stakeholders' Resistance to Reform

US reform also struggles because of the self-preservation strategies manifested by subsystems of the US educational system. Because the various US educational stakeholders are all products of the same educational culture, they jointly enable and enact the expected teaching culture; further—like a metaphorical immune response system—US educational stakeholders actively resist deviations from the norm (Hart, Alston, & Murata, 2011). Students' resistance against changes to the status quo is one example of this immune-like response. Although not finished products of the US educational system, students rapidly learn the US cultural expectations for schooling and readily recognize changes to the status quo. US students actively resist cultural changes to teaching that might require them to struggle and develop self-made strategies, rather than looking to the teacher as the knowledge authority in the classroom (Stein, Smith, Henningsen, & Silver, 2000). The unique US education culture contains low educational norms compared to those in other high-achieving nations, making any significant reform appear as a burden instead of a blessing. For example, from the US student's perspective, personal, familial, local schooling, and societal expectations are lower (Stevenson & Stigler, 1992; Wang et al., 2011). Additionally, US educational culture deemphasizes academics in favor of extra-curricular activities (Cai, 2003, 2004; Cai & Hwang, 2002). Chinese and Japanese educational cultures (part of the Confucian Heritage Culture of East Asian nations) encourage their students to excel academically (Han & Ginsburg, 2001; Leung, 2001; Siegler & Mu, 2008).

US teachers' collective response to reform demonstrates another manner in which the US educational system self-preserves its educational culture. Teachers are the foremost authority in a nation-state on the real nature of the educational culture and pedagogical practices, for they are the ones that know it best. On a daily basis they continuously re-enact the rituals that keep the culture alive. US teachers actively resist changes to the dominant pedagogical practice (Nipper et al., 2011); like US students, they expect to enact procedural-like pedagogies given to them by experts, and shrink from having to struggle on their own (Hart & Carriere, 2011). Like teachers, parents of current US students are a third level of defense against reform. Changes to the norm are quickly passed from child to parents, who then become more involved. Parents panic, for example, when students are asked to complete homework problems unfamiliar to them. US parents also form larger collective groups to maintain the dominant teaching culture and resist change politically. Chinese and Japanese parents, however, are more willing to accept teacher-made decisions about pedagogical practice, demonstrate more educational ambitions for their children, buy more school textbooks, participate in their children's academic lives more, and push their children to excel more compared to American parents (Cai, 2003, 2004). Chinese and Japanese parents do so most likely because of their own experiences as students, but also because broader Chinese and Japanese cultures emphasize the importance of education more than in the USA (House, 2006; Perry, 2000; Stigler, Lee, Lucker, & Stevenson, 1982; Stigler, Lee, & Stevenson, 1987).

The Reflective Capabilities of Chinese and Japanese Educational Systems

China and Japan have educational systems that demonstrate intellectual participation by educational stakeholders at various levels to help reform attempts succeed (Cai, 2008; Chen & Mu, 2010; Hiebert et al., 2003; Huang & Li, 2009; Li & Yang, 2008; Wang & Lu, 2008). Although every single teacher may not participate in all the forms, in general both China and Japan have more organically intertwined, better networked, and more strongly communicating educational subsystems than the USA (An & Kulm, 2010; Ricks, Lu, & Fleener, 2009). Agents in these nested subsystems are "members of multiple groups—formal, informal, and even across distance. Together these groups construct and maintain a culture of teaching which encourages teachers in particular ways" (Paine, 1997; p. 81), especially to engage in robust reflection. Professional development in these countries focuses on issues directly related to their teaching practice (An & Kulm, 2010). These activities allow Chinese and Japanese educational systems to demonstrate unified reflective thought. Four specific mechanisms particularly contribute to this pattern: lesson study, public demonstration lessons, teaching competitions, and shared office space.

Lesson study. Lesson study is a form of unified reflective activity that occurs regularly in both China and Japan (Murata, 2011; Ricks, 2003, 2012; Stafford-Plummer, 2002). Much already exists in the Western educational literature about these activities (Lewis, 2000; Liang, 2001; Sargent & Hannum, 2005; Yang & Ricks, 2011, 2012). Although the lesson study activities in each country differ, they all share several underlying structures. Briefly, lesson study is a unifying reflective activity because: (1) participants work collaboratively to develop a hypothesis lesson targeting a pressing educational issue, (2) participants jointly test this hypothesis lesson by teaching it to an actual class, (3) participants carefully observe the influence of the hypothesis lesson on students during this enactment, (4) participants modify the hypothesis lesson based on the data they collected, (5) the cycle of testing and revision often continues until practitioners are satisfied that the final product (lesson) addresses the original issue, and (6) the results of this lesson study process are shared with others (Yuan & Ricks, 2011a; Schön, 1983). Lesson study allows for the problematizing of pressing educational issues and their joint solution by sub-groups of the wider nation-state educational system. Educational dilemmas are solved in collective manners, yielding solutions above and beyond the capabilities of individual teachers or researchers (Surowiecki, 2004). The intelligence (reflective thinking) of this system is manifest by the solutions developed through unified and concerted collaboration over time.

Public demonstration lessons. A second form of unified reflective activity occurring in Chinese and Japanese educational systems is the practice of watching public demonstration lessons (Lu & Ricks, 2013; Paine, 1997). Chinese and Japanese teachers are inculcated into public educational cultures that value the public scrutiny of teaching as a form of school-based research essential for professional advancement (Huang & Bao, 2006). Chinese and Japanese teachers' instructional workload (face-to-face classroom interaction with students) is less than US teachers', allowing for greater attention to professional development (Sargent & Hannum, 2005). For example, Chinese teachers have over three times as much time to participate in professional development than US teachers (An & Kulm, 2010).

These public demonstration lessons occur in a variety of ways. One popular form is part of mandatory training for new teachers where veterans help novices plan, enact, and reflect on lessons taught publicly in their own class—like Chinese *keli*, exemplary lesson development, or teaching research groups (Chen & Mu, 2010; Yang & Ricks, 2011). Sometimes university liaisons—district supervisors with connections to the normal universities—coordinate larger district-wide public lessons. These liaisons often meet with other university liaisons to discuss the professional development efforts in their local areas, another layer of collective reflection not found in the USA. These public "master lessons" are taught by a recognized expert teacher to a random group of students; public master lessons also occur at the regional or national level at teaching conferences and are posted online with accompanying lesson plans (An & Kulm, 2010; Li & Li, 2009; Yang & Ricks, 2011, 2012).

Teaching competitions. A third form of unified reflective activity is participation in teaching competitions, where the very best regional or national teachers present their finest work for comparative public evaluation (Li & Li, 2009). At first a competitive activity might not seem to be a unifying activity, but the process teachers go through to prepare, participate in (through direct involvement or observation), and reflect on lessons presented at teaching competitions cements a common framework and educational cultural ethos stretching across the nation-state. These teaching competition lessons are "the model of the collective. They represent and embody a common orientation which all teachers are to pursue. In a sense they are like a magnetic force that attracts and calls teachers of the whole country to work together" (Paine & Ma, 1993, p. 682). Contrast this with US selection of "top teachers" via private nominations and applications for the Presidential Award for Excellence in Mathematics and Science Teaching, where the teaching knowledge of the "top teachers" is never shared with the US educational system as a whole.

Clustered office space and shared planning times. I would be remiss not to mention two other small but very important opportunities for unified reflection by Chinese and Japanese teachers: clustered office space and shared planning times (Huang & Bao, 2006; Ricks, 2012; Yang & Ricks, 2011, 2012). Sharing clustered office space with teachers that teach related curricula allows for small-scale, microlevel communication and problem-solving on a very intimate plane, something known to strengthen the intelligence of complex adaptive systems (Davis & Simmt, 2003). Overlapping teachers' planning times also provides opportunities for these spontaneous interactions to occur. Combined with the former three forms of unified reflection, this fourth form may actually be the most powerful in initiating change in individual teacher practice (Brown & Duguid, 2000).

Discussion

Space limitations prohibit more detailed discussion of other manifestations of a lack of reflective capacity by the overall US educational system or its subsystems. Follow-up studies might consider, for example, the oscillating pendulum swings in US reform approaches, the superficiality of US reform documents or calls for reform in the educational literature (many of which are never read by educators), lack of awareness of the actual implementation of reform recommendations, haphazard and spontaneous allegiance to educational fads (e.g., learning styles or differentiated learning), the relative shortness granted to US reform attempts before abandonment, the absorption of "reformed" educators into the larger system, adoption by US educational stakeholders of reform parlance without substantive changes in practice, and dissipating difference, dissonance, and diversity (Cai, 2004, 2008; Cai & Wang, 2010; Darling-Hammond, et al., 2002; Huang & Li, 2009; Kuhn, 1962; Leung, 2005; Surowiecki, 2004). Suffice it to say that the reflective capabilities of a nation-state's educational system and its subsystems are vital for the lasting integration of any reform (Lu & Ricks, 2012; Shulman, 1986). China and Japan have educational systems that generate self-reformation through collective reflective processes; by contrast, the US educational system is categorically distinct—when the intelligent thinking capabilities of the US educational system are compared to the thinking manifested by the educational systems in other high-achieving nations, there is a great "gap."

Viewing nation-state educational systems as intelligent, self-perpetuating entities manifesting different levels of reflective capabilities—at the systemic level—suggest specific strategies to improve US reform attempts (Ricks, 2011b). Because the US educational system does not yet manifest the forms of advanced systemic reflection seen in the education systems in China and Japan, I believe the primary focus of US reform should be in developing the US educational system's reflective capabilities, just as has been recommended for US students and US teachers (Artzt & Armour-Thomas, 1999). US reformers are trying to reform the US educational system in the very ways that they (US reformers) were taught as students—a method not yielding habitual reflective thinking. Instead, new reform attempts should paradigmatically shift away from attempting to impose reform, and toward a stance that considers reform as an actual *educative act* for the entire US educational system, an educative act in harmony with Dewey's ideas of increasing capacity for thought (Zeichner & Liston, 1996). The ultimate goal, of course, is to enable the US educational system to become systemically self-reflective. Reform of the actual system will happen only after the system has learned to reflect, and only then when it reforms itself through the reflective process. In other words, the US educational system should be able to reform itself as an intelligent entity once it is able to think reflectively.

How can we help an educational system like the USA's develop its reflective capabilities? Successfully addressing this question is the crux of any successful reform. If the fractal-like self-symmetries of complex systems are any guide, then the current approach to US reform—which uses the same pedagogy operating in US classrooms to teach students and professionally develop teachers (complacent procedural mimicry)—is not the answer. US reform efforts have struggled over the last century not because of lack of effort or dedication, but because the very pedagogy being used to teach students and to develop teachers is being used to attempt to change the US educational system for the better. Experts attempt to transmit direct changes for the overall US educational system to mimic, but a pedagogy of complacent procedural mimicry does not engender reflective capabilities in its target. Instead, reforming the US educational system must be thought of as engaging the overall system in a pedagogy of struggle, similar to that used by Japan and China to develop the reflective capabilities of their educational systems. It recasts reform efforts not as attempts to change the system or its educational culture directly, through a pedagogy of experts telling novices what to do, but as a mission of genuine education of the US educational system's intelligence (Hart, Alston, & Murata, 2011).

I repeat for emphasis: system-wide US educational reform currently happens in the very same pedagogical manner that US students are taught in the classroom, that US teachers are taught in professional development, and that US schools and districts are asked to implement reform—using a pedagogy of complacent procedural mimicry. The single seminal characteristic of all US education reform attempts is that they initiate from "experts" on the peripheral boundaries of the system (from educational researchers, professional organizations, funders, and other non-school-based personnel to even more distant educational influencers, like government agencies, legislators, policy-makers, and other politicians or special interest groups), rather than from the core (from teachers and other school personnel). In China and Japan, by contrast, much if not most of professional development is insider-organized (An & Kulm, 2010; Kilpatrick et al., 2001; Lewis, 2002).

Thus, in line with Stigler & Hiebert's (1999) three descriptions of US "teaching gaps," I propose a fourth level of "teaching gap" happening in the USA—the gap between the actual century-long reform pedagogy being used to attempt direct change of the US educational system and the form of struggle-based pedagogy needed to enable genuine self-reflection by the US educational system. Perhaps the emerging role of the internationalization of pedagogy can help in this respect, as it is a primary route for the US educational system to become more reflective (Wong, Han, & Lee, 2004). It is through the juxtaposition of cultures that one's own invisible culture is made more manifest. If we think of the US educational system as a giant, nation-wide *super-student* that can interact with other nations' *super-students*, the US super-student will begin to see its own invisible culture and have the opportunity to consider other options. In a word, exposure to difference through internationalization may be the first step in helping the US educational system become more reflective.

Thus, the internationalization of pedagogy holds the potential to contribute significantly to US reform attempts. The internationalization of pedagogy occurs when nation-state educational systems—through the process of globalization—become increasingly aware of and begin to study and adopt successful features of each other's systems. Nation-state educational research (as part of this complex web of interacting nation-state educational systems) will similarly benefit by this amalgamation. Internationalizing pedagogy may also phase shift international educational collaborations into a global reflective entity—the first *global super-student*.

Conclusion

This chapter reported the results of a constant comparative study comparing the US's and other high-achieving nations' educational systems, primarily China's and Japan's. This study contributes to the field of mathematics education by suggesting new avenues to consider in the reformation of US mathematics education. It also indicates particular areas in need of more systematic research, such as how the internationalization of pedagogy can best be utilized to leverage the development of a more reflective US educational system. In particular, this study detailed the concept that each nation-state's educational system can be perceived as an intelligent, self-perpetuating entity. I introduced the idea that education systems manifest a gradation of cognitive capacity, with Dewey's (1933) concept of reflection being the most powerful. I described how the USA has a less reflective educational system than countries like China or Japan which manifest more flexible educational reform mechanisms. I discussed options available to US educational reformers, including the paradigmatic shift—away from thinking about reforming the system—toward educating the system to develop its holistic, reflective, (i.e., *fully* intellectual) capacities. I concluded with recommendations for further engaging nation-state educational systems in larger cooperative endeavors.

References

An, S., & Kulm, G. (2010). How Chinese in-service elementary mathematics teachers gain knowledge from professional development. *Journal of Mathematics Education, 1*(1), 41–57.

Anderson, P. W. (1972). More is different. *Science, 177*(4047), 393–396.

Artzt, A. F., & Armour-Thomas, E. (1999). A cognitive model for examining teachers' instructional practice: A guide for facilitating teacher reflection. *Educational Studies in Mathematics, 40*(3), 211–235.

Ball, D., Lubienski, S., & Mewborn, D. S. (2001). Research on teaching mathematics: The unsolved problem of teachers' mathematical knowledge. In V. Richardson (Ed.), *Handbook of research on teaching mathematics* (pp. 433–456). Washington, DC: AERA.

Ben-Jacob, E. (1998). Bacterial wisdom, Gödel's theorem & creative genomic webs. *Physica A, 248*, 57–76.

Bloom, B. (1956). *Taxonomy of educational objectives*. New York: David McKay.

Brown, J. S., & Duguid, P. (2000). *The social life of information*. Boston, MA: Harvard Business School Press.

Cai, J. (2003). Investigating roles in students' learning of mathematics from a cross-national perspective. *Mathematics Education Research Journal, 15*(2), 87–106.
Cai, J. (2004). Why do US and Chinese students think differently in mathematical problem solving? *The Journal of Mathematical Behavior, 23*, 135–167.
Cai, J. (2008). International assessment as a means for improving students' learning. In L. D. Miller & S. R. Saunders (Eds.), *Proceedings of the US-Sino Workshop* (pp. 155–162). Murfreesboro, Tennessee.
Cai, J., & Hwang, S. (2002). Generalized and generative thinking in US and Chinese students' mathematical problem solving and problem posing. *The Journal of Mathematical Behavior, 21* (4), 401–421.
Cai, J., & Wang, T. (2010). Conceptions of effective mathematics teaching within a cultural context. *Journal of Mathematics Teacher Education, 13*(3), 265–287.
Carroll, J. B. (1963). A model of school learning. *Teachers College Record, 64*(8), 723–733.
Chen, J., & Mu, Z. (2010). The cross-national comparison of pre-service mathematics teacher education and curriculum structure. *Journal of Mathematics Education, 1*(1), 119–136.
Cogan, L., Schmidt, W., & Wiley, D. (2001). Who takes what math and in which track? Using TIMSS to characterize US students' eighth-grade mathematics learning opportunities. *Educational Evaluation & Policy Analysis, 23*, 323–341.
Cooney, T. J., Shealy, B. E., & Arvold, B. (1998). Conceptualizing belief structures of preservice secondary mathematics teachers. *Journal for Research in Mathematics Education, 29*(3), 306–333.
Darling-Hammond, L., Ancess, J., & Ort, S. W. (2002). Reinventing high school. *American Education Research Journal, 39*(3), 639–673.
Davis, B., & Simmt, E. (2003). Understanding learning systems. *Journal for Research in Mathematics Education, 34*, 137–167.
Davis, B., & Sumara, D. (2001). Learning communities: Understanding the workplace as a complex system. *New Directions for Adult and Continuing Education, 92*, 85–95.
Davis, B., Sumara, D., & Simmt E. (2003). Complexity and collectivity. In *Proceedings of the 2003 Complexity and Education Research Conference, October 16–18* (pp. 217–230). Edmonton, Canada.
Dewey, J. (1904). The relation of theory to practice. In C. A. McMurray (Ed.), *The relation of theory to practice in the education of teachers (Third Yearbook of the NSSE)* (pp. 9–30). Chicago: University of Chicago Press.
Dewey, J. (1902). *The child and the curriculum*. Chicago: University of Chicago Press.
Dewey, J. (1933). *How we think: A restatement of the relation of reflective thinking to the educative process*. Boston: D.C. Heath.
Gardner, H., & Hatch, T. (1989). Multiple intelligences go to school. *Educational Researcher, 18* (8), 4–10.
Glaser, B. J. (1965). The constant comparative method of qualitative analysis. *Social Problems, 12* (4), 436–445.
Gumbo, J. R., Ross, G., & Cloete, T. E. (2010). The isolation and identification of predatory bacteria from a Microcystis algal bloom. *African Journal of Biotechnology, 9*(5), 663–671.
Han, Y., & Ginsburg, H. P. (2001). Chinese and English mathematics language: The relation between linguistic clarity and mathematics performance. *Mathematical Thinking and Learning, 3*(2–3), 201–220.
Hart, L. C., & Carriere, J. (2011). Developing the habits of mind for a successful lesson study community. In L. C. Hart, A. Alston, & A. Murata (Eds.), *Lesson study research and practice in mathematics education* (pp. 27–38). Chicago: University of Chicago Press.
Hart, L. C., Alston, A., & Murata, A. (Eds.). (2011). *Lesson study research and practice in mathematics education*. Chicago: University of Chicago Press.
Hiebert, J., Gallimore, R., Garnier, H., Givvin, K. B., Hollingworth, H., Jacobs, J., et al. (2003). *Teaching mathematics in seven countries: Results from the TIMSS 1999 video study*. Washington, DC: NCES.

Hiebert, J., & Grouws, D. A. (2007). The effects of classroom mathematics teaching on students' learning. In F. K. Lester (Ed.), *Second handbook of research on mathematics teaching and learning* (pp. 371–404). Charlotte, NC: Information Age.
Hiebert, J., & Stigler, J. W. (2000). A proposal for improving classroom teaching: Lessons from the TIMSS video study. *The Elementary School Journal, 101*(1), 3–20.
Hiebert, J., Stigler, J. W., Jacobs, J. K., Givvin, K. B., Garnier, H., Smith, M., et al. (2005). Mathematics teaching in the United States today (and tomorrow): Results from the TIMSS 1999 Video Study. *Educational Evaluation and Policy Analysis, 27*(2), 111–132.
House, J. D. (2006). Mathematics beliefs and achievement of elementary school students in Japan and the United States: Results from the Third International Mathematics and Science Study. *The Journal of Genetic Psychology, 167*(1), 31–45.
Huang, R., & Bao, J. (2006). Towards a model for the teacher professional development in China: Introducing Keli. *Journal of Mathematics Teacher Education, 9*, 279–298.
Huang, R., & Li, Y. (2009). Pursuing excellence in mathematics classroom instruction through exemplary lesson development in China. *ZDM Mathematics Education, 41*, 297–309.
Jensen, B., & Farmer, J. (2013). *School Turnaround in Shanghai*. Center for American Progress. Retrieved July 29, 2015, from http://www.americanprogress.org/issues/education/report/2013/05/14/63144/school-turnaround-in-shanghai/
Johnson, S. (2001). *Emergence*. New York: Scribner.
Kessel, C. (1990). Why I quit my job. *UME Trends, 2*(5), 7–8.
Kilpatrick, J., Swafford, S., & Findell, B. (Eds.). (2001). *Adding it up: Helping children learn mathematics*. Washington, DC: National Research Council/National Academy Press.
Kilpatrick, J., Mesa, V., & Sloane, F. (2006, November). US algebra teaching and learning viewed internationally. In *Paper presented at the 2nd IEA International Research Conference*. Washington, DC: Brookings Institution.
Kuhn, T. S. (1962). *The structure of scientific revolutions*. Chicago: University of Chicago Press.
Lather, P. (2004). Scientific research in education: A critical perspective. *British Educational Research Journal, 30*(6), 759–772.
Leung, F. K. S. (1995). The mathematics classroom in Beijing, Hong Kong and London. *Educational Studies in Mathematics, 29*, 297–325.
Leung, F. K. S. (2001). In search of an East Asian identity in mathematics education. *Educational Studies in Mathematics, 47*, 35–51.
Leung, F. K. S. (2005). Some characteristics of East Asian mathematics classrooms based on data from the TIMSS 1999 Video Study. *Educational Studies in Mathematics, 60*(2), 199–215.
Lewis, C. (2000, April). *Lesson study: The core of Japanese professional development*. Invited address to the Special Interest Group on Research in Mathematics Education, AERA, New Orleans , LA.
Lewis, C. (2002). *Lesson study: A handbook of teacher-led instructional change*. Philadelphia: Research for Better Schools.
Li, Y., & Li, J. (2009). Mathematics classroom instruction excellence through the platform of teaching contests. *ZDM Mathematics Education, 41*(3), 263–277.
Li, S., & Yang, Y. (2008). Reform in the process of teaching development. In L. D. Miller, & S. R. Saunders (Eds.), *Proceedings of the US-Sino Workshop* (pp. 126–132). Murfreesboro, TN.
Liang, X. (2001). *China: Challenges of secondary education*. Washington, DC: World Bank.
Liao, S., & Cao, Y. (2010). The structure of mathematics lessons in China. *Journal of Mathematics Education, 1*(1), 27–40.
Lortie, D. C. (1975). *Schoolteacher: A sociological study*. Chicago: University of Chicago Press.
Lu, L., & Ricks, T. E. (2012). Chinese mathematics teaching reform in four years' time: The DJP model. In L. R. Van Zoest, J.-J. Lo, & J. L. Kratky (Eds.), *Proceedings of the 34th annual meeting PME-NA* (p. 571). Kalamazoo, MI: Western Michigan University.
Lu, L., & Ricks, T. E. (2013). Developing discourse that promotes reasoning and proof: A case study of a Chinese exemplary lesson. In S. L. Reeder & G. T. Matney (Eds.), *Proceedings of the 40th Annual Meeting of RCML* (pp. 213–221). Tulsa, OK.

Ma, L. (1999). *Knowing and teaching elementary mathematics*. Hillsdale, NJ: Erlbaum.

Meyer, R. D., & Wilkerson, T. L. (2011). Lesson study: The impact on teachers' knowledge for teaching mathematics. In L. C. Hart, A. Alston, & A. Murata (Eds.), *Lesson study research and practice in mathematics education* (pp. 15–26). Chicago: University of Chicago Press.

Ministry of Education, P. R. China. (2003). *Mathematics curriculum standards for normal senior high schools*. Beijing, China: People's Educational Press.

Morine-Dershimer, G., & Kent, T. (1999). The complex nature and sources of teachers' pedagogical knowledge. In J. Gess-Newsome & N. G. Lederman (Eds.), *Examining Pedagogical Content Knowledge* (pp. 21–50). Netherlands: Kluwer Academic Publishers.

Murata, A. (2011). Introduction: Conceptual overview of lesson study. In L. C. Hart, A. Alston, & A. Murata (Eds.), *Lesson study research and practice in mathematics education* (pp. 1–14). Chicago: University of Chicago Press.

National Research Council and Institute of Medicine. (2000). *From neurons to neighborhoods: The science of early childhood development*. Committee on Integrating the Science of Early Childhood Development. In J. P. Shonkoff & D. A. Phillips (Eds.), *Board on Children, Youth, and Families*. Washington, DC: National Academy Press.

Nipper, K., Ricks, T. E., Kilpatrick, J., Mayhew, L., Thomas, S., Kwon, N. Y., et al. (2011). Teacher tensions: Expectations in a professional development institute. *Journal of Mathematics Teacher Education, 14*, 375–392.

Paine, L. W. (1997). Chinese teachers as mirrors of reform possibilities. In W. K. Cummings & P. G. Altbach (Eds.), *The challenge of Eastern Asian education* (pp. 65–83). Albany: State University of New York Press.

Paine, L. W., & Ma, L. (1993). Teachers working together: A dialogue on organizational and cultural perspectives of Chinese teachers. *International Journal of Educational Research, 19* (8), 667–778.

Patnam, V. S. (2013). *Factors related to student achievement in mathematics and comparison of the US with other countries: A study based on TIMSS 2007 report*. Unpublished dissertation, George Mason University.

Perry, M. (2000). Explanations of mathematical concepts in Japanese, Chinese, and US first- and fifth-grade classrooms. *Cognition and Instruction, 18*(2), 181–207.

Polikoff, M. S. (2012). The redundancy of mathematics instruction in US elementary and middle schools. *The Elementary School Journal, 113*(2), 230–251.

Porter, A. C., Polikoff, M. S., & Smithson, J. (2009). Is there a de facto national intended curriculum? Evidence from state content standards. *Educational Evaluation and Policy Analysis, 31*(3), 238–268.

Ricks, T. E. (2003). *An investigation of reflective processes during lesson study by mathematics preservice teachers*. Unpublished master's thesis, Brigham Young University, Provo, UT.

Ricks, T. E. (2007). *The mathematics class as a complex system*. Unpublished dissertation, University of Georgia, Athens, GA.

Ricks, T. E. (2009a). Juxtapositional pedagogies in mathematics methods courses. In S. L. Swars, D. W. Stinson, & S. Lemons-Smith (Eds.), *Proceedings of the 31st Annual Meeting of PME-NA* (Vol. 5, pp. 1244–1252). Atlanta, GA: Georgia State University.

Ricks, T. E. (2009b). Superbugs and ailing schools: What bacterial colonies can teach us about education. *Education Week Commentary*. published on-line, July 6, 2009.

Ricks, T. E. (2010a). China, microbes, and complexity: How to improve mathematics education. In H. Zhang (Ed.), *Complexity, Chinese culture, and curriculum reform* (pp. 112–122). Shanghai, P. R. China: East China Normal University Press.

Ricks, T. E. (2010b). Juxtaposing Chinese and American mathematics education commitments. In P. Brosnan, D. B. Erchick, & L. Flevares (Eds.), *Proceedings of the 32nd Annual Meeting of PME-NA* (Vol. 6, pp. 354–362). Columbus, OH: The Ohio State University.

Ricks, T. E. (2011a). Process reflection during Japanese lesson study experiences by prospective secondary mathematics teachers. *Journal of Mathematics Teacher Education, 14*, 251–267.

Ricks, T. E. (2011b). Mathematics education as an extension of Deweyan reflective mind development. In L. R. Wiest & T. Lamberg (Eds.), *Proceedings of the 33rd Annual Meeting of PME-NA* (pp. 1485–1492). Reno, NV: University of Nevada, Reno.

Ricks, T. E. (2012). A case for Chinese lesson study. In L. R. Van Zoest, J.-J. Lo, & J. L. Kratky (Eds.), *Proceedings of the 34th Annual Meeting PME-NA* (pp. 551–554). Kalamazoo, MI: Western Michigan University.

Ricks, T. E. (2014). Why always greener on the other side? The complexity of Chinese and US mathematics education. In B. Sriraman, J. Cai, K. H. Lee, F. Lianghuo, Y. Shimuzu, L. C. Sam, & K. Subramanium (Eds.), *The first sourcebook on Asian research in mathematics education (Chapter 27)* (pp. 759–784). Charlotte, NC: Information Age.

Ricks, T. E., Lu, L., & Fleener, M. J. (2009). Understanding students' thinking from an interrelationship perspective. *Journal of Mathematics Education, 17*(6), 66–71.

Rodgers, C. (2002). Defining reflection: Another look at John Dewey and reflective thinking. *Teachers College Record, 104*(4), 842–866.

Romberg, T., & Carpenter, T. (1986). Research on teaching and learning mathematics: Two disciplines of scientific inquiry. In M. Wittrock (Ed.), Handbook of research on teaching (pp. 850–873). New York: Macmillan Publishing Company.

Sargent, T., & Hannum, E. (2005). Keeping teachers happy: Job satisfaction among primary school teachers in rural Northwest China. *Comparative Education Review, 49*(2), 173–204.

Schleppenbach, M., Flevares, L. M., Sims, L., & Perry, M. (2007). Teachers' responses to student mistakes in Chinese and US classrooms. *The Elementary School Journal, 108*(2), 131–147.

Schmidt, W. H., Cogan, L. S., Houang, R. T., & McKnight, C. C. (2011). Content coverage differences across districts/states: A persisting challenge for US education policy. *American Journal of Education, 117*(3), 399–427.

Schmidt, W. H., Wang, H., & McKnight, C. C. (2005). Curriculum coherence: An examination of US mathematics and science content standards from an international perspective. *Journal of Curriculum Studies, 37*(5), 525–559.

Schmidt, W. H., & Houang, R. T. (2012). Curricular coherence and the Common Core State Standards for mathematics. *Educational Researcher, 41*, 294–308.

Schön, D. A. (1983). *The reflective practitioner: How professionals think in action*. New York: Basic books.

Schmidt, W. H., & Maier, A. (2009). Opportunity to learn. In G. Sykes, B. Schneider, & D. N. Plank (Eds.), *Handbook of education policy research* (pp. 541–559). Routledge: American Educational Research Association.

Shulman, L. (1986). Those who understand: Knowledge growth in teaching. *Educational Researcher, 15*(2), 4–14.

Schümer, G. (1999). Mathematics education in Japan. *Journal of Curriculum Studies, 31*, 399–427.

Shapiro, J. (1998). Thinking about bacterial populations as multicellular organisms. *Annual Review of Microbiology, 52*, 81–104.

Shapiro, J. A. (2006). Genome informatics: The role of DNA in cellular computations. *Biological Theory, 1*(3), 288–301.

Siegler, R. S., & Mu, Y. (2008). Chinese children excel on novel mathematics problems even before elementary school. *Psychological Science, 19*(8), 759–763.

Stafford-Plummer, J. (2002). *An analysis of the influence of lesson study on preservice secondary mathematics teachers' view of self-as mathematics expert*. Unpublished master's thesis. Provo, UT: Brigham Young University.

Stanley, C. (1998). A framework for teacher reflectivity. *TESOL-Quarterly, 32*(3), 584–591.

Stein, M. K., Smith, M. S., Henningsen, M. A., & Silver, E. A. (2000). *Implementing standards-based mathematics instruction: A casebook for professional development*. New York: Teachers College.

Stepanek, J. (2000). Practical inquiry: Effective practices that support teaching and learning in mathematics and science. *Northwest Educational Research Laboratory* (ERIC No. ED451044)

Stevenson, H. W., & Nerison-Low, R. (1998). To sum it up: Case studies of education in Germany, Japan, and the United States. National Institute on Student Achievement, Curriculum, and Assessment. Office of Educational Research and Improvement: US Department of Education.
Stevenson, H. W., & Stigler, J. W. (1992). *The learning gap*. New York, NY: Summit Books.
Stigler, J. W., & Hiebert, J. (1999). *The teaching gap*. NY: Free Press.
Stigler, J. W., Lee, S. Y., Lucker, G. W., & Stevenson, H. W. (1982). Curriculum and achievement in mathematics: A study of elementary school children in Japan, Taiwan, and the United States. *Journal of Educational Psychology, 74*(3), 315–322.
Stigler, J. W., Lee, S. Y., & Stevenson, H. W. (1987). Mathematics classrooms in Japan, Taiwan, and the United States. *Child Development, 58*, 1272–1285.
Surowiecki, J. (2004). *The wisdom of crowds*. NY: First Anchor Books.
Thompson, T., & Preston, R. (2004). Measurement in the middle grades: Insights from NAEP and TIMSS. *Mathematics Teaching in the Middle School, 9*, 514–519.
Tienken, C. H. (2008). Rankings of national achievement test performance and economic strength: Correlation or conjecture? *International Journal of Education Policy and Leadership, 3*(4), 1–15.
Tierney, R. J. (2001). An ethical chasm: jurisdiction, jurisprudence, and the literacy profession. *Journal of Adolescent and Adult Literacy, 45*(4), 260–276.
Wang, X., Liu, C., Zhang, L., Luo, R., Glauben, T., Shi, Y., et al. (2011). College education and the poor in China. *Phi Delta Kappan, 63*(10), 673–676.
Wang, J., & Lin, E. (2005). Comparative studies on US and Chinese mathematics learning and the implications for standards-based mathematics teaching reform. *Educational Researcher, 34*(5), 3–13.
Wang, S., & Lu, S. (2008). Research process, changes and implementation of mathematics curriculum standards for senior high schools in China. In L. D. Miller & S. R. Saunders (Eds.), *Proceedings of the US-Sino Workshop on Mathematics and Science Education* (pp. 49–53). Murfreesboro, TN: Middle Tennessee State University.
Wang, T., & Murphy, J. (2004). An examination of coherence in a Chinese mathematics classroom. In L. Fan, N.-Y. Wong, J. Cai, & S. Li (Eds.), *How Chinese learn mathematics* (pp. 107–123). Singapore: World Scientific.
Weeks, R. M. (1931). Teaching the whole child. *The English Journal, 20*(1), 9–17.
Wong, N. Y., Han, J. W., & Lee, P. Y. (2004). The mathematics curriculum: Towards globalisation or Westernisation? In L. Fan, N. Y. Wong, J. Cai, & S. Li (Eds.), *How Chinese learn mathematics* (pp. 27–70). Singapore: World Scientific.
Yang, Y., & Ricks, T. E. (2011). How crucial incidents analysis support Chinese lesson study. *International Journal for Lesson and Learning Studies, 1*(1), 41–48.
Yang, Y., & Ricks, T. E. (2012). Chinese lesson study: Developing classroom instruction through collaborations in school-based Teaching Research activities. In Y. Li & R. Huang (Eds.), *How Chinese teach mathematics and improve teaching (Chapter 4)* (pp. 51–65). London: Routledge.
Yuan, Z., & Ricks, T. E. (2011a). A comparative study on Chinese and U.S mathematics teachers' classroom teaching based on mathematical problems: A case of geometric probability model. *Fujian Education, 827*(46), 34–36.
Yuan, Z., & Ricks, T. E. (2011b). *Comparing a Chinese and American lesson on geometric probability*. Paper presented at the 3rd Classroom Teaching Research Annual Group Meeting. Changshu, China.
Zeichner, K. M., & Liston, D. P. (1996). *Reflective teaching: An introduction*. Mahwah, NJ: Lawrence Erlbaum Associates.

Chapter 14
Research on Improving Teacher Knowledge and Pedagogical Approaches: From a Comparative to a Collaborative Perspective

Sandra Crespo

Abstract International studies in education have been part of the research and policy landscape for more than five decades. They are often associated with the inevitable declaration of educational winners and losers and tend to be overshadowed by the media frenzy that accompanies any research reports that highlight world rankings. However, the organizational, political, methodological, and social skills required to design, implement, and disseminate these kinds of international comparative studies cannot be underestimated. In this commentary I first provide a brief overview of salient and widely discussed challenges and benefits of international studies and consider them in relation to the focus of this section on improving teacher knowledge and pedagogical approaches. I then highlight how each of the chapters in this section attends to some of these challenges and benefits. I close with reflections and future questions for international research focused on this area of study.

Keywords International research • Collaborative perspective • Teacher education • Teacher knowledge • Teaching practice

One challenge of international comparative research is the question of whether they compare apples and oranges. This question is typically raised of studies that use comparisons among countries to produce world rankings and benchmarks with respect to valued educational outcomes. This is a fair question for the chapters in this section, as they too have to consider the vast cultural differences between the USA and the Asian countries represented here (Korea, China, and Japan). It is for example important to consider the differing cultural expectations and demands for teachers across these countries. Another layer of cross-cultural complexity is that

S. Crespo (✉)
Michigan State University, 620 Farm Lane, Erickson Hall 362, East Lansing, MI 48824, USA
e-mail: crespo@msu.edu

J.-W. Son et al. (eds.), *What Matters? Research Trends in International Comparative Studies in Mathematics Education*, Research in Mathematics Education, DOI 10.1007/978-3-319-51187-0_14

the professional status of teachers is at a record low in the USA compared to Asian countries, where teachers are well respected and given much higher status (Burns & Darling-Hammond, 2014; Goldstein, 2014; Stigler & Hiebert, 1999). The USA and Asian countries also differ in how teacher preparation, teacher induction, and professional development programs are organized and coordinated (Burns & Darling-Hammond, 2014; Feiman-Nemser, 2001). These are important issues to consider when reading the research studies in this section, as these cross-cultural differences frame and limit what it is possible to learn from these studies.

In spite of these challenges, however, there is much to learn from international studies, as they can help us to better understand the educational issues and problems we are seeking to improve in our particular local contexts. As Paine and Zeichner (2012) tell us: "Teachers, their teaching, and teacher learning are now a central conversation, not just locally but globally" (p. 569). When framed as collaborative cross-cultural exchange studies, such work can help make visible deeply entrenched cultural scripts which to insiders seem commonsensical and typical practice (e.g., Fernandez, 2002; Stigler & Hiebert, 1999; Tobin, Hsueh & Karasawa 2009). In other words, collaborative cross-cultural studies can help identify invisible roadblocks to improving the knowledge and pedagogical skills of teachers within a particular educational system. International studies like the ones reported on here can help us raise new questions about cultural patterns and norms that may be hidden in plain view to those inside that particular local context, and help imagine possible ways to redesign and restructure educational opportunities for students and their teachers. Furthermore, as Akiba and LeTendre (2009) discuss, there is much to gain from collaboratively examining the successes and failures of systemic reform efforts to benefit the international education community.

An important benefit to approaching international research studies as a collaborative rather than simply as a competitive enterprise is that the world is increasingly much more connected. Global challenges like climate change, world hunger, and income inequality are much too big to be tackled and solved by a few wealthy and high-achieving economies: addressing them will require much more collaboration and cooperation across nations than ever before. Similarly, addressing global challenges in mathematics education requires the adoption of a global perspective to improve the quality of educational opportunities for students and teachers everywhere. But more specifically to the point of this book and the chapters in this section, the entire world stands to benefit from improving teachers' knowledge and pedagogical skills. Today's teachers are not simply charged with educating the future workforce for a specific state or nation, but rather are preparing students for a world with many fewer borders, and educating future citizens of a much smaller world and more global society.

The three chapters in this section take important steps in moving the field of mathematics education towards considering a more collaborative dialogical orientation to international research studies. Collectively, these three chapters take a stance of learning from and with international partners to explore important questions about what it will take to improve the quality of teachers' knowledge and their pedagogical approaches. Although the three chapters focus on different questions and employ different methodologies, they all narrow their focus to analyzing learning cycles, activities, or processes with the goal of improving the quality of teachers' knowledge and pedagogical approaches. Furthermore, all three chapters are modest, small-scale studies designed to explore research possibilities rather than offer generalized solutions or prescriptions. Overall they help us imagine new possibilities for designing worthwhile learning opportunities for current and future teachers of mathematics.

First, in *Cross-cultural Lesson Planning between the United States and Korea,* Lim and Son report on a collaborative course assignment for Korean in-service teachers and US teacher candidates. This innovative intercultural course project focused on the processes of preparing, refining, and reflecting on lesson plans. Important benefits include the participants' reflection on strengths and weaknesses in how they represented their content knowledge in lesson plans, and how they planned to share it with their students. Most notably, the US teacher candidates noticed the consistently high expectations for student learning in the Korean teachers' lesson plans, which were rooted in deep content knowledge. The Korean teachers in turn noticed that the US teacher candidates' lesson plans included explicit attention to issues of equity, and also included connections to students' interests and everyday life. The researchers also report added benefits in helping participants broaden their ideas about spaces and communities for teacher learning, including appreciation for developing cross-cultural and reciprocal engagement and international collaborations among future and practicing teachers.

These important insights and benefits did not come without some challenges. Lim and Son had to negotiate many challenges throughout their project, such as that of deciding which language was going to be privileged in communications among the researchers, their collaborators, as well as the participants in this project. It is important to note that both authors in this chapter are fully bilingual Korean and English speakers who could navigate and bridge both cultural and linguistic borders in ways that monolingual educators and researchers could not. Another challenge is that although their study was set up as a collaboration from the start, the situation could be interpreted as more of a mentoring situation, with more experienced Korean teachers providing advice to the inexperienced teacher candidates. Notwithstanding this unequal status, however, the participants appreciated their colleagues' feedback and learned much from comparing each other's lesson planning strengths and weaknesses.

In Chapter, *The instructional quality of mathematics student teachers in the United States and Japan: The possible impact of the structure of student teaching,* Corey, Leatham, and Peterson showcase a study of how international studies can assist in program evaluation and redesign. By comparing the performance of their future teachers to that in other countries, US-based educators can judge the progress they are making towards improving their teacher preparation programs—in this case, their student teaching experience. The authors of this chapter analyzed the quality of instruction in videotaped lessons for both US and Japanese student teachers, and took steps to ensure cultural and linguistic alignment in the research process. For instance, the Japanese data was double-coded by collaborating

Japanese educators working in their own language. Corey et al. found that there were no statistically significant differences in the overall rating of the lesson quality among the two groups of student teachers. The authors make clear and discuss the limitations to their study, and they offer plausible observations but no causal claims about the positive impact of revising the student-teaching experience in US-based teacher preparation programs.

This chapter is set up as a prototypical comparative study, where the Asian model of teaching and by extension their model of student teaching is considered superior and worthy of emulation. However, this chapter goes beyond a comparative analysis of differences: the authors had spent considerable time analyzing and learning from the structural differences in the Japanese and US models for student teaching, and used this analysis to inform the revisions made to their US institution's student teaching experiences. These revisions included providing more opportunities for the US student teachers to receive more substantive feedback from their mentor teachers and peers, to inform their reflections and revisions of thoroughly planned lessons, an established practice in the Japanese model of student teaching. This project's findings counter the dominant narrative that fixing teachers and their teaching is the solution to the problem of students' underachievement and low-quality learning. The chapter offers evidence that working on improving teachers' knowledge and quality of their teaching is not enough, but rather needs to be part of a larger scope of concurrent work on improving the structures that support teachers to improve their knowledge and skills.

Finally, in *Reflective capabilities of mathematics education systems in China, Japan and the United States,* Ricks identifies systemic features that enable and those that prevent educational reform at various levels, including the national level. Considering educational systems as complex systems that are intelligent and self-perpetuating entities, the author offers a provocative analysis as to how and why reform efforts are much more challenging to implement in the US educational system than in Asian countries, such as China and Japan. He argues that reform efforts flourish in systems that have institutionalized reflective capability alongside each level in the system, which prevents the system's self-perpetuating and self-preservation goals from undermining the need for radical reform efforts. At the core of the problem, Ricks suggests, is the prevalent pedagogy of transmission at all levels in the US education system, starting in grade school and continuing into teacher preparation and professional development. In other words, the preparation of teachers mirrors the education of students and sets up teachers to reproduce the pedagogy of transmission they experienced. This cycle of reproduction is possible to break when systems have the kinds of reflective capabilities Ricks uncovers in Japan's and China's educational systems.

Ricks' analysis tells us that the US educational system privileges self-preservation and perpetuation at the expense of developing the system's capacity for innovation and improvement. In contrast, the embedded structures for collaborative and reflective practice within each layer and level of the Chinese and Japanese education systems (e.g., shared planning times and spaces, lesson study, public demonstration lessons, and teaching competitions) set them apart.

These structures are in place to promote the collective improvement of practice rather than that of individual teachers. In the USA, however, the system is designed to exalt individual exemplary teaching, and perpetuate a culture of scapegoating teachers for the system's failures to support national educational reform efforts.

Ricks' chapter offers a different approach and theory from most typical international comparative examinations of policies and system-wide cross-cultural studies, yet it reaches a similar conclusion about the challenges to national reform efforts in systems that are not set up to support the implementation of those reform ideals. One thing missing in this chapter, however, is taking a similar reflective stance towards the research enterprise, including Ricks' own research approach. This chapter could benefit from a more intercultural dialogue with international partners, in order to consider their perspective and insights on their comparative analysis of their successes and challenges in implementing national reform efforts, which could further develop the chapter's analytical framework.

In closing, I want to reiterate the important contributions these chapters make. They do not simply dwell on outlining the challenges and roadblocks to improving the quality of teaching and learning at the local, national, or international levels. Instead, they provide us with examples of how small-scale and more collaborative (vs. comparative) international studies in mathematics education can provide unique insights and perspectives. A common theme across all three chapters is the importance of reflection and collaboration in the learning process across all types of educational settings and programs, including system-wide learning. They also illustrate the ways in which these kinds of cross-cultural research projects can explore learning experiences for teacher learning that not only improve the quality of teachers' knowledge and their instructional strategies but also aim to break reproductive cultural pedagogical scripts and cycles.

These kinds of cross-cultural studies are not for the faint of heart, as they demand a great deal of skill from researchers who embark on them. They require developing and nurturing trusting and sustained relationships with international partners and collaborators, who can see local and global benefits to participating in these endeavors. I close with some questions that are worth considering when reading these chapters, and that are worth raising for these authors and others who are engaged in exploring questions about teacher learning in local and global contexts:

- What are the core arguments and goals for developing a research agenda for studying mathematics teacher learning in global and local contexts?
- How and why are international studies in mathematics education set up as competitive and/or collaborative enterprises?
- What are important challenges and demands researchers face when conducting cross-cultural international studies in mathematics education?
- Whose languages, cultures, and practices are privileged in these kinds of international studies?
- What countries and international partners are being included and/or excluded from cross-cultural research studies, and why?

References

Akiba, M., & LeTendre, G. (2009). *Improving teacher quality: The US teaching force in global context*. New York, NY: Teachers College Press.

Burns, D., & Darling-Hammond, L. (2014). *Teaching around the world: What can TALIS tell us?* San Francisco: SCOPE (Stanford Center for Opportunity Policy In Education).

Feiman-Nemser, S. (2001). From preparation to practice: Designing a continuum to strengthen and sustain teaching. Teachers College Record, 103, 1013–1055. Available at http://www.tcrecord.org/Content.asp?ContentID=10824.

Fernandez, C. (2002). Learning from Japanese approaches to professional development: The case of lesson study. *Journal of Teacher Education, 53*(5), 393–405.

Goldstein, D. (2014). *The Teacher Wars: A History of America's Most Embattled Profession*. New York, NY: Anchor Books.

Paine, L., & Zeichner, K. (2012). The local and the global in reforming teaching and teacher education. *Comparative Education Review, 56*(4), 569–583.

Stigler, J. W., & Hiebert, J. (1999). *The teaching gap: Best ideas from the world's teachers for improving education in the classroom*. New York, NY: Free Press.

Tobin, J., Hsueh, Y., & Karasawa, M. (2009). *Preschool in three cultures revisited*. Chicago, IL: University of Chicago Press.

Part IV
Cross-national Comparative Studies with Large-Scale Data

Chapter 15
Self-Concept, Self-Efficacy, and Mathematics Achievement: Students in 65 Regions Including the US and Asia

Ming Ming Chiu

Abstract This study examines whether seven well-funded and economically equal education systems in Asian regions (Japan, Singapore, South Korea, Taiwan, Hong Kong, Macau, and Shanghai) yield students with higher mathematics achievement, lower self-concept, higher mathematics self-efficacy, and higher intrinsic motivation, compared to the US and 57 other regions. Mathematics interest, mathematics self-concept, and mathematics self-efficacy were related to math achievement in all 65 regions. Compared to US students, those from these Asian regions had higher mathematics achievement and lower self-concept, suggesting a face culture modesty bias and a dignity culture enhancement bias. These Asian students' self-efficacy exceeded their self-concept, further supporting their modesty bias. In contrast, US students' self-concept exceeded their self-efficacy, evidence of their enhancement bias.

Keywords Economic growth • Economic inequality • Family • Culture • Self-concept • Self-efficacy • Mathematics achievement

Many studies have shown that perceptions of one's competence (*expectancy beliefs*) are linked to one's mathematics achievement and that they differ across regions (e.g., Chiu & Klassen, 2010; Chiu & Zeng, 2008; Choi, Choi, & McAninch, 2012), but researchers have not explained how and why they differ. This chapter discusses how economics and cultural values can explicate the different relationships among students' expectancy beliefs and mathematics achievement in 65 regions, including Japan, mainland China [Hong Kong, Macau, and Shanghai], Singapore, South Korea, Taiwan (henceforth, *Asian regions*), and the United States (US, see Fig. 15.1).

M.M. Chiu (✉)
Department of Educational Studies, Purdue University, 5156 Beering hall, West Lafayette, IN, USA
e-mail: mingmingchiu@gmail.com

J.-W. Son et al. (eds.), *What Matters? Research Trends in International Comparative Studies in Mathematics Education*, Research in Mathematics Education, DOI 10.1007/978-3-319-51187-0_15

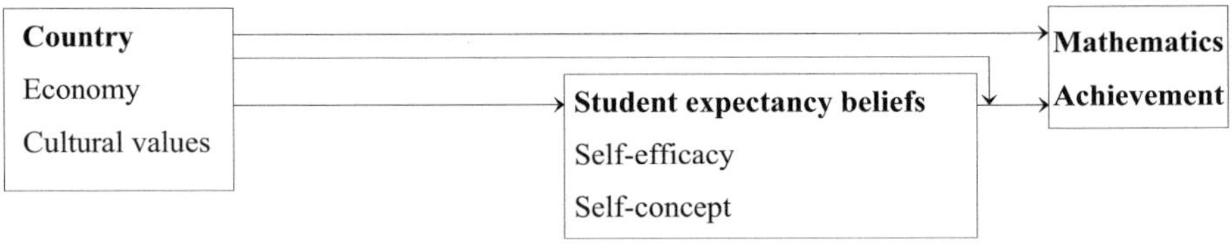

Fig. 15.1 Country attributes and expectancy beliefs can influence students' mathematics achievements

Theoretical Framework

Expectancy beliefs can affect mathematics achievement. Moreover, the economies and cultural values of Asia and the US can influence these relationships.

Social Cognitive Theory

According to social cognitive theory, students who have positive feelings about mathematics, value it, or have greater expectations of success tend to exert greater effort, learn more, and show higher mathematics performance (Pintrich & Schunk, 2002). Students' emotional reactions to a task and their task performance influence their effort, persistence, and performance (Pintrich & Schunk, 2002). Students who enjoy working on mathematics tasks ("I like playing with the colorful abacus") or relish their performance ("I love solving these math puzzles") tend to persevere and continue exerting effort, which fosters higher mathematics achievement (Chiu & Zeng, 2008).

Motivation

Students who value mathematics are more likely to exert greater effort, perceive a greater likelihood of success, persist despite difficulties, and show higher mathematics achievement (Trautwein, Lüdtke, Marsh, Köller, & Baumert, 2006). Students' beliefs about the value of mathematics consist of the reasons why students engage in learning and doing mathematics (Chiu, Pong, Mori, & Chow, 2012). For example, students who like learning mathematics ("I think mathematics is interesting," *intrinsic motivation*) often show higher mathematics achievement (Deci & Ryan, 2002), as do students who view mathematics as a useful tool for other goals ("Mathematics will help me get a good job," *instrumental motivation*) (e.g., d'Ailly, 2003; Pintrich & Schunk, 2002).

Expectancy Beliefs

Students' perceived likelihood of mathematics success (*expectancy beliefs*) includes measures of mathematics self-concept and mathematics self-efficacy. *Mathematics self-concept* reflects self-perceptions about one's abilities and competences that influence the likelihood of success in mathematics ("I'm good at math," Pintrich & Schunk, 2002). Meanwhile, *mathematics self-efficacy* is a belief in one's capabilities to execute a type of task (e.g., "I can do this math problem," Bandura, 2012). Intra-person self-concept and self-efficacy vary according to domain, with only weak correlations across them (e.g., between reading and mathematics self-concepts; Wang, 2015). When people have a positive mathematics self-concept or self-efficacy, they show more motivated behaviors and greater perseverance with challenging mathematics tasks (Pintrich & Schunk, 2002). As mathematics self-concept concerns the perception of one's general ability in contrast to mathematics self-efficacy's focus on doing mathematics tasks, comparing one's mathematics competence with that of one's schoolmates (*social comparison*) is more likely to influence self-concept than self-efficacy (Bandura, 2012; Bong & Skaalvik, 2003). Specifically, if a student has high past mathematics achievement and schoolmates with higher mathematics achievement, he or she likely has a below-average mathematics self-concept but still has high mathematics self-efficacy on basic mathematics tasks.

Student experiences, especially their past successes and failures on mathematics tasks, affect their expectancy beliefs (Pintrich & Schunk, 2002). Past successes are likely to maintain (or increase) perceived activity value and expectations, while past failures may lead students to lower their expectations and devalue the activity to protect their self-esteem from the damage of a likely future failure (Pintrich & Schunk, 2002).

Regional Differences

Regions have different economic conditions and cultural values, and both variables can influence mathematics expectancy beliefs and their relationships with mathematics achievement, as shown through studies using large-scale, international data.

Economic Differences

The US and Asia followed different paths to economic prosperity. The US economy grew fairly consistently from its founding in 1776, powered by long periods of peace, readily available human labor and physical property within the US, free markets, and investment in education (Piketty, 2013). The Atlantic and Pacific oceans isolated the US from most regions, so aside from the US Civil War, few wars were fought on US soil, unlike in Europe, Asia, and Africa. This isolation resulted in the progressive accumulation of property and human skills and also facilitated the availability of labor and property within US borders, in which businesses regularly invested to increase human productivity (Atack & Passell, 1994). Furthermore, the US's institutional infrastructure of free markets (e.g., contract law, property rights, and lender rights) protects business investors and enables a relatively equal economic field for the market to decide winners and losers, rather than relying on nepotism or crony capitalism (Jones & Romer, 2010). Lastly, the US gradually invested in education to increase its human skills, eventually resulting in universal primary education in 1918, far earlier than any other region (Pulliam & Van Patten, 2012). As a result of the US's few wars on home soil, available labor and physical property, free markets, and widespread education, its economy averaged 3.3% annual economic growth from 1781 to 2008 to become one of the richest regions in the world (Romer, 2010).

In contrast, Japan, Singapore, Taiwan, South Korea, and China were devastated by several twentieth-century wars (Stiglitz, 2013). World War II (1941–1945) ravaged Japan, Singapore, Taiwan, and South Korea; the Korean War (1950–1953) further harmed South Korea; and China fought in the Second Sino-Japanese War (1937–1945) and the Chinese Civil War (1946–1949) (Dower, 1986). Afterwards, these Asian regions all implemented free markets, growing rapidly over several decades (Stiglitz, 2013). Their governments invested heavily in schools and universal education to increase human capital (Young, 1995). Furthermore, these governments' low spending (high savings) and low interest rates spurred company investment in machines, factories, and so on (Page, 1994). Together, free markets and investments in education and machinery spurred economic growth exceeding 7% for several decades in each of these regions (Stiglitz, 2013).

As a result of sustained economic growth, the US and these Asian regions are richer than most countries in the world: in 2012, Singapore had the highest gross domestic product per capita ($54,578; measured in current US$), followed by Japan ($46,679), the US ($41,457), South Korea ($24,454), Taiwan ($21,270), and mainland China ($6265). By contrast, the median country income per capita is $5780 (World Bank, 2015). The Chinese students in this study lived in rich cities: Macau ($77,079), Hong Kong ($36,708), and Shanghai ($13,431; World Bank, 2015). Hence, Singaporean students can benefit more from its greater wealth while Shanghai students benefit less from its lesser wealth.

More resources, higher mathematics achievement. A region's economic wealth and growth can affect a student's learning, expectancy beliefs, and mathematics achievement (Chiu, Chow, & McBride-Chang, 2007; Chiu & Klassen, 2010). Richer countries typically provide more education for their people and more opportunities to earn more money, so they tend to have higher socio-economic status (SES) than people in other countries (Baker, Goesling, & LeTendre, 2002). High SES family backgrounds and better-trained teachers can increase a student's learning and mathematics achievement (Chiu & Khoo, 2005).

Specifically, students in higher-SES families often enjoy greater family investment (Chiu, 2010) in educational resources at home, which can improve a child's expectancy beliefs and mathematics achievement (Artelt, Baumert, Julius-McElvaney, & Peschar, 2003). High-SES parents typically have more years of schooling (*human capital*) and have higher income (*financial capital*; Chiu & Khoo, 2005). As they appreciate the importance of educational resources at home, they use their financial and human capital to buy suitable books, computers, and so on to create more mathematics learning opportunities for their children (Chiu, 2010). By creating stimulating learning environments, parents encourage their children to find ideas and activities that interest them, thereby enhancing their intrinsic mathematics motivation (Pintrich & Schunk, 2002). Likewise, letting them explore and succeed independently enhances their autonomy and builds their mathematics self-concept and self-efficacy (Chiu & Klassen, 2010). In short, students in richer regions tend to have higher-SES families, and their greater education investments often increase their children's learning, expectancy beliefs, and mathematics achievement.

Inequality. On the other hand, inequality at the family or school level can reduce mathematics achievement within a region. Inequality mechanisms include fewer educational resources and their inefficient allocation. Family and school inequalities may reduce students' mathematics achievement via fewer educational resources at both the national and school levels. In regions with substantial family income inequality, privileged parents often have many more educational resources (e.g., books) at home than others, and commonly send their children to private schools (Rothstein, 2000). Hence, they have a strong economic incentive to discourage public spending, for example, paying for lobbyists or personally lobbying politicians or government officials to reduce spending on public schools (e.g., Benabou, 2000). Likewise, in unequal school systems, richer schools may lobby government officials for more educational resources for their schools—and fewer educational resources for other schools. Private school officials typically favor educational vouchers for their schools over spending on public schools (e.g., Miller, 1992). For instance, the Pennsylvania Association of Private School Administrators spent \$190,000 on lobbyists during 2008–2012 (Center for Responsive Politics, 2012). Such behavior diverts education resources to lobbying and can reduce national spending on public school resources, resulting in fewer educational opportunities, less learning, and lower overall mathematics achievement (Chiu & Khoo, 2005).

Parents' lobbying behavior within schools can also reduce overall education resources. For example, some parents lobby teachers and principals to cater to the academic needs of their children, sometimes at the expense of other children (Chiu, 2009). Not only is this lobbying unfair, it also takes teacher time away from instruction, which harms all of the teacher's students. In short, fewer educational resources at both the national and school levels can reduce educational resources (whether educational materials or teacher time), which can reduce educational opportunities and student learning.

Family and school inequalities can also reduce student achievement through unequal distribution of resources, which results in diminishing marginal returns and inefficient resource allocation (Stiglitz, 2013). In a region with substantial family inequality, richer students have many more books and other educational resources than poorer students do. While having one computer can be useful, having a second computer is far less useful; this lesser value of an additional item demonstrates *diminishing marginal returns* (Chiu & Khoo, 2005). Due to this process, a richer student typically benefits less from an extra resource (e.g., computer) than a poorer student does (Demirguc-Kunt & Levine, 2009). Given a fixed amount of educational resources, distributing them less equally across families (or across schools) within a region allocates these resources inefficiently and magnifies the impact of diminishing marginal returns (Chiu, 2015). Hence, when educational resources are less equally distributed, they have less educational value for students on average, who then learn less overall and show lower mathematics achievement (Chiu, 2009). In summary, regions with greater family or school inequalities might have less efficient educational resource behaviors or diminishing marginal returns, which can reduce both overall educational resources and allocation efficiency. Both trends might reduce learning opportunities and students' mathematics achievement overall.

Regional Differences in Cultural Values

Apart from economies, cultural values differ across regions (LeTendre, Hofer, & Shimizu, 2003). The historical paths of these regions drive their different cultures.

Chinese and Korean Societies

Grounded in government exams and economic rewards, Chinese and Korean people (mainland China, Taiwan, Singapore, and South Korea) have similar collectivist cultural values that traditionally supported the education of their children (Chiu, 2013; Chiu & Joh, 2015). Beginning with the Sui dynasty, the *Keju* civil service exam system from 606 to 1905 not only selected China's government officials but also gave financial rewards, prestige, power, and fame to their extended families, thereby powering *collectivist* beliefs that value group interests over individual interests (Suen & Yu, 2006). Likewise, the *Gwageo* civil service exam in the Goryeo and Joseon dynasties (958–1894) served the same purpose for Koreans (Seth, 2010).

As a result, contemporary Chinese and Korean children often benefit from the support of extended family members (e.g., money to help pay for university studies, Chiu & Chen, 2014). The modern university exams in mainland China, Taiwan, Singapore, and South Korea serve a comparable purpose of trying to identify and reward the students with the highest academic achievement (Seth, 2010; Suen & Yu, 2006). Hence, Chinese and Korean people learn that school achievement tightly aligns with economic success (Chiu, 2014; Seth, 2010). For example, in Hong Kong's education-rewarding wage system, a high school teacher earns a manual worker's lifetime wages in 15 years while a professor earns it within 5 years (McLelland, 1991). As a result, Chinese and Korean parents, schools, and teachers view mathematics achievement as the gateway to many professions (Chiu, 2014; Seth, 2010). To meet this demand for children's education, Chinese societies devote 4–5% of their GDP to public education spending and generally fix education spending to be the same per student (e.g., Hong Kong spent US $6350 for each secondary school student in 2012).

Japan

Like South Korea and the Chinese societies, Japan highly values education, currently has a comparable examination system, and tightly aligns mathematics achievement with economic success, but it took a different path (Ishikida, 2005). Japan briefly adopted the education and Imperial examination system of China in the sixth century, but quickly returned to children inheriting titles and royal court positions from their parents (Seth, 2010). As a result, Japan also views society as an accepted hierarchy in which every person has an appropriate position (*hierarchical* cultural values), like in South Korea and in Chinese societies, but Japan is not as collectivist (Takano & Osaka, 1999).

After Japan's defeat in World War II, the US-led occupation required all Japanese children to attend middle school (Marshall, 1994). This educational investment helped spur Japan's economic growth (Stiglitz, 2013). Driven by companies' desire for high-quality, educated, and skilled labor, high schools and universities required entrance examinations to admit the highest scoring students (Seth, 2010). Companies then recruited from the high schools and universities with the highest scoring students, reinforcing the importance of these exams and aligning mathematics success with economic success (Ishikida, 2005). Facing societal demands for well-educated students, Japan spent $12,043 on each secondary student in 2012.

The United States

In contrast, the US won its independence to become an egalitarian democracy, yet has retained a frontier individualism and maintained locally controlled schools within a federal system of 50 states (Pulliam & Van Patten, 2012). Grounded in the Declaration of Independence ("all men are created equal") and the Bill of Rights, US citizens share *egalitarian* values (rather than hierarchical values) and recognize many inalienable individual rights (*individualism* rather than collectivism). Reflecting the valuation of freedom and individual choice, most schools are under local control and funded by local taxes (Chiu, 2014). Until the recent Common Core Curriculum supported by most (but not all) states, school districts or even individual schools chose their own curricula (Pulliam & Van Patten, 2012). Hence, parents can often choose a school for their children by moving into the relevant neighborhood (Andre-Bechely, 2012). As a result, US students often have strong egalitarian and individualistic beliefs. While the US spent $13,210 per student in 2012, more than any Asian region, spending varied sharply across schools and neighborhoods, yielding unequal school budgets (Chiu, 2014). Unlike cultural values, country and school inequalities are linked to mathematics achievement in international studies that test both cultural values and economic attributes (Chiu, 2007, 2010, 2015; Chiu & Chow, 2011, 2015; Chiu & Khoo, 2005). Hence, the equal schooling in the seven Asian regions is expected to yield students with higher mathematics achievement compared to students in the US's unequal school system.

Face Vs. Dignity Cultures

As a result of these different historical paths, the cultures of these Asian regions and the US differ considerably. Most significantly for the present study, the Asian regions have a face culture while the US has a dignity culture.

Face Culture

How others view a person creates that person's public self-image (*face*), which is critically important in a face culture. In such cultures, a person's status in the social hierarchy and satisfactorily fulfilling one's expected role earns him or her respect from others (Heine, 2005). Others judge whether a person behaves properly (e.g., politeness, manners) and give respect accordingly (Lee, Kam, & Bond, 2007). As a face culture is hierarchical, some people have more face than others, but everyone who performs satisfactorily has some face (Heine, 2005). In a face culture, public consensus about each person's status in the hierarchy enables harmonious functioning, so each individual is extremely attentive to others' perceptions of him or her (Kim, Cohen, & Au, 2010).

Dignity Culture

In contrast, people in the US believe in the inherent equality and primacy of the individual, a *dignity culture* (Ayers, 1984; Hofstede, Hofstede, & Minkov, 2010). Whereas others can judge a person's performance and sharply increase or decrease his or her self-worth accordingly in a face culture, they have far less influence on it in a dignity culture (Kim et al., 2010). Indeed, the US focus on liberty encourages an individual to determine his or her own self-worth rather than depending on others' judgments (Carter, 2009). In dignity cultures, people defend their agency, autonomy, and freedom against others' attempts to control them (Ayers, 1984). A person's view of himself (or herself) and others' views of that person are thus similar in Asian regions, but can differ substantially in the US (Wang, 2015).

These different cultural values can influence how adolescents estimate their mathematics self-concepts, partly by influencing the sources of mathematics expectancy beliefs and partly through differing levels of attention to in-group expectations (Oettingen & Zosuls, 2006). In face cultures, others determine one's face: as one cannot claim more face than others will give, one's own self-assessment is not relevant (Heine, 2005; Lee et al., 2007). Indeed, trying to claim too much face is considered boorish behavior that violates the rules for distributing status within a hierarchy, and thereby threatens the harmony of the hierarchy (Kim et al., 2010). Hence, a person's perception of himself or herself cannot exceed others' perceptions of him/her and should probably be a little lower, so that he or she can exercise the humility necessary not to overreach on status claims (Hamamura & Heine, 2008). Hence, students in face cultures often seek upward comparisons and have lower mathematics self-concept (*modesty bias*; Heine, 2004; White & Lehman, 2005).

In a dignity culture however, US students rely on themselves to judge their mathematics self-concept and pay little attention to others' judgments (Ayers, 1984; Hofstede et al., 2010). Given a choice of selectively comparing themselves to higher-competence classmates or lower-competence classmates, US students choose downward comparisons more often to boost their own mathematics self-concept (*self-enhancement bias*, Heine, 2004). As a result, students in the US and in other dignity cultures typically have higher mathematics self-concept than students in East Asia and in other face cultures (Chiu & Klassen, 2010).

Overall, the above research suggests how economic characteristics and cultural values together account for expectancy beliefs, mathematics achievement, and their relationships. This model yields seven specific hypotheses, which differ from culture-only theories or simple genetic claims of greater intelligence among students from Asian regions:

H1 Students in richer countries or families show greater mathematics achievement than other students.

H2 Students in countries with more income equality show greater mathematics achievement than other students.

H3 Family structures (two parents, single parent, or no parents) are more strongly linked to mathematics achievement in individualistic regions than in collectivist regions.
H4 Self-concept is greater in individualistic regions than collectivist regions.
H5 Self-efficacy is not significantly greater in individualistic regions than collectivist regions.
H6 Self-efficacy exceeds self-concept in collectivist regions but not in individualistic ones.
H7 Self-efficacy has a stronger link than self-concept to mathematics achievement in collectivist regions, but not in individualistic regions.

Methods

Multilevel analyses of 161,689 15-year-olds in 65 regions tested whether their expectancy beliefs were related to their mathematics achievement. Only one-third of the 475,760 total participants completed the survey form that included all the expectancy beliefs examined in this study, so we only used data from these 161,689 students. These included students from the following regions: Taiwan (2015 students), Hong Kong (1557), Japan (2117), Macao (1778), Shanghai (2125), Singapore (1849), South Korea (1678), and the United States (2037).

Data

The Organization for Economic Cooperation and Development's Program for International Student Assessment 2012 (OECD-PISA, 2013) assessed the mathematics achievement of students across 65 regions, and asked students and principals to fill out questionnaires. Participating students completed a 2 h mathematics test and then a 30–40-min questionnaire, from which the family variables were computed. Regions' economic data (World Bank, 2015) and cultural values data (House, Hanges, Javidan, Dorfman, & Gupta, 2004) were imported from the cited sources and merged with the PISA data.

Methodological Design

Investigating the above hypotheses requires representative sampling, precise tests and questionnaires, and suitable statistical models. In each region, OECD (2014) chose a minimum of 150 representative schools based on neighborhood SES and student intake, and sampled 35 15-year-olds from each school (stratified sampling). OECD excluded students who were mentally incapable of taking the exam, physically unable to take the exam, refused to take the exam, or did not speak the test language (less than 5% of the sample did not complete the exam). With suitable weights, OECD created representative samples of each region's schools and 15-year-olds. Missing questionnaire response data (6%) can reduce estimation efficiency, complicate data analyses, and bias results. Hence, Markov Chain Monte Carlo multiple imputation was used to estimate the values of the missing data, which addresses these missing data issues more effectively than deletion, mean substitution, or simple imputation, according to computer simulations (Peugh & Enders, 2004).

Variables

Primary variables of interest include the outcome variable *mathematics test score*, region-level variables, family variables, and student variables (see Table 15.1).

Mathematics test score. International experts from OECD and non-OECD regions defined mathematics achievement, built assessment frameworks, created test items, forward- and backward-translated them, and pilot-tested them to check their validity and reliability (for details and sample items, see OECD, 2014, and www.pisa.oecd.org). Students did not respond to all items on the entire test. Instead, they received *subtests* (overlapping subsets of all multiple choice and open-ended questions) to generate wider coverage of mathematics skills while reducing student fatigue and test-learning effects (a *balanced incomplete block* test, Baker & Kim, 2004). All data are from OECD (2014) unless otherwise noted.

Regional variables. Region-level variables include region income and income inequality. Region income is measured through gross domestic product (GDP) per capita (World Bank, 2015). Income inequality is measured with Gini. Gini has a value of 0 for perfect region income equality (everyone has the same income) and a value of 100 for perfect inequality (one person has all the income). Although a variety of other cultural values were entered, none of them were significant, so they are not discussed further. Results are available upon request.

To reduce measurement error, multiple questionnaire items were used to create several family and expectancy belief indices via Rasch models (Warm, 1989). In previous studies, the multi-group Rasch models for each item in PISA questionnaires in each region yielded similar parameters, indicating measurement equivalence across regions (May, 2006). While several studies showed consistent questionnaire responses and participant understanding across regions in earlier PISA studies (Brown, Micklewright, Schnepf, & Waldmann, 2005; Lee, 2009; Schulz, 2003), results involving cultural variables require cautious interpretation.

Family variables. *Single parent* is a dichotomous variable, indicating whether a student lives with one parent. Likewise, *living with no parents* and *foreign language spoken at home* are dichotomous variables indicating whether a student lives with no parents, or speaks a language different from the test language at home.

Table 15.1 Summary statistics ($N = 161{,}689$ students)

Variable	All 65	Hong Kong	Macau	Shanghai	Japan	South Korea	Singapore	Taiwan	USA
Mathematics achievement	470	561	541	610	536	556	569	557	487
GDP per capita (current US$)	28,985	36,708	77,079	13,431	46,679	24,454	54,578	21,270	41,457
Gini (inequality)	38	54	35	61	33	30	48	34	48
Single parent	0.12	0.12	0.13	0.10	0.12	0.09	0.08	0.12	0.19
Living with no parents	0.02	0.02	0.03	0.03	0.01	0.01	0.02	0.02	0.02
Speak foreign language (home)	0.12	0.06	0.13	0.02	0.00	0.00	0.51	0.17	0.14
Family socio-economic status (SES)	0.00	−0.43	−0.71	−0.02	0.34	0.27	−0.10	0.19	0.29
Number of books at home	2.93	2.79	2.40	2.94	3.42	3.90	3.06	3.26	2.89
Home educational resources	−0.20	−0.29	−0.29	−0.04	−0.58	−0.11	0.14	−0.24	−0.07
Girl	0.50	0.47	0.47	0.50	0.46	0.47	0.50	0.50	0.49
Relative grade	−0.13	−0.32	−0.47	−0.13	0.00	−0.06	−0.13	0.01	−0.07
Instrumental motivation	0.03	−0.30	−0.33	−0.06	−0.54	−0.43	0.33	−0.42	0.05
Mathematics interest	0.20	0.27	0.11	0.40	−0.25	−0.23	0.80	0.02	0.11
Mathematics self-concept	0.05	−0.12	−0.18	−0.01	−0.48	−0.38	0.19	−0.42	0.32
Mathematics self-efficacy	−0.03	0.20	0.13	0.91	−0.43	−0.36	0.46	0.17	0.20

Family SES is a student-level index created from the student's mother's years of schooling, father's years of schooling, and the highest job status of either parent (OECD, 2014). OECD (2014) used Ganzeboom, de Graaf, and Treiman's (1992) index to measure the highest job status among a student's parents (ranging from 16 to 90).

Number of books indicated whether a student had 0–10, 11–25, 26–100, 101–200, 201–500, or more than 500 books.

Educational resources at home is an index created for each student, indicating whether or not he or she had the following at home: a dictionary, a desk for studying, a quiet place to study, a computer for school work, educational software, books to help with school work, or technical reference books.

Student variables. Girl indicates a female student. *Relative grade* indicates whether a student was held back one grade (−1) or more (−2, −3, etc.), skipped one grade or more (1, 2, etc.), or followed the typical grade schedule (0).

Expectancy belief indices at the student level were created with Likert scales that range from 1 to 4 (disagree to agree) with the stem question, "To what extent do you agree with the following statements?" The four components for *mathematics interest* were: "I enjoy reading about mathematics," "I look forward to my mathematics lessons," "I do mathematics because I enjoy it," and "I am interested in the things I learn in mathematics."

The four components for *instrumental motivation for mathematics* were: "Making an effort in mathematics is worth it because it will help me in the work that I want to do later on," "Learning mathematics is worthwhile for me because it will improve my career prospects," "Mathematics is an important subject for me because I need it for what I want to study later on," and "I will learn many things in mathematics that will help me get a job."

The five components for *mathematics self-concept* were: "I am just not good at mathematics," "I get good grades in mathematics," "I learn mathematics quickly," "I have always believed that mathematics is one of my best subjects," and "In my mathematics class, I understand even the most difficult work."

The eight components for *mathematics self-efficacy* were: using a train timetable to work out how long it would take to get from one place to another, calculating how much cheaper a TV would be after a 30% discount, calculating how many square meters of tiles you need to cover a floor, understanding graphs presented in newspapers, solving an equation like $3x + 5 = 17$, finding the actual distance between two places on a map with a 1:10,000 scale, solving an equation like $2(x+3) = (x+3)(x-3)$, and calculating the gasoline consumption rate of a car. For hypotheses H4, H5, and H6 (relative sizes of self-concept and self-efficacy across regions), descriptive statistics of self-concept and self-efficacy in each region are examined.

This study used multilevel analysis (three levels: region, school, and student) of all the plausible values of students' mathematics test scores, which yields more precise standard errors than ordinary least squares does (Goldstein, 2011; Monseur & Adams, 2009). Plausible values are estimated values that resemble individual test scores with approximately the same distribution and yield consistent estimates of population characteristics when individuals respond to a small subset of the questions on the entire test (Monseur & Adams, 2009).

Analyses

A variance components model tested for significant differences among the three levels of region, school, and student.

$$\text{Math}_{ijk} = \beta + \text{e}_{ijk} + \text{f}_{jk} + \text{g}_{k} \tag{15.1}$$

The outcome variable Math_{ijk} of student *i* in school *j* in region *k* has a grand mean intercept β, with unexplained components (*residuals*) at the student, school, and region levels (e_{ijk}, f_{jk}, g_{k}). The above explanatory variables were entered in sequential sets into the regression models to estimate the variance explained by each set (Kennedy, 2008). Family characteristics might influence student characteristics. All continuous variables were centered on their regional means.

$$\begin{aligned}\text{Math}_{ijk} = \beta + \text{e}_{ijk} + \text{f}_{jk} + \text{g}_{k} + \beta_{uk}\mathbf{Region}_{k} + \beta_{vjk}\mathbf{Family\ structure}_{ijk} \\ + \beta_{wjk}\mathbf{Family\ resources}_{ijk} + \beta_{xjk}\mathbf{Student\ structure}_{ijk} \\ + \beta_{yjk}\mathbf{Student\ motivation}_{ijk} + \beta_{zjk}\mathbf{Student\ expectancy}_{ijk}\end{aligned} \tag{15.2}$$

To test hypotheses H1 (richer countries) and H2 (income inequality), I entered regional variables: regional income as measured by log GDP per capita (in current US$) and inequality of region as measured by Gini (**Region**). (Using the logarithm of GDP per capita reduces the impact of outlier regions with extremely large GDP per capita, and fits the data better than its linear counterpart.) To test H3 (family structure), family demographics variables were entered: student living with no parents, single parent, and foreign language spoken at home (**Family_structure**). I tested whether sets of explanatory variables were significant with a nested hypothesis test (χ^2 log likelihood, Kennedy, 2008). After entering each set of explanatory variables, nonsignificant ones were removed. To test H1 (family resources), I applied the procedure for **Family_structure** to the family resource variables: foreign language spoken at home (reflecting cultural resources), family SES, number of books at home, and home educational resources (**Family_resources**). Then, I applied this procedure to the student structure variables: gender and relative grade (**Student_structure**). Afterwards, this procedure was applied to the motivation variables: instrumental motivation for mathematics and mathematics interest (**Student_motivation**). To test hypothesis H7 (self-concept's and self-efficacy's links to mathematics achievement), this procedure was applied to expectancy belief variables: mathematics self-concept and mathematics self-efficacy (**Student_expectancy**).

An alpha level of 0.05 was used. To control for the false discovery rate (FDR), the two-stage linear step-up procedure was used, as it outperformed 13 other methods in computer simulations (Benjamini, Krieger, & Yekutieli, 2006). Control variables and standardized scores were used for further robustness tests. The small sample of regions ($N = 65$) limits identification of nonsignificant region-level results (for a 0.4 effect size at $p = 0.05$, statistical power = 0.81; Konstantopoulos, 2008). Lastly, I analyzed residuals for influential outliers.

Results

Summary Statistics

Students from the Asian regions outscored US students in mathematics overall (see Table 15.1). The US students had a mean mathematics score of 487, below the OECD mean of 500 (for all 65 regions the mean was 470). In contrast, the Asian regions had much higher mean mathematics scores: Japan (535), Macau (541), South Korea (556), Taiwan (557), Hong Kong (562), Singapore (569), and Shanghai (610). The high mathematics scores of students from Macau and Hong Kong are especially impressive, considering their low family SES (−0.71 and −0.43). Note that the mean family SES of these students can differ substantially from the overall GDP per capita, which include corporations and wealthy households without 15-year-old children. As noted earlier, most of these regions had relatively high income (GDP per capita). Meanwhile, income inequality varied widely, including a low Gini of 30 in South Korea and a high Gini of 61 in Shanghai.

Although the US has the lowest mean mathematics test score, it has the highest mathematics self-concept—but not the highest mathematics self-efficacy (supporting hypotheses H4 and H5). Furthermore, the mean self-efficacy exceeds the mean self-concept for students from the Asian regions, unlike those of the US students and students overall (supporting hypothesis H6). Other notable means are the high mathematics interest of Singapore students (0.80) and the high mathematics self-efficacy of Shanghai students (0.91).

Explanatory Model

Mathematics interest, mathematics self-concept, and mathematics self-efficacy were related to mathematics achievement in all regions (see Table 15.2). However, other explanatory variables were related to mathematics achievement in some regions but not others. All results reported are of each vector's first entry into the regression. Regions with greater income or income equality yielded higher mathematics achievement. The greater mathematics scores in richer regions suggested

Table 15.2 Results of multilevel regressions on mathematics achievement

Explanatory var.	All 65		Hong Kong		Macau		Shanghai		Japan		S. Korea		Singapore		Taiwan		USA	
Log GDP per capita (current US$)	24.61 (4.14)	***																
Gini (inequality)	−1.54 (0.53)	**																
Living with no parents	−12.30 (1.05)	***															−24.79 (11.59)	*
Single parent	−2.64 (0.47)	***													−9.49 (4.14)	*	−7.05 (2.92)	*
Foreign language spoken at home	−5.10 (0.61)	***					−54.38 (12.51)	***							−8.06 (3.62)	*		
SES	5.74 (0.20)	***			5.73 (2.26)	*	5.08 (2.16)	*	9.80 (3.28)	**	8.27 (3.51)	*	9.76 (2.15)	***	12.00 (2.38)	***	11.44 (1.85)	***
Number of books at home	9.13 (0.14)	***	6.29 (1.63)	***	3.19 (1.26)	*	4.74 (1.48)	***	5.25 (1.04)	***	12.08 (1.59)	***	8.21 (1.42)	***	9.84 (1.34)	***	14.50 (1.12)	***
Home educational resources	3.97 (0.21)	***	4.51 (2.19)	*	4.19 (2.06)	*	4.48 (1.94)	*			8.20 (2.33)	***	5.06 (1.94)	**	12.45 (2.03)	***	3.66 (1.59)	*
Girl	−18.49 (0.33)	***	−22.06 (1.36)	***	−27.63 (3.39)	***	−11.06 (3.39)	**	−16.41 (3.04)	***	−13.84 (4.11)	***			−15.59 (3.53)	***	−18.95 (2.22)	***
Relative grade	28.71 (0.30)	***	27.05 (2.78)	***	41.51 (1.77)	***	26.49 (3.51)	***					33.61 (3.87)	***	34.65 (11.10)	**	27.09 (2.22)	***
Instrumental motivation-math	4.82 (0.23)	***									5.38 (2.24)	*			()		7.09 (1.52)	***

Mathematics interest	2.08 (0.25)	***	6.00 (2.98)	*	6.39 (2.52)	*	5.87 (2.88)	*	6.29 (2.19)	**	6.57 (2.92)	*	6.47 (2.39)	**	6.71 (2.77)	*	5.86 (1.56)	***
Mathematics self-concept	23.82 (0.24)	***	17.97 (2.77)	***	11.33 (2.13)	***	27.50 (2.67)	***	10.69 (2.04)	***	17.00 (2.88)	***	18.63 (2.52)	***	23.57 (2.41)	***	27.82 (1.59)	***
Mathematics self-efficacy	23.23 (0.20)	***	26.48 (1.98)	***	28.62 (1.82)	***	29.31 (1.80)	***	21.16 (1.70)	***	35.16 (2.21)	***	36.65 (2.03)	***	36.38 (1.74)	***	29.10 (1.35)	***
Level	Variance explained at each level																	
Region	0.55																	
School	0.48		0.39		0.64		0.59		0.33		0.65		0.58		0.63		0.65	
Student	0.32		0.30		0.44		0.28		0.19		0.32		0.28		0.39		0.43	
Total variance explained	0.41		0.33		0.51		0.41		0.27		0.44		0.39		0.49		0.48	

*$p < 0.05$, **$p < 0.01$, ***$p < 0.001$

that students capitalized on their region's greater resources to learn more and attain higher mathematics scores than other students (supporting hypothesis H1). Also, students in regions with more income inequality had lower mathematics scores than other students (supporting hypothesis H2).

Family structure, family resources, and student demographics were related to mathematics achievement in some regions. Students who lived with one parent or no parents had lower mathematics achievement in most regions, but not in most of the Asian regions (except, notably, Taiwan, for single parents), largely supporting hypothesis H3. Speaking a foreign language at home was linked to lower mathematics scores overall (−5), in Shanghai (−54), and in Taiwan (−8), but not in any of the other regions. Meanwhile, family economic resources (in the form of family SES and number of books) generally had stronger positive links to mathematics scores in the US than in the Asian regions with well-funded, equal education systems (except, notably, Taiwan, for family SES), largely supporting hypothesis H3. Boys outscored girls on these mathematics tests in all regions except for Singapore. Also, students in higher grades generally outscored those in lower grades (often due to retention). As students were rarely held back in Japan or South Korea, relative grade was not significant in these regions.

Intrinsic motivation is positively related to mathematics achievement in all regions, but instrumental motivation is related to mathematics achievement only in the US (+7) and South Korea (+5). In other Asian regions, instrumental motivation is positively correlated with mathematics achievement but is not significant after mathematics interest is entered. Mathematics self-concept and mathematics self-efficacy are positively related to mathematics achievement in all regions. In the Asian regions, the mathematics self-efficacy regression coefficient is significantly higher than the mathematics self-concept one, but not in the US and not overall, supporting hypothesis H7. Other explanatory variables were not significant, and analyses of residuals showed no influential outliers.

Discussion

This study examines whether well-funded and economically equal education systems in seven collectivist Asian regions yield higher mathematics achievement, lower self-concept, and higher mathematics self-efficacy, compared to the US. Mathematics interest, mathematics self-concept, and mathematics self-efficacy were related to mathematics achievement in all regions. Compared to US students, those from the Asian regions had higher mathematics achievement and lower self-concept. These Asian students' self-efficacy exceeded their self-concept but not necessarily the US students' self-efficacy. Also, instrumental motivation for mathematics was related to mathematics achievement only in South Korea and the US.

Compared to the US, the smaller family resource regression coefficients in the Asian regions (except for Taiwan) are consistent with the view that their well-funded and equal education systems compensate for fewer family resources (Chiu,

2015). The Asian regions' schools provide educational resources that reduce the importance of family resources to students' mathematics achievement. These results are consistent with studies showing that equal school systems benefit all students, especially poorer students, and raise overall mathematics achievement (Chiu, 2015). In contrast, the US's unequal, locally funded education system yields larger budgets for superior schools in richer neighborhoods, creating school inequalities that exacerbate family inequalities and yield lower mathematics achievement overall. Future studies can examine why Taiwan's high family resource regression coefficients and high mathematics achievement does not fit this pattern.

Meanwhile, the nonsignificant family structure coefficients of most of the Asian regions are consistent with the view that extended family members in these regions substantially support students living with no parents or single parents (Chiu & Chen, 2014); as a result, these students do not have lower mathematics scores than other students. Again, Taiwan is a notable exception. In individualistic regions like the US, students living with no parents or single parents receive far less support from their extended family members, and have lower mathematics scores than other students.

The face cultures in these Asian regions help explain why they have higher mathematics scores and lower self-concept scores compared to the dignity culture of the US. These results support the modesty bias of students in the Asian face cultures, who seek upward comparisons with their higher-achieving classmates and have lower mathematics self-concepts than US students (Heine, 2004). In contrast, US students often choose downward comparisons with lower-achieving classmates to boost their own mathematics self-concept, showing self-enhancement bias (Heine, 2004). Thus, these results support prior studies showing that students in the US and in other dignity cultures typically have higher mathematics self-concept than students in Asian regions and other face cultures (e.g., Chiu & Klassen, 2010).

In all these Asian regions, self-efficacy scores exceeded self-concept scores and the former's regression coefficients were larger than the latter's, suggesting that modesty bias exerts a greater influence on self-concept than on self-efficacy. Similarly, US students' self-concepts exceeded their self-efficacy, consistent with enhancement bias. While a student often perceives his or her mathematics competence (self-concept) against the relative standard of classmates and schoolmates, his or her perceived competence on specific mathematics tasks (self-efficacy) is less likely to be affected by others' performance. Hence, researchers may find that self-efficacy is more suitable for absolute comparisons among students and for stronger links to mathematics achievement, while self-concept is more suitable for relative comparisons among students.

Finally, mathematics interest had similar, positive links to mathematics achievement in all regions, but instrumental motivation related to mathematics achievement only in the US and South Korea. Future studies can examine the determinants of these differences. Also, future studies can examine the determinants of the high mathematics interest of Singaporean students.

References

Andre-Bechely, L. (2012). *Could it be otherwise? Parents and the inequalities of public school choice*. New York: Routledge.
Artelt, C., Baumert, J., Julius-McElvaney, N., & Peschar, J. (2003). *Learners for life*. Paris: Organization for Economic Cooperation and Development.
Atack, J., & Passell, P. (1994). *A new economic view of American history: From colonial times to 1940* (2nd ed.). New York: Norton.
Ayers, E. (1984). *Vengeance and justice*. New York, NY: Oxford University Press.
Baker, D. P., Goesling, B., & LeTendre, G. K. (2002). Socioeconomic status, school quality, and national economic development. *Comparative Education Review, 46*(3), 291–312.
Baker, F. B., & Kim, S.-H. (2004). *Item response theory*. Boca Raton, FL: CRC.
Bandura, A. (2012). On the functional properties of perceived self-efficacy revisited. *Journal of Management, 38*(1), 9–44.
Benabou, R. (2000). Unequal societies: Income distribution and the social contract. *American Economic Review, 90*, 96–129.
Benjamini, Y., Krieger, A. M., & Yekutieli, D. (2006). Adaptive linear step-up procedures that control the false discovery rate. *Biometrika, 93*, 491–507.
Bong, M., & Skaalvik, E. M. (2003). Mathematics self-concept and self-efficacy. *Educational Psychology Review, 15*, 1–40.
Brown, G., Micklewright, J., Schnepf, S. V., & Waldmann, R. (2005). Cross-national surveys of learning achievement – How robust are the findings? *Journal of the Royal Statistical Society (Series A), 170*, 623–646.
Carter, I. (2009). Positive and negative liberty. In E. N. Zalta (Ed.), The Stanford encyclopedia of philosophy. Retrieved from http://plato.stanford.edu/entries/liberty–positive–negative/.
Center for Responsive Politics [CRP] (2012). Influence and lobbying. Retrieved from http://www.opensecrets.org/lobby/clientsum.php?id=D000055140&year=2012.
Chiu, M. M. (2007). Families, economies, cultures and science achievement in 41 regions. *Journal of Family Psychology, 21*, 510–519.
Chiu, M. M. (2009). Inequalities' harmful effects on both disadvantaged and privileged students: Sources, mechanisms, and strategies. *Journal of Education Research, 3*, 109–128.
Chiu, M. M. (2010). Inequality, family, school, and mathematics achievement. *Social Forces, 88* (4), 1645–1676.
Chiu, M. M. (2013). China. In J. Ainsworth & G. J. Golson (Eds.), *Sociology of education* (pp. 114–116). Thousand Oaks, CA: Sage.
Chiu, M. M. (2014). Family and school influences on students' math achievement: United States vs. China. In W. Ma (Ed.), *East meets West in teacher preparation: Crossing Chinese and American borders* (pp. 143–158). New York: Teachers College Press.
Chiu, M. M. (2015). Family inequality, school inequalities and mathematics achievement in 65 regions: Microeconomic mechanisms of rent seeking and diminishing marginal returns. *Teacher's College Record, 117*(1), 1–32.
Chiu, M. M., & Chen, G. (2014). Collectivism. In L. Ganong, M. Coleman, & G. J. Golson (Eds.), *The social history of the American family* (pp. 254–256). Thousand Oaks, CA: Sage.
Chiu, M. M., & Chow, B. W.-Y. (2011). Classroom discipline across 41 regions: School, economic, and cultural differences. *Journal of Cross-Cultural Psychology, 42*(3), 516–533.
Chiu, M. M., & Chow, B. W. Y. (2015). Classmate characteristics and student achievement in 33 regions: Classmates' past achievement, family SES, educational resources and attitudes toward reading. *Journal of Educational Psychology, 107*(1), 152–169.
Chiu, M. M., Chow, B. W.-Y., & McBride-Chang, C. (2007). Universals and specifics in learning strategies: Explaining adolescent mathematics, science, and reading achievement across 34 regions. *Learning and Individual Differences, 17*, 344–365.
Chiu, M. M., & Joh, S. W. (2015). Cross-national comparisons in education-PISA. In J. D. Wright (Ed.), *International encyclopedia of social and behavioral sciences* (Vol. 5, 2nd ed., pp. 342–348). Oxford: Elsevier.

Chiu, M. M., & Khoo, L. (2005). Effects of resources, inequality, and privilege bias on achievement. *American Educational Research Journal, 42*, 575–603.
Chiu, M. M., & Klassen, R. M. (2010). Relations of mathematics self-concept and its calibration with mathematics achievement. *Learning and Instruction, 20*, 2–17.
Chiu, M. M., Pong, S. L., Mori, I., & Chow, B. W. Y. (2012). Immigrant students' cognitive and emotional engagement at school: A multilevel analysis of students in 41 regions. *Journal of Youth and Adolescence, 41*, 1409–1425.
Chiu, M. M., & Zeng, X. (2008). Family and motivation effects on mathematics achievement. *Learning and Instruction, 18*, 321–336.
Choi, K., Choi, T., & McAninch, M. (2012). A comparative investigation of the presence of psychological conditions in high achieving eighth graders from TIMSS 2007 mathematics. *ZDM Mathematics Education, 44*, 189–199.
d'Ailly, H. (2003). Children's autonomy and perceived control in learning. *Journal of Educational Psychology, 95*(1), 84–96.
Deci, E. L., & Ryan, R. M. (2002). Overview of self-determination theory. In E. L. Deci & R. M. Ryan (Eds.), *Handbook of self-determination research* (pp. 3–33). Rochester, NY: University of Rochester Press.
Demirguc-Kunt, A., & Levine, R. (2009). *Finance and inequality: Theory and evidence* (No. w15275). Cambridge, MA: National Bureau of Economic Research.
Dower, D. W. (1986). *War without mercy: Race and power in the Pacific war*. New York: Pantheon Books.
Ganzeboom, H. B. G., de Graaf, P., & Treiman, D. J. (with de Leeuw, J.). (1992). A standard international socio-economic index of occupational status. Social Science Research, 21, 1–56.
Goldstein, H. (2011). *Multilevel statistical models*. Sydney: Edward Arnold.
Hamamura, T., & Heine, S. (2008). The role of self-criticism in self-improvement and face maintenance among Japanese. In E. Chang (Ed.), *Self-criticism and self-enhancement: Theory, research, and clinical implication* (pp. 105–122). Washington, DC: American Psychological Association.
Heine, S. J. (2004). Positive self-view. *Journal of Cultural and Evolutionary Psychology, 2*, 109–122.
Heine, S. J. (2005). Constructing good selves in Japan and North America. In R. Sorrentino, D. Cohen, J. Olson, & M. Zanna (Eds.), *Culture and social behavior* (pp. 95–116). Mahwah, NJ: Erlbaum.
Hofstede, G., Hofstede, G. J., & Minkov, M. (2010). *Cultures and organizations: Software of the mind* (3rd ed.). New York: McGraw-Hill.
House, R. J., Hanges, P. J., Javidan, M., Dorfman, P. W., & Gupta, V. (2004). *Culture, leadership, and organizations: The GLOBE study of 62 societies*. Thousand Oaks, CA: Sage.
Ishikida, M. Y. (2005). *Japanese education in the 21st Century*. Merced, CA: iUniverse, Inc.
Jones, C., & Romer, P. (2010). The new kaldor facts: Ideas, institutions, population, and human capital. *American Economic Journal: Macroeconomics, 2*(1), 224–245.
Kennedy, P. (2008). *Guide to econometrics*. New York: Wiley-Blackwell.
Kim, Y.-H., Cohen, D., & Au, W.-T. (2010). The jury and abjury of my peers: The self in face and dignity cultures. *Journal of Personality and Social Psychology, 98*(6), 904–916.
Konstantopoulos, S. (2008). The power of the test in three-level cluster randomized designs. *Journal of Research on Educational Effectiveness, 1*, 66–88.
Lee, J. (2009). Universals and specifics of math self-concept, math self-efficacy, and math anxiety across 41 PISA 2003 participating countries. *Learning and Individual Differences, 19*, 355–365.
Lee, Y. Y., Kam, C., & Bond, M. H. (2007). Predicting emotional reactions after being harmed by another. *Asian Journal of Social Psychology, 10*, 85–92.
LeTendre, G. K., Hofer, B. K., & Shimizu, H. (2003). What is tracking? *American Educational Research Journal, 40*(1), 43–89.
Marshall, B. K. (1994). *Learning to be modern: Japanese political discourse on education*. Boulder, CO: Westview Press.

May, H. (2006). A multilevel Bayesian IRT method for scaling socioeconomic status in international studies of education. *Journal of Educational and Behavioral Statistics, 31*, 63–79.
McLelland, G. (1991). Attainment targets and related targets in schools. In N. B. Crawford & E. K. P. Hui (Eds.), *The curriculum and behavior problems in Hong Kong* (Education Papers No. 11, pp. 62–86). Hong Kong: The University of Hong Kong, Faculty of Education.
Miller, J. A. (1992). Private school groups postpone quest for vouchers. *Education Week, 11*(30), 15.
Monseur, C., & Adams, R. (2009). Plausible values: How to deal with their limitations. *Journal of Applied Measurement, 10*, 320–334.
Oettingen, G., & Zosuls, K. M. (2006). Culture and self-efficacy in adolescents. In F. Pajares & T. Urdan (Eds.), *Self-efficacy beliefs of adolescents* (pp. 245–265). Greenwich, CT: Information Age.
Organization for Economic Cooperation and Development [OECD]. (2014). *PISA 2012 technical report*. Paris: OECD.
Page, J. (1994). The East Asian miracle: Four lessons for development policy. *NBER Macroeconomics Annual, 9*, 219–269.
Peugh, J. L., & Enders, C. K. (2004). Missing data in educational research. *Review of Educational Research, 74*, 525–556.
Piketty, T. (2013). *Capital in the twenty-first century*. Cambridge, MA: Harvard University Press.
Pintrich, P. R., & Schunk, D. H. (2002). Social cultural influences. In P. R. Pintrich & D. H. Schunk (Eds.), *Motivation in education* (2nd ed., pp. 190–242). Englewood Cliffs, NJ: Merrill.
Pulliam, J. D., & Van Patten, J. (2012). *The history and social foundations of American education*. New York: Pearson.
Romer, P. (2010). Which parts of globalization matter for catch-up growth? *American Economic Review, 100*(2), 94–98.
Rothstein, R. (2000). Equalizing education resources on behalf of disadvantaged children. In R. D. Kahlenberg (Ed.), *A notion at risk* (pp. 127–168). New York: Century.
Schulz, W. (2003). *Validating questionnaire constructs in international studies*. Camberwell, Victoria: Australian Council for Educational Research.
Seth, M. J. (2010). *A history of Korea: From antiquity to the present*. Lanham, MD: Rowman & Littlefield.
Stiglitz, J. E. (2013). Creating the institutional foundations for a market economy. In D. Kennedy & J. E. Stiglitz (Eds.), *Law and economics with Chinese characteristics: Institutions for promoting development in the twenty-first century* (pp. 71–111). Oxford: Oxford University Press.
Suen, H. K., & Yu, L. (2006). Chronic consequences of high-stakes testing? Lessons from the Chinese civil service exam. *Comparative Education Review, 50*, 46–65.
Takano, Y., & Osaka, E. (1999). An unsupported common view: Comparing Japan and the U.S. on individualism/collectivism. *Asian Journal of Social Psychology, 2*(3), 311–341.
Trautwein, U., Lüdtke, O., Marsh, H. W., Köller, O., & Baumert, J. (2006). Tracking, grading, and student motivation: Using group composition and status to predict self-concept and interest in ninth-grade mathematics. *Journal of Educational Psychology, 98*, 788–806.
Wang, Z. (2015). Examining big-fish-little-pond-effects across 49 regions: A multilevel latent variable modelling approach. *Educational Psychology, 35*(2), 228–251.
Warm, T. A. (1989). Weighted likelihood estimation of ability in item response theory. *Psychometrika, 54*, 427–450.
White, K., & Lehman, D. R. (2005). Culture and social comparison seeking: The role of self-motives. *Personality and Social Psychology Bulletin, 31*, 232–242.
World Bank (2015). Data: GDP per capita (current US$). Retrieved from http://data.worldbank.org/indicator/NY.GDP.PCAP.PP.CD.
Young, A. (1995). The tyranny of numbers: Confronting the statistical realities of the East Asian growth experience. *Quarterly Journal of Economics, 110*, 641–680.

Chapter 16
What Do TIMSS Studies Show About Math Achievement Inequality? A Sociological Perspective

Seong Won Han, Ji-Won Son, and Chungseo Kang

Abstract With the vast progress in international comparative achievement data collection since the early 1960s, we have witnessed the rise of research that examines why primary and secondary school students in some countries perform better than students in other countries. In addition to identifying the sources of between-country gaps in average student achievement, international data projects such as the Trends in International Mathematics and Science Study (TIMSS) have paved the way for a rich body of literature that examines achievement inequality by social background. This chapter attempts to fill a gap in the research by reviewing the comparative literature that used TIMSS data to examine math achievement inequality in Japan, Korea, Singapore, and the United States over the past two decades. We focus on two types of inequality: inequality in student test scores on the basis of socioeconomic and demographic backgrounds, as measured by achievement gaps among different groups (e.g., gender, family socioeconomic status, immigrant background, and family structure), and the relative importance of school compared to family background on mathematics achievement. In this chapter, we summarize what comparative research studies say about the relationship between social backgrounds and math achievement in primary and secondary schools. Finally, we suggest possible avenues for further research.

Keywords Mathematics achievement • Inequality • Sociological perspective • TIMSS

S.W. Han (✉) • C. Kang
Department of Educational Leadership and Policy, University at Buffalo – The State University of New York, Buffalo, NY, USA
e-mail: seongwon@buffalo.edu; chungseo@buffalo.edu

J.-W. Son
Department of Learning and Instruction, University at Buffalo – The State University of New York, Buffalo, NY, USA
e-mail: jiwonson@buffalo.edu

J.-W. Son et al. (eds.), *What Matters? Research Trends in International Comparative Studies in Mathematics Education*, Research in Mathematics Education, DOI 10.1007/978-3-319-51187-0_16

Introduction

Understanding inequality in educational achievement and attainment is an important research topic addressed in the sociology of education. Achievement studies primarily focus on academic achievement (e.g., test scores, grades) in primary and secondary school contexts, while attainment studies focus on the highest level of education that an individual has completed (e.g., completion of high school and higher education). Because educational achievement is crucial for eventual educational attainment, such as access to higher education, research on educational achievement inequality can contribute to a better understanding of educational attainment inequality. With the vast progress in large-scale international comparative achievement data collection since the early 1960s, we have witnessed the rise of comparative research that examines achievement inequality on the basis of students' social and demographic backgrounds. Among several school subjects, researchers have extensively examined mathematics achievement (Camburn & Han, 2009). This is partly attributable to the importance of mathematics in gaining access to and completing higher education in both STEM and non-STEM fields (Adelman, 1999, 2006).

International comparative studies allow researchers to examine variation across countries (National Research Council, 2003). For example, large-scale international comparative achievement data such as the Trends in Mathematics and Science Study (TIMSS) and the Program for International Student Assessment (PISA) allow researchers to examine differences between countries in the associations between student-level factors (such as gender and social background) and achievement, and the degree to which teachers and schools moderate the effects of individual-level student factors on student achievement. More importantly, these studies offer rich opportunities for researchers to examine national contexts related to differences in student achievement across countries. Accordingly, comparative research on achievement inequality can benefit when researchers focus on macro-level social, economic, and educational factors in explaining cross-national differences in achievement inequality.

These factors include, for example, characteristics of national education systems (e.g., the uniformity of curriculum and assessment nationwide); labor market conditions as arenas where workers exchange their labor power in return for wages, status, and other job rewards; shared social norms and values at national and subnational levels; and governmental policies (e.g., governmental educational policies to equalize educational spending) (Else-Quest, Hyde, & Linn, 2010; Guiso, Monte, Sapienza, & Zingales, 2008; Van de Werfhorst & Mijs, 2010). This research is important because country differences in these factors (e.g., particular structures of educational and labor force organizations) affect the patterns in the connections between family of origin and educational outcomes, and between educational attainments and labor market outcomes (Kerckhoff, 1995).

We focus on two aspects of inequality: (1) inequality in student test scores on the basis of socioeconomic and demographic backgrounds, as measured by achievement gaps among different groups (e.g., gender, family socioeconomic status, immigrant background, and family structure), and (2) the relative importance of school on student test scores compared to family background. Sociological research on these two aspects of inequality has been inspired by the *Equality of Educational Opportunity Study* (EEOS) (Coleman et al., 1966), also known as the Coleman Study (Gamoran & Long, 2008). Coleman et al. (1966) found that the impact of school resources on student achievement was modest compared to the impact of students' family backgrounds and other characteristics. This finding triggered decades of research on school effects, namely, the relative effects of school resources and family background on achievement and the impact of socioeconomic status (SES) on achievement. This line of research is important to improve equity in education.

The purpose of the current study is to review international comparative studies that examined math achievement inequality. We review those studies that used TIMSS data. TIMSS assessed mathematics and science at grade 4, 8, and 12. It also collects information on cross-national differences in mathematics curricula, as well as learning and teaching contexts, by using teacher and curriculum questionnaires. These features are key differences between TIMSS and other international data (e.g., PISA) and allow researchers to examine the degree to which mathematics classroom instruction (e.g., teaching approaches) explains mathematics achievement inequality. In this study, mathematics achievement inequality refers to mathematics achievement *gaps* on the basis of student background, such as gender, immigrant background, family structure, and school factors. In this chapter, we summarize what research studies say about the relationship between social background and mathematics achievement in primary and secondary schools across countries, and national contexts related to these between-country differences. Possible avenues for further research, particularly in mathematics education, are also suggested.

Methods

Selection and Search Procedure

The narrative research review presented here is based on an analysis of peer-reviewed research articles that used data from the TIMSS. Conducted on a regular 4-year cycle, the TIMSS has assessed mathematics and science at grades 4 and 8 in 1995, 1999, 2003, 2007, and 2011. TIMSS data have been collected twice from students at grade 12, in 1995 and 2008, which can be referred to as TIMSS Advanced. In addition to student achievement scores, the TIMSS collected contextual information about school characteristics, instruction, and home and family backgrounds through student, teacher, and school background questionnaires. The curriculum questionnaire was also completed by the National Research

Coordinators (NRCs), who were asked to supply information about their education systems, mathematics curricula, and resources for mathematics instruction in their target grades, with the assistance of their curriculum experts. Unlike other large-scale international studies such as the PISA, the TIMSS collected detailed information about implemented mathematics curricula (what teachers taught) by using teacher questionnaires. This feature of the TIMSS allows researchers to examine the degree to which inequality of mathematics achievement is associated with classrooms/teachers (e.g., instructional practices) as well as societal factors. Because the TIMSS assessed fourth and eighth graders, researchers can examine variation of math achievement inequality across grades within countries and across countries. In addition, multiple waves of TIMSS data (i.e., the 1995, 1999, 2003, 2007, and 2011 waves) allow researchers to examine changes in math achievement inequality at the national level over time. However, the cross-sectional design in TIMSS does not allow researchers to conduct longitudinal analyses (comparisons over time) at the student level.

To conduct the research review, we constructed a database of all peer-reviewed research articles that used data from TIMSS surveys to examine inequality of mathematics achievement. The database was created progressively in five steps, to narrow the articles for review: (1) Peer-reviewed journal articles using data from TIMSS surveys were selected for review; (2) research articles were selected for review when the outcome was student math achievement; (3) research articles were selected for review when the sample of countries included Japan, Korea, Singapore, or the United States; (4) research articles were selected for review when the main focus of research was sociological factors in explaining mathematics achievement inequality and gaps (e.g., family SES, gender, and immigrant background); and (5) the methodology and findings of all the articles resulting from the search in step 4 were summarized.

Multiple strategies were used to identify an exhaustive list of research articles that used TIMSS 1995, 1999, 2003, 2007, and 2011. Note that all the articles reviewed in this chapter used TIMSS data in measuring students' mathematics achievement as well as classroom and school-level factors. However, the sources of national-level factors vary from TIMSS data to external databases such as the World Bank and United Nations Educational, Scientific and Cultural Organization (UNESCO) Statistics. Research articles were identified through searches of Academic Search Elite, Google Scholar, and JSTOR. Our primary search terms were the full name of the TIMSS as well as the acronym. We identified a total of 22 research articles that met our selection criteria (see Appendix for the complete listing).

Coding Framework

For this chapter, we examined how well two aspects of math achievement inequality were covered in the literature. Beyond the coverage of different aspects of

Table 16.1 Coding framework for coverage of aspects in mathematics achievement inequality and research design

	Level of analysis			
		Between countries		
Aspects of inequality	Within countries	National education system factor	Societal factor	Policy factor
Achievement gaps on the basis of social background				
Relative importance of school compared to family backgrounds on mathematics achievement				

mathematics achievement inequality, we also reviewed the research design and analytic approaches used. Lastly, we summarized major findings about the link between macro-level societal, educational, and policy factors and two aspects of math achievement inequality. Table 16.1 presents the coding framework we used in reviewing articles. First, to evaluate the coverage of different aspects of mathematics achievement inequality, we coded whether studies focused on achievement gaps on the basis of student background (referred to as gap study) or the relative importance of school factors compared to family background on mathematics achievement (referred to as school effects). Among studies that examined achievement gaps, we coded the focus of achievement gaps (e.g., gender, family SES, immigrant background, family structure, or geographical location of schools). If a study did not focus on specific groups' achievement gap and examined mathematics achievement on the basis of several student background factors, we coded it as exploratory.

Second, to evaluate the level of analysis, we coded whether studies focused on within-country or between-country differences in mathematics inequality. Of studies that examined between-country differences, we coded what national-level factors were used in explaining cross-national variation in mathematics achievement inequality. In particular, we coded whether studies focused on national education systems, societal, or governmental policy factors. In this study, national education system factors include national-level characteristics such as the existence of a national curriculum and assessments, and centralized teacher training and hiring that differentiate education systems across countries. Societal factors include national economic development level and social inequalities, which can be called macro-level societal environment. In this study, policy factors refer to regulatory measures and funding priorities concerning education, such as public expenditure on childcare and early education. These policy factors are often promulgated by a governmental entity or its representatives.

Findings

This section consists of three parts. First, we summarize overall tendencies in research on mathematics achievement inequality by describing the coverage of different aspects of mathematics achievement inequality, the research design, and analytic approaches used in the literature. Next, we present the results of the research summary on the basis of two aspects of math achievement inequality, achievement gaps on the basis of student background, and school effects compared to family background. We then summarize math achievement inequality that is related to socioeconomic and demographic characteristics such as gender, family SES, and immigrant background. Finally, we summarize studies that investigated the relative importance of school compared to family background, known as "school effect" research.

Overall Tendencies in Research on Math Achievement Inequality

We identified 22 studies that examined mathematics achievement inequality using TIMSS data (see Appendix). These studies were analyzed with respect to their coverage of different aspects of mathematics achievement inequality (e.g., gap study, school effect research) and research design, to establish the overall tendencies in comparative research on mathematics achievement inequality.

Coverage of aspects in mathematics achievement inequality. Table 16.2 displays the frequency of studies that examined different aspects of math achievement inequality. About two-thirds of the studies (64%) focused on achievement gaps among different groups, with a focus on students' socioeconomic and demographic backgrounds (i.e., gap study research). The remaining studies (36%) focused on the relative importance of school and family background on math achievement (i.e., school effect research). The mathematics achievement gap studies were further classified into five groups based on their research focus: gender studies, immigrant status studies, family SES studies, family structure studies, and exploratory studies (see Table 16.2).

Table 16.2 Frequency of studies focused on math achievement inequality with TIMSS data

Research focus		*N*	%
Gap study	Gender	4	18.2
	Immigration	3	13.6
	Family SES	2	9.1
	Family structure	2	9.1
	Exploratory studies	3	13.6
Relative importance of school compared to family		8	36.4

Three studies explored the relative importance of several student background factors on mathematics achievement inequality using an exploratory model (see Table 16.2). The purpose of these exploratory studies was to examine whether several family background factors (e.g., parental education, parental educational expectation, and family structure) were equally related to mathematics achievement. Four studies focused on gender differences in mathematics achievement. While three studies investigated the achievement gap between immigrant and native students, two studies focused on the effects of family SES on mathematics achievement by comparing students with high-SES families to those with low-SES families. Two studies examined the effects of family structure on mathematics achievement by assessing differences between two-parent families and single-parent families in mathematics achievement (Table 16.2).

Two tendencies should be noted here. First, although we reviewed studies that were conducted over a 20-year span, only 22 studies were identified as focusing on mathematics achievement inequality using TIMSS data. This suggests that TIMSS studies have not been extensively used to examine mathematics achievement inequality. This might be partly due to the limited measure of family SES in TIMSS studies. In sociology, family SES is typically measured by three components (i.e., parents' education, parents' occupation, and family income), but the TIMSS data only includes information on parents' education and family home possessions (e.g., study desk, computer, and dictionary). However, some researchers (e.g., Carnoy & Rothstein, 2013) argued that home possessions (e.g., number of books at home) are better measures of social class than parental education and occupation because students' answers about the latter variables are not always reliable. Given that the TIMSS administered several questions about home possessions, TIMSS data is still valuable source for research on mathematics achievement inequality. Second, several researchers suggested that regional or geographic variation in achievement is one of the important factors in understanding achievement inequality (Bray & Thomas, 1995; Burt & Namgi, 2009; Hanushek & Yilmaz, 2011; Roscigno, Tomaskovic-Devey, & Crowley, 2006). However, among studies that used TIMSS data from Japan, Korea, Singapore, or the United States, we found that no study investigated achievement gaps related to geographic location (e.g., the gap between urban and rural areas) (Fig. 16.1).

We then explored these research studies with respect to the grade level (i.e., fourth grade only, eighth grade only, and both), the degree to which studies used multiwave data (i.e., single vs. multiwave), and analytic level (i.e., student-level by country studies, country-level studies, and student-level by country and country-level studies). Figure 16.1 illustrates the grade levels and aspects of mathematics achievement inequality on which the studies focused. With respect to the grade levels, about 68% of the studies used data from TIMSS Grade 8, whereas about 18% used data from TIMSS Grade 4. Only about 14% of the studies used both TIMSS Grade 4 and Grade 8 data. The limited number of studies on mathematics achievement inequality in Grade 4 was partly due to limited information on family background in the early elementary grades. For example, information on family structure was only collected in TIMSS 1995, leading to limited research on

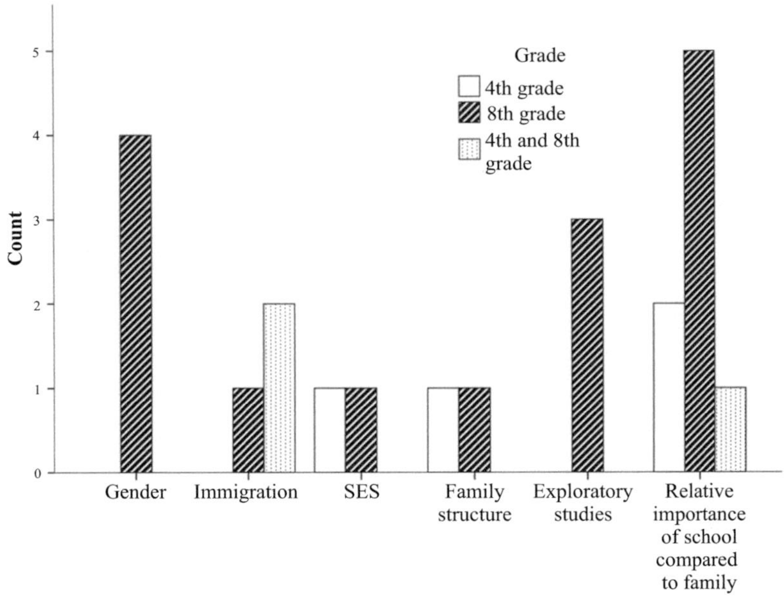

Fig. 16.1 Research focus by grade level

disparity of achievement on the basis of family structure. We also found that no study examined fourth graders' gender gap in mathematics achievement using TIMSS data from Japan, Korea, Singapore, or the United States over the past two decades. This finding suggests that more studies are needed to focus on Grade 4 in exploring mathematics achievement inequality.

With respect to the degree to which studies used multiwave data in examining mathematics achievement inequality, about two-thirds of all studies utilized only single-wave data, whereas about one-third of all studies employed more than two waves of TIMSS data (see Fig. 16.2). Despite the availability of five waves of TIMSS data, we found that the majority of studies did not take full advantage of such multiwave data. One of the advantages to using multiwave data, for example, is that differences in participating countries across TIMSS waves allow researchers to test a robustness of findings (Wößmann, 2002). As shown in Fig. 16.2, no study utilized multiwave data in examining achievement gap on the basis of family structure, although information on family structure was collected repeatedly in Grade 8 in TIMSS data. Among all studies that examined the relative importance of school factors compared to family factors, about 75% of studies (6 out of 8) utilized data from a single wave of the TIMSS, whereas about 25% of studies employed multiwave data.

Figure 16.3 illustrates the frequency of the research studies in five research foci that used different approaches to level of analysis. Twenty-two studies were grouped into three groups in terms of their analytic levels: (1) student level by country studies, (2) country-level studies, and (3) student level by country and

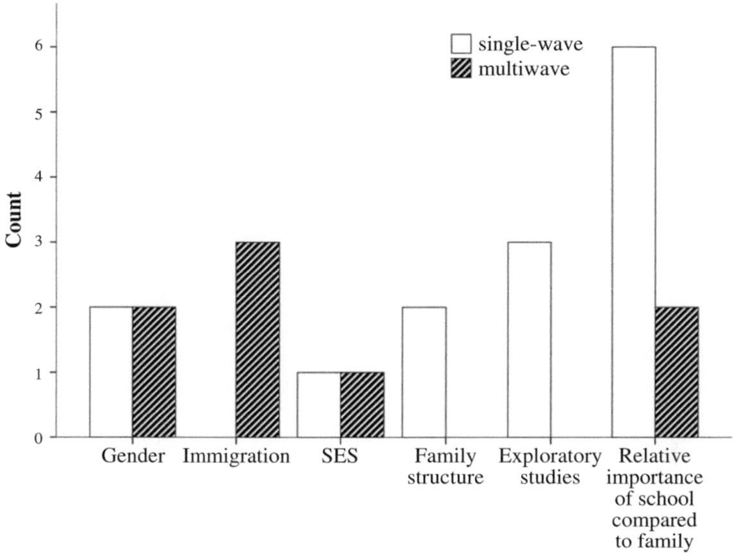

Fig. 16.2 Research focus by use of multiwave data

country-level studies. Student level by country studies examine the association between family, school, and student achievement in each country. Thus, the findings from this type of study cannot take into account the influence of national-level factors in explaining cross-national differences in mathematics achievement inequality. The second category, country-level studies, examines the link between national-level factors (e.g., GDP per capita, GINI inequality index, and labor market condition) and student achievement. Researchers adopted two different approaches when they examined this question: (a) statistical models that include student-, school-, and national-level predictors simultaneously by using data from more than two countries, and (b) multiple regression models that predict national-level student outcomes (e.g., the aggregated mathematics achievement gap between high-SES and low-SES students at the national level) with national-level predictors. The former approach estimates country effects after controlling for heterogeneity in the student, teacher or classroom, and school levels. The latter approach does not take into account heterogeneity at the student and school levels within countries, by focusing only on country-level outcomes and predictors. Finally, the third category, student level by country and country-level studies, examines the association between family, school, and student achievement by country and then investigates this within-country variation by using country-level predictors. This type of study focuses on both within-country variation and between-country variation. For example, researchers examined gender gaps in mathematics achievement in each country and then investigated the degree to which national factors were associated with gender gaps in mathematics achievement.

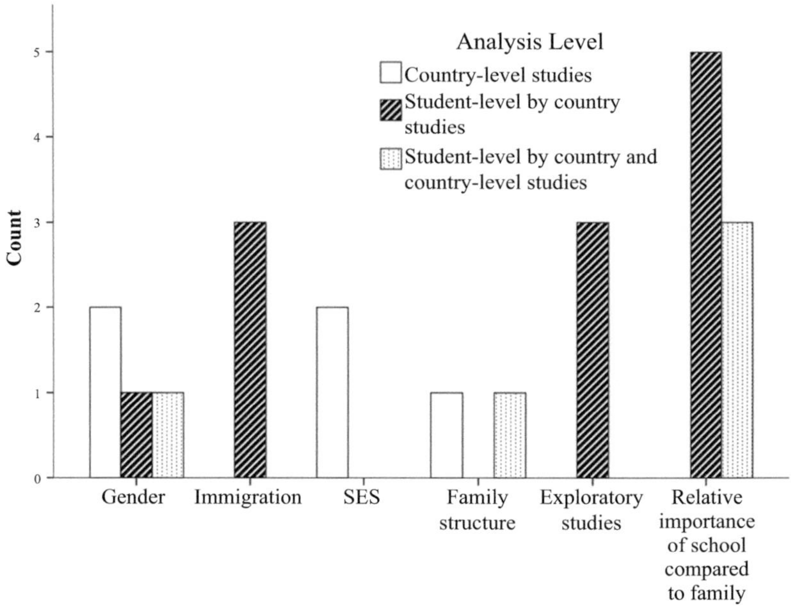

Fig. 16.3 Research focus by level of analysis

Around half of the reviewed studies (55%, 12 out of 22 studies) adopted student level by country analyses; five studies used country-level analyses; and five studies used student level by country and country-level analyses (see Fig. 16.3). It is important to note that only student level by country analyses were conducted to examine achievement gaps in both research on immigrant background and exploratory studies. In other words, no studies examined the degree to which country-level factors are associated with achievement gaps between immigrant and native students. This suggests that research has contributed to the understanding of educational inequality associated with immigrant status in individual countries, but there is a lack of understanding about national-level factors (e.g., national education systems or governmental policies) in explaining between-country differences in achievement gaps between immigrant and native students. In addition, all the studies that examined achievement gaps on the basis of family SES utilized country-level analyses. This suggests that researchers examined the degree to which national-level factors were associated with cross-national variation in achievement gaps between high- and low-SES students, but there is a lack of research on educational inequality associated with family SES in individual countries. When researchers investigated "school effects," the relative importance of school factors compared to family background, five studies utilized student level by country and three studies adopted student level by country and country-level analyses.

Mathematics Achievement Gaps on the Basis of Student Socioeconomic and Demographic Background

Of the 22 articles, 14 (64%) investigated mathematics achievement gaps on the basis of students' socioeconomic and demographic factors, such as gender, family SES, and immigrant status. Of these 14 studies, seven examined the association between family, school, and mathematics achievement by country, focusing on mathematics achievement inequality in each individual country. The remaining seven studies investigated the degree to which national factors (e.g., national education systems, labor market conditions, social norms and values, and governmental policies) are associated with mathematics achievement inequality. In the following section, we first present research studies that examine micro-level factors, followed by those considering the influence of national-level factors in explaining mathematics achievement inequality.

Micro-level factors and gap studies. Our review found that seven studies examined the disparity of mathematics achievement on the basis of student socioeconomic and demographic background within countries: one study on the achievement gap between males and females, three studies on the achievement gap between immigrant and native students, and three exploratory studies that examined the achievement gap. None of these studies took national-level factors into account in explaining cross-national differences in mathematics achievement inequality.

We found only one study that examined gender differences in mathematics achievement within countries. Neuschmidt, Barth, and Hastedt (2008) examined changes in gender differences in mathematics achievement in sixteen countries, including Japan, Korea, Singapore, and the United States, using data from TIMSS 1995, 1999, and 2003, with a focus on Grade 8 students. To test whether gender differences in mathematics achievement were statistically significant, they calculated a single number, national average mathematics test scores, by gender. They found that changes in the magnitude of gender gaps in mathematics achievement varied from country to country. For example, in the United States, they found no difference in mathematics achievement between boys and girls at the eighth-grade level in TIMSS 1995, but boys performed better than girls in TIMSS 2003. In Japan and Korea, boys performed better than girls in TIMSS 1995, but there were no gender differences in TIMSS 2003. In Singapore, there was no gender difference in mathematics achievement in TIMSS 1995, but girls performed better than boys in TIMSS 2003.

Neuschmidt et al. (2008) also extended their research on gender differences from overall achievement in mathematics to specific mathematics content areas such as algebra, geometry, and measurement. They found that changes in the gender gap varied across specific mathematics content areas within countries as well as across countries. In Singapore, there were no gender differences in algebra and geometry in TIMSS 1995, but girls performed better than boys in TIMSS 2003. In the United States, mirroring the overall mathematics gender gap trend, the gender gap has widened in geometry. In Korea, there was no change in the gender gap in geometry between TIMSS 1995 and 2003; boys performed better than girls. In Korea and Japan, the gender gap has narrowed in measurement between TIMSS 1995 and 2003. However, we claim that this study has limitations. First, the use of a single-point estimate without taking into account other student characteristics can over-simplify trends in gender differences in mathematics achievement. In addition, a single number such as an overall mean difference cannot capture different trends across levels of academic performance (Raudenbush & Kim, 2002); it is difficult to compare trends in the top distribution with those in the bottom distribution.

Of the studies we examined, three investigated mathematics achievement on the basis of immigrant status within countries: Schnepf (2007), Cheng, Wang, Hao, and Shi (2014), and Andon, Thompson, and Becker (2014). All three studies compared mathematics achievement gaps between native and immigrant students within a country or from country to country, but none tested whether national-level factors are associated with cross-national differences in immigrant students' educational disadvantage. Using data from TIMSS 1995 and 1999 as well as other international achievement surveys, Schnepf (2007) examined differences in educational achievement between immigrants and natives across ten countries, including the United States, Australia, Canada, Germany, and France. He found that the achievement gap between immigrants and natives in the United States is smaller than that of continental European countries (e.g., Germany and Switzerland), but is larger than that of other English-speaking economically developed countries (e.g., Australia and Canada). He also found that language skills explained immigrant students' educational disadvantage in the United States. A recent study by Cheng et al. (2014) extended research on immigrant students' disadvantage by examining whether immigrants' educational disadvantage was consistent across different racial immigrant groups. Using the US data from TIMSS 2007, they found that most White and Asian immigrant students outperformed their non-immigrant peers, whereas Hispanic and African American immigrants underperformed their counterparts. These two studies classified students as first-generation immigrants, second-generation immigrants, or natives. To better understand first-generation immigrants' educational disadvantage in the United States, future research needs to take into account the age of students at the time of immigration, along with race or ethnicity. A recent international study showed that children who immigrate at age 5 or older have a greater educational disadvantage than children who immigrate before age 5 (Ohinata & van Ours, 2012).

Unlike the other two studies, Andon et al. (2014) combined results from multiple studies about immigrant students' educational disadvantage. They conducted meta-analyses to compute standardized mean differences between immigrants and natives on mathematics achievement in three international student achievement studies, including the TIMSS, across all Organisation for Economic Co-operation and Development (OECD) countries. In addition to the overall mean effect size for immigrants' educational disadvantage in mathematics, they found that achievement gap between immigrants and natives was 0.64 standard deviations among fourth graders, and the gap for eighth graders was 0.403 standard deviations; younger students (fourth graders) show a larger immigrant achievement gap than older students (eighth graders), by about one-fourth of a standard deviation. Based on conventional benchmarks for interpreting effect sizes (i.e., 0.2 = small, 0.5 = medium, and 0.8 = large), this suggests that the immigrant achievement gap for fourth graders is moderately large and the gap for eighth graders is moderate. Thus, the difference between fourth graders and eighth graders in terms of the immigrant student achievement gap is small.

Taken together, these three studies about achievement gaps between immigrant and native students showed that immigrants have an educational disadvantage, but the magnitude of this disadvantage varies across immigrants' race or ethnicity, grade level, and country. Research on immigrants' educational disadvantage with TIMSS data concentrated on the United States. This is partly due to the quite small percentage of immigrant students in Japan and Korea.

The three remaining studies focusing on the mathematics achievement gap used exploratory approaches, in which researchers examined the disparities in mathematics achievement on the basis of several student background factors, but their research questions did not focus on specific groups such as immigrant students or disadvantaged students. All three studies compared findings from the United States with findings from other territories, including Hong Kong (Wang, 2004), Japan (Yoshino, 2012), and Russia, Singapore, and South Africa (Wang, Osterlind, & Bergin, 2012). Wang (2004) and Yoshino (2012) examined the associations between student characteristics and mathematics achievement without taking school characteristics into account. Using data from TIMSS 1995, Wang (2004) found different patterns in the association between parental education and student achievement in the United States and Hong Kong: in the United States, there was a positive association between parental education and mathematics achievement, while there was no association between these two factors in Hong Kong.

Using data from TIMSS 2007, Yoshino (2012) also found that parental education is positively related to mathematics achievement in both the United States and Japan. However, Wang et al. (2012), who used TIMSS 2003 data from four countries (the United States, Russia, Singapore, and South Africa), found no association between parental education and mathematics achievement in the United States and Singapore, whereas parental education was positively linked to mathematics achievement in Russia and South Africa. This contradictory finding about the link between parental education and student achievement in the United States might be partly attributable to different analytic strategies employed and variables included in statistical models. Wang (2004) and Yoshino (2012) did not take into account school characteristics, but they included several student characteristics, such as family structure and home possessions. On the other hand, Wang et al. (2012) controlled for teacher and school characteristics, but they included only student gender and parental education as student characteristics in their statistical models. In addition, Yoshino (2012) controlled for race when analyzing US data, while Wang et al. (2012) did not take race into account. Another possible explanation is that these three studies analyzed TIMSS 1995, 2003, and 2007, respectively.

National-level factors and gap study. Of the 22 articles we studied, seven investigated the degree to which national factors are associated with mathematics achievement inequality. National-level factors include national education systems, societal factors (e.g., labor market conditions, social values and norms), or government policies. Using the TIMSS curriculum survey, researchers can measure national-level characteristics of education systems, such as national or regional curriculum in fourth and eighth grades, including what is prescribed and how it is disseminated, requirements for becoming teachers, and how teachers are informed about the mathematics and science curriculum. However, in order to measure societal-level country characteristics, researchers used other data, such as data from the World Bank, the OECD, the International Labour Organization, and the World Values Survey. That is, researchers used the TIMSS data to measure students' achievement, and supplemental data from other sources to measure country-level societal factors.

Table 16.3 displays a summary of the seven studies that focused on national-level factors in examining mathematics achievement inequality on the basis of students' socioeconomic and demographic backgrounds. As shown in Table 16.3, these studies focused on mathematics achievement disparity on the basis of gender, family SES, and family structure. Research that examines the link between immigrants' educational disadvantages and national-level factors is relatively rare.

Of the three studies that examined gender gaps in mathematics achievement, two studies focused on societal factors (Nosek et al., 2009; Penner, 2008) and only one focused on national education systems: Ayalon and Livneh (2013). The former two studies examined whether gender-stratified societies (i.e., the level of inequality in society based on gender) were associated with gender gaps in mathematics achievement. Penner (2008) measured gender-stratified societies at the national level using several indicators: women's domestic duties, gender differences in secondary enrollment and attainment, labor force participation and position, and general status in 22 countries (including Australia, Canada, Germany, France, the Netherlands, and the United States). Using TIMSS 1995 data, Penner (2008) found that there were smaller female disadvantages in mathematics achievement in countries where there was greater gender equality in education, domestic duties, the labor market, and status in general. He advanced our understanding of gender gaps in mathematics achievement by examining whether the link between societal factors and gender gaps in mathematics achievement varied across the distribution of mathematics achievement. Penner found that among high academic achievers, gender differences in mathematics performance were more closely associated with labor market factors than with educational gender inequality.

Nosek et al. (2009) focused on different aspects of national-level gender inequality, such as gender-science stereotypes that "men are naturally more talented and interested in mathematics and science than women" (p. 10593), in explaining gender gaps in mathematics achievement. Nosek et al. differentiated implicit from explicit gender-science stereotypes. Explicit stereotypes were measured from self-reports on survey items, whereas implicit stereotypes were measured from the Gender Science Implicit Association Test at a laboratory. They found

Table 16.3 Summary of the findings of the studies focusing on macro-level factors in examining mathematics achievement inequality

Research focus	Author(s)	National education system factors	Societal factors	Policy factors
Gender gap	Ayalon and Livneh (2013)	Highly standardized education system and smaller gender gaps in mathematics achievement		
	Penner (2008)		Greater gender equality in education, home life, labor force position, general status and smaller female disadvantages in mathematics achievement	
	Nosek et al. (2009)		No association between national differences in *explicit* gender-science stereotypes and gender differences in mathematics achievement Positive association between *implicit* gender-science stereotypes and gender gaps in mathematics achievement	
SES gap	Akiba et al. (2007)	No association between opportunity gap in students' access to qualified teachers and achievement gaps between high- and low-SES students		
	Waldfogel and Zhai (2008)			No association between public preschool expenditures and achievement gaps between high- and low-SES students in mathematics achievement
Family structure	Schiller, Khmelkov, and Wang (2002)		Higher levels of economic development levels and achievement advantages of students from two-parent families	
	Pong, Dronkers, and Hampden-Thompson (2003)			Family policies that equalize resources between single- and two-parent families (e.g., family or child allowances) and smaller achievement gaps across different family structure

that societal indicators of explicit gender-science stereotypes were not associated with gender gaps in mathematics achievement, but higher levels of implicit gender-science stereotypes were linked to larger gender gaps in mathematics achievement. This study suggests that socially constructed, implicit male-oriented science stereotypes embedded in individual minds (though not explicitly endorsed stereotypes) can play a role in shaping gender gaps in mathematics achievement. In addition, this study suggests that self-reported responses on gender stereotypes in surveys have limitations in examining the effects of stereotypes on gender inequality in education, because few respondents explicitly endorse gender-science stereotypes on survey items.

Ayalon and Livneh (2013) focused on national education systems (measured by the level of educational standardization) as well as societal-level gender stratification in examining gender gaps in mathematics achievement. Standardization in this study refers to "the degree to which the quality of education meets the same standards nationwide" (Allmendinger, 1989, p. 233). Ayalon and Livneh found that more standardized education systems, as measured by the use of national examinations and between-teacher uniformity in covering major mathematics topics, were linked to smaller gender gaps in mathematics achievement. In other words, countries with a higher level of education system standardization showed smaller gender gaps in mathematics achievement than countries with a lower level of standardization (i.e., no national examinations and substantial variation between teachers in covering major mathematics topics). Unlike Penner's study, Ayalon and Livneh found that women's labor market participation was not associated with gender gaps in mathematics achievement. This finding is contradictory to the findings from prior research. Ayalon and Livneh speculated that this was partly due to the countries included in the analytic sample; Penner's sample concentrated on economically developed Western countries, whereas Ayalon and Livneh's sample represented a wide range of cultures. Ayalon and Livneh suggested the importance of understanding the countries included in an analysis that examines societal-level gender stratification and gender inequality in education.

Of the two studies that investigated mathematics achievement gaps between high- and low-SES students, one focused on national indicators of teacher quality and the other focused on national policy. Akiba, LeTendre, and Scribner (2007) examined the degree to which the distribution of teacher quality was associated with achievement gaps between high- and low-SES students. They measured teacher quality using several indicators: teaching certification, subject majors, and years of teaching experience. Using data from TIMSS 2003, Akiba et al. (2007) found that while the national level of teacher quality in the United States was about equal to the international average, the "opportunity gap" between wealthy and poor students in access to highly qualified teachers was the fourth largest of the 39 countries. They also found that the opportunity gap measured by the difference in the percentages of high-SES students and low-SES students taught by teachers with mathematics degrees was significantly and positively associated with the achievement gap. However, larger opportunity gaps in access to qualified teachers, measured by teacher certification, mathematics education major, and teaching experience, did not predict larger achievement gaps between high-SES and low-SES students cross-nationally. Waldfogel and Zhai (2008) examined whether public preschool expenditures were associated with achievement gaps between high- and low-SES students. They found that these expenditures were positively associated with the mathematics achievement scores of fourth graders, but there was no evidence to support that there were smaller achievement gaps between high- and low-SES students in countries that spent more on public preschool.

Of the two studies that investigated mathematics achievement gaps between two-parent families and single-parent families, Schiller et al. (2002) focused on national-level economic development while Pong et al. (2003) focused on government policies that equalize family income and parental time inputs. Both studies found that students living with two parents had higher mathematics achievement scores than students living with a single parent across countries, even after controlling for other family background indicators such as parental education. They also found that the magnitude of the achievement gap on the basis of family structure varied from country to country. Schiller, Khmelkov, and Wang analyzed data from 34 countries and found larger achievement gaps between students living with both parents and students living with only one or neither parent in economically developed countries. Pong et al. compared US students with students in nine economically developed countries (e.g., Australia, Canada, and Norway). They found smaller achievement gaps between single- and two-parent families in countries where family policies equalize financial resources and parental time inputs between single- and two-parent families. This study shows that government policies can play a role in narrowing achievement gaps between two-parent families and single-parent families.

Relative Importance of School and Family Backgrounds Research

Of the 22 articles, eight investigated the relative importance of school factors compared to family background, known as the school effect. These studies examined the degree to which national-level factors were associated with cross-national differences in the relative importance of school factors compared to family background. Since Coleman et al. (1966) found relatively modest effects of school factors on student achievement compared to the effects of student family background, researchers examined whether this finding holds true in other countries. In particular, this line of research examined the degree to which between-school differences explain the variation of mathematics achievement by country by conducting the aforementioned analyses separately for each country. Other studies investigated cross-national variation in school effects and its association with national-level factors such as economic development levels. In the next section, we present the micro-level studies and then the national-level studies that examined school effects.

Micro-level studies examining school effects. Of the eight studies that investigated the relative importance of school and family backgrounds, five examined the relative importance of school factors and family background on student achievement by country. Several studies provided evidence that the variance in student achievement explained by school factors differs across four countries (i.e., the United States, Korea, Singapore, and Hong Kong); in Korea, the lowest level of variance of student achievement was attributable to schools (Byun & Kim, 2010; Pahlke, Hyde, & Mertz, 2013; Pong & Pallas, 2001). In addition to national differences in school effects, some studies examined trends in school effects. For example, using eighth-grade student data from TIMSS 1999, 2003, and 2007, Byun and Kim (2010) found that the variation of mathematics achievement attributable to schools in Korea was 6.9%, 9%, and 9.5%, respectively, whereas the variation of mathematics achievement attributable to schools in the United States was 38.7%, 38.5%, and 34.5%, respectively. These findings suggest that the between-school variance in student achievement has significantly increased during the three cycles of TIMSS surveys in South Korea, whereas it has significantly decreased in the United States. In addition, Byun and Kim found growing educational inequality resulting from family background in Korea; that is, the effect of family background on student achievement has increased over time. Chen (2014) investigated the between-school variation of fourth-grade mathematics achievement in Hong Kong and Singapore and found that it was 29% and 23%, respectively. Because these studies examined different grade levels, we cannot compare the between-school variance across Hong Kong, Korea, Singapore, and the United States.

Unlike other studies that focused only on the relative importance of school factors, Lamb and Fullarton (2002) examined the relative importance of classroom, teacher, and school factors compared to family background in explaining the variation in student mathematics achievement in the United States and Australia. Using data from TIMSS 1995, Lamb and Fullarton found that the between-school variation of mathematics achievement in the United States was slightly higher than the between-school variation in Australia. Among several school-level variables, they found that the mean SES of the school was positively associated with student achievement in the United States but not Australia. They also found that tracking and SES composition of the classroom was positively associated with student achievement in both countries. All five of these studies investigated whether school effects varied from country to country, but they did not examine factors that were associated with national differences in school effects. Although the findings from this line of the study were informative, this analytic approach (micro-level) did not take into account national-level factors in explaining cross-national differences in school effects.

National-level studies examining school effects. Of the eight studies that investigated the relative importance of school and family backgrounds, three studies examined the degree to which national-level factors are associated with cross-national differences in school effects. After single-country case studies in developing countries found that school resources had a large effect on student achievement after controlling for family background (Heyneman, 1976), Heyneman and Loxley

(1983) tested whether national economic development levels were associated with national differences in school effects. Using data from the 1970s, Heyneman and Loxley (1983) found a large school effect on achievement, compared to family background, in low-income countries. In contrast, economically advanced nations showed stronger family background effects and weaker school resource effects on achievement, which became known as the "Heyneman-Loxley effect" (HL effect). Using recent data, several researchers tested whether the HL effect still holds true (Baker, Goesling, & LeTendre, 2002; Chudgar & Luschei, 2009; Hanushek & Luque, 2003). Analyzing data from TIMSS 1995, Baker et al. (2002) and Hanushek and Luque (2003) found no evidence to support the HL effect. They found that the relative importance of school factors and family background on achievement within countries was not associated with national income levels.

Baker et al. (2002) argued that there might be strong school effects in poor countries, but the sample of poor nations in their analyses had moved beyond the threshold at which additional input of school resources makes a substantial difference in student achievement, because of expansions of mass education and economic growth. Chudgar and Luschei (2009) extended the focus of research from national income levels to income inequality in explaining national differences in the relative importance of school and family background on student achievement. Using data from TIMSS 2003, however, Chudgar and Luschei (2009) found a stronger school effect on student achievement in poorer and more unequal countries. They argued that their findings on the link between national income levels and school effects might be due to the different approaches used in estimating school effects; prior research estimated school effects using a set of school variables in conventional regression and multilevel approaches, whereas Chudgar and Luschei estimated school effects by quantifying the variance of student achievement attributable to schools and family background.

Summary and Implications

We undertook a review of international studies that focused on mathematics achievement inequality in Japan, Korea, Singapore, and the United States over the past two decades. Before we discuss the summaries, two limitations should be noted. First, we acknowledge that some studies that were excluded from our chapter due to the source of data (e.g., PISA) may echo similar findings we have identified in this study or extend our understanding of the link between macro-level factors and math achievement inequality. For example, using PISA data, several studies examined the degree to which national income inequality explained math achievement inequality (e.g., Chiu, 2005, 2010) or the degree to which student tracking systems (e.g., academic versus vocational track schools) account for math achievement inequality (e.g., Knipprath, 2010; Marks, Cresswell, & Ainley, 2006). Although the studies that examined mathematics achievement inequality by using PISA data are valuable, we limited our review to TIMSS studies because TIMSS studies offer rich opportunities to examine how students, teachers, classrooms, schools, and national curricula all play a role in shaping mathematics student achievement. In addition, although a meta-analysis can provide a systematic quantitative summary of the relationship between students' achievements and various factors that were reported by prior studies, we cannot conduct a meta-analysis in summarizing our findings because the number of studies we reviewed is too small.

Despite these limitations, the findings of this study can be informative with respect to the relative importance of school resources and family background in mathematics achievement *gaps* on the basis of student background, such as gender and SES. Our findings on mathematics achievement inequality based on TIMSS data can be summarized as follows:

- The magnitude of the gender gap in mathematics achievement varied across countries. Both national education systems and societal factors explained between-country differences in this gender gap. In particular, a standardized education system (measured by the use of national examinations and teachers' uniformity in covering major mathematics topics) was linked to smaller gender gaps in mathematics achievement. In countries where women are more equal in education, home life, labor force position, and general status, there was a smaller gender gap in mathematics.
- The magnitude of the achievement gap between immigrants and natives varied from country to country as well as across academic grades. In the United States, there was a smaller achievement gap between immigrants and natives compared to European countries. The size of the immigrant achievement gap was larger for fourth graders than for eighth graders.
- The achievement gap between high-SES and low-SES students varied across countries. Neither national education systems (measured by the opportunity gap in students' access to qualified teachers) nor government policies (measured by the public preschool expenditure) explained the achievement gap between high- and low-SES students.
- In economically developed countries, there were larger achievement gaps between students living with both parents and students living with only one or neither parent. In countries where governmental family policies equalize resources between single- and two-parent families (e.g., child allowances), there were smaller achievement gaps between single- and two-parent families.
- The variance of student achievement explained by school factors differed across countries. However, there was no consensus whether between-country differences in the relative importance of school effects compared to family factors depend on national income levels.

We found several voids in the TIMSS-based comparative research on mathematics achievement inequality. Compared to other large-scale international achievement data, the TIMSS is unique in that it assessed several grade levels in TIMSS-participating countries. However, our analyses show that mathematics achievement inequality in Grade 4 and changes in mathematics achievement inequality between grades are under-researched areas. Although the TIMSS

Grade 4 data allow researchers to examine the gender gap in mathematics achievement, for example, no studies examined this gender inequality in mathematics achievement with TIMSS Japan, Korea, Singapore, or US data over the past two decades. TIMSS data also offer rich opportunities for researchers to examine the variation of mathematics achievement inequality across grade levels, but only the research on achievement gaps between native and immigrant students and the research on the relative importance of school compared to family backgrounds utilized data from more than one grade level. Although it is difficult to examine whether inequality has widened from Grade 4 to Grade 8 in TIMSS studies, a comparison of mathematics achievement inequality between these two grades can shed some light on the persistence of achievement gaps throughout the educational trajectory. Although students are not identical, for example, the majority of fourth graders in TIMSS 2007 had reached the eighth grade by 2011. Researchers could compare TIMSS 2007 fourth graders' mathematics achievement inequality with TIMSS 2011 eighth graders' mathematics achievement inequality, to examine changes in mathematics achievement inequality between Grades 4 and 8. This research can also be extended to examine trends over time. For instance, examining changes in mathematics inequality across grades between Grade 4 in 1995 (Grade 8 in 1999) and Grade 4 in 2003 (Grade 8 in 2007) would be also informative.

Since 1995 the TIMSS has completed five cycles of data collection and collected the sixth cycle of data in 2015. However, our analyses show that only a few studies utilized multiwave data from the TIMSS, and the majority of studies used a single wave of TIMSS data. Given that the TIMSS used a cross-sectional design, findings from a single wave of cross-sectional international data must be interpreted with caution. One possible way to assess the robustness of findings in large-scale international achievement studies is to replicate studies across different waves.

Compared to domestic research, international studies allow researchers to examine the link between society (e.g., organization and structure of the education system, cultural tradition, economic conditions, and political structure) and educational outcomes more precisely. However, our review shows that a substantial body of research on mathematics achievement inequality focused on a descriptive portrait of differences from country to country. For example, research on the disparity of achievement on the basis of immigrant status documented between-country differences in the magnitude of this gap, but no studies investigated factors that were associated with this cross-national variation in immigrants' educational disadvantage. Using PISA data, a few studies have examined the extent to which macro-level characteristics of host countries, origin countries, and communities (e.g., origin countries' national economic development levels) help explain differences in immigrant students' educational achievement (Levels, Dronkders, & Kraaykamp, 2008). Several national-level factors can play a role in the formation of achievement gaps between immigrants and natives. For instance, one of the key classifiable features of national education systems is the level of stratification, which refers to the degree to which students are sorted into different schools or programs that are differently valued in higher education and the labor market (Kerckhoff, 2001). Countries with highly stratified education systems tend to sort students into different programs at an early age. Future research can test whether stratified education systems were associated with first-generation immigrants' educational disadvantage.

We also found only one study that examined how national-level factors were associated with achievement gaps between high- and low-SES students using TIMSS data. Akiba et al. (2007) found that the "opportunity gap" in students' access to highly qualified teachers between wealthy and poor students in the United States was larger than that of other high-achieving countries such as Korea and Japan, but larger opportunity gaps in access to qualified teachers did not predict larger achievement gaps between high-SES and low-SES students cross-nationally. What is less clear, however, is what can explain cross-national variation in the distribution of teacher quality across social groups and schools. Examining how policy contexts surrounding teacher quality impact the distribution of teacher quality and students' access to high-quality instruction would be informative. In Korea, for example, public school teacher assignments are governed by the Ministry of Education and Human Resources Department and metropolitan and provincial offices of education, to ensure equitable distribution of teachers across schools (Kim & Han, 2002). This might explain how the Korean education system has achieved a more equal distribution of qualified teachers across schools than the United States has.

Our analyses showed that research on the relative importance of family backgrounds and school factors (school effect research) focused only on economic conditions such as national economic development levels and income inequality in examining cross-national variation in school effects. Future research needs to be extended to specific national education conditions or policies that redistribute unequal distribution of resources across schools. For example, some countries (e.g., Australia) provide a wide range of incentives (e.g., salary allowance, bonuses, and transportation assistance) in order to encourage experienced teachers to teach and remain in rural and remote locations (Organisation for Economic Co-operation and Development, 2004). This may help ensure that all students have access to teachers of similar quality across schools and will influence between-school variation in student achievement. Thus, it is important to examine the degree to which policies promoting equal distribution of teaching resources in relation to schools' socioeconomic background and geographic location are associated with cross-national variation in school effect.

More importantly, of the studies that investigated the degree to which national-level factors were associated with mathematics achievement inequality, the majority focused on student family background, school contexts, and national-level factors. As Bray and Thomas (1995) pointed out, this approach fails to consider important aspects for comparative education analyses such as classrooms, teaching methods, and curriculum. The TIMSS collected rich information on course content that is actually taught in classrooms, instructional practices, and textbooks, as well as intended mathematics curricula at a national level. To better understand mathematics achievement inequality, it is important to examine teaching and learning processes that mediate between upper levels (e.g., organization and structure of national education systems) and student learning outcomes.

Ayalon and Livneh (2013) examined the degree to which between-teacher uniformity in covering major mathematics topics was linked to smaller gender gaps in mathematics achievement. Yet, we highlight that this approach needs to be extended to the relationship between teaching and learning process and mathematics achievement gap between high- and low-SES students. For example, several studies found that opportunity to learn (OTL) specific mathematics topics varied across schools as well as across countries and found positive relationships between SES and OTL (Schmidt, Burroughs, Zoido, & Houang, 2015; Schmidt et al., 2001). However, it is not known how the organization and structure of national education systems (e.g., a standardized mathematics curriculum) account for the positive relationship between SES and OTL. Future research needs to examine how macro-level educational and social contexts interact with micro-level factors such as parents and teachers to exacerbate or alleviate mathematics achievement inequality. Furthermore, research examining whether the association between national-level factors and mathematics achievement inequality remains consistent across the distribution would be informative.

References

Adelman, C. (1999). *Answers in the toolbox: Academic intensity, attendance patterns, and bachelor's degree attainment*. Washington, DC: Department of Education, Office of Educational Research and Improvement.

Adelman, C. (2006). *The toolbox revisited: Paths to degree completion from high school through college*. Washington, DC: US Department of Education.

Allmendinger, J. (1989). *Career mobility dynamics: A comparative analysis of the United States, Norway, and West Germany*. Berlin: Max–Planck–Institut für Bildungsforschung.

Bray, M., & Thomas, R. M. (1995). Levels of comparison in educational studies: Different insights from different literature and the value of multilevel analyses. *Harvard Education Review, 65* (3), 472–490.

Burt, M., & Namgi, P. (2009). Education inequality in the Republic of Korea: Measurement and causes. In D. B. Holsinger & W. J. Jacob (Eds.), *Inequality in education: Comparative and international perspective* (Vol. 24, pp. 261–289). Dordrecht, Netherlands: Springer.

Camburn, E. M., & Han, S. W. (2009). Two decades of generalizable evidence on U.S. instruction from national surveys. *Teachers College Record, 113*(3), 561–610.

Carnoy, M., & Rothstein, R. (2013). *What do international tests really show about U.S. student performance?* Washington, DC: Economic Policy Institute.

Chiu, M. M. (2005). Effects of resources, inequality, and privilege bias on achievement: Country, school, and student level analyses. *American Educational Research Journal, 42*(4), 575–603.

Chiu, M. M. (2010). Effects of inequality, family and school on mathematics achievement. *Social Forces, 88*(4), 1645–1676.

Coleman, J. S., Campbell, E. Q., Hobson, C. J., McPartland, J., Mood, A. M., Weinfeld, F. D., & York, R. L. (1966). *Equality of educational opportunity*. Washington, DC: U.S. Government Printing Office.

Else-Quest, N. M., Hyde, J. S., & Linn, M. C. (2010). Cross-national patterns of gender differences in mathematics: A meta-analysis. *Psychological Bulletin, 136*(1), 103–127.

Gamoran, A., & Long, D. A. (2008). Equality of educational opportunity: A 40 year perspective. In R. Teese, S. Lamb, & M. Duru-Bellats (Eds.), *Education and equity*. International perspectives on theory and policy (Vol. 1, pp. 23–47). New York: Springer.

Guiso, L., Monte, F., Sapienza, P., & Zingales, L. (2008). Culture, gender, and math. *Science, 320* (5880), 1164–1165.

Hanushek, E. A., & Yilmaz, K. (2011). Urban education: Location and opportunity in the United States. In N. Brooks, K. Donaghy, & G.-J. Knaap (Eds.), *The Oxford handbook of urban economics and planning* (pp. 583–615). Oxford and New York: Oxford University Press.

Heyneman, S. P. (1976). A brief note on the relationship between socioeconomic status and test performance among Ugandan primary school children. *Comparative Education Review, 20*, 42–47.

Heyneman, S. P., & Loxley, W. A. (1983). The effects of primary school quality on academic achievement across twenty-nine high and low income countries. *American Journal of Sociology, 88*, 1162–1194.

Kerckhoff, A. C. (1995). Institutional arrangements and stratification processes in industrial societies. *Annual Review of Sociology, 21*, 323–347.

Kerckhoff, A. C. (2001). Education and social stratification processes in comparative perspective. *Sociology of Education, 74*, 3–18.

Kim, E., & Han, Y. (2002). *Attracting, developing and retaining effective teachers: Background report for Korea*. Seoul, Korea: Korean Educational Development Institute.

Knipprath, H. (2010). What PISA tells about the quality and inequality of Japanese education in mathematics and science. *International Journal of Science and Mathematics Education, 8*(3), 389–408.

Levels, M., Dronkders, J., & Kraaykamp, G. (2008). Immigrant children's educational achievement in western countries: Origin, destination, and community effects on mathematical performance. *American Sociological Review, 73*, 835–853.

Marks, G. N., Cresswell, J., & Ainley, J. (2006). Explaining socioeconomic inequalities in student achievement: The role of home and school factors. *Educational Research and Evaluation, 12* (2), 105–128.

National Research Council. (2003). *Understanding others, educating ourselves: Getting more from international comparative studies in education*. Washington, DC: National Academic Press.

Ohinata, A., & van Ours, J. C. (2012). *Young immigrant children and their educational attainment, IZA discussion paper*. Bonn, Germany: Institute for the Study of Labor.

Organisation for Economic Co-operation and Development. (2004). *Attracting, developing and retaining effective teachers: Background report for Australia*. Retrieved April 21, 2010 from OECD http://www.oecd.org/dataoecd/63/50/3879121.pdf.

Raudenbush, S. W., & Kim, J.-S. (2002). Statistical issues in analysis of international comparison of educational achievement. In A. C. Porter & A. Gamoran (Eds.), *Methodological advances in cross-national surveys of educational achievement* (pp. 267–294). Washington, DC: National Academic Press.

Roscigno, V. J., Tomaskovic-Devey, D., & Crowley, M. (2006). Social forces. *Education and the Inequalities of Place, 84*(4), 2121–2145.

Schmidt, W. H., Burroughs, N. A., Zoido, P., & Houang, R. T. (2015). The role of schooling in perpetuating educational inequality: An international perspective. *Educational Researcher, 44* (7), 371–386.

Schmidt, W. H., McKnight, C. C., Houang, R. T., Wang, H. C., Wiley, D. E., Cogan, L. S., et al. (2001). *Why schools matter: A cross-national comparison of curriculum and learning, The Jossey-Bass education series*. San Francisco, CA: Jossey-Bass.

Van de Werfhorst, H. G., & Mijs, J. B. (2010). Achievement inequality and the institutional structure of educational systems: A comparative perspective. *Annual Review of Sociology, 36*, 407–428.

Wößmann, L. (2002). *Central examinations improve educational performance: International evidence, Kiel discussion papers 397*. Kiel: Institute for World Economics.

Appendix: Research Articles Reviewed

Akiba, M., LeTendre, G. K., & Scribner, J. P. (2007). Teacher quality, opportunity gap, and national achievement in 46 countries. *Educational Researcher, 36*(7), 369–387.

Andon, A., Thompson, C. G., & Becker, B. J. (2014). A quantitative synthesis of the immigrant achievement gap across OECD countries. *Large-scale Assessments in Education, 2*(1), 1–20.

Ayalon, H., & Livneh, I. (2013). Educational standardization and gender differences in mathematics achievement: A comparative study. *Social Science Research, 42*(2), 432–445.

Baker, D. P., Goesling, B., & LeTendre, G. K. (2002). Socioeconomic status, school quality, and national economic development: A cross-national analysis of the "Heyneman-Loxley Effect" on mathematics and science achievement. *Comparative Education Review, 46*(3), 291–312.

Byun, S. Y., & Kim, K. K. (2010). Educational inequality in South Korea: The widening socioeconomic gap in student achievement. *Research in Sociology of Education, 17*, 155–182.

Chen, Q. (2014). Using TIMSS 2007 Data to build mathematics achievement model of fourth graders in Hong Kong and Singapore. *International Journal of Science and Mathematics Education, 12*(6), 1519–1545.

Cheng, Q., Wang, J., Hao, S., & Shi, Q. (2014). Mathematics performance of immigrant students across different racial groups: An indirect examination of the influence of culture and schooling. *Journal of International Migration and Integration, 15*(4), 589–607.

Chudgar, A., & Luschei, T. F. (2009). National income, income inequality, and the importance of schools: A hierarchical cross-national comparison. *American Educational Research Journal, 46*(3), 626–658.

Hanushek, E. A., & Luque, J. A. (2003). Efficiency and equity in schools around the world. *Economics of Education Review, 22*(5), 481–502.

Lamb, S., & Fullarton, S. (2002). Classroom and school factors affecting mathematics achievement: A comparative study of Australia and the United States using TIMSS. *Australian Journal of Education, 46*(2), 154–171.

Neuschmidt, O., Barth, J., & Hastedt, D. (2008). Trends in gender differences in mathematics and science (TIMSS 1995–2003). *Studies in Educational Evaluation, 34*(2), 56–72.

Nosek, B. A., Smyth, F. L., Sriram, N., Lindner, N. M., Devos, T., Ayala, A., et al. (2009). National differences in gender-science stereotypes predict national sex differences in science and math achievement. *Proceedings of the National Academy of Sciences, 106*(26), 10593–10597.

Pahlke, E., Hyde, J. S., & Mertz, J. E. (2013). The effects of single-sex compared with coeducational schooling on mathematics and science achievement: Data from Korea. *Journal of Educational Psychology, 105*(2), 444.

Penner, A. (2008). Gender differences in extreme mathematical achievement: An international perspective of biological and social forces. *American Journal of Sociology, 114*, S138–S170.

Pong, S. L., & Pallas, A. (2001). Class size and eighth-grade math achievement in the United States and abroad. *Educational Evaluation and Policy Analysis, 23*(3), 251–273.

Pong, S. L., Dronkers, J., & Hampden-Thompson, G. (2003). Family policies and children's school achievement in single- versus two-parent families. *Journal of Marriage and Family, 65*(3), 681–699.

Schiller, K. S., Khmelkov, V. T., & Wang, X. Q. (2002). Economic development and the effects of family characteristics on mathematics achievement. *Journal of Marriage and Family, 64*(3), 730–742.

Schnepf, S. V. (2007). Immigrants' educational disadvantage: An examination across ten countries and three surveys. *Journal of Population Economics, 20*(3), 527–545.

Waldfogel, J., & Zhai, F. (2008). Effects of public preschool expenditures on the test scores of fourth graders: Evidence from TIMSS. *Educational Research and Evaluation, 14*(1), 9–28.

Wang, D. B. (2004). Family background factors and mathematics success: A comparison of Chinese and US students. *International Journal of Educational Research, 41*(1), 40–54.

Wang, Z., Osterlind, S. J., & Bergin, D. A. (2012). Building mathematics achievement models in four countries using TIMSS 2003. *International Journal of Science and Mathematics Education, 10*(5), 1215–1242.

Yoshino, A. (2012). The relationship between self-concept and achievement in TIMSS 2007: A comparison between American and Japanese students. *International Review of Education, 58* (2), 199–219.

Chapter 17
When Knowing Basic Skills and Procedures Is Not Enough

Kyong Mi Choi and Dae S. Hong

Abstract Helping students achieve higher level cognitive processes is one of the goals of mathematics education. Studies have shown that students who engaged in high-level cognitive demand tasks were successful in mathematics performance. Based on an analysis of the TIMSS 2011 data, this chapter argues that eighth-grade US students' mathematics achievement is explained more by basic skills and procedures across ability levels. Data analyses of 8th grade students in Taiwan, Hong Kong, Korea, and Singapore indicate that success in the applying domain affects their overall scores more than the knowing domain. This pattern is consistent across all ability level groups. The findings of this study show that to be competitive internationally, US students should have more opportunities to develop higher level cognitive abilities when learning mathematics.

Keywords International comparison study • Mathematics • High-level cognitive practice • TIMSS • Applying • Reasoning

We encourage students to reach and use high-level cognitive practices so that they gain deeper understanding of mathematical concepts (Boston & Smith, 2009). Indeed, several researchers have shown that students engaged in high-level cognitive activities are more successful in mathematics performance (Boaler & Staples, 2008; Stein & Lane, 1996). The National Council of Teachers of Mathematics (NCTM) and Common Core State Standards for Mathematics (CCSSM) also support the importance of higher level cognitive practices (Common Core State Standards Initiative [CCSSI], 2010; NCTM, 2000). These two standards emphasize the development of higher level cognitive abilities such as reasoning, presenting, connecting, making and critiquing arguments, and modeling. The goal is to develop students' cognitive skills in addition to their basic skills and procedural knowledge

K.M. Choi (✉) • D.S. Hong
Department of Teaching and Learning, College of Education,
The University of Iowa, Iowa City, IA, USA
e-mail: kyongmi-choi@uiowa.edu; dae-hong@uiowa.edu

J.-W. Son et al. (eds.), *What Matters? Research Trends in International Comparative Studies in Mathematics Education*, Research in Mathematics Education, DOI 10.1007/978-3-319-51187-0_17

when teaching mathematics. Such emphasis is not limited to the US curriculum standards documents, but also appears in international assessments.

One of the largest international comparison assessments, the Trends in International Mathematics and Science Study (TIMSS), presents its assessment framework to guide expected cognitive processes that students across over 60 countries develop in three domains—knowing, applying, and reasoning (Mullis, Martin, Ruddock, O'Sullivan, & Preuschoff, 2009). It is reasonable to assume that the applying and reasoning domains are higher levels than knowing, as students use knowledge on mathematical facts, procedures, and concepts when applying and reasoning. TIMSS provides subcategories of each domain to identify what students would *do* when operating at a certain level of cognitive process: students who can *apply* are able to select appropriate strategies for problem-solving, and to generate and represent mathematical models using facts, concepts, and procedures. As reasoning involves the capacity for logical thinking, students who can *reason* should be able to analyze and make inferences from mathematical situations, generalize/specialize, integrate/synthesize mathematical knowledge, and justify their arguments using mathematical properties.

As the importance of higher level cognitive practices is emphasized, many educational researchers have examined the instructional practices using tasks with high-level cognitive demands. They argue that teachers need to provide students with learning opportunities to engage in high-level tasks, and to use pedagogical methods to engage and maintain students in high-level tasks (Boaler & Staples, 2008; Jensen, Choi, Sherry, & Kye, 2016; Stein, Grover, & Henningsen, 1996; Stein & Lane, 1996; Wilhelm, 2014). Studies found that engaging in cognitively demanding learning environments helped students develop deeper understanding of mathematical concepts (Boston & Smith, 2009), and is likely to result in larger gains in mathematical understanding and achievement (Boaler & Staples, 2008; Stein & Lane, 1996). Despite encouragement from curriculum standards documents and research studies that promote the importance of cognitively demanding practices, however, teachers experience difficulties maintaining these high levels of cognitive demand, and deprive students of these opportunities by guiding students toward a correct answer or way of solving a problem (Lithner, 2008). Facing such difficulties to challenge students cognitively, teachers may do most of the cognitive work and provide students opportunities to do little more than single-step basic arithmetic or recall of simple facts (Lithner, 2008; Yackel, Cobb, & Wood, 1998).

There is evidence that students learning mathematics with cognitively demanding tasks have deeper understanding of mathematical concepts and larger gains in achievement scores, but it is still not clear if the impact on achievement is due to their improved higher level cognitive abilities or the enhanced basic skills and procedural understandings that they attain through cognitively demanding tasks. In order to clarify the impact of high-level cognitive skills on students' performance, it is appropriate to investigate whether possessing higher level cognitive ability affects student performance more than possessing basic skills and procedural knowledge. More specifically, we would like to examine how each of the three

TIMSS cognitive stages—knowing, applying, and reasoning—contributes to overall mathematics achievement.

The TIMSS international report describes each participating country's emphasis of cognitive practices in its curriculum, which could contribute to the differences in students' cognitive abilities (Mullis, Martin, Foy, & Arora, 2012). Students who learn mathematics with simple factual and procedural knowledge may not be as good with mathematics problems that assess their application skills and higher level cognitive practices. For this reason, we examined what types of cognitive practices are emphasized in each country's curriculum. TIMSS surveyed national research coordinators on this topic to highlight what cognitive practices are emphasized in each country's intended curriculum (Mullis et al., 2012). National research coordinators rated these practices on a four-answer scale: no emphasis, very little emphasis, some emphasis, or a lot of emphasis. Table 17.1 shows the results of the five-country survey that TIMSS conducted in 2007 and 2011. We reviewed the emphases made in 2007 curricula because eighth graders in 2011 were fourth graders in 2007, and their learning practices could have influenced by the 2007 curricula. We also examined the 2011 curricula to see if there was any change during the 4-year period. When taking curriculum guidelines into consideration, it is important to remember that there is always a gap between intended curriculum and implemented curriculum, and the impact of intended curriculum on students' learning outcomes is only a conjecture.

Literature Review

Reasoning and Applying

Although mathematics educators aim for students to reach and use high-level cognitive practices so that they gain deeper understanding of mathematical concepts (Boston & Smith, 2009), evaluating these instructional goals is somewhat difficult because it is unclear what to assess for problem-solving skills and conceptual understanding (Thomson, 2006). For this reason, researchers have attempted to understand what each goal means, and to develop relevant measurement tools and comprehensive frameworks for assessment. One popular framework to consider cognitive abilities is Bloom's Taxonomy, and researchers have continuously used and revised this framework (Gierl, 1997; Krathwohl, 2002). In particular, Krathwohl's (2002) revision of Bloom's Taxonomy identified major categories of cognitive domains as the verbal aspects of learning mathematics. These categories are considered as learning goals measured in mathematics assessments, which represent what students are intended to do with mathematics contents. According to Kastberg (2003), these categories, including comprehension, application, analysis, synthesis, and evaluation, are originally based on Bloom's Taxonomy (1956). They are also used in the TIMSS (Mullis et al., 2009) as subcategories of three

Table 17.1 Emphases on cognitive domains in the intended mathematics curriculum reported by TIMSS national research coordinators (Foy & Olson, 2009; Mullis et al., 2012, p. 60)

	Country	Cognitive domain		
		Mastering basic skills and procedures	Applying mathematics in real-life contexts	Reasoning mathematically
TIMSS 2007	TWN	A lot	Some	Some
	HGK	Some	Some	Some
	KOR	A lot	A lot	Some
	SGP	A lot	A lot	A lot
	USA	A lot	A lot	Some
TIMSS 2011	TWN	A lot	Some	Some
	HGK	A lot	Some	A lot
	KOR	A lot	A lot	A lot
	SGP	A lot	A lot	A lot
	USA	A lot	Some	Some

cognitive domains (knowing, applying, and reasoning) in its assessment framework.

There has been lots of attention paid to cognitive practices (e.g., mathematical reasoning and applying) in mathematics teaching and learning because of the increased importance and usefulness of those practices in professional and everyday life (NCTM, 2000). Such emphases are reflected in the development of the CCSSM for students' college readiness as well as their opportunities to learn with a high-quality mathematics curriculum (Schmidt & Houang, 2012). The Standards for Mathematical Practices (SMP) in CCSSM identify the practices of applying and reasoning as core practices in teaching and learning. SMP emphasizes that students ought to *apply* the mathematics content to practical situations—"everyday life, society, and the workplace." Moreover, several of the eight standards address the use of reasoning: the second standard, for example, is entitled "reason abstractly and quantitatively" (CCSSI, 2010).

To explain the significance of applying and reasoning, TIMSS established its mathematics and science assessment framework in three domains: Knowing, Applying, and Reasoning (Mullis et al., 2012). As students' use of mathematics and reasoning about mathematics "depends on mathematical knowledge and familiarity with mathematical concepts" (p. 41), it is relevant for students to *recall* basic facts, *recognize* mathematical objects, *compute* algorithmic procedures, *retrieve* relevant information, *measure*, and *classify/order* objects and numbers, which are subcategories of the knowing domain. The applying domain "involves applications of mathematical tools in a range of contexts" using facts, concepts, and procedures that are routinized and familiar to the students (p. 43). The TIMSS assessment framework includes practices to *select*, *represent*, *model*, *implement*, and *solve routine problems* as subcategories of the applying domain. In the reasoning domain, there are five subcategories describing practices to *analyze* mathematical situations, *generalize/specialize* the result of mathematical thinking, *integrate/synthesize* different elements of mathematics, *justify* mathematical arguments, and *solve nonroutine problems*. The three TIMSS cognitive domains form the core of our analysis frame as they are the three variables we employ for the correlation analyses.

International Comparison Studies

There have been many efforts to learn how students in other countries learn and perform mathematics, and how diverse teaching and learning practices exist across countries. TIMSS provides rich datasets every 4 years for over 60 countries, which include various data: achievement scores—overall, content domain, and cognitive domains; students' attitudes toward mathematics; gender differences; school/national curricula; and many other educational background indices (Mullis et al., 2012). Using this information, researchers have investigated a wide range of educational practices and student performance levels, from simple comparisons of achievement scores and various related aspects that affect student performance, such as attitudes toward mathematics, gender, structures of educational systems, and details of student achievement (content and/or cognitive domains), to addressing measurement issues and analyzing the data using complex psychometric approaches for additional information (e.g., Andrews, Ryve, Hemmi, & Sayers, 2014; Choi, Choi, & McAninch, 2012; Choi, Lee, & Park, 2015; Eklöf, 2007; Lee, Park, & Taylan, 2011; Mere, Reiska, & Smith, 2006; Papanastasiou & Papanastasiou, 2006; Shen & Tam, 2008; Thomson, 2006; Wilkins, 2004; Yoshino, 2012).

By analyzing students' responses to TIMSS questionnaires, past studies have found that students with positive attitudes toward mathematics, in general, have higher achievement in mathematics (Mullis et al., 2012); however, there are countries whose students did not show the same patterns (Choi et al., 2012; Eklöf, 2007; Shen & Tam, 2008; Yoshino, 2012; Wilkins, 2004). Researchers suspected that different educational cultures and teaching practices may have resulted in the different patterns and influenced how students perceived mathematics and themselves, as well as how they respond to questionnaires about themselves, since the TIMSS data on attitudes toward mathematics are mostly based on students' self-reports. For example, one of the attitude survey question on students' self-perception was "I usually do well in mathematics," and students' responses to this question did not completely align with their mathematics performance (Wilkins, 2004). Choi et al. (2012) also described how students in a few higher performing countries have less positive attitudes toward mathematics than students in lower performing countries. Although it seems natural that self-perception and attitude toward a discipline affect achievement, given these patterns and results it is logical that there must be other factors that affect students' attitudes and achievement. One such factor could be how students use their cognitive abilities when learning and doing mathematics.

To address measurement issues or provide different perspectives to understand the TIMSS results, researchers used novel approaches. Lee and colleagues' study in 2011 was one of the first research studies conducted using the Cognitive Diagnosis Modeling approach, which allowed educators to understand if TIMSS students used important mathematics concepts when taking the assessment rather than simply reporting students' scores. This study opened up the possibility of using large-scale or standardized assessments to improve instruction and understand what students really learned or didn't. Choi et al. (2015) took a similar approach, using the TIMSS 2003 eighth-grade mathematics assessment to compare what mathematics students in different countries attained. This could help educators in each country examine how much content knowledge their students have, and reflect on their practices and systems.

Cognitive domains are also of interest to many mathematics educators. The TIMSS cognitive domains of knowing, applying, and reasoning describe what cognitive practices students use when answering mathematics problems, in addition to four content domains: Number, Algebra, Geometry, and Data and Chance (Mullis et al., 2012). Using the cognitive domains, TIMSS designed assessment items that require students to use a particular cognitive skill when solving them. With such information, it became possible to understand if students possessed the skillsets to use appropriate cognitive practices, and to examine students' cognitive practices in international perspectives. Studies on TIMSS cognitive domains mostly reported the average scores of each domain along with achievement scores, and compared them with the international average or averages of other countries (Thomson, 2006). Along with this information, some researchers investigated other factors that influence students' cognitive abilities, such as SES (Mere et al., 2006). However, the relationship between cognitive skills and overall achievement scores—how students' cognitive practices impact achievement scores—has not yet been investigated, despite the crucial importance of students' cognitive processes in learning mathematics.

The Purpose of This Study

Based on previous research on students' mathematics achievement and higher level cognitive practices (Boaler & Staples, 2008; Stein & Lane, 1996), we attempted to investigate how the development of higher level cognitive ability (e.g., applying and reasoning) impacts one's overall score when compared with basic skills and procedural knowledge. Specifically, the purpose of this study is to explore if eight-grade students in the USA and four other high-performing countries (Korea, Hong Kong, Taiwan, and Singapore) rely on the knowing stage or on higher level cognitive stages to achieve certain mathematics scores. If the mathematics achievement of the high-performing countries is explained by the knowing domain (i.e., the knowing domain is the most contributing cognitive stage) more than the higher cognitive level domains, then we can conclude that it may be enough that students achieve basic knowledge and procedural ability to score well on international assessments like TIMSS. Although scoring higher is not the goal of mathematics education, it is one indicator of where a country is placed in an international perspective. We also partitioned each country's sample down to ability levels, called TIMSS international benchmark levels, to compare groups with similar mathematics performance.

Methods

Data

We used the TIMSS 2011 mathematics assessment data of 30,627 students from five countries: Taiwan (TWN), Hong Kong (HKG), Singapore (SGP), South Korea (KOR), and the USA. Table 17.2 describes the data and the average scores of eighth graders in each country, and in each benchmark. Among the five plausible values that TIMSS provides, we used the first plausible value of mathematics to represent students' mathematics achievement, a method that Von Davier, Gonzales and Mislevy (2009) suggested as a way to reduce the bias of "variance and percentile estimates" (p. 23). TIMSS administered multiple sets of assessment items at the same time, and in order to obtain unbiased estimations of students' proficiency (since not all students answered the same set of problems), plausible values were calculated and determined based on students' responses to the item set that they received, combined with other relevant background information (Mislevy, 1991). Also, we collected the first plausible value for each cognitive domain in mathematics—knowing (KNO), applying (APP), and reasoning (REA). In TIMSS 2011, 35%, 40%, and 25% of mathematics assessment items were in the knowing, applying, and reasoning domains, respectively.

TIMSS provides five benchmarks based on standardized scores: <400: below, 400–475: low, 475–550: intermediate, 550–625: high, and >625: advanced (Mullis et al., 2012; p. 87). We compare students in each benchmark from the five countries that we selected for this study. In this way, we can look into students' mastery of each cognitive domain and its effect on achievement scores among similar-performance-level groups to compare them.

Analysis

We employed a linear regression model with three dependent domain-variables (KNO, APP, and REA) and one independent variable of mathematics achievement (ACH). Initially, we examined the correlation between one variable and the achievement (KNO and ACH; APP and ACH; and REA and ACH); however,

Table 17.2 Descriptive statistics of the TIMSS 2011 data by the international benchmarks

		International benchmark (TIMSS 2011)										Total	
		Below (<400)		Low (400–475)		Intermediate (475–500)		High (550–625)		Advanced (>625)			
	Domain	Mean	*N* (%)	Mean	*N* (%)	Mean	*N* (%)	Mean	*N* (%)	Mean	*N* (%)	Mean	*N* (%)
TWN	ACH	357.54	166 (3.29)	441.10	406 (8.05)	514.55	710 (14.08)	590.81	1242 (24.63)	694.65	2518 (49.94)	612.20	5042 (100)
	KNO	368.84		426.72		509.93		596.24		698.63		614.08	
	APP	393.40		447.86		522.38		600.12		692.60		616.29	
	REA	397.01		447.87		519.50		594.08		685.30		610.87	
HKG	ACH	362.85	122 (3.04)	442.65	293 (7.30)	517.45	742 (18.48)	589.88	1488 (37.06)	669.30	1370 (34.12)	585.95	4015 (100)
	KNO	381.73		451.58		526.68		598.06		667.26		591.22	
	APP	386.18		450.55		524.31		594.87		663.84		588.49	
	REA	367.52		431.45		510.63		587.57		661.71		580.57	
KOR	ACH	364.94	70 (1.36)	445.74	300 (5.81)	516.91	821 (15.89)	590.38	1563 (30.26)	688.02	2412 (46.69)	612.84	5166 (100)
	KNO	382.24		447.32		519.68		597.79		688.63		616.13	
	APP	400.06		459.00		525.01		597.04		687.16		616.98	
	REA	406.43		460.52		522.58		593.96		680.08		612.53	
SGP	ACH	378.94	72 (1.21)	443.31	405 (6.83)	519.06	930 (15.69)	589.65	1809 (30.52)	677.56	2711 (45.74)	606.22	5927 (100)
	KNO	411.46		462.15		532.18		599.32		676.02		612.21	
	APP	397.07		448.45		523.14		594.57		674.52		607.55	
	REA	357.31		425.25		503.18		584.02		674.31		599.03	
USA	ACH	364.58	836 (7.98)	442.39	2547 (24.31)	512.82	3945 (37.65)	581.19	2516 (24.01)	655.64	633 (6.04)	508.92	10,477 (100)
	KNO	395.27		455.46		520.96		587.15		658.10		519.19	
	APP	377.19		438.18		504.54		572.57		645.45		503.09	
	REA	378.78		438.17		505.53		572.86		646.78		503.74	

Note. *ACH* the first plausible value of mathematics achievement, *KNO* the first plausible value of mathematics knowing, *APP* the first plausible value of mathematics applying, and *REA* the first plausible value of mathematics reasoning

these relationships were too strong as one might expect. Thus, we decided to construct a regression model that included a linear combination of KNO, APP, REA, and the error term, because we fixed the intercept term to zero. It is reasonable to assume that the 0 point for any cognitive domain score would yield nothing for overall mathematics achievement. Thus, the full regression equation was expressed as follows:

$Y = B_1(\text{KNO}) + B_2(\text{APP}) + B_3(\text{REA}) + E$where E is the error and Y represents the first plausible value of mathematics achievement.

We analyzed the data with SPSS version 21 in two phases. In the first phase, we investigated a linear equation explaining the entire sample of each country. However, as mentioned previously, we questioned the fairness of comparing students with large achievement gaps, and thus in the second phase we split each country's sample into five groups of students based on the TIMSS international benchmark of mathematics achievement (see Mullis et al., 2012, p. 87). We applied the regression model again for each benchmark group in each country. This resulted in a total of 25 linear regression models in the second phase.

In each phase of data analysis, based on the regression equations, we conducted hypothesis testing for each correlation first, which would answer whether or not each cognitive domain significantly contributes to mathematics achievement, when controlling for the other two cognitive domains in the sample. This means that we investigated expected changes of mathematics achievement scores associated with a unit change in each cognitive domain when controlling the effects of the two other cognitive domains. We then compared these expected changes in mathematics achievement per unit change in the considered cognitive domain across five countries, including the USA.

Results

We report the first phase of analysis (country-level comparison) to describe each country's eighth graders' development on three cognitive levels in mathematics, and its effects on achievement scores. This will help us understand if development of the knowing level is enough to achieve highly in mathematics, or if further development to the applying or reasoning level is beneficial for eighth graders' mathematics performance. For the second-phase analysis, we investigated if students in a similar achievement level group show similar development of cognitive levels. This will test whether the development of knowing explains students' achievement more than the development of applying or reasoning for each of the five benchmarks across the five countries.

Country-Level Comparison

As shown in Table 17.3, the most prominent cognitive domain that explains American eighth-grade students' mathematics achievement was the knowing

Table 17.3 B coefficients in the regression models using the TIMSS 2011 international benchmarks

Country	Cognitive domains	Group by international benchmark					Whole sample
		Low (<400)	Below (400–475)	Intermediate (475–550)	High (550–625)	Advanced (>625)	
TWN	KNO	-0.124^{***}	-0.032^{***}	0.091^{***}	0.067^{***}	0.116^{***}	0.181^{***}
	APP	0.551^{***}	0.550^{***}	0.532^{***}	0.490^{***}	0.472^{***}	0.429^{***}
	REA	0.455^{***}	0.458^{***}	0.362^{***}	0.429^{***}	0.415^{***}	0.384^{***}
HKG	KNO	0.502^{***}	0.318^{***}	0.347^{***}	0.340^{***}	0.367^{***}	0.346^{***}
	APP	0.393^{***}	0.792^{***}	0.618^{***}	0.525^{***}	0.616^{***}	0.548^{***}
	REA	0.045^{***}	-0.140^{***}	0.017^{***}	0.124^{***}	0.021^{***}	0.100^{***}
KOR	KNO	-0.206^{***}	0.236^{***}	0.207^{***}	0.199^{***}	0.120^{***}	0.205^{***}
	APP	0.909^{***}	0.526^{***}	0.521^{***}	0.455^{***}	0.556^{***}	0.509^{***}
	REA	0.185^{***}	0.208^{***}	0.256^{***}	0.333^{***}	0.325^{***}	0.280^{***}
SGP	KNO	0.586^{***}	0.642^{***}	0.524^{***}	0.427^{***}	0.383^{***}	0.389^{***}
	APP	0.534^{***}	0.341^{***}	0.360^{***}	0.471^{***}	0.475^{***}	0.443^{***}
	REA	-0.213^{***}	-0.020^{***}	0.100^{***}	0.090^{***}	0.144^{***}	0.164^{***}
USA	KNO	0.683^{***}	0.649^{***}	0.530^{***}	0.452^{***}	0.445^{***}	0.505^{***}
	APP	0.184^{***}	0.204^{***}	0.315^{***}	0.373^{***}	0.350^{***}	0.335^{***}
	REA	0.059^{***}	0.125^{***}	0.149^{***}	0.176^{***}	0.209^{***}	0.153^{***}

$^{***}p < 0.01$, $^{**}p < 0.05$, and $^{*}p < 0.1$

domain ($B_{\text{USA-KNO}} = 0.505$), followed by the applying ($B_{\text{USA-APP}} = 0.335$) and reasoning ($B_{\text{USA-REA}} = 0.153$) domains. This implies that US eighth graders' achievement in TIMSS 2011 was uniquely explained by knowing ability rather than the other higher cognitive domains. One the other hand, the other four countries we evaluated present the applying domain as the most significant. For Taiwan and Korea, the reasoning domain was the second-most significant stage to explain students' achievement, while in Hong Kong and Singapore applying was followed by the knowing domain. The effect of reasoning ability on US eighth graders' overall achievement was smaller than their counterparts in all four countries, except for Hong Kong.

As seen in Table 17.1, the large influence of the knowing domain on US students is consistent with the emphasis on basic skills and procedures in their intended curriculum from 2007 and 2011. According to the national research coordinator, applying mathematics in real-life context was emphasized a lot in the 2007 curriculum, but the emphasis was reduced in the 2011 curriculum. Students in Hong Kong, Korea, and Singapore may have paid more attention to applying and reasoning practices as their curricula placed a lot of emphasis on higher level cognitive practices. The impacts of applying and reasoning on their students' mathematics achievement could be explained through their learning experiences utilizing higher level cognitive practices.

Benchmark-Level Comparisons

In the TIMSS 2011 report, there are five benchmarks based on standardized scores: <400: below; 400–475: low; 475–550: intermediate; 550–625: high; and >625: advanced (Mullis et al., 2012, p. 87). All the analysis results referred to in this section are in Table 17.3. The high and advanced benchmark groups are combined as their coefficients have similar patterns.

Below the benchmark. Among eighth graders in Hong Kong, Singapore, and the USA, having a knowing ability below the benchmark tended to contribute the most toward overall achievement. However, for students in Hong Kong and Singapore, their applying skills ($B_{\text{HKG-APP}} = 0.393$ and $B_{\text{SGP-APP}} = 0.534$) explain their overall achievement more than their American peers ($B_{\text{USA-APP}} = 0.184$). Analyses of student data from Taiwan and Korea did not provide statistically significant results for the knowing domain, but their applying domain significantly explains Taiwanese and Korean students' overall achievement ($B_{\text{TWN-APP}} = 0.551$ and $B_{\text{KOR-APP}} = 0.909$).

Low benchmark. For the students whose overall scores are between 400 and 475, the pattern is slightly different from the group of students below the benchmark. The achievement of students in Hong Kong is now explained more by the applying domain ($B_{\text{HKG-APP}} = 0.792$), while the knowing domain was dominant previously. Singaporean students' knowing ability still contributes the most to their overall achievement, even more strongly than their peers in the group below the benchmark. Similarly, US eighth graders' knowing ability ($B_{\text{USA-KNO}} = 0.649$) still explains their overall achievement the most, representing about three times the contribution of the applying domain ($B_{\text{USA-APP}} = 0.204$). Similar to the group below the benchmark, for Korean and Taiwanese students in the low benchmark, the applying domain is the most effective domain. It is notable that Taiwanese students' reasoning ability ($B_{\text{TWN-REA}} = 0.458$) explains their achievement very strongly.

Intermediate. Students in the intermediate benchmark, although their knowing effect has been reduced and applying contributed more than students in the lower benchmark groups, still show the dominant effect of knowing on US and Singaporean students' overall achievement. The Singaporean students' trend indicates a similar pattern to US eighth graders, while their peers in Taiwan, Korea, and Hong Kong showed that their applying ability impacted achievement at least twice as much as their knowing ability.

High and advanced. US students' trend of dominant knowing-impact extended to the two highest benchmark groups (High and Advanced). For the two US groups, correlation coefficients of applying have increased slightly ($B_{\text{high}} = 0.373$ and $B_{\text{advanced}} = 0.35$) from the three lower benchmark groups; however, they were still lower than students in the other four countries. Singaporean students, whose trends in the three lower benchmark groups were similar to their US peers', now changed their pattern: the applying domain became more influential than the knowing domain in explaining their overall achievement. One notable change for US eighth graders in these two higher benchmark groups was that their reasoning ability was more influential than it was for their counterparts in Hong Kong and Singapore. For Singaporean and Hong Kong students in the two groups, the reasoning domain never exceeded 0.144, while American students in the Advanced benchmark showed that reasoning and applying together ($B_{\text{USA-REA}} = 0.35$ and $B_{\text{USA-APP}} = 0.445$) explained achievement more so than the knowing domain ($B_{\text{USA-KNO}} = 0.445$).

Discussion and Conclusion

In general, US eighth graders' mathematics achievement was explained with their basic skills and procedure level ability than applying or reasoning abilities across advanced, high, intermediate, or lower achieving groups. In comparison to eighth-grade peers in high-achieving countries, the US trend of strong knowing effects was distinctive, as applying was the most dominant effect in the other four countries overall, especially for higher achieving students. While knowing explained the overall achievement for lower achieving groups in Singapore and Hong Kong, the effect of the applying domain become stronger in their higher achieving groups. For Taiwanese and Korean eighth graders, the effect of the applying domain is very strong from their below-benchmark groups to their advanced groups. The coefficients of the applying domain are larger than 0.5 unless the reasoning effect is strong.

At the beginning of the study, we assumed hierarchical relationships among the three cognitive domains—knowing, applying, and reasoning. The results do not mean that students in the four Asian countries do better at applying than knowing. The way we read and understand the results is that mathematics assessment problems in the applying domain contribute more to their overall scores than those in the knowing domain. In other words, correct responses to knowing items do not contribute to their overall score increments as much as correct responses to applying items do.

As presented in Table 17.1, in 2007 knowing and applying were emphasized a lot in the US intended curriculum, based on the national research coordinator's report. In 2011, on the other hand, US national research coordinators reported that American mathematics curriculum emphasized basic skills and procedures a lot while there was only some emphasis on the applying and reasoning processes. The shift of emphasis on applying practices from *a lot* in 2007 to *some* in 2011 might explain why American eighth graders' mathematics achievement dominantly relied on basic knowledge and procedural practices. We should not forget that there is always a gap between intended curriculum and implemented curriculum: The emphasis made on applying and reasoning practices in the 2007 intended mathematics curriculum may not have extended to the classroom practices to become a fully implemented curriculum.

In the USA, there have been curriculum reform movements in last couple decades that seem to emphasize students' critical thinking skills and promote students' engagement in real-word problems using mathematical reasoning (Schoen & Hirsch, 2003). Reform-based textbooks provide activities to promote students' active participation using higher level thinking skills. Despite 20 years of such reform, however, we do not see students' use of higher level cognitive practices in the TIMSS data. These movements thus seem to be ineffective, based on the findings of this study.

For the four Asian countries we examined in the study, our findings could be partly explained by previous curriculum and textbook studies: Korean, Singaporean, and Chinese textbooks provide more or widespread opportunities for students to engage in multistep tasks and tasks requiring explanation, compared to both traditional and reform-based American textbooks (Fan & Zhu, 2007; Son & Senk, 2010). Similarly, Taiwanese textbooks include more challenging problems and problems requiring explanations (Charalambous, Delaney, Hui-Yu, & Mesa, 2010). These practices give students more experience and engagement in cognitively challenging tasks. They could have helped students in high-achieving Asian countries to develop higher level cognitive abilities and, in turn, led them to perform higher on international assessments.

We would like to provide a few explanations for the misalignment between recent curricular movements and students' practices. First, teachers may not have received proper training when they receive new reform-based textbooks. After all, "reform" means something different from the previous familiar practices that teachers have trained with. Without knowing how to use a new type of textbook and curriculum effectively, teachers either struggle to develop their own methods, which may take a lot of time, or stick with an old style of teaching that is mismatched with the new textbook. Researchers suggest that proper training programs for in-service teachers as well as preservice teachers will provide a great opportunity to prepare teachers to help young students develop and use higher level cognitive practices in mathematics (Franke, Kazemi, & Battey, 2007; Jensen et al., 2016). In such programs, teachers learn methods to provide students with opportunities to learn and engage in high cognitive demand tasks, but also with opportunities to understand the importance of developing higher level cognitive abilities such as mathematical reasoning and applying processes. In turn, teachers will understand why it is important for students to learn mathematics using higher level cognitive practices, and how they can help students accomplish this.

Another explanation would be a bit more complicated than the teacher training issue: although many teachers receive proper training to teach students with new curricula and are familiar with new ideas in mathematics instruction, a shift in instruction also takes time. Teachers must adjust their instructional methods, and students need time to get used to student-centered learning environments. Sometimes students are resistant to play an active role in learning, as they are accustomed to sitting in the classroom and listening, or watching what teachers show them. As Emanuelsson and Sahlstrom (2008) recognized that teachers tended to lower cognitive demand required for students to engage when students did not immediately respond to the teachers' prompts, it takes time for students to engage actively in a cognitively challenging environment. However, school administrators and parents will not wait for this shift, and want to see immediate "results" such as achievement scores. To respond to such requests, teachers may decide to go back to the old and comfortable methods. Teaching for basic skills and procedures is what they are accustomed to teach, and it is likely for students to see these elements in standardized tests. Although teachers' perceptions of reform-based curricula are positive, for these reasons, teachers may stay with basic knowledge and procedural skills and students would likewise emphasize these practices.

School administrators and other stakeholders do need to understand the importance of students' learning and using higher level cognitive practices such as applying and reasoning, and what it looks like when students use these practices. At the same time, they need to understand that it takes time to shift instructional practices, and that what the new generation learns would be very different from what they have learned and have been seeing. To help in this process, we could share information about how high-level cognitive mathematical practices (e.g., applying and reasoning) helped students in other high-achieving countries achieve advanced performance.

In this study, the investigation was limited to eighth-grade students in the five selected countries. Based on the results of this study, we would like to explore if this trend persists or if basic skills and procedural knowledge is enough for younger high performers, as TIMSS also offers assessment data from fourth-grade students in many countries. We would also like to expand this study to include more countries. It would give us a comprehensive picture of the relationship between cognitive abilities and achievement. Also, we would like to conduct a large-scale study on how teachers perceive and implement reform-based curricular materials that are designed to promote students' use of higher level cognitive practices when learning and doing mathematics. Theoretically, we have assumed a hierarchical relationship among the three cognitive domains such that the knowing domain is cognitively lower than the applying and reasoning domains (Bloom, 1956). It would be meaningful to empirically verify this theory using Confirmatory Factor Analysis and Structural Equation Modeling to see if students' performance also indicates the hierarchy we assumed for this study.

American eighth graders' mathematics achievement has been significantly and gradually improved since 1995 when TIMSS was first conducted (Mullis et al., 2012). Our students moved from slightly below the international average to significantly higher than the international average. To become competitive in an international perspective, our students first showed compatibility in international scenes such as international comparison studies like TIMSS. Boaler and Staples (2008) argue for the importance of higher level cognitive activities in successful outcomes in mathematics learning. To be able to compete with students in Asian high-achieving countries, we now need to emphasize higher level cognitive practices, not just basic knowledge and procedures.

References

Andrews, P., Ryve, A., Hemmi, K., & Sayers, J. (2014). PISA, TIMSS and Finnish mathematics teaching: An enigma in search of an explanation. *Educational Studies in Mathematics, 87*, 7–26.

Bloom, B. S. (Ed.). (1956). *A taxonomy of educational objectives: Handbook I, the cognitive domain*. New York, NY: Longman.

Boaler, J., & Staples, M. (2008). Creating mathematical futures through an equitable teaching approach: The case of Railside School. *Teachers College Record, 110*, 608–645.

Boston, M. D., & Smith, M. S. (2009). Transforming secondary mathematics teaching: Increasing the cognitive demands of instructional tasks used in teachers' classrooms. *Journal for Research in Mathematics Education, 40*(2), 119–156.

Charalambous, C., Delaney, S., Hui-Yu, H., & Mesa, V. (2010). A comparative analysis of the addition and subtraction of fractions in textbooks from three countries. *Mathematical Thinking and Learning, 12*, 117–151.

Choi, K., Choi, T., & McAninch, M. (2012). A comparative investigation of the presence of psychological conditions in high achieving eighth graders from TIMSS 2007 Mathematics. *ZDM, 44*(2), 189–199.

Choi, K., Lee, Y. S., & Park, Y. S. (2015). What CDM can tell about what students have learned: An analysis of TIMSS eighth grade mathematics. *Eurasia Journal of Mathematics, Science & Technology Education, 11*(6), 1563–1577.

Common Core State Standards Initiative [CCSSI]. (2010). *Common core state standards for mathematics*. Retrieved from http://www.corestandards.org/the-standards/mathematics.

Eklöf, H. (2007). Self-concept and valuing of mathematics in TIMSS 2003: Scale structure and relation to performance in a Swedish setting. *Scandinavian Journal of Educational Research, 51*(3), 297–313.

Emanuelsson, J., & Sahlström, F. (2008). The price of participation: Teacher control versus student participation in classroom interaction. *Scandinavian Journal of Educational Research, 52*(2), 205–223.

Fan, L., & Zhu, Y. (2007). Representation of problem–solving procedures: A comparative look at China, Singapore, and US mathematics textbooks. *Educational Studies in Mathematics, 66*, 61–75.

Foy, P., & Olson, J. F. (Eds.). (2009). *TIMSS 2007 international database and user guide*. Chestnut Hill, MA: TIMSS & PIRLS International Study Center, Boston College.

Franke, M. L., Kazemi, E., & Battey, D. (2007). Mathematics teaching and classroom practice. *Second Handbook of Research on Mathematics Teaching and Learning, 1*, 225–256.

Gierl, M. J. (1997). Comparing cognitive representations of test developers and students on a mathematics test with Bloom's taxonomy. *Journal of Educational Research, 91*(1), 26–32.

Jensen, J. L., Kye, A., Sherry, M., & Choi, K. (2016). *Maintaining high level of cognitive demand through student silence*. Paper presented at the National Council of Teachers of Mathematics Research Conference, San Francisco, CA.

Kastberg, S. E. (2003). Using Bloom's Taxonomy as a framework for classroom assessment. *Mathematics Teacher, 96*(6), 402–405.

Krathwohl, D. R. (2002). A revision of Bloom's Taxonomy: An overview. *Theory into Practice, 41*(4), 212–218.

Lee, Y. S., Park, Y. S., & Taylan, D. (2011). A cognitive diagnostic modeling of attribute mastery in Massachusetts, Minnesota, and the US national sample using the TIMSS 2007. *International Journal of Testing, 11*(2), 144–177.

Lithner, J. (2008). A research framework for creative and imitative reasoning. *Educational Studies in Mathematics, 67*(3), 255–276.

Mere, K., Reiska, P., & Smith, T. M. (2006). Impact of SES on Estonian students' science achievement across different cognitive domains. *Prospects, 36*(4), 497–516.

Mislevy, R. J. (1991). Randomization-based inference about latent variables from complex samples. *Psychometrika, 56*(2), 177–196.

Mullis, I. V. S., Martin, M. O., Foy, P., & Arora, A. (2012). *TIMSS 2011 international results in mathematics*. Chestnut Hill, MA: TIMSS & PIRLS International Study Center, Lynch School of Education, Boston College and International Association for the Evaluation of Educational Achievement (IEA).

Mullis, I. V. S., Martin, M. O., Ruddock, G. J., O'Sullivan, C. Y., & Preuschoff, C. (2009). *TIMSS 2011 assessment frameworks*. Chestnut Hill, MA: TIMSS & PIRLS International Study Center, Lynch School of Education, Boston College.

National Council of Teachers of Mathematics [NCTM]. (2000). *Principals and standards for mathematics*. Reston, VA: NCTM.

Papanastasiou, C., & Papanastasiou, E. C. (2006). Modeling mathematics achievement in Cyprus. In S. J. Howie & T. Plomp (Eds.), *Contexts of learning mathematics and science: Lessons learned from TIMSS* (pp. 113–125). New York, NY: Routledge.

Schoen, H. L., & Hirsch, C. R. (2003). The core-plus mathematics project: Perspectives and student achievement. In S. Senk & D. Thompson (Eds.), *Standards-oriented school mathematics curricula: What are they? What do students learn?* (pp. 311–344). Hillsdale: Lawrence Erlbaum Associates.

Shen, C., & Tam, H. P. (2008). The paradoxical relationship between student achievement and self-perception: A cross-national analysis based on three waves of TIMSS data. *Educational Research and Evaluation, 14*(1), 87–100. doi:10.1080/13803610801896653.

Schmidt, W. H., & Houang, R. T. (2012). Curricular coherence and the common core state standards for mathematics. *Educational Researcher, 41*(8), 294–308.

Son, J., & Senk, S. (2010). How standards curricula in the USA and Korea present multiplication and division of fractions. *Educational Studies in Mathematics, 74*, 117–142.

Stein, M. K., Grover, B., & Henningsen, M. (1996). Building student capacity for mathematical thinking and reasoning: An analysis of mathematical tasks used in reform classrooms. *American Educational Research Journal, 33*(2), 455–488.

Stein, M. K., & Lane, S. (1996). Instructional tasks and the development of student capacity to think and reason: An analysis of the relationship between teaching and learning in a reform mathematics project. *Educational Research and Evaluation: An International Journal on Theory and Practice, 2*(1), 50–80.

Thomson, S. (2006). Australian students achievement in the TIMSS 2002 mathematics cognitive domains. *TIMSS Australia Monograph Series, 2*.

Von Davier, M., Gonzalez, E., & Mislevy, R. (2009). What are plausible values and why are they useful. *IERI Monograph Series, 2*, 9–36.

Wilhelm, A. G. (2014). Mathematics teachers' enactment of cognitively demanding tasks: Investigating links to teachers' knowledge and conceptions. *Journal for Research in Mathematics Education, 45*(5), 636–674.

Wilkins, J. L. M. (2004). Mathematics and science self-concept: An international investigation. *Journal of Experimental Education, 72*(4), 331–346. doi:10.3200/JEXE.72.4.331-346.

Yackel, E., Cobb, P., & Wood, T. (1998). The interactive constitution of mathematical meaning in one second grade classroom: An illustrative example. *The Journal of Mathematical Behavior, 17*(4), 469–488.

Yoshino, A. (2012). The relationship between self-concept and achievement in TIMSS 2007: A comparison between American and Japanese students. *International Review of Education, 58* (2), 199–219.

Chapter 18
The WIFI Study: Students' Valuing of Mathematics Learning in Hong Kong and Japan

Wee Tiong Seah, Takuya Baba, and Qiaoping Zhang

Abstract This chapter introduces the reader to the *What I Find Important (in my mathematics learning)* study (WIFI), conducted by a consortium of 21 research teams from 18 economies. It uses the same questionnaire to assess what students value in their respective mathematics education experiences. Two case economies, Hong Kong and Japan, provide the context for the discussion. This provides a reference point for analyzing four significant themes: the affordance to identify and define cultures and subcultures, the documenting and comparing of espoused and enacted valuing, the triangulation of survey responses, and the culturally situated labelling of values and valuing.

Keywords Values • Valuing • Questionnaire • Hong Kong • Japan • WIFI

Introduction

The *What I Find Important (in my mathematics learning)* study (WIFI), based at The University of Melbourne, is a consortium of 21 research teams from 18 economies across 5 continents around the world that gather, analyze, and compare information relating to what school students value in their respective mathematics

W.T. Seah (✉)
Melbourne Graduate School of Education, The University of Melbourne, Melbourne, VIC, Australia
e-mail: wt.seah@unimelb.edu.au

T. Baba
Graduate School for International Development and Cooperation, Hiroshima University, Higashihiroshima, Japan
e-mail: takuba@hiroshima-u.ac.jp

Q. Zhang
Department of Curriculum and Instruction, Faculty of Education, The Chinese University of Hong Kong, Sha Tin, NT, Hong Kong SAR, China
e-mail: qpzhang@cuhk.edu.hk

J.-W. Son et al. (eds.), *What Matters? Research Trends in International Comparative Studies in Mathematics Education*, Research in Mathematics Education, DOI 10.1007/978-3-319-51187-0_18

educations. This is one of a few large-scale, international studies which have been designed to survey and understand an aspect of students' volition. The findings are intended to complement those from other studies which examine cognitive and affective variables, to allow for a more holistic understanding of how we can facilitate optimal levels of mathematics learning in schools across different cultures.

Large-scale, cross-national educational research studies such as TIMSS, PISA, and TEDS-M have received widespread attention in the media and in mathematics education scholarship in recent years. These tests do not measure cognitive performance alone. For example, PISA (Organization for Economic Cooperation and Development (OCED), 2014) surveyed affective variables such as student perseverance, anxiety, and self-concept in addition to student knowledge and performance. None of the studies, however, assess what students or teachers value and find important in mathematics pedagogy. In this context, the WIFI Study presents us with a great opportunity to understand better what students value, and thus what drives student performance and/or enhances teachers' and students' mathematics teaching/learning experience. However, values are culturally referenced (Bishop, 1988). Thus, for example, findings from Zhang et al. (2016) which showed that junior secondary students in mainland China valued *achievement* more than anything else should be understood in the context of that country's history and culture. In more ways than one, this cautions us against expecting to replicate the PISA 2012 success of the Chinese city of Shanghai simply by copying its mathematics education model.

Investigating how students in different countries value the learning and teaching of mathematics has its own set of issues as well. This chapter attempts to highlight what these challenges are, in addition to the affordances. To contextualize this discussion, we will begin with a review of related research that sheds light on the role that values/valuing plays in the learning and teaching of mathematics. Specifically, we will introduce Alan Bishop's framework of values in the "Western mathematics" classroom. The methodology will be described next, including an introduction to and justification for the two case economies (Hong Kong and Japan) that make up the focus of this chapter. Next, we share the results of the questionnaire data analysis, and discuss our findings according to the following four themes: the affordance to identify and define cultures and subcultures, the documenting and comparing of espoused and enacted valuing, the triangulation of survey responses, and the culturally situated labelling of values and valuing.

The Value of Values/Valuing in Mathematics Learning

Not only are students in East Asian contexts performing very well in international comparative tests such as TIMSS and PISA, but studies such as those by Byun and Park (2012) as well as Wei and Eisenhart (2011) have also reported that Asian students, especially East Asian students, in "Western" education systems also perform better than their peers in school mathematics. These (East) Asian students attended the same schools as their peers, so they would have been taught by the same teachers, performed similar activities during mathematics lessons, attempted the same homework, and completed the same assessment tasks. They would also have experienced the same classroom learning environments and conditions. Given these same opportunities to learn (at school), then, why do East Asian students perform better in school mathematics? Lee and Zhou's (2015) analysis of the mathematics performance of migrant children in the USA painted the same picture, thus raising the question: all else being equal, how might the Asian ethnicity of these students explain their relative superior achievement in the American education system?

Several reports (e.g., Leung, 2006; Wei & Eisenhart, 2011) have made reference to culturally based values in mathematics education. Askew, Brown, Rhodes, Johnson, and William (1997) might have stopped short of naming "values" as the factor associated with the "effective" teaching they observed, although they wrote about these teachers "believing in the importance of" (p. 4) particular pedagogical practices in their mathematics teaching repertoire. Later on, Askew, Hodgen, Hossain, and Bretscher (2010) wrote that

> one of the most striking things the review has shown is that high attainment may be much more closely linked to cultural values than to specific mathematics teaching practices. This may be a bitter pill for those of us in mathematics education who like to think that how the subject is taught is the key to high attainment. [...] Being born into a culture that highly values success in mathematics establishes a 'virtuous cycle' of continuing success. (p. 12)

Would it not be more empowering (for teachers and students alike) if we identify the values that underlie such classroom activities, and apply these values in other classrooms through culturally appropriate adaptations?

Values/Valuing in Mathematics Education

We adopt Seah and Andersson's (2015) definition of values/valuing in the context of mathematics learning and teaching:

> Values are the convictions which an individual has internalised as being the things of importance and worth. What an individual values defines for her/him a window through which s/he views the world around her/him. Valuing provides the individual with the will and determination to maintain any course of action chosen in the learning and teaching of mathematics. They regulate the ways in which a learner's/teacher's cognitive skills and emotional dispositions are aligned to learning/teaching in any given educational context. (p. 169)

According to Krathwohl, Bloom, and Masia's (1964) "Affective Taxonomy of Educational Objectives", the internalized nature of valuing makes it extremely stable. Drawing on McLeod's (1992) conception, this stability reflects a high investment of cognition and a correspondingly low involvement of affect. Similarly, the valuing process proposed by Raths, Harmin, and Simon (1987) demonstrates the extent to which thinking and reasoning are a part of coming to value particular attributes. Such internalization and stability have contributed in part to the appeal of values/valuing as being valid and reliable, and may well explain why they are associated with the characteristics of will and determination. This in no way implies that values or valuing need to be directly observable. Although values might be regarded as beliefs in action (Clarkson, Bishop, FitzSimons, & Seah, 2000), Munir Fasheh's observation at the 2015 Mathematics and Society Conference—that values are what we do not violate in action—reflects another way in which valuing can manifest itself less directly.

Regardless of what observable valuing might look like, Krathwohl et al. (1964) Taxonomy suggests that values can modify related beliefs and other emotional constructs. Here, we emphasize the distinction between values and beliefs: whereas values demonstrate what are considered to be important, beliefs are expressions of truth (Seah, 2013). Thus, a culture may value *communication*, and in so doing it expresses the importance of such a form of interaction, but this valuing is no indication of the extent to which communication is held to be correct or wrong. On the other hand, a related belief might be "student-student communication enhances student understanding of mathematical concepts," which is a statement declaring some perceived truth in mathematics pedagogy. Value's interaction with Bloom, Englehart, Furst, Hill, and Krathwohl (1956) Taxonomy also suggests that what we value can affect our choice of mental strategies, reasoning, and decisions deployed to "do mathematics." In other words, what is valued regulates both cognitive processes and affective modes.

Research into the role of values and valuing in mathematics learning and teaching began with Alan Bishop's proposal of three pairs of complementary values relating to "Western" mathematics (Bishop, 1988). These are convictions in the discipline of mathematics that are taught in contemporary schooling: *rationalism* and *objectism*, *control* and *progress*, as well as *mystery* and *openness* (Bishop, 1988). Bishop (1996) later proposed that these mathematical values constitute one of three categories of valuing that are often expressed in the mathematics classroom. One of the two other categories captures the mathematics educational values which are reflected in the pedagogical practices of school mathematics. The range of these values can be extensive, and examples include *information and communication technology [ICT]*, *practice*, *ability*, and *effort*.

Both mathematical and mathematics educational values have the potential to affect the quality of a student's mathematics learning experience. The mathematical values that are passed on to students carry with them the message of what is regarded as important in the discipline and in the practice of "doing mathematics". Similarly, the sorts of mathematics educational values that are shared and potentially embraced relay messages to the students regarding the norms and practices in mathematics pedagogy, suggesting to them what it takes to learn mathematics well. Bishop's (1996) third category of values in the mathematics classroom, general educational values, refers to the sorts of values which educational systems expect to inculcate in students through the school subjects. They do not directly affect mathematics performance (if at all), and thus they will not be discussed in this chapter.

Outside the mathematics education area, the empirical work of Dutch social scientist Geert Hofstede in more than 50 countries during the 1970s led to his formulation (Hofstede, 1997) of five value continua. In his view, each culture can be uniquely defined by this set of five value continua: power distance (from small to large), collectivism–individualism, femininity (concern and preservation)–masculinity (being assertive and ambitious), uncertainty avoidance (from weak to strong), and orientation in life (from short term to long term). These value continua have been considered in our study reported here as general educational values, and are assessed in our data collection instrument.

Methodology

The WIFI Study has been designed to document quantitatively what students value in their respective mathematics education experiences. It is an 18-economy international research study, which features local research teams from Australia, Brazil, mainland China, Germany, Ghana, Greece, Hong Kong SAR, Indonesia, Japan, Korea, Malaysia, South Africa, Spain, Sweden, Taiwan, Thailand, Turkey, and the United Kingdom. The Australian team also coordinates the study's research design, data collection, data analysis, and dissemination. One innovation of the WIFI Study is its attempt to design and use a questionnaire to assess what students broadly value in mathematics learning. Whereas qualitative methods might have traditionally been used to find out what students and teachers value in mathematics education (see, for examples, Keitel, 2003; Tan & Lim, 2013), the questionnaire allows for quick and efficient collection of data from a large participant pool. This means that the findings are generalizable, thus facilitating comparative studies between and amongst participating economies.

Several values questionnaires (see below) have been constructed in mathematics education research, though they only assessed specific categories of values. The WIFI questionnaire, on the other hand, seeks to paint a holistic picture of what the student respondent values in his/her mathematics learning experience. Of course, values questionnaires also exist in other research areas not related to mathematics education. Examples include the "survey for terminal and instrumental values" (Rokeach, 1973) and the "survey for personal and work values" (Senge, Kleiner, Roberts, Ross, & Smith, 1994). The WIFI questionnaire, however, has been designed specially to survey broadly what students find important—that is, value—with regard to mathematics, to mathematics learning, and to being educated. This schema corresponds to Bishop's (1996) categories of mathematical, mathematics educational, and general educational values.

The WIFI questionnaire items were drawn from four existing relevant sources, namely the "values and mathematics project" (Clarkson et al., 2000), a questionnaire used by Alan Bishop in a Thai workshop, the "mathematics education values

Section A

For each of the items below, tick a box to tell us how **important** it is to you when you learn mathematics.

	Absolutely important	Important	Neither important nor unimportant	Unimportant	Absolutely unimportant
1. Investigations	☐	☐	☐	☐	☐
2. Problem-solving	☐	☐	☐	☐	☐
3. Small-group discussions	☐	☐	☐	☐	☐

Fig. 18.1 Examples of the Likert-scale items in the WIFI questionnaire

questionnaire" (Dede, 2011), and the findings of a prior international collaboration study in the "Third Wave Project" (see Seah, 2011). Given that values are the invisible threads of culture (Henderson & Thompson, 2003), and that individuals may not be aware of what they themselves value (Clarkson et al., 2000), the questionnaire does not ask respondents to list what they value. Rather, they are asked to indicate the extent to which they find particular learning activities (e.g., small-group discussions) important, in the form of 64 five-choice Likert-scale items (see Fig. 18.1). The other sections of the WIFI questionnaire are not relevant to the focus of this chapter, and thus they will not be discussed here.

A feature of the WIFI questionnaire aimed at optimizing its validity is that multiple items are asked of the respondents for each of the intended values. For example, the following items were selected for inclusion in the questionnaire to test for the valuing of *process* (versus *product*): working systematically (item 6), explaining my solutions to the class (item 19), writing the solutions step by step (item 33), and understanding why my solution is incorrect (item 64). In this manner, triangulating the answers can help eliminate other possible valuing which might manifest in the same action descriptions.

Given that several different languages (e.g., Chinese, German, Japanese, Spanish, and Swedish) constitute a variety of mediums of instruction in the participating schools, across the 18 economies, it was expected that the WIFI questionnaire would be translated from the English language to the students' respective languages of instruction. The original, English-language questionnaire was validated with more than 1000 primary and secondary school students, as reported by Seah (2013).

The Cases of Hong Kong and Japan

The cases of Hong Kong and Japan provide the participant context for this chapter. Our intention was to select two of the several high-performing economies in the PISA assessment tests so that we could explore the ways in which values and valuing contribute to mathematical performance. Indeed, Hong Kong and Japan were ranked third and seventh, respectively, in PISA 2012, out of 65 participating economies. That both these economies are East Asian was also a selection factor, since it was desirable to keep the two comparison economies as culturally similar as possible, so that the attributes being valued across them would be comparable. In this research, the student participants were drawn from different parts of Hong Kong and five prefectures in Japan. Amongst the 1081 student respondents in Hong Kong, 367 were in primary schools, and the other 714 were in secondary schools. Japan's 3818 students, on the other hand, were made up of 1631 and 2187 primary and secondary school students, respectively.

Both the Hong Kong and Japanese research teams conducted initial data screening exercises to test for univariate normality (Tabachnick & Fidell, 1996). Descriptive statistics normality tests (normal probability plot, detrended normal, skewness, and kurtosis) showed that assumptions of univariate normality were not violated. Principal component analysis (PCA) with a varimax rotation and Kaiser normalization was used to examine items 1–64 of the questionnaire. The significance level was set at 0.05, while a cutoff criterion for component loadings of 0.45 was used in interpreting the solutions. Variables loading on more than one component were eliminated.

Results

Appendices 1 and 2 show the respective rotated component matrices for Hong Kong and Japan. The Hong Kong and Japanese research teams analyzed their own matrices to propose a set of valuing each, which reflects what students in their respective economies valued in mathematics learning. In this context, Hong Kong students valued these items in their mathematics learning: *explorations, alternative approaches, effort, (mathematics) identity, recall, ICT, feedback, applications,* and *exposition.* Their peers in Japan, however, were found to value *wonder, creativity, results, others' involvement, know-how, ICT, discussion, reality,* and *mystery.*

The factor structures for the Hong Kong and Japan component matrices were then compared using the simple Pearson *r* to test for correlations between pairs of factor vectors between the Hong Kong and Japanese data. This "methodology is concerned with calculations made directly on the factor loading vectors alone, immaterial of the relationships between the factors within any single study" (Barrett, 1986, p. 328). As shown in Appendix 3, none of the Hong Kong students' valuing and the Japanese students' valuing is correlated, $p < 0.01$. Overall, this means that the students in the two places valued different attributes in their respective mathematics learning experiences. This is despite the fact that our Hong Kong and Japanese researchers had assigned the same value name to their respective Component 6, *ICT*.

Our work analyzing the collected data in the WIFI Study has facilitated national reporting and international comparisons (e.g., Zhang et al., 2016). Additionally, this quantitative cross-cultural research study has led us to reflect on several themes, which will be elaborated below: the affordance to identify and define cultures and subcultures, the documenting and comparing of espoused and enacted valuing, the triangulation of survey responses, and the culturally situated labelling of values and valuing.

Discussion

The Affordance to Identify and Define Cultures and Subcultures

Culture is "an organised system of values which are transmitted to its members both formally and informally" (McConatha & Schnell, 1995, p. 81). Research designs that facilitate comparative studies across countries tend to collect large-scale data, which in turn has meant that group patterns can be made visible, thus enabling the definitions of subcultures within populations. For instance, subsets of populations can be selected and tested through factor analyses, to identify components that represent what is valued collectively by such groups, and possibly leading to the definitions of subgroups in the society. In the context of school (mathematics) teaching and learning, Stigler and Hiebert (1999) acknowledged that "the national patterns of teaching that we have observed must arise out of a knowledge base that is widely shared by teachers within each culture" (p. 83). These patterns can also be confirmed in terms of the characteristics of the curriculum and textbooks.

Furthermore, established subcultures such as gender and grade levels can be investigated through such processes as factor analysis and MANOVA tests for evidence of values-based differences. In our case studies of Hong Kong and Japan, we have been able to learn more than just what school students valued in their mathematics learning. In fact, each of the nine attributes that were valued by students in Japan was significantly influenced by school level. That is to say, amongst the student respondents to the WIFI questionnaire in Japan, the primary school and secondary school students differently valued *wonder, creativity, results, others' involvement, know-how, ICT, discussion, reality,* and *mystery* to an extent that is statistically significant, with three values (*wonder*, *results*, and *discussion*) exhibiting medium effect sizes. Given that primary school students in Hong Kong and Japan move on to different secondary schools (rather than to the same ones), it would have been difficult to assess their valuing again when they are in their secondary school years. As such, the approach we adopted (i.e., surveying different groups of students in primary and secondary schools at the same time) appears to be the optimal one with which to examine how values change across school systems.

By giving the different research teams the opportunity to assign a value name to each component in its own PCA matrix, the WIFI Study capitalizes on the multinational nature of its research capacity, to facilitate culturally situated interpretations of what students in each culture and subculture value. For example, we can see from Appendices 1 and 2 that students in Hong Kong and Japan regarded the following three activities as being of importance: stories about mathematics (item 17), stories about recent developments in mathematics (item 18), and stories about mathematicians (item 61). Yet our research team members from Hong Kong and Japan, acting as "cultural insiders," interpreted these phenomena differently. The Hong Kong researchers emphasized the practicality mindset of the Hong Kong population, in which decisions and efforts with regard to school education are almost always directed at optimizing one's own assessment scores, thus arriving at the conclusion that these stories were valued for their potential to provide students with alternative, more efficient solutions to mathematics questions. On the other hand, the same types of stories (as described in the three questionnaire items listed above) were shared with children in Japan not for the purpose of seeking alternative and more efficient solutions, but for the purpose of generating a sense of wonder in the young minds of the students.

Thus, we can see how the same classroom phenomena may be valued by students in different cultures for different reasons, reflecting different values/valuing. This has been facilitated by a research group whose members represent these different cultures. It is doubtful that had the research group been staffed by Australian researchers only, for example, that we would have been able to identify from the component matrices alone what was valued similarly and differently.

Documenting and Comparing Espoused and Enacted Valuing

Qualitative methods such as (lesson) observations and interviews allow for in-depth observation and clarification, respectively, of the valuing process. What students (and teachers) valued in a lesson would have been observed and earmarked during lesson visits, while post-lesson interviews would facilitate explanations and justifications of what appeared to be valued in class. Whereas such qualitative methods might be useful in revealing both espoused and enacted valuing, the same cannot be said of quantitative methods such as the questionnaire. This and other large-scale survey methods are known to be useful in generating information about what participants espouse, and yet we are aware of the gaps that can exist between

Section C

Imagine that we are going to produce a magic pill.
Anyone who takes this magic pill will become very good at mathematics!

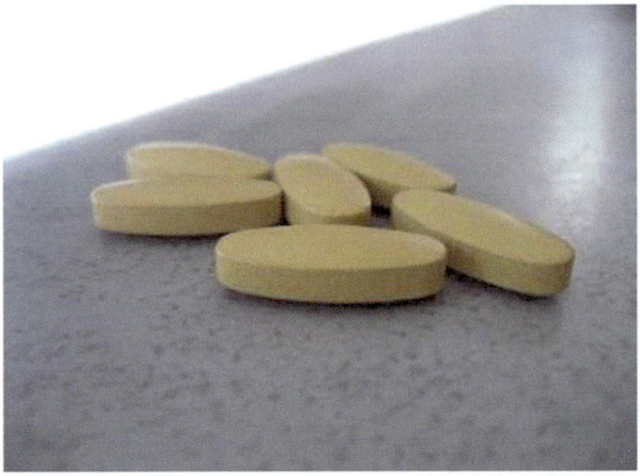

What will you choose to be the **top 3 main ingredients** of this magic pill?

Fig. 18.2 A contextualized item in the WIFI questionnaire

espoused and enacted valuing (Howell, Kirk-Brown, & Cooper, 2012). This thus poses threats to the validity of findings derived from questionnaires and other large-scale survey methods.

The design of the WIFI questionnaire, however, aimed to minimize errors regarding such construct validity. In responding to the survey items, student respondents were encouraged to relate to their own respective mathematics learning experiences. In addition, one of the sections was contextualized in a scenario, with student respondents invited into the "story" to elicit better open-ended responses that should probe for what respondents espouse as being of value. This is demonstrated in Fig. 18.2.

The Triangulation of Survey Responses

The provision of such contexts to the rest of the survey items, however, would lead to a questionnaire that is too long, which can affect either the quality of responses (as a result of respondent fatigue) or the proportion of valid responses. The fact that qualitative methods such as observations and interviews establish the context within which responses can be interpreted should be useful for researching values, because it can be difficult to identify what exactly is being valued based on a survey item response alone. For example, a student's expressed enthusiasm for the use of iPads (or other digital devices) might reflect several possible valuing priorities, such as *fun*, *efficiency* (e.g., "the software program processed the data efficiently"), and *accuracy* (e.g., "the digital devices compute the answers accurately each and every time"). The nature of large-scale survey methods such as the questionnaire, however, means that we do not have access to this rich background information. As an attempt to compensate for this, multiple items were asked of the respondents for each of the valuing elements targeted in the WIFI questionnaire, as explained earlier in this chapter.

The Culturally Situated Labelling of Values

The main concern facing the various research teams in the WIFI Study, however, has been that the same valuing might be named differently—and/or that different valued attributes might be named similarly—in and by different cultures. That is, the same questionnaire items may load onto a component in each economy but be allocated different labels, thereby suggesting that the students in the different economies (and thus cultures) were valuing different attributes of mathematics learning. Or, even though two sets of questionnaire items that cluster together in two component matrices may not be identical, different cultures may assign the same value name to them.

In our case example here of Hong Kong and Japan, the students valued different attributes in their respective mathematics learning experiences. This is despite the fact that both economies are generally considered to be East Asian, that both societies are regarded as embracing the Confucian Heritage Culture, and that they are ranked in the top seven performing mathematics education systems in PISA 2012. This suggests that subcultures may exist in ways which define how mathematics is learnt effectively, regulating students' valuing in the course of doing so.

Likewise, the questionnaire items which made up Component 6 in both the Hong Kong and Japanese component matrices are not identical, yet they were both regarded in the respective economies as being reflective of valuing *ICT*. In the Hong Kong data, there were five items which loaded onto the component (items 4, 22, 23, 24, 27), whereas in the Japan data only four items loaded onto it (items 22, 23, 24, 25). Three questionnaire items (i.e., items 22, 23, 24) were common between the two economies, and their nature probably led to the two separate components being named similarly. However, item 27 (being lucky at getting the correct answer), which loaded onto the Hong Kong data, and item 25 (mathematics games), which loaded onto the Japan data, can hardly be seen to demonstrate student valuing of *ICT* in their respective math learning experiences. It does appear that different labels ought to be proposed in both Hong Kong and Japan, to account for the sorts of things that are valued by students in Hong Kong and Japan, respectively. Indeed, *ICT* may not have been an appropriate label for both economies in the first place, and the Japanese data might well reflect student valuing of, say, *fun*, which may or may not be in the form of electronic or online delivery.

This example highlights that even though two components from the factor analyses of two economies/cultures may be similar, in that they are constituted by similar but not identical survey items, it need not imply that the same attributes (of mathematics learning) are being valued between or across the cultures. When a survey respondent answers any of the WIFI questionnaire items, this apparently straightforward exercise is associated with a culturally based interpretation of what the item means in his/her socio-historical space. Perhaps the "average" Hong Kong student regards the calculator as a convenient tool with which the correct answer to a mathematics question may be obtained/checked, a gadget to which a set of steps/ procedures can be applied to derive/check this correct answer, to the extent where luck plays a role. On the other hand, the student in Japan may embrace the calculator as an instrument with which mathematics games and recreational activities may be performed, given that there is also the valuing of mathematics games (item 25) in the Japanese component.

Thus, what we are seeing here is the embodiment of the cultural nature of values and valuing. Each of the WIFI questionnaire items potentially reflect different values/valuing according to the culture within which it is posed. How these items come together in any subsequent factor analysis process defines for a cultural setting its unique value/valuing with regard to mathematics, to its pedagogy, and to school education. It is in this manner that although Hong Kong students and their peers in Japan consider it important to use the calculator to check answers (item 22), and to learn mathematics with the computer (item 23) and on the internet (item 24), these commonly valued activities do not automatically imply the same underlying values as they relate to mathematics, its learning, and education more generally. There are certainly lessons here too for exercising caution when interpreting values/ valuing based on what is observable in practice.

To this end, the pursuit of inter-rater agreement/consistency between cultures should not apply when values/valuing are being investigated in comparative studies. Implicit in this discussion, too, is the belief that the culturally referenced values and valuing within particular cultures are best interpreted by "cultural insiders." Any of us who have not had personal life experiences in Hong Kong or Japan would not have been able to interpret the respective components in ways which capture and account for local ways and means of reasoning and acting.

Concluding Ideas

The quantitative, cross-cultural WIFI Study is an example of research efforts that have been invested into looking at ways in which we can better facilitate the learning and teaching of mathematics in schools, using the construct of values/valuing. As a volitional variable, values are regarded as having cognitive and affective components (Seah, 2004), as well as sociocultural (Seah & Andersson, 2015) components. While earlier research on values in the context of mathematics education research has largely been qualitative in nature, a desire to understand what and how bigger groupings of students (and teachers) value aspects of mathematics education more recently had led to the conceptualization and design of the WIFI Study, which allows for research—and comparisons—across cultures of a large number of participants at any one time. This has been especially meaningful, though challenging, given the sociocultural nature of values/valuing.

In using the WIFI questionnaire to collect values/valuing data across 18 economies located in five continents (Africa, Asia, Australia, Europe, and South America), we have identified some affordances and barriers that are unique to the large-scale, quantitative research approach. It has allowed us to collect and work with large quantities of data so that generalizations about what students in individual economies and across regions (e.g., Zhang et al., 2016) value and find important can be proposed. Indeed, the large-scale design of the WIFI Study—and the statistical testing which it afforded—made it possible for cultures and subcultures to be identified and defined in terms of what they value (Hofstede, 1997) in the mathematics learning process.

On the other hand, there have also been challenges in ascertaining the validity of our interpretations, given the constraints posed by the nature of the research design. As discussed in the section above, these included the collection of data pertaining to espoused values rather than enacted ones, the need for triangulation, and the culturally situated labelling of what is valued. The WIFI Study's design had taken these potential issues into account such that survey participants were encouraged to respond to the items in context, and multiple items were included to assess particular values. Similarly, local researchers assigned value names, reflecting our recognition that as "cultural insiders" they were best placed to make sense of and to interpret what students in their own cultures valued in school mathematics education. In this way, the WIFI Study has brought into sharper focus the constituents of each value as it is understood and emphasized in different cultures, highlighting the differences between what might appear as similar values across different cultures. In this manner, we hope that the WIFI Study will continue to deepen our collective understanding of (student) values/valuing as socially situated constructs that facilitate the mathematics learning process. These findings will present us with the first steps towards more meaningfully facilitating the learning of mathematics for the next generations of students, both to cultivate mathematical ways of sense-making and to improve the quality and quantity of citizens involved in the various STEM careers in the twenty-first century.

Appendix 1: Rotated component matrix for Hong Kong data

	1	2	3	4	5	6	7	8	9
Component 1: Exploration									
Q56 Knowing the steps of the solution	0.748								
Q54 Understanding concepts / processes	0.708								
Q55 Shortcuts to solving a problem	0.703								
Q51 Learning through mistakes	0.655								
Q58 Knowing which formula to use	0.604								
Q63 Understanding why my solution is incorrect or correct	0.593								
Q50 Getting the right answer	0.588								
Q59 Knowing the theoretical aspects of mathematics	0.564								
Q49 Examples to help me understand	0.546								
Q2 Problem-solving	0.530								
Q47 Using diagrams to understand maths	0.491								
Q53 Teacher use of keywords	0.490								
Component 2: Alternative approaches									
Q17 Stories about mathematics		0.760							
Q61 Stories about mathematicians		0.754							
Q18 Stories about recent developments in mathematics		0.696							
Q34 Outdoor mathematics activities		0.666							

(continued)

	1	2	3	4	5	6	7	8	9
Q25 Mathematics games		0.559							
Q52 Hands-on activities		0.555							
Component 3: Effort									
Q37 Doing a lot of mathematics work			0.849						
Q36 Practicing with lots of questions			0.822						
Q57 Mathematics homework			0.732						
Q62 Completing mathematics work			0.690						
Q43 Mathematics tests examinations			0.519						
Component 4: (Mathematics) identity									
Q30 Alternative solutions				0.687					
Q21 Students posing maths problems				0.601					
Q31 Verifying theorems hypotheses				0.593					
Q29 Making up my own maths questions				0.585					
Q19 Explaining my solutions to the class				0.487					
Component 5: Recall									
Q28 Knowing the times tables					0.629				
Q14 Memorizing facts					0.570				
Q38 Given a formula to use					0.548				
Q13 Practicing how to use maths formulae					0.517				
Q32 Using mathematical words					0.513				
Component 6: ICT									
Q22 Using the calculator to check the answer						0.802			

(continued)

	1	2	3	4	5	6	7	8	9
Q23 Learning maths with the computer						0.760			
Q4 Using the calculator to calculate						0.724			
Q24 Learning maths with the internet						0.692			
Q27 Being lucky at getting the correct answer						0.560			
Component 7: Feedback									
Q45 Feedback from my friends							0.666		
Q44 Feedback from my teacher							0.646		
Q46 Me asking questions							0.485		
Component 8: Applications									
Q10 Relating mathematics to other subjects in school								0.636	
Q12 Connecting maths to real life								0.553	
Q11 Appreciating the beauty of mathematics								0.549	
Q8 Learning the proofs								0.485	
Component 9: Exposition									
Q5 Explaining by the teacher									0.550
Q7 Whole-class discussions									0.493
Q6 Working step by step									0.485

Extraction Method: Principal Component Analysis; Rotation Method: Varimax with Kaiser Normalization

[a]Rotation converged in ten iterations

Appendix 2: Rotated component matrix for Japan data

	1	2	3	4	5	6	7	8	9
Component 1: Wonder									
61. Stories about mathematicians	0.760								
17. Stories about mathematics	0.755								
18. Stories about recent developments in mathematics	0.732								
11. Appreciating the beauty of mathematics	0.682								
60. Mystery of mathematics	0.649								
39. Looking out for math in real life	0.620								
40. Explaining where the rules/formulae came from	0.619								
34. Outdoor mathematics activities	0.568								
10. Relating mathematics to other subjects in school	0.529								
12. Connecting math to real life	0.517								
21. Students posing math problems	0.481								
20. Mathematics puzzles	0.481								
Component 2: Creativity									
30. Alternative solutions		0.691							
15. Looking for different ways to find the answer		0.665							
16. Looking for different possible answers		0.655							
31. Verifying theorems/hypotheses		0.546							
37. Doing a lot of mathematics work		0.515							

(continued)

	1	2	3	4	5	6	7	8	9
63. Understanding why my solution is incorrect or correct		0.486							
36. Practicing with lots of questions		0.460							
8. Learning the proofs		0.460							
29. Making up my own math questions		0.455							
Component 3: Results									
14. Memorizing facts (e.g., Area of a rectangle = length × breadth)			0.712						
13. Practicing how to use math formulae			0.639						
2. Problem-solving			0.526						
28. Knowing the times tables			0.522						
43. Mathematics tests/examinations			0.497						
32. Using mathematical words			0.470						
Component 4: Others' involvement									
44. Feedback from my teacher				0.645					
41. Teacher helping me individually				0.601					
45. Feedback from my friends				0.553					
46. Me asking questions				0.473					
35. Teacher asking us questions				0.472					
5. Explaining by the teacher				0.458					
Component 5: Know-how									
55. Shortcuts to solving a problem					0.633				
56. Knowing the steps of the solution					0.597				

(continued)

	1	2	3	4	5	6	7	8	9
64. Remembering the work we have done					0.471				
54. Understanding concepts/processes					0.451				
Component 6: ICT									
23. Learning math with the computer						0.854			
24. Learning math with the internet						0.844			
22. Using the calculator to check the answer						0.637			
25. Mathematics games						0.578			
Component 7: Discussion									
7. Whole-class discussions							0.693		
3. Small-group discussions							0.660		
9. Mathematics debates							0.493		
Component 8: Reality									
48. Using concrete materials to understand								0.474	
Component 9: Mystery									
27. Being lucky at getting the correct answer									0.565

Extraction Method: Principal Component Analysis; Rotation Method: Varimax with Kaiser Normalization

[a]Rotation converged in 19 iterations

Appendix 3: Correlations between all pairs of factor vectors in the Hong Kong and Japanese data

		JPN_C1	JPN_C2	JPN_C3	JPN_C4	JPN_C5	JPN_C6	JPN_C7	JPN_C8	JPN_C9
HKG_C1	Pearson correlation	[a]	[a]	[a]	[a]	0.232	[a]	[a]	[a]	[a]
	Sig. (2-tailed)					0.851				
	N	0	1	1	0	3	0	0	0	0
HKG_C2	Pearson correlation	0.849	[a]	[a]	[a]	[a]	[a]	[a]	[a]	[a]
	Sig. (2-tailed)	0.151								
	N	4	0	0	0	0	1	0	0	0
HKG_C3	Pearson correlation	[a]	1.000**	[a]	[a]	[a]	[a]	[a]	[a]	[a]
	Sig. (2-tailed)									
	N	0	2	1	0	0	0	0	0	0
HKG_C4	Pearson correlation	[a]	0.949	[a]	[a]	[a]	[a]	[a]	[a]	[a]
	Sig. (2-tailed)		0.205							
	N	1	3	0	0	0	0	0	0	0
HKG_C5	Pearson correlation	[a]	[a]	0.001	[a]	[a]	[a]	[a]	[a]	[a]
	Sig. (2-tailed)			0.999						
	N	0	0	4	0	0	0	0	0	0
HKG_C6	Pearson correlation	[a]	[a]	[a]	[a]	[a]	−0.765	[a]	[a]	[a]
	Sig. (2-tailed)						0.446			
	N	0	0	0	0	0	3	0	0	1
HKG_C7	Pearson correlation	[a]	[a]	[a]	0.787	[a]	[a]	[a]	[a]	[a]
	Sig. (2-tailed)				0.423					
	N	0	0	0	3	0	0	0	0	0
HKG_C8	Pearson correlation	−0.479	[a]	[a]	[a]	[a]	[a]	[a]	[a]	[a]
	Sig. (2-tailed)	0.682								
	N	3	1	0	0	0	0	0	0	0
HKG_C9	Pearson correlation	[a]	[a]	[a]	[a]	[a]	[a]	[a]	[a]	[a]
	Sig. (2-tailed)									
	N	0	0	0	1	0	0	1	0	0
	Sig. (2-tailed)									
	N	0	0	0	0	0	0	0	0	1

**Correlation is significant at the 0.01 level (2-tailed)

[a]Cannot be computed because at least one of the variables is constant

References

Askew, M., Brown, M., Rhodes, V., Johnson, D., & William, D. (1997). *Effective teachers of numeracy: Final report*. London: King's College.

Askew, M., Hodgen, J., Hossain, S., & Bretscher, N. (2010). *Values and variables: Mathematics education in high-performing countries*. London: Nuffield Foundation.

Barrett, P. (1986). Factor comparison: An examination of three methods. *Personality and Individual Differences, 7*(3), 327–340.

Bishop, A. J. (1988). *Mathematical enculturation: A cultural perspective on mathematics education*. Dordrecht, The Netherlands: Kluwer Academic Publishers.

Bishop, A. J. (1996). *How should mathematics teaching in modern societies relate to cultural values–some preliminary questions*. Paper presented at the Seventh Southeast Asian Conference on Mathematics Education, Hanoi, Vietnam.

Bloom, B., Englehart, M., Furst, E., Hill, W., & Krathwohl, D. (1956). *Taxonomy of educational objectives: The classification of educational goals. Handbook I: Cognitive domain*. New York: Longmans, Green.

Byun, S. Y., & Park, H. (2012). The academic success of East Asian American youth. *Sociology of Education, 85*(1), 40–60.

Clarkson, P., Bishop, A. J., FitzSimons, G. E., & Seah, W. T. (2000). Challenges and constraints in researching values. In J. Bana & A. Chapman (Eds.), *Mathematics Education Beyond 2000: Proceedings of the Twenty-third Annual Conference of the Mathematics Education Research Group of Australasia Incorporated held at Fremantle, Western Australia, 5–9 July, 2000* (Vol. 1, pp. 188–195). Perth, Australia: Mathematics Education Research Group of Australasia Incorporated.

Dede, Y. (2011). Mathematics education values questionnaire for Turkish preservice mathematics teachers: Design, validation, and results. *International Journal of Science and Mathematics Education, 9*, 603–626.

Fasheh, M. (2015). Over 68 years with mathematics: My story of healing from modern superstitions and reclaiming my sense of being and well-being. In S. Mukhopadhyay & B. Greer (Eds.), *Proceedings of the Eighth International Mathematics Education and Society Conference* (Vol. 1, pp. 33–60). Portland, OR: MES8.

Henderson, M., & Thompson, D. (2003). *Values at work: The invisible threads between people, performance and profit*. Auckland, New Zealand: Harper Collins.

Hofstede, G. (1997). *Cultures and organizations: Software of the mind* (Revised ed.). New York, NY: McGraw-Hill.

Howell, A., Kirk-Brown, A., & Cooper, B. (2012). Does congruence between espoused and enacted organizational values predict affective commitment in Australian organizations? *The International Journal of Human Resource Management, 23*(4), 731–747.

Keitel, C. (2003). *Values in mathematics classroom practice: The students' perspective*. Paper presented at the Conference of the Learners' Perspective Study International Research Team, Melbourne, Australia.

Krathwohl, D. R., Bloom, B. S., & Masia, B. B. (1964). *Taxonomy of educational objectives: The classification of educational goals (Handbook II: Affective domain)*. New York, NY: David McKay.

Lee, J., & Zhou, M. (2015). *The Asian American achievement paradox*. New York, NY: Russell Sage Foundation.

Leung, F. K. S. (2006). Mathematics education in East Asia and the West: Does culture matter? In F. K. S. Leung, K.-D. Graf, & F. J. Lopez-Real (Eds.), *Mathematics education in different cultural traditions: A comparative study of East Asia and the West* (pp. 21–46). New York, NY: Springer.

McConatha, J. T., & Schnell, F. (1995). The confluence of values: Implications for educational research and policy. *Educational Practice and Theory, 17*(2), 79–83.

McLeod, D. B. (1992). Research on affect in mathematics: A reconceptualization. In D. A. Grouws (Ed.), *Handbook of research on mathematics teaching and learning* (pp. 575–596). New York, NY: MacMillan.

Organization for Economic Cooperation and Development. (2014). *PISA 2012 results: Ready to learn: Students' engagement, drive and self-beliefs* (Vol. 3). Paris, France: OECD.

Raths, L. E., Harmin, M., & Simon, S. B. (1987). Selections from 'values and teaching'. In J. P. F. Carbone (Ed.), *Value theory and education* (pp. 198–214). Malabar, FL: Robert E. Krieger.

Rokeach, M. (1973). *The nature of human values*. New York, NY: The Free Press.

Seah, W. T. (2004). Shifting the lens of inquiry into the socialisation of mathematics teachers: Nature of value differences. In I. Putt, R. Faragher, & M. McLean (Eds.), *Proceedings of the 27th Annual Conference of the Mathematics Education Research Group of Australasia* (Vol. 2, pp. 501–508). Townsville, Australia: James Cook University.

Seah, W. T. (2011). Effective mathematics learning in two Australian primary classes: Exploring the underlying values. In B. Ubuz (Ed.), *Proceedings of the 35th Conference of the International Group for the Psychology of Mathematics Education* (Vol. 4, pp. 129–136). Ankara, Turkey: PME.

Seah, W. T. (2013). Assessing values in mathematics education. In A. M. Lindmeier & A. Heinze (Eds.), *Proceedings of the 37th Conference of the International Group for the Psychology of Mathematics Education* (Vol. 4, pp. 193–200). Kiel, Germany: PME.

Seah, W. T., & Andersson, A. (2015). Valuing diversity in mathematics pedagogy through the volitional nature and alignment of values. In A. Bishop, H. Tan, & T. Barkatsas (Eds.), *Diversity in mathematics education: Towards inclusive practices* (pp. 167–183). Switzerland: Springer.

Senge, P., Kleiner, A., Roberts, C., Ross, R., & Smith, J. (1994). *The fifth discipline fieldbook. Strategies and tools for building a learning organisation*. NY: Currency, Doubleday.

Stigler, J. W., & Hiebert, J. (1999). *The Teaching Gap: Best ideas from the world's teachers for improving education in the classroom*. NY: Free Press.

Tabachnick, B. G., & Fidell, L. S. (1996). *Using multivariate statistics* (3rd ed.). New York, NY: Harper Collins.

Tan, S. F., & Lim, C. S. (2013). Effective mathematics lesson from the eyes of primary pupils. In M. Inprasitha (Ed.), *Proceedings of the 6th East Asia Regional Conference on Mathematics Education* (Vol. 2, pp. 184–193). Khon Kaen, Thailand: Centre for Research in Mathematics Education.

Wei, M.-H., & Eisenhart, C. (2011). Why do Taiwanese children excel at math? *The Phi Delta Kappan, 93*(1), 74–76.

Zhang, Q., Barkatsas, T., Law, H. Y., Leu, Y. C., Seah, W. T., & Wong, N. Y. (2016). What primary students in the Chinese Mainland, Hong Kong and Taiwan value in mathematics learning: A comparative analysis. *International Journal of Science and Mathematics Education, 14*(5), 907–924.

Chapter 19
Examining the Association Between Teacher Feedback and Mathematics Instruction in Japan, Korea, Singapore, and the United States

Seong Won Han, Ji-Won Son, and Chungseo Kang

Abstract Recent research demonstrates that professional learning experiences and consistent evaluation and feedback systems significantly improve teachers' instructional practices and students' learning. This chapter examines the role of teacher evaluation and feedback systems in supporting instructional change with four other factors: professional development, collaboration, teacher beliefs about constructive pedagogy, and teacher–student relationships. Data came from lower secondary school mathematics teachers' responses to the Teaching and Learning International Study (TALIS) 2013, and focuses on schools in Japan, Korea, Singapore, and the United States. Descriptive analyses illustrate how teacher evaluation and feedback vary across these countries, while regression analyses examine the degree to which teacher feedback is associated with mathematics instruction. In particular, the study explores factors that mediate the impact of teacher feedback on mathematics instruction. The results of this study provide comparative insights into how to use teacher evaluation and feedback systems effectively in improving mathematics instruction.

Keywords Teacher feedback • Professional development • Mathematics instruction • TALIS

The first two authors contributed equally to this work.

S.W. Han (✉) • C. Kang
Department of Educational Leadership and Policy, University at Buffalo - The State University of New York, Buffalo, NY, USA
e-mail: seongwon@buffalo.edu; chungseo@buffalo.edu

J.-W. Son
Department of Learning and Instruction, University at Buffalo - The State University of New York, Buffalo, NY, USA
e-mail: jiwonson@buffalo.edu

J.-W. Son et al. (eds.), *What Matters? Research Trends in International Comparative Studies in Mathematics Education*, Research in Mathematics Education, DOI 10.1007/978-3-319-51187-0_19

Introduction

Researchers and policymakers have an ongoing interest in understanding the ways in which teachers contribute to students' learning and academic achievement, both in the United States and in many other countries, including Japan, Korea, and Singapore (Organisation for Economic Co-operation and Development (OECD), 2005; United Nations Educational, Scientific and Cultural Organization (UNESCO) Institute for Statistics, 2006). In the United States, for example, prompted by federal investments in education, the Race to the Top program was authorized under sections 14005 and 14006 of the American Recovery and Reinvestment Act of 2009. It places specific emphasis on improving teacher quality; as a result, researchers, policymakers, and practitioners have become increasingly interested in the interplay between teacher quality and student achievement. With the Common Core State Standards Initiative similarly emphasizing students' mathematical proficiency, teachers' continued learning opportunities (i.e., professional development) and teacher appraisal and feedback have been used as essential mechanisms to enhance teacher quality, teaching practices, and, ultimately, student achievement (Youngs, 2013).

However, as with many constructs in education, teacher quality is a multidimensional construct that includes not only teacher qualifications but also their knowledge, self-efficacy, beliefs, attitudes toward students and teaching, and instructional practices (Darling-Hammond & Youngs, 2002; Goe, 2007). Educational researchers have attempted to validate empirically which strategies are effective for improving teacher quality. One of the key levers for improving teacher quality is professional development, but recently much attention has been extended from professional development to teacher evaluation. Unlike research on professional development that identified several features of effective professional development for enhancing teacher quality (Yoon, Duncan, Lee, Scarloss, & Shapley, 2007), only a few studies have examined the link between teacher appraisal and feedback and teaching practices (OECD, 2009). Overall, there is little consensus on the specific teacher evaluation and feedback types that are consistently associated with instructional practices.

The purpose of this research is to examine the role of teacher evaluation and feedback systems in supporting effective mathematics teaching, when combined with other factors such as professional development and collaboration. In doing so, we first describe patterns in teachers' instructional practices in lower secondary mathematics classrooms in four countries that participated in the Teaching and Learning International Study (TALIS) 2013: Japan, Korea, Singapore, and the United States. Although TALIS was administered to nationally representative samples of all lower secondary-level teachers in the participating countries, we focused our analyses on mathematics teachers. Next, we examine the association between teachers' instructional practices and feedback types in connection with professional development and other teacher factors. In particular, this chapter compares and contrasts the extent to which the effect of teacher evaluation and feedback on math instruction is mediated by teachers' professional development learning, collaboration with peers, teacher beliefs about constructive pedagogy, and teacher–student relationships in the examined countries. This research also highlights the challenges of using data from international cross-sectional comparative studies to link teachers' instructional practices with feedback types and professional development.

In the next section, we first define the characteristics of good teaching that were used to measure teaching practices in the four countries, and then present a literature review on the ways in which teacher evaluation and professional development affect teachers' instructional practice.

What Constitutes Good Teaching Practices in the Teaching of Mathematics

Many US-based and international research studies have demonstrated that the quality of instruction is fundamental to student learning (Rivkin, Hanushek, & Kain, 2005; Schmidt, McKnight, & Raizen, 1997). Prior research has identified several empirically validated, effective strategies for promoting student understanding and fluency in mathematics, including motivating students with meaningful contexts and visual representation (Gersten & Clarke, 2007), using cognitively engaging or challenging instruction (Stigler & Hiebert, 2004), providing students with clear learning objectives and probing questions (Stein, Grover, & Henningsen, 1996), providing students with opportunities to explain and discuss alternative strategies (Hamilton & Martínez, 2007), using interactive or hands-on teaching practices (Baker, Gersten, & Lee, 2002), and using a variety of assessment strategies including formative assessment (Gersten et al., 2009). This research points to several characteristics of high-quality mathematics instruction. For example, in effective classrooms, teachers:

- Get students to believe they can do well in schoolwork and value learning mathematics.
- Incorporate students' backgrounds, interests, real-life situations, and meaningful contexts in instruction to motivate students who show low interest in schoolwork.
- Create lessons that allow students to think critically and participate in activities in which they understand that learning is a process and mistakes are a natural part of the learning.
- Link mathematics activities and problems to students' prior learning experiences and understanding, using multiple representations, relevant examples, and clear explanations.
- Teach students to express their understanding of how mathematical concepts and key ideas are connected, and provide alternative explanations when students are confused.
- Implement alternative instructional strategies by making connections between and among disciplines, and show how mathematics is a part of other major subjects by working with other teachers.

- Use probing questions to foster student understanding with a variety of assessment strategies.

However, the effectiveness of these teaching approaches depends on the quality of the teachers (Grouws, 2004). For example, small-group instruction will benefit students only if a teacher knows when and how to use this teaching practice. To develop students' mathematical skills effectively, teachers must themselves be effective, which requires continued learning opportunities (i.e., professional development) and teacher appraisal and feedback. Building on prior research, we identified several TALIS survey items that would capture the characteristics of good teaching practices, as shown in Appendix 1. We then measured the extent to which teachers create such teaching practices, and explored links between mathematics instructional practices, teacher feedback types, professional development, and other teacher factors in the four countries.

Classroom Instruction, Professional Development, and Teacher Evaluation

Highly effective mathematics instruction that is central to raising student achievement begins with highly qualified teachers, but prior research has documented the challenges teachers face in creating cognitively demanding instruction (e.g., Stein et al., 1996; Stigler & Hiebert, 2004). This is partly attributable to teachers' limited knowledge of teaching mathematics (e.g., Ball, Hill, & Bass, 2005; Hill et al., 2008; Ma, 1999; Son, 2013). Accordingly, professional development, teacher appraisal, and subsequent teacher feedback have been used as essential mechanisms for raising teacher quality and improving teaching practices (Ball & Cohen, 1999; Cocoran, Shields, & Zucker, 1998; Darling-Hammond & McLaughlin, 1995; Youngs, 2013).

Links Between Professional Development and Student Achievement

Providing continuous professional learning opportunities to teachers is considered one of the critical ways to improve their instructional practices, and thus improve students' learning. The effect of professional development on student achievement is mediated through the following steps: "first, professional development enhances teacher knowledge and skills; second, better knowledge and skills improve classroom teaching; and third, improved teaching raises student achievement" (Yoon et al., 2007, p. 4). Research has shown that professional development plays an important role in changing teachers' practices and that these changes have a positive impact on students' learning (Borko & Putnam, 1995; Desimone, Porter,

Garet, Yoon, & Birman, 2002; Supovitz, Mayer, & Kahle, 2000). These studies suggest that effective professional development on improving teachers' knowledge and instruction tends to focus on enhancing specific subject-related content and pedagogical knowledge, understanding how students acquire specific content knowledge and skills as well as the specific learning difficulties they may encounter, providing sufficient time for significant learning and mentoring support, and organizing collective participation from the same school (Cohen & Hill, 2000; Darling-Hammond, Wei, Richardson, Andree, & Orphanos, 2009; Desimone et al., 2002; Garet, Porter, Desimone, Birman, & Yoon, 2001).

Research on professional development in mathematics also provides supporting evidence for a positive connection between professional development and student outcomes. Prior research found that, compared to programs concerned with teacher behaviors that can be applied to all subjects, programs focusing on teachers' knowledge of a specific subject and the curriculum have larger positive effects on student achievement outcomes (Kennedy, 1998; Yoon et al., 2007). They also found that programs deepening teachers' understanding of how students learn the subject demonstrated larger influences on student learning than did programs simply offering courses in mathematics (Kennedy, 1998; Yoon et al., 2007). Furthermore, prior research has shown that professional development is most effective when it offers sustained, active engagement to improve instruction and student achievement in collaborative professional communities (Darling-Hammond et al., 2009). In countries like Japan and China, teachers routinely work with their colleagues on developing curriculum, observing one another's teaching, participating in study groups, and conducting research on teaching (Darling-Hammond, 1997). Evidence in the United States has suggested that teacher collaboration likewise influences classroom practices and contributes to improved student learning (Supovitz, Sirinides, & May, 2010).

However, some studies have reported no impact of professional development on teachers' instructional practices or student achievement (Isenberg et al., 2009; Yoon et al., 2007). Although experimental or quasi-experimental studies examining the effects of professional development found increased teacher knowledge and classroom practice promoted by professional development interventions, these rigorous studies did not find that this teacher knowledge resulted in gains in student learning outcomes or sustainable changes in practice over time (Borman, Gamoran, & Bowdon, 2008; Garet et al., 2008; Grigg, Kelly, Gamoran, & Borman, 2012). One possible way to improve the effectiveness of professional development is to incorporate teacher evaluation results (Coggshall, Rasmussen, Colton, Milton, & Jacques, 2012; Goe, Biggers, & Croft, 2012). If the results of teacher evaluations are used to provide effective professional learning opportunities for teachers, this can lead to the improvement of teaching and, consequently, student learning. Because professional development "takes place in the context of high standards, challenging curricular, system-wide accountability, and high-stakes assessments" (Yoon et al., 2007, p. 4), it is important to investigate how teacher evaluation interacts with professional development in improving teachers' instructional practices.

Links Between Teacher Evaluation and Student Achievement

Teacher evaluation is often used both for the improvement of teaching and learning and for accountability purposes (Baratz-Snowden, 2009). The Race to the Top Fund, part of the American Recovery and Reinvestment Act, specifically includes teacher evaluation as a key element of its approach to improve teacher quality. With the recent Common Core State Standards Initiative, increasing attention is being paid to the effects of teacher evaluations on teacher improvement and student learning. In particular, international comparative research has extended its focus from how education systems prepare and support a high-quality teaching force through rigorous recruitment policies and professional development, to how they support effective teaching through appraisal and feedback (OECD, 2009, 2014a).

A recent international study found between- and within-country differences in teacher appraisal and feedback (OECD, 2009). Using data from the TALIS 2008, which was administered through surveys of nationally representative samples of all lower secondary-level teachers across all subjects, the report showed that of the various aspects of teacher appraisal and feedback, the greatest emphasis was placed on relations with students in TALIS-participating countries (OECD, 2009). This was followed by teacher knowledge and understanding of instructional practice, and then classroom management. Another area of relatively high importance in teacher evaluation was direct appraisal of classroom teaching. Comparatively less importance was placed on teaching students with special needs, the retention and pass rates of students, and teaching in a multicultural setting in assessing teaching and teachers' work. Of the four countries analyzed in this chapter, only Korea participated in the TALIS 2008 study. On average, about 80% of lower secondary teachers in TALIS 2008-participating countries reported that emphasis was placed on knowledge of their subject when they received feedback, whereas about 64% of teachers in Korea reported such emphasis. In Korea, the greatest emphasis was placed on classroom management in assessing teaching and teachers' work (75%); this was followed by relations with students (70%) and direct appraisal of classroom teaching (68%).

A more recent international report (OECD, 2014a) showed that there was a change in the focus of teacher feedback between 2008 and 2013. In TALIS 2013-participating countries, about 88% of teachers reported that the greatest emphasis was placed on student performance, whereas about 67% of teachers reported a strong emphasis on student performance in TALIS 2008 countries. In TALIS 2013, about 87% of teachers reported that the feedback they received emphasized student behavior and classroom management, along with pedagogical competencies in teaching the subject field(s). This was followed by student assessment practice (83%) and knowledge and understanding of the subject fields (83%). In the United States, the greatest emphasis was placed on student performance in assessing teachers' work (92%); this was followed by student behavior and classroom management (82%) and student assessment practices (81%).

Prior research also showed that appraisal and feedback have a positive impact on teaching (OECD, 2009). Using data from the TALIS 2008, the OECD (2009) study conducted descriptive analyses on the percentage of lower secondary school teachers who reported that the appraisal and feedback they received led directly to changes in their practices. Across TALIS-participating countries, the researchers found that emphasis placed on student test scores in teacher appraisal and feedback had the greatest impact on teachers' emphasizing improved student test scores in their teaching. In addition, this study found impacts of appraisal and feedback on classroom management practices, understanding of instructional practices, knowledge, and a teacher development or training plan to improve teaching. In Korea, teachers reported that the appraisal and feedback they received led to the greatest changes in establishing a teacher development or training plan to improve teaching. About half of Korean teachers also reported that the appraisal and feedback they received led to moderate or large changes in their knowledge or understanding of instructional practices (48%) and in student discipline and behavior problems (47%).

To better understand the link between teacher evaluation and teachers' work, the TALIS study conducted a path analysis (OECD, 2009). In all TALIS 2008-participating countries, this study found that more changes take place in teachers' knowledge and understanding of their main subject fields and instructional practices when greater emphasis is placed on these areas in teacher appraisal and feedback. The findings from this study are informative. However, prior research did not focus on teachers' specific instructional practices, such as how to pose questions and use assessment strategies. Moreover, it is unclear whether these findings remain consistent across different subject fields. Although associations between teacher appraisal and feedback and teaching practices overlap across subjects, some practices may be specific to particular subjects. Given the importance of professional development in improving teacher quality in some countries, it is important to clarify the role of appraisal and feedback not only in identifying development needs, but also in assessing the relative influence of teacher appraisal and professional development on instruction.

Teacher Feedback Systems and Professional Development in the Four Countries

We acknowledge existing differences in teacher feedback systems and professional development in the four countries. While the United States has no uniform policy on the organization of professional development and teacher evaluation, Japan, Korea, and Singapore provide highly structured professional development, organized at the national level, throughout teachers' professional lives (Akiba & LeTendre, 2009). In the United States, policies concerning professional development and teacher evaluation vary by school district (i.e., local municipal education agencies) (OECD, 2005). Although teacher evaluation has become a central driving educational reform in recent years, approaches to teacher evaluations also vary across states in the United States (Doherty & Jacobs, 2013; Hull, 2013).

In contrast, in-service teacher training is structured and centralized in Korea (OECD, 2005). The in-service training aimed at professional development for currently employed teachers can be divided into: (a) new teacher training, (b) qualification training for higher teacher certificates (e.g., from a level 2 teacher certificate to a level 1 teacher certificate,[1] or from vice principal to principal), and (c) general training for professional development. Recently, the Korean Ministry of Education announced a new teacher evaluation system that is required for all teachers. The old evaluation system relied on principals' judgement of teacher performance, whereas the new system consists of peer evaluations (from a panel of a principal, a master teacher, and at least three other teachers), student surveys (grades 4–12), and parent surveys. These multiple evaluators assess teachers' instructional practices and student advisement. Teachers who receive low ratings are required to attend an intensive program of professional development at a designated teacher training institution or the National Training Institute of Education, Science, and Technology.

Similarly, in Singapore, the Ministry of Education developed a comprehensive human resource system, including the recruitment of academically talented students to the teaching profession, coherent training, and ongoing support to improve teacher quality (Steward, 2012). All teachers in Singapore are required to take at least 100 h of professional development per year to improve their teaching practices (OECD, 2011). Using the Enhanced Performance Management System (EPMS), Singapore clearly articulated expectations for teachers in their chosen fields of excellence by level (e.g., beginning teacher, senior teacher, and master teacher) (Kaur, 2010). To be fair, teacher performance is evaluated relative to a teacher's substantive level. Performance is appraised annually by a number of people (including the teachers' supervisors) on student performance outcomes as well as child character development, collaboration with parents, and contribution to their colleagues and the school community (OECD, 2010).

In Japan, the Ministry of Education requires all first-year and tenth-year teachers to participate in required professional development activities at the school and prefectural education centers (Shimahara, 2002). Each prefecture (equivalent to a "state" in the USA) has a board of education, which is responsible for actually conducting these activities. Some of the professional development takes place under the guidance of administrators or "master teachers." As part of the required professional development, a teacher may be asked to teach a research lesson that is observed by other teachers from the same school. As for teacher appraisal

[1]University graduates who complete teacher training courses are conferred the level 2 teacher certificate through a non-examination authorization procedure. Level 2 teachers who work for 3 years are entitled to obtain the level 1 teacher certificate by taking a required in-service training course.

(evaluation), it is typically done by a principal with the help of assistant principals and master teachers in the school.

Indeed, the aforementioned differences in the organization of professional development and teacher evaluation can cause challenges when using data from international cross-sectional comparative studies to link teachers' instructional practices with feedback types and professional development. Yet we view such differences as opportunities to learn, by finding better ways of improving teaching practices and student learning. In particular, prior research conducted nationally and internationally has shown that both professional development and teacher appraisal and feedback have a positive impact on teaching, but there is little consensus on the specific teacher evaluation and feedback types consistently associated with instructional practices (OECD, 2009). Thus, it would be informative to analyze differences in the associations between teacher feedback, professional development, and math instruction within and across countries.

Research Questions

The purpose of this study was to examine the degree to which different types of teacher feedback are associated with teachers' instructional practices in four countries: Japan, Korea, Singapore, and the United States. The study specifically focused on four types of feedback: (a) emphasis placed on student performance level, (b) knowledge and understanding of subject field(s), (c) collaboration or work with other teachers, and (d) student assessment practices. Several research questions guided this study:

First, to what extent are the four types of teacher feedback associated with math teachers' instructional practices, after controlling for other factors known to be related to teaching practices (i.e., teacher professional development, teacher–student relationships, teacher collaboration, and teacher beliefs about constructive pedagogy) and teacher backgrounds (i.e., gender and years of teaching experience)?

Second, what types of feedback are associated with good teaching practices across Korea, Japan, Singapore, and the United States, when we take into account differences in teacher characteristics in the four countries?

Third, when there is a positive association between teacher feedback and mathematics instruction, what factors mediate the impact of teacher feedback types on teaching practices? Are the mediating factors between feedback and teaching practices consistent across the four countries?

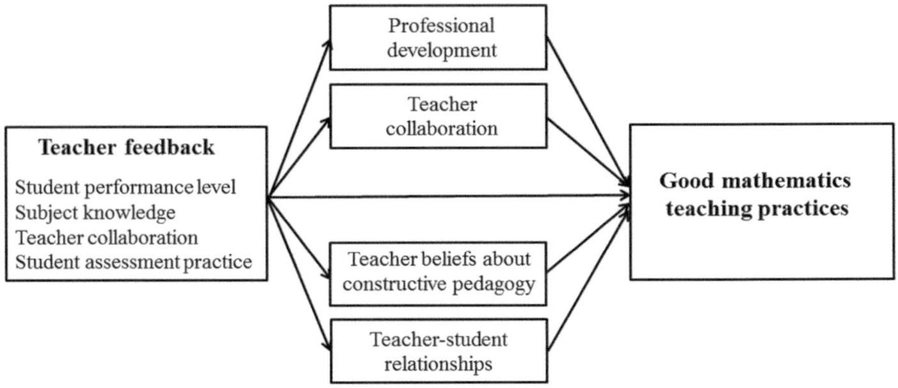

Fig. 19.1 Conceptual framework for the study. Note: The proposed framework controls for teacher characteristics (i.e., gender and teaching experiences)

Study Design and Conceptual Framework

In this study, we explored how teacher feedback is associated with teaching practices. We investigated two different pathways: direct and indirect. In addition to the direct effect of teacher feedback on teaching, teacher feedback can affect teaching practices *indirectly* through several factors. This is called "the Mediation Effect." Figure 19.1 illustrates the conceptual framework used in this study. We hypothesized that teacher feedback can affect teaching practices indirectly through four different factors: professional development, teacher collaboration, teacher beliefs about constructive pedagogy, and teacher–student relationships. This framework allows researchers to compare the direct effect of teacher feedback on teaching practice with the indirect effects of teacher feedback through these four factors. In addition, researchers can examine which factors are most significant in mediating between teacher feedback and teaching. For example, they can assess whether professional development plays a more important role than teacher collaboration in mediating the effect of teacher feedback on teaching. Note that teachers' gender and teaching experience were controlled (not shown in Fig. 19.1).

Data and Methods

The study used data from the TALIS 2013, which was administered by the OECD. Conducted on a 5-year cycle since 2008, the OECD TALIS 2013 is an international, large-scale survey that focused on the working conditions of teachers and the learning environments at lower secondary schools in 34 countries, including Japan, Korea, Singapore, and the United States. The TALIS 2013 data oversampled mathematics teachers, allowing researchers to conduct subsample analyses of nationally representative mathematics teachers in each country. In the TALIS teacher questionnaire, teachers reported the subject they taught during the school year. Using teachers' self-reported responses on the subject they taught, we limited our sample to mathematics teachers by including 455 teachers from Japan, 319 teachers from Korea, 414 teachers from Singapore, and 282 teachers from the United States. Note that, although the United States response rate on the TALIS 2013 survey did not meet the international technical standards for TALIS, the National Center for Education Statistics (NCES), which conducted an initial nonresponse bias analysis, concluded that the US TALIS data were sufficiently high-quality enough to be nationally representative of US teachers (NCES, 2014).

Dependent Variable

The outcome measure for this study was the extent to which teachers create good mathematics teaching practices, based on self-report teacher surveys. Teachers were asked to indicate the extent to which they can implement the following teaching practices on a four-point Likert scale (i.e., not at all, to some extent, quite a bit, and a lot): promotion of student motivation and academic engagement, posing good questions for students, development of student critical thinking, implementation of alternative instructional strategies, and use of a variety of assessment strategies. A detailed description of the outcome measure is presented in Appendix 1. Each teacher's responses to eight items were accumulated and then standardized to create an index for each teacher's teaching practices. Several studies highlighted the presence of systematic cross-cultural differences in response style on Likert-type data (Buckley, 2009; Chen, Lee, & Stevenson, 1995). Therefore, we created separate measures of teaching practice in each country, indicating that the reference group for each teacher's responses was fellow teachers in the same country. This suggests that the level of teacher instructional practices cannot be compared across four countries. We applied this approach to all our independent and control variables that are discussed below.

Independent Variable

The key independent variables were four different types of teacher feedback: (a) emphasis placed on student performance level, (b) knowledge and understanding of subject field(s), (c) collaboration or work with other teachers, and (d) student assessment practices (see Appendix 1 for details). Participants were asked to indicate whether the emphasis was placed on these elements when they received each type of feedback. Similar to the teaching practice measure, each teacher's response to the four feedback types was standardized to create an index for teacher feedback. That is, we created four separate measures of teacher feedback types in each country. We included the four independent variables separately in regression and multiple mediator analyses: one model that included only feedback on student performance level, one model that included feedback on knowledge, one model that included feedback on teacher collaboration, and one model that included feedback on student assessment practices.

Other Control Variables

We also included four variables found to be related to good mathematics teaching practices: (a) professional development, (b) teacher collaboration, (c) teachers' beliefs about constructive pedagogy, and (d) teacher–student relationships. Note that teacher collaboration as a control variable indicates the degree to which teachers collaborated with their colleagues, whereas feedback on teacher collaboration indicates the degree to which emphasis was placed on teacher collaboration when the teacher received feedback. We simply accumulated individual scores on the items and then standardized them. A detailed description of the measures is presented in Appendix 1.

We also included two teacher background variables, gender and years of teaching experience, as control variables. Appendix 2 shows descriptive statistics of the variables used in our statistical models for Japan, Korea, Singapore, and the United States. Note that all continuous variables (including an outcome measure and key independent variables) were standardized (mean = 0, SD = 1) before we ran regression analyses. Consequently, the coefficients can be interpreted as standardized regression coefficients. We checked correlations among dependent, independent, and control variables by each country. In Korea, Singapore, and the United States, the correlation coefficients ranged from −0.20 to 0.34. In Japan, all variables used in the analyses showed low to modest correlations (i.e., −0.20 to 0.48). Specifically, the correlation between two mediator variables, professional development participation and teacher collaboration, was 0.48. Thus, we assessed multicollinearity by using the variance inflation factor (VIF) test (Zurr, Ieno, & Elphick, 2010). The results provided no evidence that all variables in the Japan dataset have multicollinearity.

Analytic Strategy

We first conducted descriptive statistics analyses to show between-country differences in good mathematics teaching practices and focus on teacher feedback among mathematics teachers. To obtain the estimates for mathematics teachers, we selected mathematics teachers and used the final teacher weight variable (TCHWGT). We also used balanced replication weights (BRR) to take into account the clustered nature of the TALIS data and obtain unbiased estimates for standard errors (OECD, 2014b).

To examine the degree to which the four different types of teacher feedback were associated with mathematics teachers' instructional practices, we used an ordinary least-squares (OLS) regression model for each country. In Model 1, we examined the association between teacher feedback and teaching practices, without controlling for any other factors or teachers' backgrounds. In Model 2, we included four factors (i.e., professional development, teacher–student relationships, teachers' beliefs about constructive pedagogy, and teacher collaboration) and the teacher background variables in addition to the feedback variables. The results of Model 2 showed whether teacher feedback was linked to good teaching practice, after taking into account other factors such as professional development and collaboration. To examine the variation in the association between feedback types and teaching across the four countries, we ran models separately by the four different types of feedback as well as by the countries. The OLS model is as follows:

$$S_i = \beta_0 + \beta_t X_i + \beta_c C_i + e,$$

where S_i represents a teaching practice score of an individual teacher i, X_i indicates a teacher feedback factor, C_i indicates a vector of control variables related to an individual teacher i, and e is a random error term.

If there was a positive association between the types of teacher feedback and teaching, we wanted to explore factors that mediate this positive association. Because the purpose of our study was exploratory, we employed a multiple mediator model analysis with control variables. This method is useful to assess possible pathways from a predictor to a dependent variable indirectly through multiple simultaneous mediators. For the multiple mediation analysis we utilized the MEDIATE macro, which is a regression-based path analysis macro for SPSS that has the advantage of providing inferential statistical methods for estimating indirect effects, including bootstrap confidence levels. This method of bootstrap confidence interval (CI) has the advantage that it does not impose the assumption of normality of the sampling distribution when testing mediation (Preacher & Hayes, 2008). Using this method, we explored the degree to which the association between the type of feedback and teaching practice was mediated by professional development, teacher–student relationships, teachers' beliefs about constructive pedagogy, and teacher collaboration simultaneously, after controlling for teacher backgrounds. Thus, the indirect effects of each teacher feedback type on the degree of good teaching practice were estimated as the product of the coefficients for the variables linking "teacher feedback" to good teaching practice through four specific mediators.

Findings

We began by examining between-country differences in mathematics teaching practices and the emphasis placed on the four different types of feedback teachers received. Next, we present the regression analysis results for the association between teacher feedback and good mathematics teaching practices, after taking into account the other four factors linked to instructional practices, such as PD and collaboration. Finally, we present the multiple mediated analysis results, to examine factors that mediate the impacts of teacher feedback on good mathematics teaching practices in the four countries.

Teacher Reports on Mathematics Classroom Instruction and the Focus of Teacher Feedback

Figure 19.2 presents the average composite scores on good teaching practices. Eight items were used to measure the extent to which teachers can implement teaching practices aligned with the recommendations, on a four-point scale. Thus, composite scores range from 8 (one on all eight items) to 32 (four on all eight items). Figure 19.2 shows the differences in teaching practices across the four countries. However, we cannot compare the level of good teaching practices in the United States to teaching practices in Japan, Korea, and Singapore because there are cross-cultural differences in response styles on Likert type of survey items. Nevertheless, we found that the Korean, Singaporean, and the US data showed similar standard deviation values in the good teaching practice index scores, which

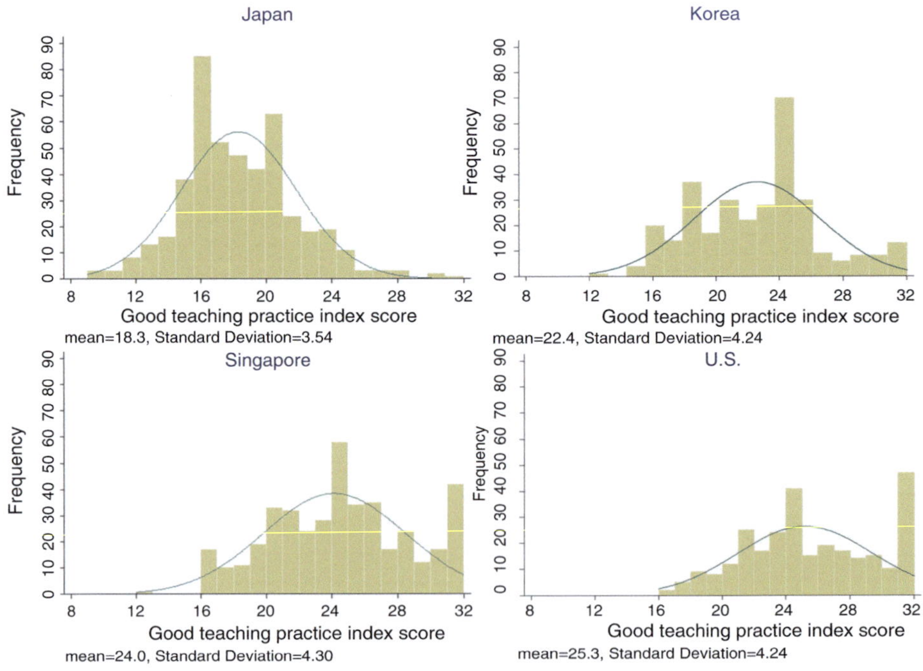

Fig. 19.2 Distribution of good teaching practice index score

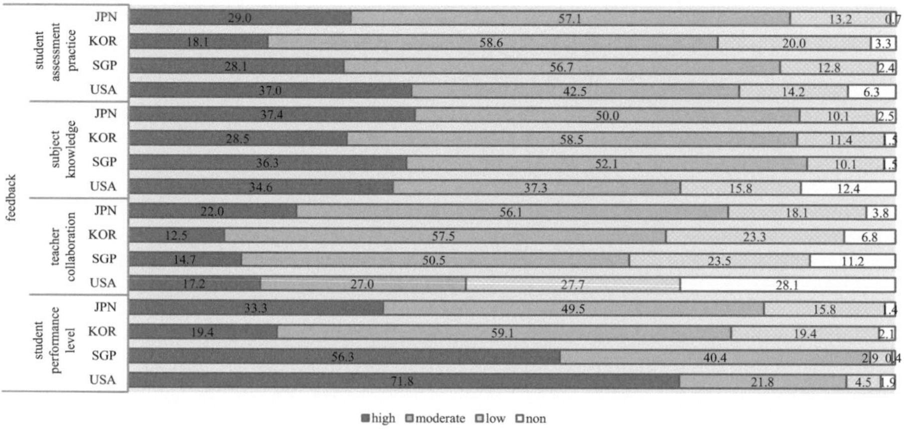

Fig. 19.3 Emphasis in feedback

ranged from 4.24 to 4.30. The Japanese dataset showed a 3.54-standard-deviation value, which is moderately lower than the other three countries. This shows that the variations of good teaching practice index scores are similar within countries.

Next, we examined between-country differences in the emphasis in received feedback. As shown in Fig. 19.3, there were several noteworthy trends in teacher feedback practices across the countries. First, on average, about 45% of mathematics teachers across the four countries reported a high importance placed on student performance levels in the feedback they received. However, feedback on student performance showed substantial variation across the four countries: about 72% of mathematics teachers in the United States reported a high importance placed on student performance levels in the feedback they received, whereas only about 20% of Korean mathematics teachers reported a similar emphasis. Singapore (56%) and Japan (35%) fell between these two extremes.

The percentage of teacher feedback emphasizing teacher collaboration also showed substantial differences across the four countries. On average, about 64% of mathematics teachers reported a moderate or high importance placed on teacher collaboration in the feedback they received. In Japan, about 78% of mathematics teachers reported that a moderate or high importance was placed on teacher collaboration, while about 44% of mathematics teachers in the United States reported a moderate or high importance. Feedback emphasizing teacher subject knowledge showed different patterns between top-performing Asian countries and the United States. In Japan, Korea, and Singapore, about 87% or 88% of mathematics teachers reported a moderate or high importance placed on subject knowledge in the feedback they received, whereas about 72% of mathematics teachers in the United States reported a moderate or high importance of subject knowledge.

There were no substantial differences in feedback that emphasized student assessment practices across the four countries. On average, about 82% of mathematics teachers reported a moderate or high importance placed on student assessment practices in the feedback they received. In Japan, about 86% of mathematics teachers reported that a moderate or high importance placed was placed on student assessment practices, while about 77% of mathematics teachers in Korea reported a moderate or high importance of student assessment practices. Mathematics teachers in Singapore (85%) and the United States (80%) fell between these two points.

In sum, Japanese and Korean mathematics teachers tended to report that strong emphases were placed on all four different types of feedback, with the highest emphasis on subject knowledge. In the United States and Singapore, mathematics teachers reported that the highest level of emphasis was placed on student performance level in the feedback they received, whereas the lowest emphasis was placed on teacher collaboration. However, the level of emphasis placed on teacher collaboration was much higher in Singapore than in the United States: 44% of teachers in the United States versus 65% of teachers in Singapore reported a moderate or high importance placed on teacher collaboration. In the next section, we present the degree to which feedback types were associated with teaching practices in the four countries.

Results from Regression Analyses

Table 19.1 presents the results from a series of OLS regression analyses that examined the degree to which feedback types were associated with teaching practices. The null model included only a feedback variable, and the full model included the feedback variable as well as other factors known to be related to good teaching practices (i.e., teacher professional development, teacher–student relationships, teacher collaboration, and teacher beliefs about constructive pedagogy) and teacher backgrounds (i.e., gender and years of teaching experience).

The results revealed positive associations between teacher feedback types and teaching practices across the four countries. As shown in the null models in Table 19.1, the emphases of the feedback placed on knowledge and understanding of subject field(s), collaboration or work with other teachers, and feedback on student assessment practices were found to be positively associated with good teaching practices in all four countries. However, the association between feedback on student performance level and teaching practices varied from country to country. While there was a positive association between teacher feedback on performance level and teaching practices in Japan and Singapore, no such association existed in Korea or the United States.

In addition, we found that positive associations between teacher feedback types and teaching practices remained even after controlling for other factors known to be related to teaching practices (see full models in Table 19.1). In the United States, feedback emphases on teacher knowledge, collaboration, and assessment practices were positively associated with teaching practices, whereas an emphasis placed on student performance level was not linked to teaching practices. In the United States, for example, a 1-standard-deviation increase of feedback emphasizing teacher collaboration was linked to a 0.27-standard-deviation increase in teaching

Table 19.1 OLS regression results: association between teacher feedback and teaching practices across the four countries

		Feedback on knowledge				Feedback on collaboration				Feedback on performance level				Feedback on assessment practice			
		Null model		Full model		Null model		Full model		Null model		Full model		Null mode		Full model	
		B	S.E	*B*	S.E	*B*	S.E	*B*	S.E	*B*	S.E	*B*	S.E	*B*	S.E1	*B*	S.E
Japan	Corresponding feedback	.15*	(.06)	.10	(.06)	.26**	(.05)	.12†	(.06)	.09†	(.05)	.03	(.05)	.22**	(.05)	.11†	(.05)
	Professional development			.02	(.06)			.02	(.06)			.04	(.06)			.01	(.06)
	Constructive belief			.19**	(.07)			.20**	(.07)			.20**	(.06)			.19**	(.05)
	Teacher collaboration			.23**	(.05)			.22**	(.06)			.25**	(.06)			.23**	(.05)
	Teacher-student relationships			.11	(.08)			.10	(.07)			.09	(.07)			.12	(.05)
Korea	Corresponding feedback	.14†	(.07)	-.01	(.07)	.21**	(.06)	.08	(.06)	.02	(.07)	-.03	(.05)	.17*	(.06)	.09	(.06)
	Professional development			.19**	(.06)			.17**	(.06)			.19**	(.06)			.19**	(.06)
	Constructive belief			.07	(.05)			.06	(.05)			.07	(.05)			.05	(.07)
	Teacher collaboration			.23**	(.07)			.22**	(.07)			.23**	(.07)			.22**	(.07)
	Teacher-student relationships			.17**	(.06)			.16**	(.06)			.17**	(.06)			.15*	(.07)
Singapore	Corresponding feedback	.14**	(.05)	.06	(.05)	.23**	(.05)	.14**	(.05)	.19**	(.05)	.10*	(.05)	.21**	(.05)	.14**	(.06)
	Professional development			.02	(.06)			.01	(.06)			.20	(.06)			.00	(.06)
	Constructive belief			.07	(.05)			.09†	(.05)			.08†	(.05)			.10	(.06)
	Teacher collaboration			.04	(.05)			.01	(.06)			.04	(.05)			.04	(.05)
	Teacher-student relationships			.31**	(.05)			.30**	(.05)			.30**	(.05)			.30*	(.08)
USA	Corresponding feedback	.18*	(.07)	.13†	(.07)	.32**	(.08)	.27**	(.09)	.08	(.06)	.09	(.06)	.33**	(.06)	.26**	(.06)
	Professional development			.05	(.07)			.072	(.07)			.07	(.07)			.05	(.05)
	Constructive belief			.16*	(.07)			.13†	(.07)			.13†	(.07)			.14*	(.05)
	Teacher collaboration			.18*	(.08)			.13†	(.08)			.21**	(.07)			.17*	(.07)
	Teacher-student relationships			.19**	(.07)			.20**	(.07)			.20**	(.07)			.17*	(.06)

Note: Teacher gender and teaching year are controlled. *B* = coefficient, *S.E.*= standard errors. $^{\dagger}p < 0.10$, $^{*}p < 0.05$, $^{**}p < 0.01$. The results were derived from BRR weighting. *Highlighted cells with grey color* indicate that the association between feedback and teaching practice is statistically significant, even after controlling for four factors that are known to be associated with teaching practices

practices, and a 1-standard-deviation increase of feedback on student assessment practices was associated with a 0.26-standard-deviation increase in teaching practices. Given that our estimated coefficients are standardized, these results suggest that in the United States feedback on collaboration and student assessment practices had stronger associations with good mathematics teaching practices than feedback on teacher knowledge. We found the same patterns in other countries; these two types of feedback were also positively linked to teachers' good teaching practices in Japan and Singapore.

In addition to feedback on collaboration and student assessment practices, in Singapore, we found a positive association between an emphasis on student performance level and teaching practices. In Singapore, a 1-standard-deviation increase in feedback emphasis on student performance level was linked to a 0.10-standard-deviation increase in teaching practices. Compared to the other three countries, Korea was exceptional in that none of the feedback types were associated with teaching practices after controlling for other teacher factors (i.e., teacher professional development, teacher–student relationships, teacher collaboration, and teacher beliefs about constructive pedagogy).

Furthermore, it is important to note that there was no common feedback type positively associated with teaching practices across the four countries. It is also noteworthy that the association between feedback types and teaching practices varied across countries. Only in the Unites States was there a positive association between an emphasis on teacher knowledge and teaching practices, while there was no association between them in the other three countries. In the United States, a 1-standard-deviation increase in feedback emphasizing teacher knowledge was linked to a 0.13-standard-deviation increase in teaching practices. Only in Singapore, we found a positive association between an emphasis on student performance level and teaching practices, whereas there was no association between them in Japan, Korea, or the United States. Feedback emphasizing both collaboration and assessment practices was positively associated with teaching practices in Japan, Singapore, and the United States.

Korea showed a very different pattern in the association between teacher feedback types and teaching practices. In the null model that included only one teacher feedback variable, emphasis placed on knowledge and understanding of subject field(s), collaboration or work with other teachers, and student assessment practices were positively associated with teaching practices. However, in the full model that included other factors known to be related to teaching practices, none of the teacher feedback types were associated with teaching practices, while professional development, teacher collaboration, and teacher–student relationships were positively associated with teaching practices. This finding suggests that the positive association between teacher feedback and teaching practices in Korea might be mediated by other factors such as professional development and teacher collaboration. This finding is consistent with the design of teacher feedback that provides teachers with opportunities for professional development plans (Seo, 2012). In the next section, we report the factors that mediate the positive associations between teacher feedback types and teaching practices.

Results from Multiple Mediator Model Analyses with Control Variables

To explore the factors that mediate positive associations between teacher feedback types and teaching practices and to evaluate the mediated pathways, we conducted multiple mediator model analyses with control variables (i.e., teacher professional development, teacher–student relationships, teacher collaboration, and teacher beliefs about constructive pedagogy). We report the full results in Appendix 3. Given the importance of teacher knowledge for teaching and student learning (Hill, Rowan, & Ball, 2005), the example of feedback on teacher knowledge was chosen to visualize these results (see Fig. 19.4).

As shown in Fig. 19.4, we found that factors mediating positive associations between teacher feedback types and teaching practices varied across countries. In the United States, the effect of emphasis placed on knowledge and understanding of subject field(s) on teaching practices was mediated indirectly only by teacher collaboration; the other three mediators—teacher–student relationships, professional development, and teacher beliefs about constructive pedagogy—did not mediate the association between teacher feedback and teaching practices. We also found that teacher collaboration plays an important role in mediating between other types of teacher feedback and teaching practices in the United States; the effects of feedback with emphasis placed on collaboration and student assessment practices on good mathematics teaching practice were also mediated by teacher collaboration (see Appendix 3).

In Korea, the effect of feedback emphasizing knowledge and understanding of subject field(s) on teaching practices was mediated indirectly by three variables: teacher–student relationships, professional development, and teacher collaboration. In particular, professional development was the strongest mediator between teacher feedback on subject knowledge and teaching practices in Korea. This finding may suggest that, unlike in the United States, professional development is an important factor that mediates between teacher feedback and teaching practices in Korea. We also found the same pattern of mediation in other types of teacher feedback; in Korea, three variables (i.e., teacher–student relationships, professional development, and teacher collaboration) were important factors that mediate between good teaching practices and feedback on teacher collaboration and student assessment practices (see Appendix 3).

The other two Asian countries, Japan and Singapore, showed different patterns of mediation from the Korean pattern. In Singapore, the effect of feedback stressing knowledge and understanding of subject field(s) on teaching practices was mediated indirectly only by teacher–student relationships; the other three variables did not mediate the association. The same pattern of mediation processes emerged for the other three types of feedback: teacher collaboration, student performance level, and assessment practices (see Appendix 3). In Japan, the effect of emphasis on knowledge and understanding of subject field(s) on teaching practices was mediated indirectly by teacher–student relationships, teacher beliefs about constructive

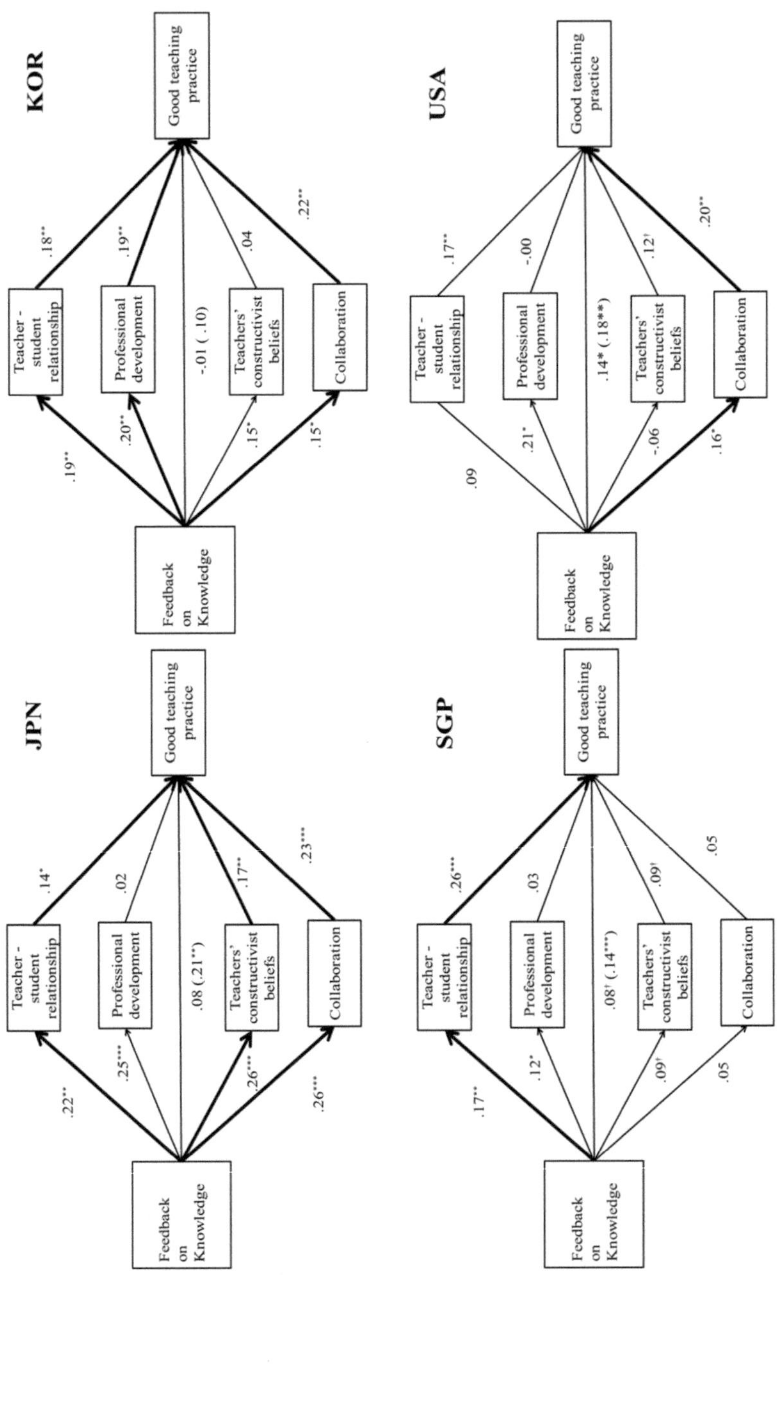

Note: † p<.10, * p<.05, ** p<.01.

Fig. 19.4 Multiple mediator analysis: emphasis of teacher feedback placed on knowledge and understanding of subject field

pedagogy, and particularly teacher collaboration. We also found the same pattern of mediation for other types of teacher feedback; in Japan, three variables (i.e., teacher–student relationships, teacher beliefs about constructive pedagogy, and teacher collaboration) were important factors that mediate between good teaching practices and feedback on teacher collaboration and student assessment practices (see Appendix 3).

In sum, we found between-country differences in factors that mediated between teacher feedback and good mathematics teaching practices: in the United States and Japan, teacher collaboration was the strongest mediator between teacher feedback and mathematics teaching practices; in Korea, professional development; and in Singapore, teacher–student relationships. We also found that the same factor mediated between different types of feedback and teaching practices within countries.

Discussion and Implications

Over the past few years, educational researchers and policymakers have become increasingly interested in teacher appraisal and feedback as a way to improve teachers' teaching practices. Recent international studies found that greater emphasis on specific aspects of teacher feedback was more likely to change teachers' practices (OECD, 2009, 2014a). However, there is a lack of empirical research on the degree to which this finding holds in different subject fields. The main focus of the present study was to assess the degree to which teacher feedback was associated with mathematics teachers' instructional practices in four countries: Japan, Korea, Singapore, and the United States. We found that, consistent with previous studies (OECD, 2009, 2014a), teacher feedback was positively associated with teaching practices, but this positive association between teacher feedback and teaching practices varied across countries. More importantly, we found that factors mediating the positive associations between teacher feedback types and teaching practices varied across countries.

In the United States, teacher collaboration was an important mediator between emphases on specific aspects of teacher feedback and teaching practices. Prior research in the United States showed that teacher collaboration can enhance teacher quality (Cochran-Smith & Lytle, 1999; Goddard, Goddard, & Tschannen-Moran, 2006; McLaughlin & Talbert, 2006). The findings of our study support the claim that it is important to improve teacher collaboration and build strong teacher learning communities in order to make teacher feedback effective, and thus to improve teachers' mathematics instruction. Given the fact that teacher collaboration varies across schools (Goddard et al., 2006), school administers and policymakers in the United States acknowledge that it is important to implement teacher feedback with strong teacher collaboration and professional learning communities, particularly at disadvantaged schools, in order to close the mathematics teaching gaps across schools.

Interestingly, in the United States, professional development was neither associated with teachers' good teaching practices nor did it mediate the effect of teacher feedback on teaching practices. In contrast, professional development in Korea was associated with teachers' good teaching practices and significantly mediated the link between feedback types and teaching practices when teacher feedback was focused on teacher knowledge. Indeed, there is a systemic difference in providing professional development between Korea and the United States. In Korea, teacher evaluation and feedback have been designed to provide opportunities for professional development at the national level. However, in the United States, there is no strong institutionalized link between teacher feedback and professional development. This finding seems to suggest that policymakers and school administrators in the United States need to make a systematic effort to coordinate teacher evaluation and feedback with teachers' professional development.

In addition to professional development and teacher collaboration, teacher–student relationships were found be an important factor that mediates the positive associations between teacher feedback types and teaching practices in Korea. This was also an important mediating factor in Singapore. In Singapore's case, there was a positive association between feedback on teacher knowledge and professional development (see Fig. 19.4). However, similar to the United States, there was no significant association between professional development and good teaching practices in Singapore. This finding suggests that policymakers and school administrators in Singapore need to reconsider teachers' professional development and find a better way to link teacher evaluation and feedback with teachers' professional development systematically, which can lead to quality teaching.

In Japan, three factors—teacher–student relationships, teachers' constructivist beliefs, and collaboration among teachers—were identified as mediating between feedback on teacher knowledge and good teaching. However, similar to the United States and Singapore, no significant association existed between professional development and good teaching practices based on teacher reports. This finding also calls for policymakers and school administrators in Japan to reconsider the link between teachers' professional development, teacher evaluation and feedback, and teaching practices.

Our study focused on the role of teacher feedback on mathematics teaching. Future research is needed to examine whether there is a differential effect of teacher feedback on mathematics teaching across the distribution of teaching experience. In particular, future research needs to examine whether specific types of teacher feedback are effective at improving inexperienced teachers who have less than 3–5 years teaching experience. This research will contribute to reducing the teaching gap between inexperienced and experienced teachers. Furthermore, given that this study identified mediating factors based on four different aspects of teacher feedback, future research should investigate other types of teacher feedback, as well as the factors that mediate the positive associations between teacher feedback types and teaching practices. Because improving teacher practices through teacher feedback aims to raise student learning, examining the degree to which specific types of teacher feedback are associated with student mathematics achievement across countries would be also informative.

Appendix 1

Items used to measure outcome and key independent variables

Variable name	Description	Reliability
Teacher's good teaching practices	In your teaching, to what extent can you do the following? (1) Get students to believe they can do well in schoolwork (2) Help my students value learning (3) Craft good questions for my students (4) Motivate students who show low interest in schoolwork (5) Help students think critically (6) Use a variety of assessment strategies (7) Provide an alternative explanation for example when students are confused (8) Implement alternative instructional strategies in my classroom	Japan $\alpha = 0.849$ Korea $\alpha = 0.904$ Singapore $\alpha = 0.894$ USA $\alpha = 0.878$
Teacher feedback types	In your opinion, when you receive this feedback, what is the emphasis placed on the following areas?	
(a) Student performance level	Emphasis placed on student performance level	
(b) Subject knowledge	Emphasis placed on (1) Knowledge and understanding of my subject field (2) Pedagogical competencies in teaching my subject field(s)	Japan $\alpha = 0.849$ Korea $\alpha = 0.904$ Singapore $\alpha = 0.894$ USA $\alpha = 0.878$
(c) Teacher collaboration	Emphasis placed on (1) Collaboration or working with other teachers (2) The feedback I provide to other teachers to improve their teaching	Japan $\alpha = 0.627$ Korea $\alpha = 0.695$ Singapore $\alpha = 0.584$ USA $\alpha = 0.760$
(d) Student assessment practices	Emphasis placed on student assessment practices	
Professional development	Considering the professional development activities you took part in during the last 12 months, to what extent have they included the following? (1) A group of colleagues from my school or subject group (2) Opportunities for active learning methods (not only listening to a lecturer) (3) Collaborative learning activities or research with other teachers (4) An extended time-period (several occasions spread out over several weeks or months)	Japan $\alpha = 0.623$ Korea $\alpha = 0.790$ Singapore $\alpha = 0.741$ USA $\alpha = 0.726$

(continued)

Variable name	Description	Reliability
Teacher beliefs about constructive pedagogy	We would like to ask about your personal beliefs on teaching and learning. Please indicate how strongly you agree or disagree with each of the following statements. (1) My role as a teacher is to facilitate students' own inquiry. (2) Students learn best by finding solutions to problems on their own. (3) Students should be allowed to think of solutions to practical problems themselves before the teacher shows them how they are solved. (4) Thinking and reasoning processes are more important than specific curriculum content.	Japan $\alpha = 0.801$ Korea $\alpha = 0.862$ Singapore $\alpha = 0.824$ USA $\alpha = 0.884$
Teacher collaboration	On average, how often do you do the following in this school? (1) Teach jointly as a team in the same class (2) Observe other teachers' classes and provide feedback (3) Engage in joint activities across different classes and age groups (e.g., projects) (4) Exchange teaching materials with colleagues (5) Engage in discussions about the learning development of specific students (6) Work with other teachers in my school to ensure common standards in evaluations for assessing student progress (7) Attend team conferences (8) Take part in collaborative professional learning	Japan $\alpha = 0.701$ Korea $\alpha = 0.815$ Singapore $\alpha = 0.668$ USA $\alpha = 0.748$
Teacher–student relationships	How strongly do you agree or disagree with the following statements about what happens in this school? (1) In this school, teachers and students usually get on well with each other. (2) Most teachers in this school believe that the students' well-being is important. (3) Most teachers in this school are interested in what students have to say. (4) If a student from this school needs extra assistance, the school provides it.	Japan $\alpha = 0.829$ Korea $\alpha = 0.799$ Singapore $\alpha = 0.793$ USA $\alpha = 0.890$

Appendix 2 Descriptive statistics (mean and standard deviation) of variables used in analyses

Variable	Description	Descriptive statistics			
		Japan	Korea	Singapore	USA
Sample size		455	319	414	282
Outcome variable					
Teacher's good teaching practice	Standardized person score for teachers' good teaching practice in classroom	0.00 (1.00)	0.00 (1.00)	0.00 (1.00)	0.00 (1.00)
Independent variables					
Teacher feedback type					
(a) Student performance level	Standardized person score for feedback emphasis placed on student performance level	0.00 (1.00)	0.00 (1.00)	0.00 (1.00)	0.00 (1.00)
(b) Subject knowledge	Standardized person score for feedback emphasis placed on subject knowledge	0.00 (1.00)	0.00 (1.00)	0.00 (1.00)	0.00 (1.00)
(c) Teacher collaboration	Standardized person score for feedback emphasis placed on teacher collaboration	0.00 (1.00)	0.00 (1.00)	0.00 (1.00)	0.00 (1.00)
(d) Student assessment practices	Standardized person score for feedback emphasis placed on student assessment practices	0.00 (1.00)	0.00 (1.00)	0.00 (1.00)	0.00 (1.00)
Mediator variables					
Professional development	Standardized person score for participation in professional development	0.00 (1.00)	0.00 (1.00)	0.00 (1.00)	0.00 (1.00)
Teacher beliefs about constructive pedagogy	Standardized person score for teacher beliefs about constructive pedagogy	0.00 (1.00)	0.00 (1.00)	0.00 (1.00)	0.00 (1.00)
Teacher collaboration	Standardized person score for teacher collaboration	0.00 (1.00)	0.00 (1.00)	0.00 (1.00)	0.00 (1.00)
Teacher–student relationships	Standardized person score for teacher–student relationships	0.00 (1.00)	0.00 (1.00)	0.00 (1.00)	0.00 (1.00)
Control variables					
Female	Dummy variable coded 1 if the teacher is female	0.23 (0.42)	0.68 (0.47)	0.60 (0.49)	0.61 (0.49)
Years of experience	Standardized number of years of experience the teacher has in any school	16.80 (11.00)	15.94 (9.84)	8.06 (8.39)	13.11 (9.79)

Note: weighted by BRR weighting technique. Standard deviation in parenthesis

Appendix 3 Indirect effects of emphasis placed on teacher feedbacks on teaching practices

	Japan			Korea			Singapore			United States		
		95% CI			95% CI			95% CI			95% CI	
	β	Lower	Upper	β	Lower	Upper	β	Lower	Upper	β	Lower	Upper
Panel A												
Feedback on subject knowledge												
via teacher–student relationships	0.03*	0.00	0.07	0.03*	0.01	0.08	0.04*	0.02	0.08	0.02	−0.00	0.05
via professional development	0.01	−0.02	0.04	0.04*	0.01	0.08	0.00	−0.01	0.02	0.00	−0.03	0.03
via teacher beliefs about constructive pedagogy	0.04*	0.01	0.10	0.01	−0.01	0.03	0.01	−0.00	0.03	−0.01	−0.04	0.01
via teacher collaboration	0.06*	0.03	0.11	0.03	0.01	0.08	0.00	−0.00	0.02	0.03*	0.01	0.08
Panel B												
Feedback on teacher collaboration												
via teacher–student relationships	0.03*	0.00	0.07	0.02*	0.00	0.06	0.04*	0.02	0.08	0.02	−0.00	0.05
via professional development	0.01	−0.02	0.03	0.05*	0.01	0.11	0.00	−0.01	0.02	0.00	−0.01	0.03
via teacher beliefs about constructive pedagogy	0.04*	0.01	0.09	0.01	−0.01	0.04	0.00	−0.01	0.02	0.01	−0.00	0.04
via teacher collaboration	0.07*	0.03	0.11	0.05*	0.02	0.11	0.00	−0.02	0.03	0.05*	0.01	0.10
Panel C												
Feedback on student performance level												
via teacher–student relationships	0.01	−0.00	0.04	0.02	−0.00	0.06	0.04*	0.02	0.08	0.01	−0.01	0.04
via professional development	0.00	−0.00	0.02	0.00	−0.02	0.03	0.00	−0.00	0.02	0.00	−0.02	0.02
via teacher beliefs about constructive pedagogy	0.03*	0.01	0.07	0.01	−0.01	0.03	0.00	−0.01	0.02	0.01	−0.00	0.04
via teacher collaboration	0.02	−0.01	0.06	0.00	−0.02	0.03	0.00	−0.00	0.02	0.01	−0.01	0.05
Panel D												
Feedback on student assessment practice												
via teacher–student relationships	0.02*	0.00	0.05	0.03*	0.01	0.07	0.03*	0.01	0.07	0.02*	0.00	0.07
via professional development	0.00	−0.03	0.03	0.02*	0.00	0.06	0.00	−0.02	0.02	0.00	−0.03	0.02
via teacher beliefs about constructive pedagogy	0.03*	0.01	0.08	0.01	−0.01	0.03	0.01	0.00	0.03	0.01	−0.00	0.04
via teacher collaboration	0.05*	0.02	0.10	0.03*	0.01	0.08	0.00	−0.00	0.02	0.03*	0.01	0.08

Each column in each panel reports results from one regression. All regressions controlled for teacher's gender and years of teaching experience
β = coefficient. CI = confidence interval. $*p \leq 0.05$

References

Akiba, M., & LeTendre, G. K. (2009). *Improving teacher quality: The U.S. teaching force in global context*. New York: Teachers College Press.

Baker, S., Gersten, R., & Lee, D.-S. (2002). A synthesis of empirical research on teaching mathematics to low-achieving students. *The Elementary School Journal, 103*(1), 51–73.

Ball, D. L., & Cohen, D. K. (1999). Developing practice, developing practitioners: Toward a practice-based theory of professional education. In G. Sykes & L. Darling-Hammond (Eds.), *Teaching as the learning profession: Handbook of policy and practice* (pp. 3–32). San Francisco: Jossey-Bass.

Ball, D. L., Hill, H. C., & Bass, H. (2005). Knowing mathematics for teaching: Who knows mathematics well enough to teach third grade, and how can we decide? *American Educator, 29* (1), 14–17 20–22, 43–46.

Baratz-Snowden, J. (2009). *Fixing tenure: A proposal for assuring teacher effectiveness and due process*. Washington, DC: Center for American Progress.

Borko, H., & Putnam, R. T. (1995). Expanding a teacher's knowledge base: A cognitive psychological perspective on professional development. In T. R. Guskey & A. M. Huberman (Eds.), *Professional development in education: New paradigms and practices* (pp. 35–65). New York: Teachers College Press.

Borman, G., Gamoran, A., & Bowdon, J. (2008). A randomized trial of teacher development in elementary science: First-year achievement effects. *Journal of Research on Educational Effectiveness, 1*, 237–264.

Buckley, J. (2009). *Cross-national response styles in international educational assessments: Evidence from PISA 2006*. Paper presented at the NCES Conference on the Program for International Student Assessment (PISA): What We Can Learn from PISA, Washington, DC.

Chen, C., Lee, S.-Y., & Stevenson, H. W. (1995). Response style and cross-cultural comparisons of rating scales among East Asian and North American students. *Psychological Science, 6*(3), 170–175.

Cochran-Smith, M., & Lytle, S. (1999). Teacher learning communities. *Review of Research in Education, 24*, 24–32.

Cocoran, T. B., Shields, P. M., & Zucker, A. A. (1998). *Evaluation of NSF's Statewide System Initiatives (SSI) program: The SSIs and professional development for teachers*. Menlo Park, CA: SRI International.

Coggshall, J. G., Rasmussen, C., Colton, A., Milton, J., & Jacques, C. (2012). *Generating teaching effectiveness: The role of job-embedded professional learning in teacher evaluation*. Washington, DC: National Comprehensive Center for Teacher Quality.

Cohen, D. K., & Hill, H. (2000). Instructional policy and classroom performance: The mathematics reform in California. *Teachers College Record, 102*(2), 294–343.

Darling-Hammond, L. (1997). *Doing what matters most: Investing in quality teaching*. New York: National Commission on Teaching & America's Future.

Darling-Hammond, L., & McLaughlin, M. (1995). Policies that support professional development in an era of reform. *Phi Delta Kappan, 76*(8), 597–604.

Darling-Hammond, L., Wei, R. C., Richardson, N., Andree, A., & Orphanos, S. (2009). *Professional learning in the learning profession: A status report on professional development in the U.S. and abroad*. Washington, DC: National Staff Development Council.

Darling-Hammond, L., & Youngs, P. (2002). Defining "highly qualified teachers": What does "scientifically-based research" actually tell us? *Educational Researcher, 31*(9), 13–25.

Desimone, L. M., Porter, A. C., Garet, M. S., Yoon, K. S., & Birman, B. F. (2002). Effects of professional development on teachers' instruction: Results from a three-year longitudinal study. *Educational Evaluation and Policy Analysis, 24*(2), 81–112.

Doherty, K. M., & Jacobs, S. (2013). *State of the states 2013: Connect the dots: Using evaluations of teacher effectiveness to inform policy and practice*. Washington, DC: National Council on Teacher Quality.

Garet, M. S., Cronen, S., Eaton, M., Kurki, A., Ledwig, M., Jones, W., et al. (2008). *The impact of two professional development interventions on early reading instruction and achievement*. Washington, DC: Institute of Education Sciences.

Garet, M. S., Porter, A. C., Desimone, L., Birman, B. F., & Yoon, K. S. (2001). What makes professional development effective? Results from a national sample of teachers. *American Educational Research Journal, 38*(4), 915–945.

Gersten, R., Chard, D., Jayanthi, M., Baker, S., Morphy, P., & Flojo, J. (2009). *A meta-analysis of mathematics instructional interventions for students with learning disabilities: A technical report*. Los Alamitos, CA: Instructional Research Group.

Gersten, R., & Clarke, B. S. (2007). *Effective strategies for teaching students with difficulties in mathematics*. NCTM: National Council of Teachers of Mathematics.

Goddard, Y. L., Goddard, R. D., & Tschannen-Moran, M. (2006). A theoretical and empirical investigation of teacher collaboration for school improvement and student achievement in public elementary schools. *Teachers College Record, 109*(4), 877–896.

Goe, L. (2007). *The link between teacher quality and student outcomes: A research synthesis*. Washington, DC: National Comprehensive Center for Teacher Quality.

Goe, L., Biggers, K., & Croft, A. (2012). *Linking teacher evaluation to professional development: Focusing on improving teaching and learning*. Washington, DC: National Comprehensive Center for Teacher Quality.

Grigg, J., Kelly, K. A., Gamoran, A., & Borman, G. (2012). Effects of two scientific inquiry professional development interventions on teaching practice. *Educational Evaluation and Policy Analysis, 35*(1), 38–56.

Grouws, D. (2004). Mathematics. In G. Cawelti (Ed.), *Handbook of research on improving student achievement* (pp. 160–178). Arlington, VA: Educational Research Service.

Hamilton, L. S., & Martínez, J. F. (2007). What can TIMSS surveys tell us about mathematics reforms in the United States during the 1990s. In T. Loveless (Ed.), *Lessons learned: What international assessments tell us about math achievement* (pp. 127–174). Washington, DC: Brookings Institution Press.

Hill, H. C., Blunk, M. L., Charalambos, C. Y., Lewis, J. M., Phelps, G. C., Sleep, L., et al. (2008). Mathematical knowledge for teaching and the mathematical quality of instruction: An exploratory study. *Cognition and Instruction, 26*(4), 1–81.

Hill, H. C., Rowan, B., & Ball, D. L. (2005). Effects of teachers' mathematical knowledge for teaching on student achievement. *American Educational Research Journal, 42*(2), 371–406.

Hull, J. (2013). *Trends in teacher evaluation: How states are measuring teacher performance*. Alexandria, VA: Center for Public Education.

Isenberg, E., Glazerman, S., Bleeker, M., Johnson, A., Lugo-Gil, J., Grider, M., et al. (2009). *Impacts of comprehensive teacher induction: Results from the second year of a randomized controlled study (NCEE 2009-4072)*. Washington, DC: National Center for Education Evaluation and Regional Assistance, Institute of Education Sciences, U.S. Department of Education.

Kaur, B. (2010). Towards excellence in mathematics education—Singapore's experience. *Procedia—Social and Behavioral Science, 8*, 28–34.

Kennedy, M. M. (1998). *Form and substance in in-service teacher education*. Arlington, VA: National Science Foundation.

Ma, L. (1999). *Knowing and teaching elementary mathematics: Teachers' understanding of fundamental mathematics in China and the United States*. Mahwah, NJ: Lawrence Erlbaum.

McLaughlin, M. W., & Talbert, J. E. (2006). *Building school-based teacher learning communities: Professional strategies to improve student achievement*. New York: Teachers College Press.

National Center for Education Statistics (2014). *Teaching and Learning International Survey (TALIS) 2013: U.S. technical report*. Washington, DC: National Center for Education Statistics.

Organisation for Economic Co-operation and Development (2005). *Teachers matter: Attracting, developing and retaining effective teachers*. Paris: OECD Publishing.

Organisation for Economic Co-operation and Development (2009). *Creating effective teaching and learning environments: First results from TALIS*. Paris: OECD Publishing.
Organisation for Economic Co-operation and Development (2010). *Strong performers and successful reformers in education: Lessons from PISA for the United States*. Paris: OECD Publishing.
Organisation for Economic Co-operation and Development (2011). *Building a high-quality teaching profession: Lessons from around the world*. Paris: OECD Publishing.
Organisation for Economic Co-operation and Development (2014a). *TALIS 2013 results: An international perspective on teaching and learning*. Paris: OECD Publishing.
Organisation for Economic Co-operation and Development (2014b). *TALIS 2013 technical report*. Paris: OECD Publishing.
Preacher, K. J., & Hayes, A. F. (2008). Asymptotic and resampling strategies for assessing and comparing indirect effects in multiple mediator models. *Behavior Research Methods, 40*, 879–891.
Rivkin, S. G., Hanushek, E. A., & Kain, J. F. (2005). Teachers, schools, and academic achievement. *Econometrica, 73*(2), 417–458.
Schmidt, W. H., McKnight, C. C., & Raizen, S. A. (1997). *A splintered vision: An investigation of U.S. science and mathematics education*. Dordrecht; Boston: Kluwer Academic Publishers.
Seo, K. (2012). Lessons from Korea. *Educational Leadership, 70*(3), 75–78.
Shimahara, N. K. (2002). *Teaching in Japan: A cultural perspective*. New York: Routledge Farmer.
Son, J. (2013). How preservice teachers interpret and respond to student errors: Ratio and proportion in similar rectangles. *Educational Studies in Mathematics, 84*(1), 49–70.
Stein, M. K., Grover, B., & Henningsen, M. A. (1996). Building student capacity for mathematical thinking and reasoning: An analysis of mathematical tasks used in reform classrooms. *American Educational Research Journal, 33*(2), 455–488.
Steward, V. (2012). *A world-class education: Learning from international models of excellence and innovation*. Alexandria, VA: Association for Supervision & Curriculum Development.
Stigler, J. W., & Hiebert, J. (2004). Improving mathematics teaching. *Educational Leadership, 61* (5), 12–16.
Supovitz, J. A., Mayer, D. P., & Kahle, J. B. (2000). Promoting inquiry-based instructional practice: The longitudinal impact of professional development in the context of systemic reform. *Educational Policy, 14*(3), 331–356.
Supovitz, J. A., Sirinides, P., & May, H. (2010). How principals and peers influence teaching and learning. *Educational Administration Quarterly, 46*, 31–56.
United Nations Educational, Scientific and Cultural Organization (UNESCO) Institute for Statistics (2006). *Teachers and educational quality: Monitoring global needs for 2015*. Montreal, Quebec: UNESCO Institute for Statistics.
Yoon, K. S., Duncan, T., Lee, S. W.-Y., Scarloss, B., & Shapley, K. L. (2007). *Reviewing the evidence on how teacher professional development affects student achievement*. Washington, DC: U.S. Department of Education, Regional Educational Laboratory Southwest.
Youngs, P. (2013). *Using teacher evaluation reform and professional development to support Common Core assessments*. Washington, DC: Center for American Progress.
Zurr, A. F., Ieno, E. N., & Elphick, C. S. (2010). A protocol of data exploration to avoid common statistical problems. *Methods in Ecology and Evolution, 1*, 3–13.

Chapter 20
Large-Scale International Datasets—What We Can and Cannot Learn from Them, and How We Could Learn More

Sarah Theule Lubienski

Abstract The chapters in this section focus on a variety of important ideas, including mathematical affect, values, forms of mathematics knowledge, and teacher evaluation. Socioeconomic equity themes are woven throughout. Several chapters provide models of ways to be sensitive to national contexts when conducting international comparisons, such as attending to how countries' histories and political systems might affect mathematics outcomes. This crosschapter analysis highlights these and other strengths of the five chapters and also points to ways in which analyses of international datasets could be improved, including greater attention to multiple comparisons, caution in interpreting self-reported data, and a more consistent reporting of effect sizes to clearly communicate results. Future mathematics education research could benefit from more collaboration across data collection efforts, as well as the establishment of a longitudinal international dataset.

Keywords TIMSS • PISA • TALIS • Values • Culture • Self-concept • Self-efficacy • Mathematics achievement • Mathematics instruction • Longitudinal datasets • Equity • Socioeconomic status

Chapters 15–19 focus on large-scale, international datasets, namely TIMSS (Han, Son, & Kang, Chap. 16; Choi & Hong, Chap. 17), PISA (Chiu, Chap. 15), TALIS (Han, Son & Kang, Chap. 19), and the WIFI dataset (Seah, Baba, & Qiaoping, Chap. 18). These chapters focus on a variety of important issues related to mathematics education, including affect, values, equity, forms of mathematics knowledge, and teacher evaluation. Each study is grounded in relevant literature and pursues interesting questions that are worthy of our attention.

S.T. Lubienski (✉)
Department of Curriculum and Instruction, University of Illinois at Urbana-Champaign, Champaign, IL 61801, USA
e-mail: stl@illinois.edu

J.-W. Son et al. (eds.), *What Matters? Research Trends in International Comparative Studies in Mathematics Education*, Research in Mathematics Education, DOI 10.1007/978-3-319-51187-0_20

Despite the many strengths of these studies, one limitation they share is that they are based on datasets that are cross-sectional, as opposed to longitudinal. That is, instead of tracking a sample of students over time, they provide a single snapshot of student data at one time point. Cross-sectional datasets within a single country, such as the National Assessment of Educational Progress (NAEP) in the USA, clearly have their limits. For example, NAEP data have revealed that US eighth graders who use calculators more often have higher mathematics achievement (Braswell et al., 2001). This correlation might lead some to conclude that calculator use *causes* higher achievement. The problem with this logic is the old adage, "correlation does not imply causation." While it might be true that calculator use boosts student achievement, the correlation might instead be due to higher-achieving or otherwise more advantaged students having greater access to calculators in their math classrooms (Lubienski, 2006). But to realize how likely this is, one must understand the ways in which US eighth graders tend to be tracked for mathematics, as well as the ways in which instruction and school resources vary by race and family socioeconomic status in the USA.

Making good use of cross-sectional data when comparing trends across countries is even more challenging, as the factors that confound causal claims are even more complex and difficult to identify. It is relatively easy to draw questionable claims from current international datasets, and much more difficult to use those data with appropriate care. For example, an overly simplistic study using international data might compare math achievement across countries and conclude that whichever instructional methods are used relatively often in the highest-scoring countries must boost achievement, and should therefore be emulated in other countries. But again, there may be many other explanations for such correlations. Indeed, there are myriad variables that muddy cross-national comparisons, including differences in culture, history, politics, and the structure of schooling. It is impossible for a single researcher to understand or statistically account for all of these subtle differences. Hence, it is a true challenge to untangle how much school or teacher variables really matter when compared to other factors, such as inter-country differences in education funding models, expectations of children and families, or the use of additional instruction or tutoring outside the regular school day. Given these concerns, what can we possibly learn from current international datasets? And how can we analyze them well? Luckily, Chaps. 15–19 help provide some answers to these questions.

What We Can Learn from International Datasets

Chapters 15–19 each make a unique contribution to what we know about mathematics education in Asia and the USA. Additionally, the cross-cutting theme of educational equity—most notably, associations between family socioeconomic status and educational opportunities—merits special attention.

Highlights of Individual Chapter Contributions

Self-concept versus self-efficacy. In exploring differences in mathematics self-efficacy and self-concept, Chiu (Chap. 15) reveals several interesting patterns. He finds, for example, that self-efficacy exceeds self-concept for students from several Asian countries, while the reverse is true of US students. Chiu's findings are consistent with literature suggesting that US students tend to compare themselves to lower-achieving peers, while students in "face cultures" compare themselves to higher-achieving peers to avoid the possibility of claiming more status than others would give them. One of the many strengths of this chapter is the clear, operationalized definitions of self-efficacy and self-concept, along with the implication for researchers that self-efficacy may be better for predicting math achievement and for comparing students in more absolute terms, while self-concept is better for relative comparisons among similar students.

Taking stock of what's out there. In Chap. 16, Han, Son, and Kang take stock of the TIMSS studies examining math achievement and equity that were published over the past 20 years. They found 22 such studies, and through analyzing each of those studies using a framework they developed, they point to areas in which the research is lacking. For example, they found no study using TIMSS fourth-grade data to examine gender and math, and they argue that more studies should make use of multiple waves of data. They point to the limits of socioeconomic status measures in TIMSS as a possible reason why there have not been more studies of equity in TIMSS data.

Unpacking "mathematics achievement." Choi and Hong (Chap. 17) go beyond the typical, single measure of mathematics achievement and closely examine patterns in students' proficiency in "Knowing," "Applying," and "Reasoning," including how these patterns vary across countries and by achievement level. One contribution of this chapter is its empirical investigation of the assumed hierarchy of these knowledge forms, as well as attempts to link patterns in the data to shifts in education policy within each country.

Measuring what students value in mathematics class. Seah, Baba, and Qiaoping (Chap. 18) introduce readers to the WIFI study, which focuses on students' values and valuing in mathematics. This researcher-driven effort spans 18 countries and 5 continents. The study focuses on what students value during the mathematics learning process, including factors that help or hinder their learning (e.g., investigations, small-group discussions). The take-aways from this chapter fit solidly under the question of how to conduct international comparisons well, and will therefore be discussed further in the next section.

Teacher evaluation across countries. In Chap. 19 Han, Son, and Kang dig into the TALIS dataset, illuminating the emphasis placed on various types of evaluative feedback given to teachers. For example, they find that the emphasis on student achievement was particularly high in the evaluations of teachers in Singapore and the USA. They examine correlations between specific forms of teacher feedback and teachers' self-reported beliefs.

Guilt by Association

Educational equity was thematically woven throughout Chaps. 15–19, serving as a focus of analysis in some chapters and an important backdrop in others. One important pattern that emerged is the variation in association between socioeconomic status and educational opportunities across countries. More specifically, Chaps. 15 and 16 noted that within the USA, student SES is a larger predictor of whether students will have qualified teachers (Chap. 16) and high achievement (Chaps. 15 and 16) than in other countries studied. Despite the limits of international datasets in supporting causal claims, this relatively large association between socioeconomic status and educational opportunities in the USA is striking, and should be enough to cause US policymakers to stop and consider how our localized funding and control of education may allow the "rich to get richer."

This showcases international data comparisons at their finest—illuminating the ways that detrimental practices viewed as "normal" or inevitable within a country may not be. These chapters make an important contribution by subtly raising the question of how and why the USA allows such egregious inequities in school funding and opportunities to continue. Furthermore, the chapters point to potential policy-driven solutions, noting, for example, the ways in which teachers are assigned to schools in Korea (Chap. 16).

Analyzing Large-Scale, International Datasets Well

The studies in Chaps. 15–19 also provide models of ways to be sensitive to inter-country differences, both through narrative discussions of relevant historical, cultural, and political factors within each country and through the selection of the researchers and research methods involved.

Tying History, Culture, and Policy to Analyses of Mathematics Education Data

All five chapters emphasize the importance of interpreting large-scale international comparisons while considering micro- and macro-level factors within each country. Policies related to teacher distribution and evaluation, for example, become salient for understanding data on inequitable student outcomes. In addition to understanding current policies, some chapters delve even further, considering how countries' histories, economies, and/or political systems affect mathematics education.

It is rare for mathematics education researchers to possess deep knowledge about a society's history, its effect on current culture, and how that impacts patterns in large-scale international datasets, but in Chap. 15, Chiu provides a model of how this can be done. Specifically, in his discussion of face culture versus dignity culture, he ties the history of regions to the evolution of education, economics, and values of countries, and ultimately how these factors shape current students' math self-perceptions. What is perhaps most striking is the even-handed ways in which these issues are discussed, understanding each cultural orientation on its own terms without placing a higher value on one orientation or another.

One interesting side note mentioned by Chiu is that several variables designed to detect differences in cultural values were not significant in his statistical models and were therefore not discussed in the chapter. This lack of significance highlights the difficulty of statistically accounting for subtle cultural differences among countries (and the difficult task undertaken by the WIFI study), and it suggests that more qualitative, narrative considerations of these variables, such as those provided by Chiu and other authors in this section, are necessary when interpreting large-scale international comparisons. Additionally, methods of analysis that can account for inter-country differences in other ways are needed—the topic of the next section.

Within-Country Analysts and Analyses

Several chapters in this section present analyses conducted with unusually careful focus *within* countries, in conjunction with cross-national comparisons. Some chapters utilized international author teams, which allowed for both depth and breadth of relevant expertise. As a prime example, Seah, Baba, and Qiaoping focus specifically on Hong Kong and Japan, with a research team from each country working on the analysis and interpretation of data from their own country. In order to see how the various value-related survey items cohered within each country, they had each within-country team run a factor analysis on their own data and create the factors that made the most sense in their context. While this can pose challenges for cross-country comparisons (since the composite variables being examined differ by country), this method emphasizes the importance of creating variables that make optimal sense within a specific cultural context, and avoiding faulty assumptions about the consistency of relationships among variables across countries. Ultimately, Seah, Baba, and Qiaoping find differences in how researchers in different countries interpreted similar findings. For example, students in both Hong Kong and Japan placed importance on hearing stories about mathematicians. However, the research team in Hong Kong attributed this to students' desire to learn how to be successful in math, while the Japanese research team felt this was connected to students' sense of wonder about mathematics. Hence, having cultural insider knowledge can lead to very different interpretations of patterns in large-scale, international data.

The methods used in other chapters likewise revealed sensitivity to national contexts in additional ways, including standardizing teacher variables within each country (Han, Son, and Kang in Chap. 19), examining reported curricular emphases and considering students at various levels of achievement within each country

(Choi and Hong), and distinguishing between within-country and between-country analyses when taking stock of the literature (Han, Son, and Kang in Chap. 16).

Future Directions for Large-Scale International Datasets

Improving Analyses of Current Data

As noted above, Chaps. 15–19 provide some important models of ways to integrate historical, cultural, and political sensitivities into large-scale analyses of international mathematics education data, thereby providing food for thought as researchers consider ways to improve analyses of those data. Additionally, other aspects of the chapters point to methodological issues that should be considered in future large-scale studies, including cautions related to the data, methods, and the reporting of results.

Beware of international, self-reported data. First, fundamental questions about the reliability of self-reported data, particularly in international comparisons, lurks behind several of the studies. For example, Han, Son, and Kang (Chap. 19) utilized teachers' self-reports about their instruction with care, noting that there can be cross-cultural differences in survey responses. Chiu's findings illustrate those differences vividly, showing that Asian students are more modest than US students when rating their own mathematics abilities in comparison to others. Similarly, US teachers may rate their teaching practices higher than is warranted (for example, in Chap. 19 US teachers rated their quality of instruction much higher than did Japanese teachers). Hence, these chapters highlight the importance of being extremely cautious when comparing self-reported survey data across countries. Standardizing variables within countries (as done by Han, Son, and Kang) is one way to address this issue, although this might not be appropriate for all studies. At the very least, this issue should be seriously considered in any cross-national study that makes use of self-reported survey data.

Multiple comparisons. A second issue lurking in several chapters is that of multiple comparisons. Given the many different countries examined, models run, and variables considered, the statistical claims of several studies would be stronger if an adjustment for multiple comparisons was considered. This becomes particularly important if researchers move toward running statistical analyses within each country, as opposed to building a single, multilevel model.

That said, multiple comparisons can also add important richness to analyses and actually help test the robustness of findings. As Han, Son, and Kang argue in Chap. 16, examining only means can mask important trends that vary across the achievement distribution. Indeed, this has become clear within research on gender in the USA and around the world—boys, on average, outscore girls most prominently at the top of the distribution, where future mathematicians and scientists tend to reside (OECD, 2015; Robinson & Lubienski, 2011). Researchers using international datasets should consider examining the robustness of trends throughout the distribution when possible, as Choi and Hong do. Han, Son, and Kang also encourage researchers to test the robustness of findings by confirming analyses with multiple waves of data. Examining multiple waves can also allow researchers to track changes in age cohorts over time. Hence, making better use of multiple waves of data is another potential avenue to enhance the quality and richness of research using international datasets. As our analyses become more multilayered, adding dimensions of time or within-subgroup analyses, we must strive to use the most appropriate methods, including those that help us account for multiple comparisons.

Clear communication of results. Third, researchers using established international datasets such as TIMSS and PISA should ensure that they communicate results as clearly as possible. For example, readers may not be familiar with the achievement scales used in these international datasets, and so consistent inclusion of standard deviations and/or effect sizes would make the meaning of various disparities between groups clearer for readers.

Better coordination across datasets? Fourth, the fact that four different international data collection efforts were discussed in these chapters (TIMSS, PISA, TALIS, and WIFI) raises the question of whether these efforts could be better coordinated. For example, what are the unique contributions of TIMSS and PISA? Are they coordinating in optimal ways? Are there ways in which they are duplicating efforts? Similarly, if the WIFI study identifies variables that reliably measure important differences in students' values, could such variables be added to one of the larger existing datasets? Streamlining existing cross-sectional data collection efforts could perhaps make room for a longitudinal study—the topic of the next section.

Is It Time for a Longitudinal, International Dataset?

As mentioned at the start of this commentary, although cross-sectional datasets within a single country have their limits, the limits for international datasets are even more severe. Researchers must be ever vigilant about keeping causal claims in check when working with the cross-sectional, international datasets. As just one example, Chap. 19 sets out to examine how teacher collaboration might mediate the relationship between professional development and teaching practices. However, given that these variables were all measured at the same time, and given that a confounder such as school socioeconomic status might predict greater frequency of professional development, collaboration, and better teaching practices (at least within some countries, including the USA), the conclusions that can be drawn from this analysis are far more limited than they would be if variables could be measured over time. A longitudinal dataset could, for example, allow an analysis of whether professional development or collaboration predicts a later increase in the quality of teaching practices (if teachers were followed over time), or it could allow for analyses of whether teachers who use particular teaching practices promote greater increases in students' scores (if students were followed over time).

However, a longitudinal dataset is only as good as its variables, and so the points raised earlier about self-reported measures, as well as ways to combine or better leverage the variables in existing datasets would still stand. In fact, one fundamental question in the interest of efficiency is whether there is a way for various countries' individual datasets (such as NAEP or the Early Childhood Longitudinal Study in the USA) to tie into international data collection efforts better. For example, could countries identify a set of common items they could agree to administer within their own countries, if not to replace some current data collection efforts, then at least to allow national samples to link to and/or augment the international samples? This could allow richer analyses with larger samples within each country, and might even move us toward tracking some cross-national samples longitudinally (although issues of population sampling and attrition would certainly need to be carefully considered). As a first step toward a longitudinal data system that would leverage individual countries' ongoing data collection efforts, could we agree on a core set of socioeconomic measures, a small set of mathematics items, and a core set of student and teacher survey variables that get at what is most important within the mathematics education community?

In truth, perhaps not. It is difficult enough for education leaders within a single country to agree on such things, and there are important differences between countries in terms of their goals for mathematics teaching and learning, which make sense in light of each country's history and culture. Given these differences, it is certainly impressive that so many nations have come together to be part of the data collection efforts discussed in this volume. Overall, Chaps. 15–19 contribute to our understanding of what can be gained from existing international datasets, provide models of ways to improve our analyses of those data, and allow us to start envisioning alternatives that could move large-scale international studies of mathematics education forward in the decades to come.

References

Braswell, J. S., Lutkis, A. D., Grigg, W. S., Santapau, S. L., Tay-Lim, B., & Johnson, M. (2001). *The nation's report card: Mathematics 2000*. Washington, DC: U.S. Department of Education, National Center for Education Statistics.

Lubienski, S.T. (2006). Examining instruction, achievement, and equity with NAEP mathematics data. *Education Policy Analysis Archives, 14*(14). Retrieved from http://epaa.asu.edu/epaa/v14n14.

Organisation for Economic Co-Operation and Development. (2015). *The ABC of gender equality in education: Aptitude, behaviour, confidence*. Paris: OECD Publishing.

Robinson, J. P., & Lubienski, S. T. (2011). The development of gender achievement gaps in mathematics and reading during elementary and middle school: Examining direct cognitive assessments and teacher ratings. *American Educational Research Journal, 48*(2), 268–302.

Part V
Final Commentary

Chapter 21
Reflections on Research Trends in International Comparative Studies in Mathematics Education

Gabriele Kaiser and Xinrong Yang

Abstract The chapter reflects on research trends in international comparative studies in mathematics education based on a short description of the historical development and its most recent discussions. As commentary paper it integrates the chapters in the book into this overall framework in order to discuss possible prospects of international comparative studies.

Keywords History of comparative education • Goals of comparative education • Limitations of comparative studies • Comparison between East and West • Qualitative and quantitative empirical studies

Introduction: Setting the Scene

Two famous and often quoted characterizations of comparative education set the scene for this volume. First, Thut and Adams (1964) remarked: "To study education well is to study it comparatively" (Back cover). However, Husén—the founder of the IEA and chair of the First International Mathematics Study (FIMS)—stated, "Comparing the outcomes of learning in different countries is in several respects an exercise in comparing the incomparable" (1983, p. 455).

Despite these challenges, comparative education has a long tradition. In his introduction to *The Encyclopedia of Comparative Education and National Systems of Education*, a benchmark book in the field, Postlethwaite (1988) identified several phases in the historical development of comparative education. In the first phase,

G. Kaiser (✉)
Faculty of Education, University of Hamburg, Hamburg, Germany
e-mail: gabriele.kaiser@uni-hamburg.de

X. Yang
Faculty of Education, University of Hamburg, Hamburg, Germany

School of Mathematics and Statistics, Southwest University, Chongqing, China
e-mail: xinrong.yang@uni-hamburg.de

J.-W. Son et al. (eds.), *What Matters? Research Trends in International Comparative Studies in Mathematics Education*, Research in Mathematics Education, DOI 10.1007/978-3-319-51187-0_21

which Postlethwaite labeled "Travellers' Tales," Greek and Roman travelers described the education of young persons they observed abroad, a practice later adopted by Marco Polo and Alexis de Tocqueville. The next developmental phase started in the beginning of the 1800s with the systematic collection of data about education in different countries, spearheaded by Marc Antoine Jullien, who is thus called the father of comparative education. Many scholars, including the Russian novelist Leo Tolstoy, visited other countries to observe the organization of education there and speculate on their underlying educational principles. The feasibility of borrowing ideas from one educational system and transferring them to another was intensively discussed and often rejected. Especially important was a 1900 article by Sir Michael Sadler of England, who had visited Germany, especially Prussia: "How far can we learn anything of practical value from the study of foreign systems of education?" Sadler (1900) pointed out that specific elements of a foreign system of education are not detachable parts; however, he saw the benefit of studying foreign educational systems in order to better understand his own country's system. He warned against the misconception "that all other nations have better systems of education than we have. It is a great misunderstanding to think, or imply, that one kind of education suits every nation alike" (p. 231). Sadler also described the difficulties of learning from other educational systems by simply transferring foreign components into one's own system:

> In studying foreign systems of education we should not forget that the things outside the schools matter even more than the things inside the schools, and govern and interpret the things inside. We cannot wander at pleasure among the educational systems of the world, like a child strolling through a garden and pick off a flower from one bush and some leaves from another and then expect that if we stick what we have gathered into the soil at home, we shall have a living plant. A national system of education is a living thing, the outcome of forgotten struggles and 'of battles long ago'. It has in it some of the secret workings of national life (p. 310).

Such sustained discussions about borrowing were followed by a phase emphasizing international cooperation, in which scholars aimed to identify the forces influencing educational systems at an international level. These studies featured a more analytical understanding of the relationship between society and education, and of the political systems in which education was embedded. The post-Second World War era can be characterized by the usage of social-science methods and both quantitative and qualitative data to examine the effect of various factors on educational developments. These studies estimated the strengths of specific variables or constructs on other variables, describing, for example, the efficiency of educational systems. Due to the inherent measurement problems, many scholars rejected this empirical approach and pleaded for a historical or hermeneutic approach. While the advantages of empirical approaches are no longer in doubt, many problems surrounding adequate measurement are still unsolved.

Given these problems, what do we mean by comparative education? Postlethwaite (1988) offers this definition: "Strictly speaking, to 'compare' means to examine two or more entities by putting them side by side and looking for similarities and differences between or among them. In the field of education, this can apply both to comparisons between and comparisons within systems of education" (p. xvii). However, many international studies are not comparative studies in this strict sense, but they do describe, analyze, or make proposals about particular aspects of education in countries other than the author's own country. Still, Postlethwaite (1988) asserted that "[w]hen well done, comparative education can deepen our understanding of our own education and society, it can be of assistance to policy makers and administrators, and it can be a valuable component of teacher education programmes" (p. xix).

In his comprehensive overview Postlethwaite distinguished four major aims of comparative education:

(a) *Identifying what is happening elsewhere that might help improve our own system of education...*
(b) *Describing similarities and differences in educational phenomena between systems of education and interpreting why they exist...*
(c) *Estimating the relative effects of variables (thought to be determinants) on outcomes (both within and between systems of education)...*
(d) *Identifying general principles concerning educational effects* (pp. xix–xx, italics in original).

Later researchers stressed the contribution of comparative studies in understanding their own cultures. For instance, Stigler and Perry (1988) argued, "[c]ross-cultural comparison also leads researchers and educators to a more explicit understanding of their own implicit theories about how children learn mathematics. Without comparison, we tend not to question our own traditional teaching practices and we may not even be aware of the choices we have made in constructing the educational process" (p. 199).

According to these goals, two different kinds of comparative studies can be identified:

- Country studies, in which the educational system of a country is described.
- Thematically oriented studies within a country or between several countries, focusing on the economics of education, or pedagogically oriented studies such as comparative achievement studies.

According to Postlethwaite (1988) emphasis had shifted in the 1970s and 1980s from country studies to thematically oriented studies, using more and more sophisticated methods from social sciences, especially quantitative methods. However, even then critics had already pointed out the methodological constraints and limitations of comparative work, partly unsolved until today. In his seminal analysis, Hilker (1962) analyzed the missing normative strength of comparative studies, which cannot create the norms of education needed for comparative purposes by themselves. The *tertium comparationis* needed as a benchmark for judging comparative studies can only be created outside of comparative education, based on educational philosophy. Eckstein (1988) similarly complained that especially large-scale quantitative studies conceptually assume "that educational and social phenomena are results of multiple causes, that they are regularities or tentative laws of input and outcomes (cause and effect), and that these are discoverable through systematic collection and analysis of the relevant facts" (p. 9).

Both Stenhouse (1979) and Crossley and Vulliamy (1984) criticized the limitations of quantitatively oriented research, aiming for macro-level education with an experimental sample paradigm. They called for practical meaningful case studies based on the careful study of particular settings close to school practice and relevant to day-to-day educational reality. Crossley and Vulliamy developed the concept of the "ecological validity of the data," which reflects the extent to which observations made in one cultural context are valid for other contexts. They pointed out that due to the ethnographical references of case studies, they seem to be adequate to secure high ecological validity.

Apart from these fundamental problems, comparative studies also face problems such as the costs and difficulty of sampling data from foreign sources, a lack of comparability of the data sampled, problems concerning the validity and reliability of the data sampled (especially if the data were not collected for the study's specific purposes), and problems constructing valid scales. The central problem however is an "ethnocentric bias in defining the topic to be investigated, establishing the bases for classifying data, drawing inferences, and making policy recommendations" (Noah, 1988, p. 10).

To summarize these different positions and paradigms, it is apparent that comparative education tries to identify general patterns in order to enhance the mutual understanding of different educational systems. On the other hand, these comparisons cannot lead to far-reaching recommendations for changes to educational systems due to their cultural dependency. Alexander (1999) describes this dichotomy as follows:

> I argue that the educational activity which we call pedagogy ... is a window on the culture of which it is a part, and on that culture's underlying tensions and contradictions as well as its publicly declared educational policies and purposes. Second, ... I argue that the comparative perspective is an important and necessary part of the quest to understand and improve the science, art or craft of teaching, and to enable us to distinguish those aspects of teaching which are generic and cross international boundaries from those which are culture-specific (p. 149).

The more recent debate on comparative mathematics education is strongly influenced by the already long-standing outperformance of Western students by their Eastern counterparts, which is apparent in the large-scale PISA and TIMSS studies. TIMSS, originally the Third International Mathematics and Science Study, but now renamed the Trends in International Mathematics and Science Study, was carried out by the IEA as an independent organization, while PISA was carried out on behalf of the OECD. Both studies point out that students from Shanghai (China), Korea, Japan, Singapore, Taiwan, and Hong Kong mathematically outperform the students of most Western countries, with the exception of Finland and Switzerland. These results can be seen as a continuation of the First International Mathematics Study (FIMS; Husén, 1967), in which only Japan represented East Asia, but still placed in the top-achieving group. More East Asian countries participated in later versions of PISA and TIMSS, yielding a clear performance gap between East Asian and Western students.

Specific studies comparing East Asian students and American students have been carried out on smaller scales since the 1980s. In 1983, the *Dallas Times-Herald Survey*—inspired by the report *A Nation at Risk*—compared the mathematical achievements of students from Dallas with those from Australia, Canada, England, France, Japan, Sweden, and Switzerland. Tests in mathematics, science, and geography ranked the Dallas students as lowest in mathematics, better in science and geography, but still under the international median (Robitaille & Travers, 1992). The so-called Japan-Illinois study similarly explored patterns of mathematics achievement and mathematical motivation for high school students from the state of Illinois and Japan. In all age cohorts (15-, 16-, and 17-year-olds) Japanese students significantly outperformed their Illinois counterparts, with strong gender differences in the Japanese sample (Harnisch, Walberg, Shiow-Ling, Takahiro, & Fyans, 1985). Song and Ginsburg (1987) analyzed mathematical thinking in Korean and American children at ages 4–8. They identified slight advantages for American children between ages 4 and 6 in informal mathematics, whereas no differences could be identified in formal mathematics. At ages 7 and 8, Korean children showed superior achievement on both informal and formal mathematics.

The highly discussed and well-documented superior mathematics achievement of Japanese students at the secondary level, compared to their US counterparts, provided the background for the so-called Michigan Studies, carried out at the University of Michigan by Harold Stevenson and colleagues. Their research focused on a cross-cultural study in mathematical achievements by Japanese, Taiwanese, and American students at grades 1 and 5, carried out in two cycles from 1979–1980 and 1985–1986. The studies showed the clear superiority of the Japanese and the Taiwanese students already at the beginning of schooling, with the differences becoming greater at the later wave (Stevenson & Bartsch, 1992; Stigler & Perry, 1988). The achievement tests covered not only computational proficiency but creativity as well.

Classroom observations and textbook analyses carried out simultaneously pointed out significant differences concerning time devoted to mathematics instruction, patterns of teaching and learning formats, and differences in the role of verbal communication, which was significantly higher in Japanese classrooms than in the American and Taiwanese classrooms. The textbook analyses revealed that important mathematical topics such as decimals and fractions were introduced earlier in Japanese textbooks than in their US counterparts. Furthermore, problem-solving was emphasized more strongly and earlier in Japanese textbooks. The US textbooks contained much more repetition than the Japanese textbooks, which was interpreted as reflecting a spiral curriculum in the United States (Stevenson & Bartsch, 1992). The higher importance of problem-solving activities in Japanese classrooms was confirmed by studies by Becker and colleagues, who emphasized the high relevance of open-ended problem-solving in Japanese classrooms (Becker, Sawada, & Shimizu, 1999).

These studies provided the background for a video study carried out within the frame of the Third International Mathematics and Science Study (known as TIMSS 1995) which was conducted at five grade levels in more than 40 countries (the third, fourth, seventh, and eighth grades, and the final year of secondary school). The video study concentrated on students from Germany, Japan, and the USA, and collected more than one hundred lessons per country (Stigler & Hiebert, 1999). A continuation, the TIMSS 1999 Video Study, was conducted in Australia, the Czech Republic, Hong Kong, the Netherlands, Switzerland, and the USA, again using the Japanese videos. The study revealed significant differences between the countries, identifying specific country profiles. However, the findings are not inconsistent with each other; for example, it was reported that a relatively small percentage of mathematics problems (and lessons) in countries other than Japan involved proofs. The same trend held for mathematical reasoning (Hiebert et al., 2003).

These studies have raised intensive discussions about the underlying reasons for these discrepancies. In his seminal work on the search for an East Asian identity in mathematics education, Leung (2001) identified several relevant differences—described as dichotomies—between the East Asian and the Western traditions in mathematics education, which may provide explanations for these achievement differences. He distinguished product (content) as being more strongly emphasized in East Asian classrooms, while process, i.e., doing mathematics, was of higher importance in Western countries. Furthermore, the difference between rote learning versus meaningful learning is of importance: rote learning and memorization are a necessary way of learning in Eastern countries, whereas Western cultures emphasize the necessity of understanding the phenomenon before it can be memorized. The third dichotomy, studying hard versus pleasurable learning, refers to traditional views in East Asian countries that studying is a serious endeavor relying on hard work, in contrast to many Western views which emphasize that children have to enjoy the learning process by making it meaningful for themselves.

The fourth dichotomy presented by Leung (2001) refers to extrinsic versus intrinsic motivation. He argued that Western educators mainly value intrinsic motivation in learning mathematics while their Eastern counterparts emphasize the relevance of extrinsic motivation, especially in terms of high-stakes tests. The fifth dichotomy relates to a different understanding of the role of the teacher: whole-class teaching with the teacher as the role model is regarded as highly important in East Asian countries, in contrast to the stronger focus on independent and individualized learning in Western countries. Finally, the sixth dichotomy contrasts East Asian expectations for teachers to be scholars and role models with profound subject-matter knowledge, whereas Western teachers are expected to possess profound pedagogical competencies. In sum, Leung (2001) characterized Western approaches to mathematics education as student-centered, while East Asian approaches utilized a tripartite emphasis on the student, the teacher, and the subject matter. Leung (2001) hypothesized that this model represented the essence of East Asian identity, an idea in line with other approaches, e.g., the concept of learning in a Confucian heritage culture (CHC). According to Wong, Wong, and Wong (2012), CHCs are characterized by:

> 1. A high degree of parental involvement in and commitment to the education of children; 2. A basic eagerness to learn and a positive attitude toward school on the part of children; 3. High status for teachers. A strong commitment on teachers' part to teaching and to involve in their students' overall development; 4. The premise of egalitarian access to the rewards of successful learning; 5. The assumption that it is effort rather than innate ability which yields rewards in schooling; 6. The occupational system values education as appropriate preparation for work. ... Through education, Confucian ideals are passed from generation to generation and have become a cultural heritage. (p. 11)

Similar to Leung (2001), Wong (2004) described an orientation towards social or collective achievement as a central feature of CHC learning, in contrast to an orientation towards individual achievement in Western cultures—including an emphasis on diligence, an attribution of success to effort, a competitive spirit, and a high relevance of practice. So, there seems to be some consensus that a kind of joint identity within East Asian or Western learning traditions exists. However, Wong et al. (2012) point out that there are also other prominent philosophical traditions and values important in East Asia, and it is too simplistic to attribute the superior academic performance only to the role of Confucian values.

To connect this historical context to the present book, we now turn to the most recent survey on international comparative studies in mathematics education: Cai, Mok, Reddy, and Stacey's (2016) article, "International Comparative Studies in Mathematics: Lessons for Improving Students' Learning." They took the strong position that the main purpose of educational research is to improve student learning, which includes international comparative studies. Accordingly, they identified four lessons to be learned from international comparative studies:

- Understanding students' thinking: focusing on narrow in-depth studies, they pointed out how the variety of students' problem-solving methods, as well as students' individual misconceptions, influence test scores, although quantitative studies can often not identify these problems.
- Promoting students' mathematical literacy: referring to the PISA study, they highlighted the potential of evaluating specific country profiles of mathematical literacy, which allows insight into students' dispositions towards formal and applied mathematics.
- Changing Classroom Instruction: taking into account the 1995 and 1999 TIMSS Video Studies, along with the Learner's Perspective Study (Clarke, Keitel, & Shimizu, 2006), they described the potential of these kinds of studies, in which the influence of cultural traditions on classroom teaching allowed scholars to form country profiles for educational practices.
- Making Global Research Locally Meaningful: based on the findings of TIMSS in South Africa, they pointed out how these studies can be used for local policy reforms, especially since the contextual factors influencing students' achievements are of high importance, in order to develop measures to overcome educational inequality.

Cai et al. (2016, p. 31) summarized,

> Over the last three decades, international comparative studies have completely transformed the way we see mathematics education. For example, because of the very high ranking of some Asian countries, the field of mathematics education has become interested in mathematics education in Asian countries. We used to think that there was one basic way of teaching mathematics; international comparative studies, however, showed us many different ways of teaching mathematics in the classroom. We also learned that some student background variables (e.g., attitudes, gender) operate in different ways for students in different countries. (p. 31)

They conclude:

> [W]e would like to emphasize the complementary roles of small-scale and large-scale international comparative studies. ... Small-scale studies tend to be used towards the beginning of a research program to explore new phenomena. Large-scale studies, in contrast, tend to be employed after such methods or instruments have been piloted and their use justified, and the phenomena to which they apply have been adequately defined. (p. 31)

The survey points out that large-scale international comparative studies such as TIMSS and PISA report many variables, including achievement variables, school variables, system variables, and student variables, based on representative samples. However, despite resource and time constraints, small-scale international comparative studies can offer advantages too, as they allow in-depth analysis of the issues being studied. Large- and small-scale studies can also be combined or conducted concurrently, and their findings and analysis could complement each other to understanding and improving students' learning. Alternatively, scholars could conduct in-depth analyses of specific issues based on large-scale data, e.g., analyzing PISA's three processes of Formulate, Employ, and Interpret based on the results of students from Australia, China, Singapore, and the USA. This analysis shows different patterns: Australian and US-American students have relatively better performance on interpreting than on the other two processes, while Shanghai and Singapore students did a better job on formulating than on the other two processes. "With in-depth analysis from large-scale data, we can find such subtle, but important differences, but we need a range of other studies to understand why this might be the case" (Cai et al., 2016, p. 33).

This Book's Contributions

The present book, *What Matters? Research Trends in International Comparative Studies in Mathematics Education*, departs from the work described above in important ways, but also takes many important elements into account. It is a comparative study focusing on six countries, one Western (the United States of America) and five high-achieving East Asian countries (Japan, Korea, China, Taiwan, and Singapore). These East Asian countries can be characterized by commonalities in their educational philosophy, such as the Confucian Heritage

Culture model summarized above. The studies collected here focus strongly on the educational differences between "East and West." Twelve out of the 16 papers compare the American educational system to at least one of the five East Asian countries; three of these studies compare the USA to one East Asian country, while the other nine compare it to multiple countries. The remaining four papers deal with comparisons within East Asian educational systems; half the papers compare two countries and the other half compare several. Overall, a comparison between East and West remains the dominant focus of the book.

Concerning the content of the book, four main research perspectives inform these comparative studies: (a) research on curriculum's influence on student learning, (b) research on institutional systems of mathematics teacher education, (c) research on improving teacher knowledge and pedagogical approaches, and (d) research using large-scale data, as a kind of a meta-perspective. Each of these perspectives is supplemented by a commentary focusing on the papers in that part of the book. Therefore, our comments here will focus on the overall structure of the book and its novelty.

Compared to the overview of the state-of-the-art ICME-13 Topical Survey by Cai et al. (2016), described above, this book is characterized by three distinct features:

- A strong emphasis on teacher education and teacher knowledge in addition to students' learning
- A more systemic view of comparative studies, including curricular aspects
- The East-West-paradigm as a guideline for the comparative studies, although comparative studies between East Asian countries go beyond the East-West-paradigm

More specifically, Part 1 presents research on curricular influence on students' mathematics learning. Son and Diletti's paper covers textbook analysis from a more general perspective, while Watanabe, Lo, and Son focus on the intended curricular treatments of fractions and fraction operations in Japan, Korea, and Taiwan. Next, Zhu analyzes the difficulty level of mathematics textbooks in China, the USA, Korea, Singapore, and Japan. The results confirm previous findings that Chinese textbooks are the most difficult, while American textbooks are the easiest. However, the fact that the analyzed Japanese textbooks involve the least amount of mathematics and are at a similar level to the American textbooks is more than surprising. Unfortunately, due to resource restrictions only one textbook series per country could be analyzed in both studies, which strongly restricts the generalizability of the results. Zhu also examines achievement differences between East Asian and American students through a secondary analysis of the PISA data, showing inconsistencies between the students' cognitive and the affective attainments in the participating countries and strong differences in the opportunities to learn between the various educational systems. Finally, Shimizu challenges the notion of "Asian" in international comparative studies by using data from the Learners' Perspective Study, pointing out significant differences in instructional patterns within various East Asian educational systems. Silver's (2017) commentary on these papers emphasizes that there is probably more than one simple explanation for the higher performance of East Asian students, and that therefore "those who seek to improve mathematics teaching and learning in the United States and other lower performing countries would be wise to recognize the value in not only recognizing the differences across education systems but also in trying to understand how it is that different educational practices arise and flourish within countries."

The second set of papers deals with research on institutional systems of mathematics teacher education. First, Kim and Ham analyze the knowledge expectations for mathematics teachers in Korea and the USA, showing that there exist transnational commonalities and national differences concerning the social expectations for teachers' knowledge. Along similar lines, Koyaman and Lew examine preservice teacher education in Japan and Korea. The final paper in this section, by Wang and Hsieh, is based on data from the IEA large-scale study on teacher education TEDS-M, and analyzes the teaching readiness of future mathematics teachers. The study confirms the well-known importance of intrinsic motivation among future teachers. To summarize, as Kilpatrick (2017) pointed out in his commentary these three chapters have "the limitations of cross-sectional data, the use of limited measures, and insufficient phases of analysis." However, they still "constitute an important first step in establishing a research basis for studying the effectiveness of mathematics teacher education programs around the world."

The third section is devoted to research on improving teacher knowledge and pedagogical approaches. The first paper, by Lim and Son, describes a cross-cultural collaborative project in which Korean and US-American teachers shared feedback on writing lesson plans. The second paper (Corey, Leatham, and Peterson) examines the instructional quality of future mathematics teachers in the USA and Japan using a video coding protocol, and points out that US future teachers can implement lessons of similar quality as Japanese future teachers do, if they receive adequate support. Finally, Ricks' paper focuses on the macro-level of education. He analyzes the reflective capabilities of mathematics education systems in China, Japan, and USA and points out the higher reform resistance of the US mathematics education system compared to the other two systems. In her section commentary, Crespo (2017) poses several challenging questions for the readers of these papers, including: "Whose language, cultures, and practices are privileged in these kinds of international studies?"

The final set of essays departs from the East vs. West theme, and instead concentrates on large-scale studies and their specific contribution to comparative studies. In the first paper, Chiu examines the relationships among self-concept, self-efficacy, and mathematics achievement using PISA data from 65 regions, including the USA and Asia, and points out strong discrepancies between mathematical achievements and self-concept. Based on data from TIMSS over the past two decades, Han, Son, and Kang focus on inequality, specifically the relationship between social background and mathematics achievement in Japan, Korea, Singapore, and the USA. The next paper, by Choi and Hong, refers to TIMSS data too, emphasizing Taiwan, Hong Kong, Korea, and Singapore, and points out significant differences between American and East Asian students' success in mathematics, arguing that they are strongly influenced by applying basic and procedural skills versus applying higher level cognitive abilities. Next, based on data from the WIFI study, Seah, Baba, and Zhang examine what students from Hong Kong and Japan value in mathematics education; the study also highlights the cultural dependence of values and their differences, even between students from similar cultures.

The final paper in this section (Han, Son, and Kang) refers to data from the TALIS project and examines the relation between teacher feedback and mathematics instruction in Japan, Korea, Singapore, and the USA, describing the variation among these countries. Theule Lubienski's (2017) section commentary identifies several restrictions of these studies, such as the fact that their cross-sectional methods do not allow scholars to trace the development of specific students' cohorts. However, she also highlights that the papers "contribute to our understanding of what can be gained from existing international datasets, provide models of ways to improve our analyses of those data, and allow us to start envisioning alternatives that could move large-scale international studies of mathematics education forward in the decades to come."

Overall, this book presents many new insights into the strength of international comparative studies, especially those focusing on Eastern and Western countries. However, two main problems, both raised earlier in the history of systematic comparative research, still remain. First, to what extent can we learn anything of practical value from the study of foreign systems of education? Alexander (1999, p. 158) described this as the "the 'so what?' problem in educational research" and called for culturally sensitive studies with practical insight. The essays printed here neither challenge nor really discuss the approach of borrowing good aspects from one educational system and transferring them to another. So, despite the rich database provided by international comparative studies, questions remain about their relevance and potential consequences. In light of the power of international comparative studies this question is especially pressing, but still unanswered.

Secondly, the problem of the missing *tertium comparationis* as a benchmark is not dealt with. Apart from the PISA study, no explicit theoretical framework with which to judge the outcomes of comparative studies is developed here, though its necessity is discussed broadly. Specifically, Noah (1988) called attention to the "ethnocentric bias in defining the topic to be investigated," which is outside the focus of most of these studies. In contrast, data from large-scale studies are interpreted within a Western paradigm and referring to Western concepts, without questioning whether these concepts are adequate for Eastern educational approaches. Prominent examples are the Western concepts of self-efficacy and self-confidence in mathematics, which are used in many large-scale studies whether or not they are adequate for students from different educational backgrounds, including Confucian Heritage Culture countries. In particular, the distinction between "Eastern" and "Western" cultures is not challenged, although it is obvious that strongly different philosophical backgrounds exist: in addition to CHC, there is the continental European didactic tradition referring to John Amos Comenius and the Great Didactic (Hudson & Meyer, 2011). The question of what the East and the West can learn from each other is still unsolved (Kaiser & Blömeke, 2013).

Concerning the consistently outstanding performance of East Asian students in recent international comparative studies (e.g., TIMSS and PISA), most studies have focused on factors such as number systems, cultural contexts (e.g., parental expectations and beliefs in ability), school organizations, and mathematics curricula. However, so far few studies have focused on explaining these achievement differences in terms of the professional competencies of the teacher and their teaching-and-learning processes. In particular, very few scholars have examined these relationships over a certain period of time in different social and cultural contexts. Further research is necessary to evaluate these trends in terms of the differences between East Asian and Western traditions in mathematics education, as discussed earlier.

This distinction can be further explored with grand theories from cultural psychology or sociology, such as the famous cultural-psychological framework developed by Hofstede (1986). Departing from a definition of culture as shared motives, values, beliefs, identities, and interpretations or meanings of significant events, which results from common experiences of members of collectives that are transmitted across generations, Hofstede (1986) concluded that through socialization processes a country's culture has an impact on the preferred modes of learning. Within this framework, the collectivism vs. individualism dimension refers to the extent to which the individuals of a society are perceived as autonomous. This criterion seems to be particularly relevant in explaining differences between East Asian and Western teaching and learning processes and the associated differences in mathematics achievement. Triandis (1995) transferred this framework to education, describing learning in individualistic countries as activities of autonomous subjects, who are not obliged to learn due to societal norms. Failure is attributed to context conditions such as poor explanations by the teacher, and changes therefore focus on changing instructional quality, but not on more efforts by the students. In contrast, in collectivistic countries students' learning is seen as a commitment to their teachers, their families, and their society. Failure is attributed to lacking efforts and should result in more effort to learn.

These different cultural orientations also influence thinking on teacher education and teacher expertise. Kaiser and Li (2011) found that Eastern educators tend to view teacher expertise in a more holistic manner, which aims to make systematic changes to in-school teaching-and-learning processes by encouraging teachers to be researchers, and to develop their expertise in doing scientific work. This holistic view refers to the public valuing of expert teachers, who according to Yang (2014) play multiple roles, including expert role models for students and colleagues. Kaiser and Li (2011) describe the Western perspective, in contrast, as "clearly focused on the teaching-and-learning process within the classroom, where experienced teachers shall display their expertise especially in interactions with the students. Characteristic for the Western approach to expertise is the focus on the individual student, who is put into the centre of reflections and actions; the promotion of learning processes of individual students is a major goal of the classroom activities" (p. 349). Based on these theoretical frameworks, future studies should examine whether the gap identified in student mathematics achievements is valid for the professional competencies of practicing teachers as well. This could help alleviate the limitations of cross-sectional (as opposed to longitudinal) datasets, and provide a new perspective to interpret differences in Western and Eastern students' mathematics achievements.

The TEDS-M study has already confirmed that achievement gaps between Eastern and Western future teachers at the school level persist at the tertiary level. Based on follow-up studies to TEDS-M (for an overview, see Kaiser et al., 2016), an ongoing project by Yang will discuss what the East and the West can potentially learn from each other, and whether cultural influences on the teacher's role and function, especially with respect to teacher knowledge, are decisive for these achievement differences at the school level. Leung and Park (2002) had already hypothesized that competent students might become competent teachers and in turn produce competent students. They see in this competence cycle one reason for the outstanding performance of East Asian students. However, few studies provide similar empirical evidence for teachers in other countries, apart from the German study COACTIV (Kunter et al., 2013) and the American Michigan studies (Hill, Rowan, & Ball, 2005). In addition, no previous studies compare East Asian and European countries in this manner. Accordingly, Yang's work will analyze at a general level whether studies developed in the West, namely TEDS-M and its follow-up studies, can be transferred to the East. Are the theoretical framework and the instruments culturally insensitive enough to be used in China too? Can they yield valid results? At a more concrete level the study will examine the relationship between teachers' competencies and students' progress, taking into account instructional quality, a construct mainly developed in Europe. Results of the study are expected in 2017, and will describe new chances and limitations of international comparative studies.

References

Alexander, R. (1999). Culture in pedagogy, pedagogy across cultures. In R. Alexander, P. Broadfoot, & D. Philips (Eds.), *Learning from comparing: New directions in comparative educational research* (Vol. 1, pp. 149–180). Oxford: Symposium Books.

Becker, J. P., Sawada, T., & Shimizu, Y. (1999). Some findings of the U.S.–Japan cross–cultural research on students' problem solving behaviours. In G. Kaiser, E. Luna, & I. Huntley (Eds.), *International comparisons in mathematics education* (pp. 121–139). London: Falmer Press.

Cai, J., Mok, I. A. C., Reddy, V., & Stacey, K. (2016). International comparative studies in mathematics: Lessons for improving students' learning. In *ICME–13 Topical Survey*. Dordrecht: Springer.

Clarke, D., Keitel, C., & Shimizu, Y. (2006). *Mathematics classrooms in twelve countries: The insiders' perspective*. Rotterdam: Sense.

Crespo, S. (2017). Research on improving teacher knowledge and pedagogical approaches—From a comparative to a collaborative perspective. In J.-W. Son, T. Watanabe, & J.-J. Lo (Eds.), *Research trends in international comparative studies in mathematics education in six countries*. New York, NY: Springer.

Crossley, M., & Vulliamy, G. (1984). Case-study research methods and comparative education. *Comparative Education, 20*(2), 193–207.

Eckstein, M. A. (1988). Concepts and theories in comparative education. In T. N. Postlethwaite (Ed.), *The encyclopedia of comparative education and national systems of education* (pp. 7–10). Oxford: Pergamon.

Harnisch, D. L., Walberg, H. J., Shiow-Ling, T., Takahiro, S., & Fyans, L. R. (1985). Mathematics productivity in Japan and Illinois. *Evaluation in Education: An International Review Series, 9* (3), 277–284.

Hiebert, J., Gallimore, R., Garnier, H., Bogard Givvin, K., Hollingsworth, H., Jacobs, J., Miu-Ying Chui, A., Wearner, D., Smith, M., Kersting, N., Manaster, A., Tseng, E., Etterbeek, W., Manaster, C., Gonzales, P., & Stigler, J. (2003). *Teaching Mathematics in Seven Countries: Results from the TIMSS 1999 Video Study*. Washington: U.S. Department of Education, National Center for Education Statistics.

Hilker, F. (1962). *Vergleichende Pädagogik. Eine Einführung in ihre Geschichte, Theorie und Praxis*. München: Hueber.

Hill, H. C., Rowan, B., & Ball, D. L. (2005). Effects of teachers' mathematical knowledge for teaching on student achievement. *American Education Research Journal, 42*(2), 371–406.

Hofstede, G. (1986). Cultural differences in teaching and learning. *International Journal of Intercultural Relations, 10*, 301–320.

Hudson, B., & Meyer, M. A. (2011). *Beyond fragmentation: Didactics, learning and teaching in Europe*. Opladen: Barbara Budrich.

Husén, T. (Ed.). (1967). *International study of achievement in mathematics* (Vols. I, II). Stock–holm: Almqvist & Wiksell.

Husén, T. (1983). Are standards in U.S. schools really lagging behind those in other countries? *Phi Delta Kappan, 64*, 455–461.

Kaiser, G., & Blömeke, S. (2013). Learning from the Eastern and the Western debate: The case of mathematics teacher education. *ZDM Mathematics Education, 45*(1), 7–19.

Kaiser, G., & Li, Y. (2011). Reflections and future prospects. In Y. Li & G. Kaiser (Eds.), *Expertise in mathematics instruction: An international perspective* (pp. 343–353). New York, NY: Springer.

Kaiser, G., Blömeke, S., König, J., Busse, A., Döhrmann, M., & Hoth, J. (2016). Professional competencies of (prospective) mathematics teachers—cognitive versus situated approaches. *Educational Studies in Mathematics* (10.1007/s10649-016-9713-8; Erratum:10.1007/s10649-016-9724-5).

Kilpatrick, J. (2017). Similarities and differences in programs for prospective secondary mathematics teachers. In J.-W. Son, T. Watanabe, & J.-J. Lo (Eds.), *Research trends in international comparative studies in mathematics education in six countries*. New York, NY: Springer.

Kunter, M., Baumert, J., Blum, W., Klusmann, U., Krauss, S., & Neubrand, M. (2013). *Cognitive activation in the mathematics classroom and professional competence of teachers: Results from the COACTIV project*. New York, NY: Springer.

Leung, F. K. S. (2001). In search of an East Asian identity in mathematics education. *Educational Studies in Mathematics, 47*(1), 35–51.

Leung, F. K. S., & Park, K. M. (2002). Competent students, competent teachers? *International Journal of Educational Research, 37*(2), 113–129.

Noah, H. J. (1988). Methods of comparative education. In T. N. Postlethwaite (Ed.), *The encyclopedia of comparative education and national systems of education* (pp. 10–13). Oxford: Pergamon.

Postlethwaite, N. (1988). *The encyclopedia of comparative education and national systems of education*. Oxford: Pergamon.

Robitaille, D. F., & Travers, K. J. (1992). International studies of achievement in mathematics. In J. S. Grouws (Ed.), *Handbook of research on mathematics learning and teaching* (pp. 687–709). New York, NY: Macmillan.

Sadler, M. (1900). How far can we learn anything of practical value from the study of foreign systems of education? Comparative Education Review, 7, 307–314. Reprinted from: *Surrey Advertiser*, December 1, 1900.

Silver, E. A. (2017). Toward understanding the influence of curriculum resources on students' mathematics learning: Cross-national perspectives on what matters where, when and for whom. In J.-W. Son, T. Watanabe, & J.-J. Lo (Eds.), *Research trends in international comparative studies in mathematics education in six countries*. New York, NY: Springer.

Song, M. J., & Ginsburg, H. (1987). The development of informal and formal mathematical thinking in Korean and U.S. children. *Child Development, 58*, 1286–1296.

Stenhouse, L. (1979). Case study in comparative education: Particularity and generalisation. *Comparative Education, 15*(1), 5–10.

Stevenson, H., & Bartsch, K. (1992). An analysis of Japanese and American textbooks in mathematics. In R. Leestma & H. J. Walberg (Eds.), *Japanese educational productivity* (pp. 103–133). Ann Arbor: University of Michigan, Michigan Papers in Japanese Studies.

Stigler, J. W., & Hiebert, J. (1999). *The teaching gap: Best ideas from the world's teachers for improving education in the classroom*. New York, NY: The Free Press.

Stigler, J. W., & Perry, M. (1988). Cross cultural studies of mathematics teaching and learning: Recent finding and new directions. In D. A. Grouws, T. J. Cooney, & D. Jones (Eds.), *Perspectives on research on effective mathematics teaching* (pp. 194–223). Reston: National Council of Teachers of Mathematics.

Theule Lubienski, S. (2017). Part IV commentary on chapters 12–16: Large-scale international datasets: What we can and cannot learn from them and how we could learn more. In J.-W. Son, T. Watanabe, & J.-J. Lo (Eds.), *Research trends in international comparative studies in mathematics education in six countries*. New York, NY: Springer.

Thut, I. N., & Adams, D. (1964). *Educational patterns in contemporary societies*. New York, NY: McGraw-Hill.

Triandis, H. C. (1995). *Individualism and collectivism*. San Francisco, CA: Westview Press.

Wong, N.-Y. (2004). The CHC learner's phenomenon: its implications on mathematics education. In L. Fan, N. Y. Wong, J. Cai, & S. Li (Eds.), *How Chinese learn mathematics: Perspectives from insiders* (pp. 503–534). Singapore: World Scientific.

Wong, N.-Y., Wong, W. Y., & Wong, E. W. Y. (2012). What do the Chinese value in (mathematics) education? *ZDM Mathematics Education, 44*(1), 9–19.

Yang, X. (2014). *Conception and characteristics of expert mathematics teachers in Mainland China*. Wiesbaden: Springer Spektrum (PhD thesis, University of Hong Kong).

Chapter 22
The Missing Link—Incorporating Opportunity to Learn in Educational Research Analyses

William H. Schmidt, Leland S. Cogan, and Michelle L. Solorio

Abstract International comparative education has developed a rich and complex history to address a variety of questions about how education systems are similar or different in their organization, operation, and outcomes. Despite different approaches and the exploration of different research questions, a common challenge is making these comparisons both trustworthy and credible. In the language of research methodology these are issues of reliability and validity, and are of particular interest as researchers attempt to compare what some consider to be the incomparable (Husén, 1983). Our contention is that whatever approach, method, or question informs the research, a critical element to be considered is the subject matter that is the focus of the educational enterprises being compared.

Keywords Opportunity to learn • Statistical bias • Student performance

International comparative education has developed a rich and complex history to address a variety of questions about how education systems are similar or different in their organization, operation, and outcomes. Despite different approaches and the exploration of different research questions, a common challenge is making these comparisons both trustworthy and credible. In the language of research methodology these are issues of reliability and validity, and are of particular interest as researchers attempt to compare what some consider to be the incomparable (Husén, 1983). Our contention is that whatever approach, method, or question informs the research, a critical element to be considered is the subject matter that is the focus of the educational enterprises being compared.

W.H. Schmidt (✉)
Education Policy Center, Michigan State University, East Lansing, MI, USA

Center for the Study of Curriculum Policy, Michigan State University, East Lansing, MI, USA
e-mail: bschmidt@msu.edu

L.S. Cogan • M.L. Solorio
Center for the Study of Curriculum Policy, Michigan State University, East Lansing, MI, USA

J.-W. Son et al. (eds.), *What Matters? Research Trends in International Comparative Studies in Mathematics Education*, Research in Mathematics Education, DOI 10.1007/978-3-319-51187-0_22

A hypothetical example may help clarify our contention. Imagine two photographs arranged side by side, both featuring students clad in white coats and engaged in measuring, mixing, and heating various mixtures. In one picture these activities are being conducted in a large room that has a number of stoves, ovens, and common kitchen pots and pans; in the other there are a number of open-flame "cooking stations" surrounded by various glass cylinders, bowls, and cups. Comparing the laboratories in the two pictures seems a bit odd if we assume all the students are studying the same thing. However, the differences become immediately meaningful once we realize the first is a laboratory kitchen for future chefs while the other is a laboratory for organic chemistry students. Although a picture may well be worth a thousand words, these pictures require a few words for a faithful and reasonable comparison.

The need to provide important contextual information about the subject matter in comparative education research is not limited to any particular set of research methods, goals, or questions. Indeed, such studies encompass a variety of approaches and methodologies that have been strongly influenced by the disciplinary traditions informing the research: sociology, psychology, philosophy, economics, political science, and of course various traditions within education. This history reveals a changing focus on various goals and accompanying methods. Although not strictly chronological there is a sense of development, with contemporary research typically embracing more than a single goal and often reflecting a multidisciplinary or interdisciplinary orientation.

Among the earliest comparisons of education were descriptions made during the Greek and Roman eras that documented differences among foreign—"xeno" or "barbarous"—peoples. This goal of rich description of differences continues to find expression in ethnographies and in the compendia of education indicators produced, for example, by UNESCO and OECD. A second goal has been to examine foreign education systems and practices specifically to discover novel practices, approaches, or structures that could be employed in one's own context. Sometimes studying the mundane in an unfamiliar context, i.e., a different culture or social setting, can spark insights and perspectives that would otherwise remain tacit and unexamined. The World Bank and other international agencies often examine educational practices with the explicit humanitarian goal of improving education in order to improve people's well-being and the overall economies in developing countries. As these goals have been pursued, an additional goal arose: to examine specific factors thought to shape education. The advent of relatively cheap and powerful computers has led many scholars to elucidate more specifically quantitative explanations, and to examine causal relationships among many different education resources, practices, and products.[1]

[1]The broad goals summarized here have been gleaned from histories of international comparative education by Hans (1949), Noah and Eckstein (1969), Cowen (1996), Mitter (1997), and Lundgren (2011).

The root of current large-scale comparative studies sponsored by the International Association for the Evaluation of Educational Achievement (IEA) and OECD, such as TIMSS and PISA, can be found in the interest in the late 1950s among university research professors and education ministry officials to investigate education practices and outcomes in a systematic manner. The initial 12-country pilot administered in 1960 (Foshay, Thorndike, Hotyat, Pidgeon, & Walker, 1962) was the result of a consensus in this group of "the need to introduce into comparative educational studies established procedures of research and quantitative assessment" (Husén, 1967, p. 13). Their goals included providing rich qualitative descriptive data situating education in its social, cultural, and political context, but also moving beyond this to provide insight into possible causal relationships between educational inputs and outputs.

Benjamin Bloom was a member of this group and was selected to lead the initial pilot study. To move beyond mere description required a theoretical or conceptual model that would identify constructs of interest and would inform the creation of instruments. Carroll and Bloom were an integral part of the early discussions. Consequently, the constructs embedded in Bloom's (1974) mastery learning model and Carroll's (1963) model of school learning served central roles in the research. More specifically, in thinking about influences on student achievement to include in the instrumentation, "one of the factors which may influence scores on the achievement examination was whether or not the students had an opportunity to study a particular topic of how to solve a particular type of problem" (Husén, 1967, pp. 162–163). This opportunity to learn (OTL) construct, termed "time actually spent on learning" in Carroll's model, was conceptualized at the student level as his was a psychological model. Given the practical challenges of a large-scale research endeavor, the decision was made to measure OTL through a teacher survey rather than burdening students with greater response time. Measuring OTL at the classroom level through teachers' survey responses has been a hallmark of large-scale comparative surveys. Most recently, PISA 2012 for the first time included an OTL measure and it was measured through student responses (Cogan & Schmidt, 2015).

Conducted appropriately, comparisons can lead to deep insight into a researcher's own education system. Done poorly, however, researchers and others are left with shallow observations regarding superficial differences and similarities, observations that fail to provide insights that may be leveraged to make sense of the resultant data. Common to all of the research goals in comparative education identified earlier, implicitly if not explicitly, is a desire to learn about different education systems in order to gain insights toward potentially improving one's own education system. The consistency of this foundational purpose of comparative education is striking and important, and underscores the important question posed earlier: "how have researchers made their comparisons both trustworthy and credible?" Throughout the evolution of international assessment studies, a problematic issue has been the tendency for researchers, policy makers, and others to use country means from the assessments to create an unsubstantiated ranking system, also referred to as a league table or a cognitive Olympics (Burstein, 1993; Husén, 1979a, 1979b). The danger is that this simplistic compilation may be used to leverage policy objectives based on comparisons of countries' mean scores alone, while failing to take into account differences in critical educational factors—educational structures, cultures, and student learning. Those researchers who developed and analyzed SIMS acknowledged this issue explicitly: "We cannot escape the ideological use and misuse of cross-national data for political purposes. We can only hope to overwhelm the most base misrepresentations with the wealth of knowledge and understanding international studies can provide" (Burstein, 1993, p. xxxi).

Whether a comparative study uses large-scale international data to look at multiple systems of education or focuses in on one or two education systems, what actually makes comparative work meaningful and useful, is a true exploration of the learning experiences students have, which provide them the opportunity to learn the material represented on assessments researchers use to compare education system outcomes. In order to make sense of comparisons that large assessments and other comparative methods allow for, researchers must pay attention to the content and substance of the education being communicated to the students whose education is being measured. Broadly speaking, we are discussing the opportunity to learn construct as a measure of the implemented curriculum, which allows for these meaningful comparisons. Referring again to the researchers who developed the IEA and the early large-scale quantitative studies, they recognized that comparison was not possible when the absence of curricular commonality existed or was adjusted for, and as the assessments were developed the opportunity to learn construct evolved:

> But the early leaders were not so naïve as to think that wishing for equity made it so. Rather they were prescient enough to introduce what may be IEA's most powerful contribution of all to the literature on educational achievement surveys; namely, the measurement of **opportunity to learn (OTL)**. (Burstein, 1993, p. xxxiii, emphasis added)

By choosing to look at educational attainment or achievement on international assessments through the lens of opportunity and with an exploration of differences in curriculum, comparisons can make sense and shed light onto why different systems of education have a different distribution of scores. That is not to say that examining different curriculums will make comparison between education systems simple; rather, the comparisons will be more meaningful if we are able to see what students have been given the opportunity to learn through intentional studies of differences in curriculum. Comparing educational opportunity itself is complicated due to the nature of the work required, and is made all the more complicated by the diverse meanings attributed to the concept of "curriculum" by educators, researchers, and teachers. In different contexts, "curriculum" can refer to textbooks, lesson plans, education frameworks, national guidelines, educational expectations, classroom activities, and a number of other attributes that make up a single system of education.

In comparative studies that incorporate factors related to opportunity to learn, "curriculum" that directly influences OTL refers to the content presented to students, the instructional opportunities that students experience, or in technical

TIMSS terms the "implemented" part of the tripartite model of curriculum. This model is comprised of the intended curriculum (what students are expected to learn as stated in national or regional goals, written frameworks, and standards), the implemented curriculum (what happens in the classroom), and the attained curriculum (what students learn). Textbooks and other learning materials comprise the potentially implemented curriculum, creating items that students will potentially have the opportunity to be exposed to, thus adding yet another element to the delicate yet complex understanding of a students' exposure to educational opportunity.

What we have learned through the phases of international assessment work, particularly through researcher-driven developments to include information about what happens inside schools through opportunity to learn measures and other practice-relevant data points, is that what makes comparisons useful is an understanding of what material students have been exposed to, in what ways, and how often. These internal workings of education get at the heart of learning, and remain the foundation for differences in student outcomes, education systems, and educational similarities and differences. Without knowing what happens in school, comparative work is reduced to meaningless numbers in a formula or words on a page, with no foundation upon which to derive understanding. Just like mathematics requires units of measure to define the meaning of a value, comparative education requires educational opportunities and exposure to subject matter to define an outcome.

Ultimately, despite the challenges faced in comparative education studies, the subject matters (Stodolsky & Grossman, 1995). Consequently, researchers need to attend to the learning opportunities that differ across education systems, classroom practices (pedagogies), and school activities in order to draw trustworthy and credible comparisons. Although Hans (1949) argued that "the application of the findings of these studies [of comparative education] is outside the scope of Comparative Education proper and belongs in its theory to the philosophy of education and in its practice to the administration and organization of education" (p. 11), we submit that the framework informing the comparative exercise is within the scope of comparative education and that this plays a central role in the proper interpretation of the research. The substance of the education enterprise—the focus and content of the curriculum—can only be excluded from consideration to the peril of the reliability and validity of the comparisons in view.

In truth, this threat to education research does not exist solely for cross-national or cross-cultural comparisons. One of the major insights from the 1995 TIMSS curriculum analysis was the great variation in what passes for eighth-grade (13-year-olds') mathematics across countries (Schmidt, McKnight, Valverde, Houang, & Wiley, 1997). Mathematics is studied around the world in every school system yet it is not all the same; math is not math. Much greater specification is needed to balance that equation. Our analysis of the US data underscored this, as we realized the great variation in what students studied in eighth-grade mathematics in our country was every bit as great as the variation across all participating TIMSS countries (Cogan, Schmidt, & Wiley, 2001).

From the disciplinary viewpoint of statistics, ignoring subject matter introduces bias. Many different studies have documented a relationship between students' motivation and their academic performance, and many investigate this relationship specifically in the context of mathematics. Regression analysis yields a numerical estimate of this relationship. However, motivation may well be related to the specific mathematics studied, i.e., students' mathematics OTL, as well as to students' achievement. In this case, if OTL is left out of the analysis model the estimate of the strength of the relationship between motivation and performance will be biased by the indirect effect. Consequently, one of the most critical contextual issues to be addressed in any piece of educational research is the substance (subject matter) that is the focus of what teachers are teaching and students are expected to learn.

The issue of bias can be framed mathematically for greater clarity. Assume that the following model defines the true relationship between two variables—OTL and student motivation, for example—and mathematics achievement:

$$y = \beta_0 + \beta_1 x_1 + \beta_2 x_2 + e$$

where x_1 is a measure of mathematics content coverage of (OTL) and x_2 is another variable describing a different aspect of schooling such as motivation or teacher quality.

Now imagine that the researcher does not have a measure of OTL and as such analyzes the data using the following model:

$$y = \beta_0 + \beta_2 x_2 + e$$

The consequence of this, given the true relationship as described in the previous equation, is that in reality:

$$\beta_2 = \beta_2 + \beta_1 \frac{\sigma_{x_1 x_2}}{\sigma_{x_2}^2}$$

where $\frac{\sigma_{x_1 x_2}}{\sigma_{x_2}^2}$ indicates the bias that results if x_2 is related to x_1 (e.g., student motivation is related to OTL) and that OTL is related to academic achievement in mathematics, which has been well established in the literature (Schmidt & Maier, 2009; Schmidt et al., 2001; Schmidt, Burroughs, Zoido, & Houang, 2015).

Furthermore, it is our contention that content coverage in mathematics is very likely related to most other school, teacher, and student characteristics, which are also related to learning. If this is the case, then most data analyses relating those characteristics to outcome measures without the inclusion of a measure of content coverage (OTL) will produce biased estimates of the relationships of those variables to student outcomes. The direction and magnitude of the bias, however, is not known.

This suggests two important roles that the measurement of content coverage plays in educational research related to practice and policy. First, it can be conceived of as an important outcome in and of itself. The first four chapters of this book have such a focus as they characterize country differences in textbook and classroom coverage. This coverage reflects the policies of the country as to what content should be covered in what grades, and differences can be used to inform potential policy or practice reforms. Such characterizations of content coverage are outcomes of educational policy, and many countries monitor this coverage as they do other achievement measures. Part II of the book demonstrates the same use of content coverage as an important indicator of schooling, but in the context of teacher preparation.

The other major use of OTL measures goes to their relationship to academic achievement. International studies have a long tradition of measuring student achievement in mathematics, and the results of TIMSS and PISA testing provide a rich source for country comparisons toward learning "what works"—or more precisely, determining the important variables that are related to achievement both across and within countries. Additional variables are included in such studies to characterize countries, schools, classrooms, teachers, and students. Many research studies have been published using this data, as well as those from TEDS-M and TALIS. We also find analyses in this book using all of these international data sets. But here we also find a shortcoming prevalent in the research literature: most of the authors do not control their analyses for differences across countries in terms of content coverage. This is especially true in Part IV of the book, as Sarah Lubienski's comments on Chaps. 15–19 confirm; she discusses the limitations of cross-sectional data sets, especially in terms of confounding variables that are not measured or are ignored in the analyses.

This is a serious limitation of the studies reported on in Chaps. 15–19. Without adequate measures of OTL we do not know if the relationships described are characterizing the variables identified or are biased coefficients resulting from no accurate control of the variation in the content coverages, both within but especially across countries where we know how different content coverage can be (Schmidt et al., 2001).

TIMSS has always had measures of OTL but unfortunately they have become less specific and are not as detailed as in the original 1995 TIMSS. PISA in 2012 had OTL measures for the first time in mathematics. In general, like much of the educational research of this sort, the studies included in this book do not include these measures of OTL either, with the notable exceptions of Chap. 9 (using TEDS-M data) and Chap. 5 (using PISA data).

Despite the limitations of the studies reported on in this book, the book does make very visible the use of mathematics content coverage in international comparative research focusing on determining differences in content coverage as an important policy variable, as well as its use in reducing the potential bias associated with characterizing the relationship between various other schooling variables and academic performance.

References

Bloom, B. S. (1974). Time and learning. *American Psychologist, 29*(9), 7.
Burstein, L. (1993). *The IEA study of mathematics III: Student growth and classroom processes* (L. Burstein, Ed.). Oxford, England: Pergamon Press.
Carroll, J. B. (1963). A model of school learning. *Teachers College Record, 64*(8), 723–733.
Cogan, L. S., Schmidt, W. H., & Wiley, D. E. (2001). Who takes what math and in which track? Using TIMSS to characterize U.S. students' eighth-grade mathematics learning opportunities. *Educational Evaluation and Policy Analysis, 23*(4), 323–341. doi:10.3102/01623737023004323.
Cogan, L. S., & Schmidt, W. H. (2015). The concept of opportunity to learn (OTL) in international comparisons of education. In K. Stacey & R. Turner (Eds.), *Assessing mathematical literacy: The PISA experience* (pp. 207–216). Dordrecht: Springer. doi:10.1007/978-3-319-10121-7.
Cowen, R. (1996). Last past the post: Comparative education, modernity and perhaps post-modernity. *Comparative Education, 32*(2), 151–170. doi:10.1080/03050069628812.
Foshay, A. W., Thorndike, R. L., Hotyat, F., Pidgeon, D. A., & Walker, D. A. (1962). *Educational achievements of thirteen-year-olds in twelve countries*. Hamburg: UNESCO Institute for Education. Retrieved from http://unesdoc.unesco.org/images/0013/001314/131437eo.pdf.
Hans, N. (1949). *Comparative education: A study of educational factors and traditions*. London: Routledge & Kegan Paul Limited.
Husén, T. (1967). *International study of achievement in mathematics, a comparison of twelve countries* (Vol. II). (T. Husén, Ed.). New York: Wiley.
Husén, T. (1979a). General theories in education: A twenty-five year perspective. *International Review of Education, 25*(2/3), 325–345.
Husén, T. (1979b). An international research venture in retrospect: The IEA surveys. *Comparative Education Review, 23*(3), 371–385. Retrieved from http://www.jstor.org/stable/1187603?seq=1#page_scan_tab_contents.
Husén, T. (1983). Are standards in U.S. schools really lagging behind those in other countries? *The Phi Delta Kappan, 64*(7), 455–461.
Lundgren, U. P. (2011). PISA as a political instrument. In M. A. Pereyra, H.-G. Kotthoff, & R. Cowen (Eds.), *Pisa under examination: Changing knowledge, changing tests, and changing schools* (pp. 17–30). Rotterdam: SensePublishers.
Mitter, W. (1997). Challenges to comparative education: Between retrospect and expectation. *International Review of Education, 43*(5–6), 401–412. doi:10.1023/A:1003084402042.
Noah, H. J., & Eckstein, M. A. (1969). *Toward a science of comparative education*. New York: The Macmillan Company.
Schmidt, W. H., & Maier, A. (2009). Opportunity to learn. In G. Sykes, B. Schneider, & D. N. Plank (Eds.), *Handbook of education policy research* (pp. 541–559). New York: American Educational Research Association, Rutledge.
Schmidt, W. H., Burroughs, N. A., Zoido, P., & Houang, R. T. (2015). The role of schooling in perpetuating educational inequality: An international perspective. *Educational Researcher, 44* (7), 371–386. doi:10.3102/0013189X15603982.
Schmidt, W. H., McKnight, C. C., Houang, R. T., Wang, H. A., Wiley, D. E., Cogan, L. S., et al. (2001). *Why schools matter: A cross-national comparison of curriculum and learning*. San Francisco: Jossey-Bass.
Schmidt, W. H., McKnight, C., Valverde, G. A., Houang, R. T., & Wiley, D. E. (1997). *Many visions, many aims, Volume I: A cross-national investigation of curricular intentions in school mathematics*. Dordrecht: Kluwer.
Stodolsky, S. S., & Grossman, P. L. (1995). The impact of subject matter on curricular activity: An analysis of five academic subjects. *American Educational Research Journal, 32*(2), 227–249.
Travers, K. J., & Westbury, I. (1989). *The IEA study of mathematics I: Analysis of mathematics curricula* (K. J. Travers & I. Westbury, Eds.). Oxford, England: Pergamon Press.

Index

A
Achievement gaps
cognitive achievement (*see* Cognitive achievement)
East Asian *vs.* American students, 95
OTL
vs. academic achievement, 108–109
challenging curriculum and pedagogy, clearer vision for, 98
content-related OTL, 105
differential OTL levels, 98–99
vs. disposition, 110
indicator system, 98
vs. motivation, 109
vs. participation, 110–111
PISA 2012 indices, 104
school-by-school accountability, 98
vs. self-beliefs, 109–110
teaching practices, 106–107
teaching quality, 105–106
research methods
data process and analysis, 100
data sources, 99
measures, 99–100
research questions and purposes, 99
test scores and attitudes, 107–108
Addition and subtraction
Japanese series
introductory problem addition, 46, 47
word problems, 46, 47, 54
Korean series
area model, 47, 48
word problems, 46, 47, 54
non-unit fractions, 53–54
Taiwanese series
linear model, 47, 48
subtraction with like denominators, 47, 48
word problems, 46, 47, 54
treatment of, 41, 42
Affective characteristics, motivation, 175, 190
individual characteristics, 178
measures, 175
American Recovery and Reinvestment Act of 2009, 356, 360
Attained curriculum, 415

B
Balanced incomplete block test, 277
Basic skills and procedural knowledge. *See* Knowing, applying, reasoning and mathematics achievement, TIMSS
Bloom's Taxonomy, 317

C
CCSSI. *See* Common Core State Standards Initiative (CCSSI)
CCSSM. *See* Common Core State Standards for Mathematics (CCSSM)
CHC model. *See* Confucian heritage culture (CHC) model
China
cultural values, 272–273
reflective capabilities, educational systems
clustered office space and shared planning times, 249
lesson study, 248

J.-W. Son et al. (eds.), *What Matters? Research Trends in International Comparative Studies in Mathematics Education*, Research in Mathematics Education, DOI 10.1007/978-3-319-51187-0

China (*cont.*)
public demonstration lessons, 248–249
teaching competitions, 249
Cognitive achievement
East Asian students, 97–98
intelligence quotient (IQ), 96
single-minded focus, 96
Cognitive and noncognitive achievement
cluster analysis, 104
East Asia *vs*. Europe, 96
East Asia *vs*. USA, 99
international rankings of indices, 101, 103
mathematical content, 100–101
mathematical contexts, 101
mathematical processes, 101
students' mathematics performance, eight education systems, 101, 102
Cognitive characteristics, entrance quality, 169–170, 181, 190
individual characteristics, 178
measures, 174
Cognitive Diagnosis Modeling approach, 320
Common Core State Standards for Mathematics (CCSSM)
fraction arithmetic, 39
higher level cognitive practices, 315
SMP, 318
Common Core State Standards Initiative (CCSSI), 356, 360
Common Core State Standards movement, 239
Complexity, 238–239
Confucian heritage culture (CHC) model, 400–403
Constant comparative method, 239–240
Content analysis
allocation of content, 14–15
Asian *vs*. US education systems, 15, 24
introduction and development of mathematical concepts and procedures, 12–14
macroanalysis, 7, 8
subcomponents, 7, 8, 11
textbook size, 14
topic placement, 11–12
Content knowledge (CK), 133
Cooperating teachers (CTs), 215–216
Cross-cultural lesson planning, United States and South Korea
cross-cultural engagement, 210
data collection and analysis, 204
implications, 211–212
literature review, 200–201
mathematics educators community, 209–210
new technologies and digital learning environments, 200
participants, 202
pedagogical cultures, 199
perceived growth, teaching mathematics
application and problem-solving, 207
content and pedagogy, 209
curriculum design and content delivery, 205
international curriculum learning, 207
Kolb and Fry's learning cycle, 205–207
perceived differences, 207–209
project
assignment, 202–204
description, 202
outcomes, 205
theoretical perspectives, 201
Curriculum, East Asian students
classroom instruction, 115, 117
definition, 116
as enacted, 116
fraction
operations, 118
subconstructs, 118
treatment of, 118
and whole numbers, 117, 118
implemented curriculum, 116–117
intended curriculum, 116–117
learning mathematics, 115
as planned, 116
superior performance, 116, 117
textbooks, 117
cross-national comparative research, 119
difficulty ratings of exercises and worked examples, 118
at junior secondary level, 118
LPS data, 119
vs. United States, 118–120

D

Dichotomies
achievement differences, mathematics education, 400
East *vs*. West Asian cultures, 85–86
Division, fractions
common aspect, 47–48
invert-and-multiply algorithm, 51–52, 57

Japanese series
area model and number line diagram, 53, 55
double-number line diagrams, 53
Grade 5, 51
opening problem, 53, 54
partitive division situation, 52, 53
partitive word problem, 52
Korean series
bar diagrams, 53
fraction division problem, 51, 52
Grade 5, 51
quotitive word problems, 52
limitations, 57–58
Taiwanese series
bar diagrams, 53
Grade 6, 51
quotitive word problems, 52
word problem, 51

E
Educational culture, 241, 244, 246–248
Educational equity, 386, 388
Educational reform mechanisms, 238, 242, 252
Education Resource and Information Center (ERIC), 6–7
Egalitarian democracy, 273
Enhanced Performance Management System (EPMS), 362
Equality of Educational Opportunity Study (EEOS), 291
Explicit stereotypes, 302

F
False discovery rate (FDR), 281
Field experience (FLD), 133
Financial capital, 271
First International Mathematics Study (FIMS), 98, 395, 398
Fractions, mathematics curricula
CCSSM approaches, 39
conceptualization, 38
depth of knowledge, 35
East Asian countries *vs.* United States, 34
informal understanding, 38
Japan, Korea, and Taiwan textbook series
addition and subtraction, 41, 42, 46–48
CCSSM, 56
content analysis, 40
grades, 40–41
intended learning progression, 42–43
multiplication and division, 47–57
rational number subconstructs, 56
sequence and grade placement, 41
subconstructs integration, 43–46
limitations and future research, 57–58
misconceptions, 37
multiplication and division, 39
national curriculum standards, 40, 41
partitioning strategies, development of, 38
part-whole relationships, 38
research questions, 40
student performance, TIMSS 2011 Grade 8 problems, 34, 35
subconstructs, 37, 38

G
Gender Science Implicit Association Test, 302
General knowledge (GK), 133
Gross domestic product (GDP), 277

H
Heyneman-Loxley effect (HL effect), 307
Hong Kong
knowing, applying, reasoning and mathematics achievement, TIMSS
benchmark-level comparisons, 325–327
country-level comparison, 323, 325
data, international benchmarks, 321, 322
linear regression model, 321, 323, 324
plausible values, 321
reform-based textbooks, 327
students' values/valuing in mathematics education, WIFI study
affordance to identify, 340–341
correlations, 339–340, 351–352
culturally situated labelling, 343–344
cultures and subcultures, definitions of, 340–341
espoused and enacted valuing, 341–342
primary and secondary school students, 339
rotated component matrices, 339–340, 346–348
survey responses, triangulation of, 342–343
univariate normality, data screening exercises, 339
Human capital, 271

I
Implemented curriculum, 116–117, 415
Implicit stereotypes, 302
Inequality mechanisms, 271–272
Instructional quality
 characteristics, 219
 Japan
  data collection, 222–223
  lesson, simultaneous equations, 228–230, 232
  MET study, 225
  MQI coding protocol, 223–224
  participants and context, 221–223
  POM, 225
  quantitative results, 226–228
  vs. United States, 216, 218–219
 mathematical work, 219–220
 opportunity to learn, 220
 students' interaction, 219
 United States
  data collection, 223
  vs. Japan, 216, 218–219
  lesson, similar figures, 230–232
  MET study, 225
  MQI coding protocol, 223–224
  participants and context, 221–223
  POM, 225
  quantitative results, 226–228
Instrumental motivation, 268, 280, 284
Intended curriculum, 116–117, 415
International Association for the Evaluation of Educational Achievement (IEA), 166, 398, 404, 413, 414
International comparative studies
 CHCs, 400–401
 Confucian values, 401
 cross cultural comparison, 397
 cross-sectional datasets, limitations of, 404, 407
 cultural-psychological framework, 406
 dichotomies, 400
 diversification, 84
 East-West-paradigm, 403
 ethnocentric bias, 398
 FIMS, 395, 398
 foreign educational systems, 396
 globalization, 84
 goals, 397
 historical/hermeneutic approach, 396
 history, 405
 intrinsic motivation, 404
 Japan-Illinois study, 399
 large-and small-scale studies, 402
 LPS
  data collection, three-camera approach, 87
  eighth-grade classrooms, 87–88
  lesson events, 88
  Shanghai classrooms, summing up in, 91
  Tokyo classrooms, Matome in, 89–90
 mathematics curricula and classroom practices, 84
 problem-solving, 399
 quantitative and qualitative data, 396
 research perspectives, 403
 similarities and differences
  cultural activity, teaching, 86–87
  cultural and social environments, 86
  teachers' behaviors, 91–93
 social-science methods, 396
 students learning, 401
 teacher education programmes, 397
 teacher feedback and mathematics instruction, 405
 teacher knowledge and pedagogical approaches, 404
 teaching-and-learning process, 406
 TEDS-M study, 407
 tertium comparationis, 397, 405
International pedagogical approaches, 240
Intrinsic motivation, 175, 268, 400

J
Japan
 cultural values, 273
 reflective capabilities, educational systems
  clustered office space and shared planning times, 249
  lesson study, 248
  public demonstration lessons, 248–249
  teaching competitions, 249
 students' values/valuing in mathematics education, WIFI study
  affordance to identify, 340–341
  correlations, 339–340, 351–352
  culturally situated labelling, 343–344
  cultures and subcultures, definitions of, 340–341
  espoused and enacted valuing, 341–342
  primary and secondary school students, 339
  rotated component matrices, 339, 349–351
  survey responses, triangulation of, 342–343
  univariate normality, data screening exercises, 339
 teacher employment examination, 151–152

teacher feedback and teaching practices
control variables, 366
dependent variable, 365
descriptive statistics analyses, 366, 379
emphasis in feedback, 369–370, 380
good teaching practice index scores, 368–369
independent variables, 365–366, 377–378
multiple mediator model analyses results, 373–375
OLS regression model, 367, 370–372
professional development activities, participation in, 362–363
research questions, 363
study design and conceptual framework, 364
TALIS 2013 data, 356, 364–365
teacher–student relationships, 376
teaching certificates
Hiroshima Prefecture Board of Education, 149
prescribed conditions for secondary school teaching certificates in Japan, 147
pre-service teacher training, 146
Teaching Profession courses, 148
undergraduate curriculum framework, 147–148
Japan, Korea, and Taiwan textbook series, fractions
addition and subtraction
area model, Korean series, 47, 48
introductory problem addition, Japanese series, 46, 47
linear model, Taiwanese series, 47, 48
non-unit fractions, 53–54
subtraction with like denominators, Taiwanese series, 47, 48
treatment of, 41, 42
word problems, 46, 47, 54
CCSSM, 56
content analysis, 40
division
area model and number line, Japanese series, 53, 55
bar diagrams, Taiwanese series, 53
common aspect, 47–48
double-number line diagrams, Japanese series, 53
fraction division problem, Korean series, 51, 52
Grade 5 and 6, 51
invert-and-multiply algorithm, 51–52, 57
limitations, 57–58
opening problem, Japanese series, 53, 54
partitive division situation, Japanese series, 52, 53
partitive word problem, Japanese series, 52
quotitive word problems, Korean series, 52
quotitive word problems, Taiwanese series, 52
word problem, 51
grades, 40–41
intended learning progression, 42–43
multiplication
area diagrams, Taiwanese series, 53
common aspect, 47–48
conceptualizing fraction multiplication, Taiwanese series, 48, 49
initial problem, Japanese series, 49, 50
measure subconstruct, 56
operator subconstruct, 56
sequence of multiplication-related topics, 49, 50
unit fractions, 49, 53
rational number subconstructs, 56
sequence and grade placement, 41
subconstructs integration
Grade 3 problems, 43, 44
introductory word problem, 45, 46
multiplication, 44–45
operator subconstruct, 45–46
part-whole and measure subconstructs, 43–44
quotient subconstruct, 44
ratio subconstruct, 44
Junior secondary school level
analyzed topics, 66
cross-national exploration, 64
nations, 64–65
textbook difficulty
Bao's model, 67
content breadth, 69–70, 76–77
content depth, 70–71, 78
difficulty level measurement, 67–69
elements, 67
four-factor model, 67–68
information technology development, 79–80
Nohara's model, 66–67
overall difficulty level, 74–76
practical applications of mathematics, 79
regional features, 78
teaching to assisting learning, 78–79

Junior secondary school level (*cont.*)
values and rankings, 74, 75
weight coefficients, 68
worked examples and exercises, difficulty level of, 71–74, 77–78
textbooks, 65
year level, 65

K
Knowing, applying, reasoning and mathematics achievement, TIMSS, 328–329, 387
Bloom's Taxonomy, 317
cognitively demanding tasks, students' engagement in, 316
cognitive skills and overall achievement scores, 320
intended curriculum, national research coordinators report on, 317, 318, 327
mathematics and science assessment framework, 318
purpose of study, 320–321
SMP in CCSSM, 318
subcategories, 316, 318–319
Taiwan, Hong Kong, Singapore, South Korea and USA eighth-grade students
benchmark-level comparisons, 325–327
country-level comparison, 323, 325
curriculum reform movements, 327
data, international benchmarks, 321, 322
linear regression model, 321, 323, 324
plausible values, 321
reform-based textbooks, 327
Knowledge achievements, teaching readiness, 171
institutional characteristics, 179
measures, 174–175
Korea
cultural values, 272–273
knowledge expectations
course classification, 188
key ideas, 188
scholars and policymakers, 188–189
program requirements, 189
qualified teachers, equal distribution of, 310
teacher employment examination, 152–153
teacher feedback and teaching practices
control variables, 366
dependent variable, 365
descriptive statistics analyses, 366, 379
emphasis in feedback, 369–370, 380
good teaching practice index scores, 368–369
independent variables, 365–366, 377–378
in-service teacher training, 362
multiple mediator model analyses results, 373–375
new teacher evaluation system, 362
OLS regression model, 367, 370–372
research questions, 363
study design and conceptual framework, 364
TALIS 2013 data, 356, 364–365
teacher–student relationships, 376
teaching certificates
actual teaching practice, 150
course categories for secondary school teaching, 149
pre-service teacher training, 146
subject matter knowledge, 149–150
undergraduate curriculum framework, 150–151

L
Large-scale international datasets
clear communication of results, 391
coordination, 391
guilt by association, 388
history, culture, and policy, 388–389
international, self-reported data, beware of, 390
longitudinal dataset, 391–392
multiple comparisons, 390–391
self-concept *vs.* self-efficacy, 387
students values/valuing, WIFI study, 387
teacher evaluation, 387
TIMSS studies, taking stock of, 387
unpacking mathematics achievement, 387
within-country analysts and analyses, 389–390
Learner's Perspective Study (LPS), 403
curriculum, 119
data collection, three-camera approach, 87
eighth-grade classrooms, 87–88
lesson events, 88
Shanghai classrooms, summing up in, 91
Tokyo classrooms, Matome in, 89–90
Likert scale, 190, 279, 338
Linear regression model, 321, 323, 324
Longitudinal datasets, 391–392
LPS. *See* Learner's Perspective Study (LPS)

M

Markov Chain Monte Carlo multiple imputation, 277
Master of Arts in Teaching (MAT) program, 202
Mathematical Quality of Instruction (MQI) video coding protocol, 219
  categories, 223–225
  MET comparison study, 225
  scores
    US and Japan STs, 233
    US mathematics teachers, sample of, 227, 228
Mathematics instruction, teacher feedback on. *See* Teacher feedback and mathematics instruction
Mathematics self-concept and self-efficacy, 387
  analyses, 280–281
  cultural values
    Chinese and Korean societies, 272–273
    Japan, 273
    United States, 273–274
  data, 276
  dignity culture, 275–276
  economic differences
    higher mathematics achievement, 270–271
    inequality, 271–272
  expectancy beliefs, 269
  explanatory model, 281–284
  face culture, 274
  methodological design, 276–277
  methods, 276
  motivation, 268
  social cognitive theory, 268
  summary statistics, 281
  variables
    educational resources at home, 279
    family SES, 279
    family variables, 277
    mathematics test score, 277
    number of books, 279
    regional variables, 277
    student variables, 279
Mathematics teacher training
  Japan
    Korea, differences and similarities with, 161–162
    teacher employment examination, 151–152
    teaching certificates, 146–149
    undergraduate curricula, 153–154
  Korea
    Japan, differences and similarities with, 161–162
    teacher employment examination, 152–153
    teaching certificates, 146, 149–151
    undergraduate curricula, 154–155
Matome, Tokyo classrooms, 88–90
Measures of Effective Teaching (MET) study, 225
Mediation Effect, 364
Middle School Curriculum Design, 202
Modesty bias, 275
Multiplication, fractions
  common aspect, 47–48
  Japanese series
    initial problem, 49, 50
    unit fractions, 49
  Korean series
    area diagrams, 53
    sequence similarities, 49
    unit fractions, 49
  measure subconstruct, 56
  operator subconstruct, 56
  sequence of multiplication-related topics, 49, 50
  Taiwanese series
    area diagrams, 53
    conceptualizing fraction, 48, 49
    sequence similarities, 49
    unit fractions, 49

N

National Assessment of Educational Progress (NAEP), 386
National Center for Education Statistics (NCES), 365
National Council of Teachers of Mathematics (NCTM)
  content analysis, 66
  higher level cognitive practices, 315
National Institute of Education, 190
National Research Coordinators (NRCs), 291–292
Nation-state educational systems, 239, 241–242
  complex-oriented theory, 240
  reflective capabilities
    US educational stakeholders resistance to reform, 246–247
    US educational structures limit reflection, 244–246
    US pedagogy limits reflection, 242–244

Natural-selection-like process, 241
NCTM. *See* National Council of Teachers of Mathematics (NCTM)
NRCs. *See* National Research Coordinators (NRCs)

O
Opportunity to learn (OTL), 311
 vs. academic achievement, 108–109
 challenging curriculum and pedagogy, clearer vision for, 98
 comparative education, 413–415
 content coverage measurement, 416
 content-related OTL, 105
 differential OTL levels, 98–99
 disciplinary traditions, 412
 vs. disposition, 110
 ethnographies, 412
 implemented curriculum, 414
 indicator system, 98
 instructional quality, 220
 learning experiences, 414
 mastery learning model, 413
 model of school learning, 413
 motivation, 416
 vs. motivation, 109
 vs. participation, 110–111
 PISA 2012, 104, 413, 417
 reliability and validity, 411
 school-by-school accountability, 98
 vs. self-beliefs, 109–110
 subject matter, 411, 412
 teaching practices, 106–107
 teaching quality, 105–106
 teaching readiness, 170, 181–182
  individual characteristics, 178–179
  institutional characteristics, 179
  measures, 174
 TEDS-M and TALIS, 417
 TIMSS, 415
Ordinary least-squares (OLS) regression model, 367, 370–372
Organisation for Economic Co-operation and Development (OECD), 276, 300
OTL. *See* Opportunity to learn (OTL)

P
Pedagogical content knowledge (PCK), 133, 153
Pedagogical knowledge (PK), 133
PISA. *See* Program for International Student Assessment (PISA)
Post-secondary teacher training programs, 241
Preparation programs
 knowledge expectations
  course classification, 188
  key ideas, 188
  national culture, 187
  non-pedagogy-related knowledge, 188
  pedagogy-related knowledge, 188
  scholars and policymakers, 188–189
  transnational discourse, 187
 requirements, 189
 research directions, 193–194
 research issues
  cross-sectional data, 191–192
  insufficient analysis phases, 192–193
  limited measures, 192
 teaching readiness, 190–191
Principal component analysis (PCA), 339
Problem analysis
 cognitive demand, 19–20
 cognitive expectations, 20–21
 contextual features, 17–18
 depth of knowledge, 21–22
 mathematical features, 16–17
 microanalysis, 7
 relevance of non-textual elements, 22–23
 response types, 18–19
 standards-based materials, 16
 subcomponent, 7, 8, 15, 16
 traditional curricular materials, 16
 US *vs.* Korean textbooks, 23
Professional development and teacher evaluation. *See* Teacher feedback and mathematics instruction
Program for International Student Assessment (PISA), 290, 334, 339
 clear communication of results, 391
 coordination, 391
 implemented mathematics curricula, 292
 origin countries' national economic development levels, 309
Proportional Odds Model (POM), 225

R
Race to the Top program, 356, 360
Rasch models, 277
Reflective capabilities
 Chinese and Japanese educational systems
  clustered office space and shared planning times, 249
  lesson study, 248
  public demonstration lessons, 248–249
  teaching competitions, 249
 complacent procedural mimicry, 251
 complex-oriented theory, 240
 intelligence, 241–242

international cross-cultural comparative study, 240
methods, 239–240
nation-state educational systems
US educational stakeholders resistance to reform, 246–247
US educational structures limit reflection, 244–246
US pedagogy limits reflection, 242–244
pedagogy, internationalization of, 251
self-perpetuation, 241
self-reformation, 250
teaching gap, 237, 238
theoretical framework, 238–239
Reflective thinking, 242

S

"School effect" research, 294
Self-concept and self-efficacy. *See* Mathematics self-concept and self-efficacy
Self-enhancement bias, 275, 285
Self-perpetuation, 241
SES. *See* Socio-economic status (SES)
Singapore
knowing, applying, reasoning and mathematics achievement, TIMSS
benchmark-level comparisons, 325–327
country-level comparison, 323, 325
data, international benchmarks, 321, 322
linear regression model, 321, 323, 324
plausible values, 321
reform-based textbooks, 327
teacher feedback and teaching practices
control variables, 366
dependent variable, 365
descriptive statistics analyses, 366, 379
emphasis in feedback, 369–370, 380
good teaching practice index scores, 368–369
human resource system, development of, 362
independent variables, 365–366, 377–378
multiple mediator model analyses results, 373–375
OLS regression model, 367, 370–372
research questions, 363
study design and conceptual framework, 364
TALIS 2013 data, 356, 364–365
teacher–student relationships, 376
teaching readiness
career-based occupation, 168
Confucian cultures, 167
data analysis, 175–176
degree of Westernization, 180
entrance quality, 170
individual characteristics, 176–179
institution characteristics, 179–180
knowledge achievements, 171
measures, 173–175
motivation, affective characteristic, 171, 181
multivariate linear modeling, 190
OTL, 170, 181–182
participants and data collection process, 173
sociodemographic characteristics, 169
teacher education quality, 172
Social comparison, 269
Social expectations for teacher knowledge, 124
core knowledge, 129
preparation and certification, institutional arrangements for, 128
in sociohistorical contexts, 129–130
teacher preparation programs
conceptual categories and courses, 133, 134
content knowledge (CK), 133
curricular structures, 133, 134
curriculum models, 136–138
differences and similarities, 124–125
educational aims, 134–136
epistemic models, 134
equity in students' learning opportunities, 124
field experience (FLD), 133
general knowledge (GK), 133
key embedded ideas, 132, 133
pedagogical content knowledge (PCK), 133
pedagogical knowledge (PK), 133
policy and practice implications, 138–140
preparation gap, 124
sampling strategy, 131–132
undergraduate-level, 132
Sociodemographic characteristics, teaching readiness, 169, 190
individual characteristics, 176, 178
measures, 174
Socio-economic status (SES), 270, 291, 295, 388
South Korea
cross-cultural lesson planning (*see* Cross-cultural lesson planning, United States and South Korea)

South Korea (*cont.*)
knowing, applying, reasoning and mathematics achievement, TIMSS
benchmark-level comparisons, 325–327
country-level comparison, 323, 325
data, international benchmarks, 321, 322
linear regression model, 321, 323, 324
plausible values, 321
reform-based textbooks, 327
teacher preparation programs (*see* Teacher preparation programs)
Standards for Mathematical Practices (SMP), 318
Stratified sampling, 276
Student teaching
Japan
instructional quality (*see* Instructional quality)
structure, 216, 218, 233
Teaching Gap study, 233
United States
CTs, 215–216
instructional quality (*see* Instructional quality)
structure, 217, 233
traditional experience, 215
Summing up lesson, Shanghai classrooms, 88, 91

T
Taiwan
knowing, applying, reasoning and mathematics achievement, TIMSS
benchmark-level comparisons, 325–327
country-level comparison, 323, 325
data, international benchmarks, 321, 322
linear regression model, 321, 323, 324
plausible values, 321
reform-based textbooks, 327
teaching readiness
career-based occupation, 168
Confucian cultures, 167
data analysis, 175–176
degree of Westernization, 180
entrance quality, 170
individual characteristics, 176–179
institution characteristics, 179–180
knowledge achievements, 171
measures, 173–175
motivation, affective characteristic, 171, 181
OTL, 170, 181–182
participants and data collection process, 173
self-reported OTL, 190
sociodemographic characteristics, 169
teacher education quality, 172
two-level hierarchical linear modeling, 190
Teacher Education and Development Study in Mathematics (TEDS-M), 166, 194, 334, 404, 407, 417
Teacher Employment Test (TET), 152–153
Teacher feedback and mathematics instruction, 387
instructional practices, characteristics of, 357–358
in Japan, Korea, Singapore, and United States
control variables, 366
dependent variable, 365
descriptive statistics analyses, 366, 379
emphasis in feedback, 369–370, 380
good teaching practice index scores, 368–369
human resource system, development of, 362
independent variables, 365–366, 377–378
in-service teacher training, 362
multiple mediator model analyses results, 373–375
new teacher evaluation system, 362
OLS regression model, 367, 370–372
professional development activities, participation in, 362–363
research questions, 363
study design and conceptual framework, 364
TALIS 2013 data, 356, 364–365
teacher–student relationships, 376
student achievement, 358–361
teacher quality, 356
Teacher knowledge
comparative perspectives
conceptual reflection, 126
knowledge bases, 126–127
and pedagogical conceptualizations, 125
cross-cultural studies, 260, 263
cultural and linguistic alignment, 261
cultural patterns and norms, 260
cultural scripts, 260
international research studies, 260
learning process, reflection and collaboration in, 263
national reform efforts, 263

and pedagogical skills, 260
program evaluation and redesign, 261
self-preservation and perpetuation, 262
social expectations for, 124
core knowledge, 129
preparation and certification, institutional arrangements for, 128
in sociohistorical contexts, 129–130
teacher preparation programs (*see* Teacher preparation programs)
student teaching, 262
system-wide learning, 263
Teacher preparation programs
curriculum models, 136–138
data and methods
conceptual categories and courses, 133, 134
content knowledge (CK), 133
curricular structures, 133, 134
epistemic models, 134
field experience (FLD), 133
general knowledge (GK), 133
key embedded ideas, 132, 133
pedagogical content knowledge (PCK), 133
pedagogical knowledge (PK), 133
sampling strategy, 131–132
undergraduate-level, 132
differences and similarities, 124–125
educational aims, 134–136
equity in students' learning opportunities, 124
policy and practice implications, 138–140
preparation gap, 124
Teacher quality, 172
institutional characteristics, 179–180
measures, 174
student achievement, 165–166
TEDS-M, 166–167
Teaching and Learning International Study (TALIS)
knowledge, 360
OECD, 361
student performance, 360
teacher feedback and professional development, 356, 365
teaching practices, 358
Teaching certificates
Japan
Hiroshima Prefecture Board of Education, 149
prescribed conditions for secondary school teaching certificates in Japan, 147
pre-service teacher training, 146
Teaching Profession courses, 148
undergraduate curriculum framework (*see* Undergraduate curricula)
Korea
actual teaching practice, 150
course categories for secondary school teaching, 149
pre-service teacher training, 146
subject matter knowledge, 149–150
undergraduate curriculum framework (*see* Undergraduate curricula)
Teaching readiness, Singapore and Taiwan *vs.* United States
career-based occupation, 168
Confucian cultures, 167
data analysis, 175–176
definition, 166
degree of Westernization, 180
entrance quality, cognitive characteristics, 169–170, 181, 190
Greek/Latin/Christian traditions, 167
knowledge achievements, 171
limitations, 180
MCK and MPCK, 168
motivation, affective characteristic, 171, 181, 190
OTL, 170, 181–182
position-based occupation, 168
predictors
individual characteristics, 176–179
institution characteristics, 179–180
research method
measures, 173–175
participants and data collection process, 173
SES, 180
social status and context, 167
sociodemographic characteristics, 169, 190
teacher education quality, 172
TEDS-M, 166–167
Textbook analysis
content analysis
allocation of content, 14–15
Asian *vs.* US education systems, 15, 24
introduction and development of mathematical concepts and procedures, 12–14
macroanalysis, 7, 8
subcomponents, 7, 8, 11
textbook size, 14
topic placement, 11–12
cross-national textbook analysis, 36
curriculum materials, 4–5

Textbook analysis (*cont.*)
definitions
intended curriculum, 6
potentially implemented curriculum, 6
textbook series, 5–6
fractions, 10
implications and future directions, 27–28
international assessment studies, 3–4
macroanalysis, 36
microanalysis, 36
performance expectations, 5
probabilistic *vs.* deterministic opportunities, 5
problem analysis
cognitive demand, 19–20
cognitive expectations, 20–21
contextual features, 17–18
depth of knowledge, 21–22
mathematical features, 16–17
microanalysis, 7
relevance of non-textual elements, 22–23
response types, 18–19
standards-based materials, 16
subcomponent, 7, 8, 15, 16
traditional curricular materials, 16
US *vs.* Korean textbooks, 23
representations of mathematical concepts, 27
selection search procedures, 6–7
student learning, impact factors, 4
survey frequency, education systems, 8, 9
textbook signatures, 24–26
type of analysis, 8–10
Textbook difficulty, junior secondary school level
Bao's model, 67
content breadth
calculation, 69–70
in five nations, 76–77
number of mathematics knowledge points, 69
relative content breadth, 69, 70
results, 70
content depth, 70–71, 78
difficulty level measurement, 67–69
difficulty level of worked examples and exercises
findings, 77–78
level of cognitive requirements, 71–72
level of contexts (BJ), 72–73
number of mathematics knowledge points included (ZS), 72
overall difficulty level (N), 73–74
elements, 67
four-factor model, 67–68
information technology development, 79–80
Nohara's model, 66–67
overall difficulty level
in five nations, 75–76
weighted mean values, 74, 75
practical applications of mathematics, 79
regional features, 78
teaching to assisting learning, 78–79
values and rankings, 74, 75
weight coefficients, 68
Third Wave Project, 338
TIMSS. *See* Trends in International Mathematics and Science Study (TIMSS)
Trends in International Mathematics and Science Study (TIMSS)
clear communication of results, 391
coding framework, 292–293
coordination, 391
findings, 294
implications
comparative education analyses, 310
cross-national variation, school effects, 310
gender and SES, 308
Grade 4 data, 309
native and immigrant students, 309
opportunity gaps, 310
PISA data, 307
SES and OTL, 311
stratified education systems, 309, 310
knowing, applying, reasoning and mathematics achievement, 328–329, 387
benchmark-level comparisons, 325–327
Bloom ' s Taxonomy, 317
cognitively demanding tasks, students' engagement in, 316
cognitive skills and overall achievement scores, 320
country-level comparison, 323, 325
curriculum reform movements, 327
data, international benchmarks, 321, 322
intended curriculum, national research coordinators report on, 317, 318, 327
international comparison studies, 319–320
linear regression model, 321, 323, 324
mathematics and science assessment framework, 318
plausible values, 321
purpose of study, 320–321

reform-based textbooks, 327–328
SMP in CCSSM, 318
subcategories, 316, 318–319
teacher training programs, 328
mathematics achievement inequality
cross-national variation, 298
family SES, 295
gap study research, 294
gender gaps, 297
grade levels, 296
multiwave data, 296
national-level factors, 297
regional/geographic variation, 295
school and family backgrounds research
cross-national variation, 305
micro-level studies, 306
national-level studies, 306–307
selection and search procedure, 291–292
STEM and non-STEM fields, 290
students' socioeconomic and demographic factors
micro-level factors and gap studies, 299–301
national-level factors and gap study, 302–305
teacher and curriculum questionnaires, 291

U

Undergraduate curricula
Japan
Courses of Study, 153
credits required for program completion, 147, 148
lower secondary schools, 154
mathematical activities, 154
mathematics, 156, 157
mathematics education, 155, 156
school-based work, 159–160
teaching profession, 157–159
upper secondary schools, 154
Korea
credits required for graduation, 150–151
lower secondary school, 154
mathematics, 156, 157
mathematics education, 155, 156
school-based work, 159–160
teaching profession, 157–159
upper secondary school, 154–155
United Nations Educational, Scientific and Cultural Organization (UNESCO), 292
United States
Common Core State Standards movement, 239
cross-cultural lesson planning (*see* Cross-cultural lesson planning, United States and South Korea)
cultural values, 273–274
educational reform, 251
educational stakeholders resistance to reform, 246–247
educational structures limit reflection
incoherent curricula, textbooks and lessons, 246
lack of time, 245–246
pervasive isolation, 244–245
knowing, applying, reasoning and mathematics achievement, TIMSS
benchmark-level comparisons, 325–327
country-level comparison, 323, 325
data, international benchmarks, 321, 322
linear regression model, 321, 323, 324
plausible values, 321
reform-based textbooks, 327
knowledge expectations
course classification, 188
key ideas, 188
scholars and policymakers, 188–189
pedagogy limits reflection
classroom pedagogy, 243
professional development pedagogy, 243–244
reform pedagogy, 244
Race to the Top program, 356
teacher feedback and teaching practices
control variables, 366
dependent variable, 365
descriptive statistics analyses, 366, 379
emphasis in feedback, 369–370, 380
good teaching practice index scores, 368–369
independent variables, 365–366, 377–378
multiple mediator model analyses results, 373–375
OLS regression model, 367, 370–372
policies, 361–362
research questions, 363
study design and conceptual framework, 364
TALIS 2013 data, 356, 364–365
teacher collaboration, 375
teacher preparation programs (*see* Teacher preparation programs)

United States (*cont.*)
teaching readiness
data analysis, 175–176
entrance quality, 170
Greek/Latin/Christian traditions, 167
individual characteristics, 176–179
institution characteristics, 179–180
knowledge achievements, 171
measures, 173–175
motivation, affective characteristic, 171, 181
OTL, 170, 181–182
participants and data collection process, 173
position-based occupation, 168
self-reported OTL, 190
sociodemographic characteristics, 169
teacher education quality, 172
two-level hierarchical linear modeling, 190

V
Values/valuing in mathematics education, WIFI study, 344–345, 387
"Affective Taxonomy of Educational Objectives", 335, 336
Alan Bishop's framework, 336
communication, 336
cultural values, 335
definition, 335
East Asian students, 334–335
Hong Kong and Japanese students, 389
affordance to identify, 340–341
correlations, 339–340, 351–352
culturally situated labelling, 343–344
cultures and subcultures, definitions of, 340–341
espoused and enacted valuing, 341–342
primary and secondary school students, 339
rotated component matrices, 339, 346–351
survey responses, triangulation of, 342–343
univariate normality, data screening exercises, 339
local research teams, 337
migrant children in USA, analysis of, 335
questionnaires, 337–339
thinking and reasoning, 336
value continua, 337
values *vs.* beliefs, 336

W
What I Find Important (in my mathematics learning) study (WIFI). *See* Values/valuing in mathematics education, WIFI study

MIX
Papier aus verantwortungsvollen Quellen
Paper from responsible sources
FSC® C105338

If you have any concerns about our products,
you can contact us on
ProductSafety@springernature.com

In case Publisher is established outside the EU,
the EU authorized representative is:
Springer Nature Customer Service Center GmbH
Europaplatz 3, 69115 Heidelberg, Germany

Printed by Libri Plureos GmbH
in Hamburg, Germany